Lecture Notes in Computer Science 13058

More information about this subseries at http://www.springer.com/series/7409

Nora Reyes · Richard Connor · Nils Kriege ·
Daniyal Kazempour · Ilaria Bartolini ·
Erich Schubert · Jian-Jia Chen (Eds.)

Similarity Search and Applications

14th International Conference, SISAP 2021
Dortmund, Germany, September 29 – October 1, 2021
Proceedings

Editors
Nora Reyes
National University of San Luis
San Luis, Argentina

Richard Connor
University of St Andrews
St Andrews, UK

Nils Kriege
University of Vienna
Vienna, Austria

Daniyal Kazempour
Kiel University
Kiel, Germany

Ilaria Bartolini
University of Bologna
Bologna, Italy

Erich Schubert
TU Dortmund University
Dortmund, Germany

Jian-Jia Chen
TU Dortmund University
Dortmund, Germany

ISSN 0302-9743 ISSN 1611-3349 (electronic)
Lecture Notes in Computer Science
ISBN 978-3-030-89656-0 ISBN 978-3-030-89657-7 (eBook)
https://doi.org/10.1007/978-3-030-89657-7

LNCS Sublibrary: SL3 – Information Systems and Applications, incl. Internet/Web, and HCI

This Springer imprint is published by the registered company Springer Nature Switzerland AG
The registered company address is: Gewerbestrasse 11, 6330 Cham, Switzerland

Preface

This volume contains the papers presented at the 14th International Conference on Similarity Search and Applications (SISAP 2021) held between September 29 and October 1, 2021. The conference was hosted by TU Dortmund, Germany. Due to the COVID-19 pandemic and international travel restrictions around the globe, SISAP 2021 was planned as a "hybrid or virtual" event, and in August 2021 it was decided that it would be held as an online conference only due to rapidly increasing incidences in Germany despite a good vaccination rate.

SISAP is an annual forum for researchers and application developers in the area of similarity data management. It focuses on the technological problems shared by numerous application domains, such as data mining, information retrieval, multimedia, computer vision, pattern recognition, computational biology, geography, biometrics, machine learning, and many others that make use of similarity search as a necessary supporting service.

From its roots as a regional workshop in metric indexing, SISAP has expanded to become the only international conference entirely devoted to the issues surrounding the theory, design, analysis, practice, and application of content-based and feature-based similarity search. The SISAP initiative has also created a repository[1] serving the similarity search community, for the exchange of examples of real-world applications, source code for similarity indexes, and experimental testbeds and benchmark data sets.

SISAP 2021 continued the two-year tradition of the SISAP Doctoral Symposium, for which a technical program was assembled to give PhD students an opportunity to present their research ideas in an international research venue. The Doctoral Symposium provides a forum that facilitates interaction among PhD students and stimulates feedback from more experienced researchers. This year's SISAP also included a single special session, on the topic of search in graph-structured data. Again in keeping with previous years, the reviewing process for the special session was integrated with the main conference program to ensure the same quality of acceptance.

The call for papers welcomed full research papers and short research papers, as well as position and demonstration papers, with all manuscripts presenting previously unpublished research contributions.

We received 44 submissions from authors based in 16 different countries. The Program Committee (PC) was composed of 55 members from 20 countries. Each submission received at least four reviews, and the papers and reviews were thoroughly discussed by the chairs and PC members. Based on the reviews and discussions, the PC chairs accepted 23 full papers and 5 short papers, resulting in an acceptance rate of 52% for the full papers and a cumulative acceptance rate of 64% for full and short papers. These rates are a little higher than usual, however the PC chairs are confident that this does not reflect a drop in standards, but rather is an artifact of the context of the COVID-19 pandemic. After a separate review by the Doctoral Symposium Program

[1] https://www.sisap.org/.

Committee members, three Doctoral Symposium papers, giving a clear sample of emerging topics in similarity search and applications, were accepted for presentation and included in the program and proceedings.

The proceedings of SISAP are published by Springer as a volume in the Lecture Notes in Computer Science (LNCS) series. For SISAP 2021, as in previous years, extended versions of selected excellent papers were invited for publication in a special issue of the journal Information Systems. The conference also conferred a Best Paper Award, a Best Student Paper Award, and a Best Doctoral Symposium Paper Award, as judged by the PC chairs and the Steering Committee.

We would like to thank all the authors who submitted papers to SISAP 2021. We would also like to thank all members of the PC and the external reviewers for their effort and contribution to the conference. We want to extend our gratitude to the members of the Organizing Committee for the enormous amount of work they have done, and our sponsors and supporters for their generosity. Finally, we thank all the participants in the online event, who make up the thriving SISAP community.

September 2021

Nora Reyes
Richard Connor
Nils Kriege
Daniyal Kazempour
Ilaria Bartolini
Erich Schubert
Jian-Jia Chen

Organization

General Chairs

Erich Schubert	TU Dortmund University, Germany
Jian-Jia Chen	TU Dortmund University, Germany

Program Committee Chairs

Richard Connor	University of St Andrews, UK
Nora Reyes	Universidad Nacional de San Luis, Argentina

Doctoral Symposium Chair

Ilaria Bartolini	University of Bologna, Italy

Publication Chair

Daniyal Kazempour	Christian-Albrechts-Universität zu Kiel, Germany

Publicity Chair

Peer Kröger	Christian-Albrechts-Universität zu Kiel, Germany

Steering Committee

Laurent Amsaleg	CNRS-IRISA, France
Edgar Chávez	CICESE, Mexico
Michael E. Houle	National Institute of Informatics, Japan
Pavel Zezula	Masaryk University, Czech Republic

Program Committee

Giuseppe Amato	ISTI-CNR, Italy
Laurent Amsaleg	CNRS-IRISA, France
Fabrizio Angiulli	University of Calabria, Italy
Ilaria Bartolini	University of Bologna, Italy
Christian Beecks	University of Münster, Germany
Panagiotis Bouros	Johannes Gutenberg University Mainz, Germany
Benjamin Bustos	University of Chile, Chile

K. Selcuk Candan	Arizona State University, USA
Edgar Chavez	CICESE, Mexico
Alan Dearle	University of St Andrews, UK
Vlastislav Dohnal	Masaryk University, Czech Republic
Vladimir Estivill-Castro	Universitat Pompeu Fabra, Spain
Rolf Fagerberg	University of Southern Denmark, Denmark
Fabrizio Falchi	ISTI-CNR, Italy
Karina Figueroa	Universidad Michoacana de San Nicolas de Hidalgo, Mexico
Claudio Gennaro	ISTI-CNR, Italy
Magnus Lie Hetland	Norwegian University of Science and Technology, Norway
Thi Thao Nguyen Ho	Aalborg University, Denmark
Michael E. Houle	National Institute of Informatics, Japan
Daniyal Kazempour	Christian-Albrechts-Universität zu Kiel, Germany
Nils Kriege	University of Vienna, Austria
Peer Kröger	Christian-Albrechts-Universität zu Kiel, Germany
Yusuke Matsui	University of Tokyo, Japan
Vladimir Mic	Masaryk University, Czech Republic
Luisa Micó	University of Alicante, Spain
Lia Morra	Politecnico di Torino, Italy
Henning Müller	HES-SO, Switzerland
Deepak P.	Queen's University Belfast, UK
Rodrigo Paredes	Universidad de Talca, Chile
Marco Patella	University of Bologna, Italy
Oscar Pedreira	Universidade da Coruna, Spain
Miloš Radovanović	University of Novi Sad, Serbia
Marcela Ribeiro	Federal University of São Carlos, Brazil
Kunihiko Sadakane	University of Tokyo, Japan
Maria Luisa Sapino	Universita' di Torino, Italy
Erich Schubert	TU Dortmund University, Germany
Tetsuo Shibuya	University of Tokyo, Japan
Tomas Skopal	Charles University in Prague, Czech Republic
Nenad Tomasev	Google DeepMind, UK
Caetano Traina	University of São Paulo, Brazil
Goce Trajcevski	Iowa State University, USA
Lucia Vadicamo	ISTI-CNR, Italy
Takashi Washio	Osaka University, Japan
Pascal Welke	University of Bonn, Germany
Kaoru Yoshida	Sony Computer Science Laboratories, Inc., Japan
Pavel Zezula	Masaryk University, Czech Republic
Kaiping Zheng	National University of Singapore, Singapore
Arthur Zimek	University of Southern Denmark, Denmark
Andreas Züfle	George Mason University, USA

Additional Reviewers

Franka Bause	University of Vienna, Austria
Andre Droschinsky	TU Dortmund University, Germany
Erik Thordsen	TU Dortmund University, Germany
Florian Kurpicz	Karlsruhe Institute of Technology, Germany
Lukas Miklautz	University of Vienna, Austria
Lutz Oettershagen	University of Bonn, Germany
Till Schulz	University of Bonn, Germany

Additional Reviewers

Contents

Similarity Search and Retrieval

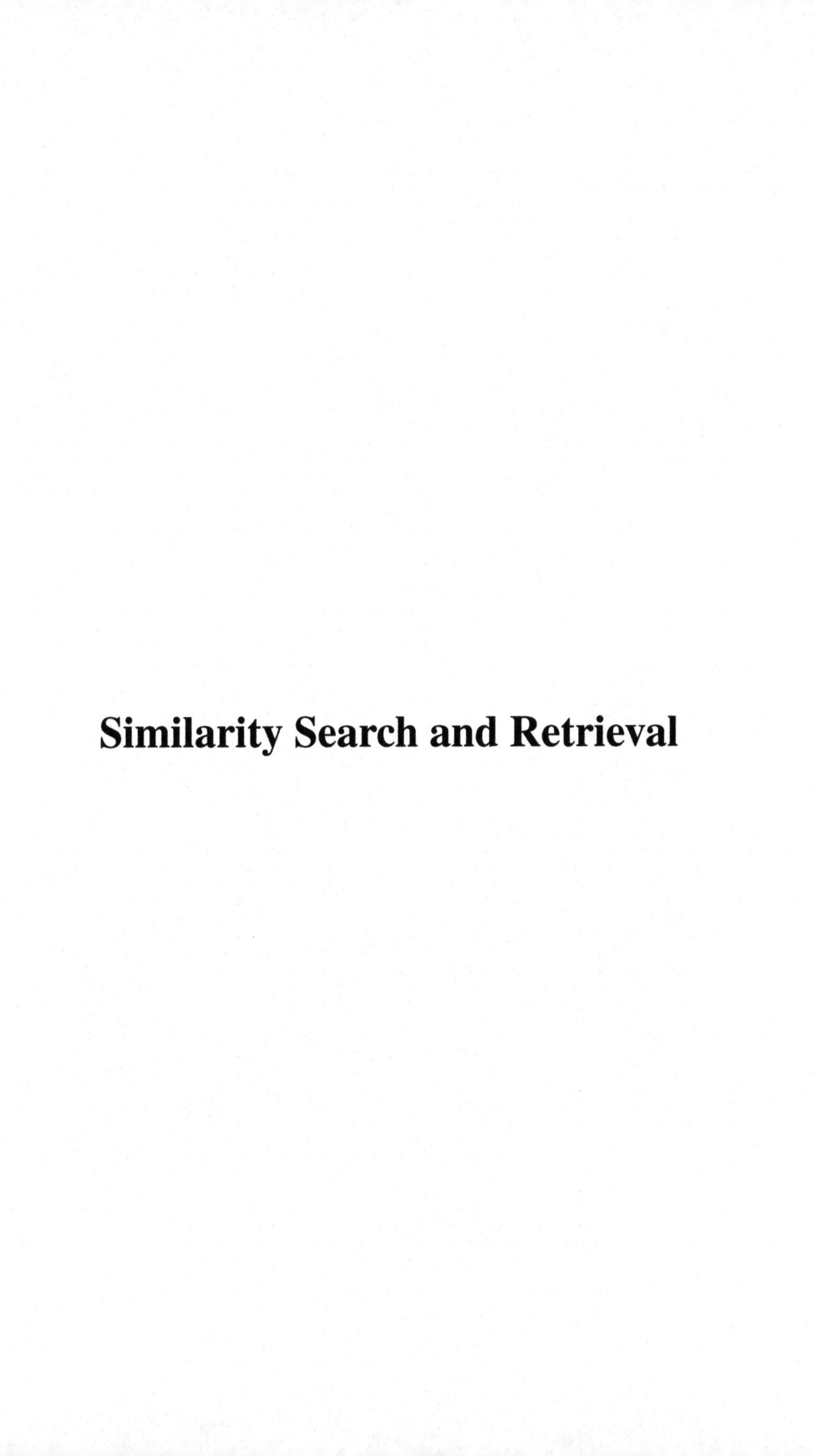

Similarity Search and Retrieval

Organizing Similarity Spaces Using Metric Hulls

Miriama Jánošová, David Procházka, and Vlastislav Dohnal(✉)

Faculty of Informatics, Masaryk University, Brno, Czech Republic
{x424615,xprocha6,dohnal}@fi.muni.cz

Abstract. A novel concept of a metric hull has recently been introduced to encompass a set of objects by a few selected border objects. Following one of the metric-hull computation methods that generate a hierarchy of metric hulls, we introduce a metric index structure for unstructured and complex data, a Metric Hull Tree (MH-tree). We propose a construction of MH-tree by a bulk-loading procedure and outline an insert operation. With respect to the design of the tree, we provide an implementation of an approximate kNN search operation. Finally, we utilized the Profimedia dataset to evaluate various building and ranking strategies of MH-tree and compared the results with M-tree.

Keywords: Metric-hull tree · Metric hull · Index structure · Nearest-neighbor query · Similarity search

1 Introduction

Content-based retrieval systems have become often applied to complement traditional retrieval systems. Such systems allow processing complex data, such as photos, medical images, protein sequences or audio recordings, and support similarity queries. Such search requests compare data items based on the similarity of their content or descriptors extracted from the content rather than the identity of data. The challenge is managing the ever-growing complex data efficiently and evaluating the similarity queries faster than by the sequential scan. Many indexing structures were proposed ranging from clustering-based ones [7,22], space-partitioning methods [4,6] to transformation techniques [1].

Complex data are thus expressed as descriptors capturing important features from their content, e.g., color histogram, texture, shape [21] or more profound vectors computed by convolutional networks [10]. Thus, the descriptors are often high-dimensional spaces[1] [18]. The problem of *dimensionality curse* then arises [5]. It leads to visiting many data partitions by an index due to frequent overlaps among them, whereas useful information is contained in a few of them.

[1] Or even distance spaces where no implicit coordinate system is defined.

The publication of this paper and the follow-up research was supported by the ERDF "CyberSecurity, CyberCrime and Critical Information Infrastructures Center of Excellence" (No. CZ.02.1.01/0.0/0.0/16_019/0000822).

N. Reyes et al. (Eds.): SISAP 2021, LNCS 13058, pp. 3–16, 2021.
https://doi.org/10.1007/978-3-030-89657-7_1

So the index must employ further filtering constraints to make query evaluation efficient [12,20].

A novel concept of metric hulls has been introduced recently [2]. The purpose of the metric hull is to embrace a set of metric objects. The metric hull is defined as a set of objects selected out of the set to encompass. We build upon this concept to create a hierarchical search structure where a metric hull represents each node. Since the authors also provide a test of whether a metric object is part of the hull or not, such a structure is viable. We perceive the metric hull as an alternative to the metric ball used by M-tree [7] or Slim-tree [22]. However, it bounds data much tighter without any additional information. As a result, node overlaps can be reduced. The issue of intersecting balls surrounding Voronoi cells is studied in VD-tree [13].

This paper proposes a metric access method that organizes data in metric hulls and addresses the issue of large node overlaps without the need for external pivots, as was applied in Pivoting M-tree [19]. We take advantage of algorithms to construct metric hulls incrementally [2] to build a hierarchy of metric hulls. Next, the issue of comparing and ordering metric hulls with respect to a similarity query is studied here in this paper. We test different variants of such and evaluate the performance of approximate k-nearest neighbors search.

The remaining parts of the paper are structured as follows. In the next section, there is a concise summary of metric space indexing and similarity queries, and more importantly the concept of metric hulls. Related work of indexing structures is surveyed in Sect. 3. The core of this paper is the proposal of Metric Hull Tree, presented in Sect. 4. Performance evaluation on a real-life high-dimensional data is described in Sect. 5. Contributions of this paper and possible future extensions are summarized in the last section.

2 Preliminaries

A *metric space* M is a pair $\mathcal{M} = (\mathcal{D}, d)$, where $\mathcal{D}$ is a domain of objects, and d is a distance function (metric) $d : \mathcal{D} \times \mathcal{D} \to \mathbb{R}_0^+$ satisfying metric postulates, namely non-negativity, the identity of indiscernibles, symmetry, and triangle inequality. A set of data objects to be queried, so-called *database*, is denoted as $X \subseteq \mathcal{D}$.

We distinguish two common retrieval operations, specifically, the *range query* ($range(q, r)$) – returning database objects, such that their distance to q is smaller than the distance r; and the *k-nearest neighbors query* (kNN(q)) – retrieving k database objects closest to the query object q; when there are more objects at the distance of the k-th nearest neighbor, the ties are solved arbitrarily. Nowadays, approximate evaluation of similarity queries (e.g., approximate kNN(q)) loosens the restrictions on returning the genuine answer at much lower search costs. Such evaluation can be implemented by an early termination strategy that stops the search when a predefined number of data objects are visited. The identification of the most relevant data parts is thus the center of interest.

A more advanced data processing technique is the Similarity Group By (SGB) query [11]. It groups data by respecting similarity constraints, e.g., distance threshold. However, the disadvantage of such queries is that the obtained groups are mere lists of objects. Thus, there is no compact representation of such groups. Hence, the objective of [2] was to examine properties of objects' groups' representations, where the *hull representation* proved to be the most compact.

2.1 Hull Representation

Let C be a group of objects from database $C \subseteq X$. Formally, the **hull representation** [2] is defined as $\mathcal{H}(C) = \{p_i \mid p_i \in C\}$ and any other object $o \in C$ is **covered** by hull. Each p_i corresponds to a boundary object of C referred to as *hull object.*

Let $\mathcal{H}$ be a hull representation $\mathcal{H} = \{p_1, \ldots, p_h\}$ and an object $o \in \mathcal{D}$. Assume p_{NN} to be the nearest hull object of $\mathcal{H}$ to o, i.e., $NN = \text{argmin}_{i=1..h}(d(o, p_i))$. We say the object o is **covered** by $\mathcal{H}$ if and only if

$$\sum_{i=1..h, i \neq NN} d(p_i, o) \leq \sum_{i=1..h} d(p_i, p_{NN}). \tag{1}$$

By the original definition, the smallest hull consists of three objects. If $|\mathcal{H}| < 3$, the hull objects are the only objects covered.

Antol et al. [2] proposed two algorithms for hull computation for a set of objects C. First, the *Basic Hull Algorithm* starts with selecting the furthest object in O and gradually adds additional objects that are furthest objects from the already selected ones. E.g., the third object has the maximum sum of distances to the previous two. This procedure terminates when the whole set C is covered by $\mathcal{H}$. In the worst case, each $o \in C$ becomes a hull object. Second, the *Optimized Hull Algorithm* is an improvement to the basic one, which reduces the number of hull objects. After selecting the initial three hull objects, the procedure is modified: instead of adding the furthest not-yet-covered object o_f to $\mathcal{H}$, the algorithm tries to replace some existing hull object with o_f to increase the coverage of C. This leads to fewer hull objects without compromising the fact that each object of C is covered by the resulting $\mathcal{H}$.

3 Existing Metric Access Methods

In this section, we overview existing metric indexes relevant to this work. We start with structures organizing objects into metric balls. The first disk-oriented and dynamic structure built in the bottom-up fashion is the M-tree [7]. Data objects are grouped into leaf nodes that are, in turn, represented by metric balls (i.e., a routing object and a covering radius). Further levels group metric balls into larger ones ending with l entries in the root node. The disadvantage is major overlaps among such ball regions. A slim-down algorithm in Slim-trees [22] has later optimized such an issue. The tree compactness is measured by the fat factor there. Additional objects are included in internal nodes to further split balls into hyperplanes in M$^+$-tree [25] and BM$^+$-tree [26]. Pivoting M-tree [19] selects a fixed set of pivots that are globally used to define ranges on distances within which objects reside – spherical cuts. This resembles Linear AESA principle [24], which recomputes distances to fixed pivots and stores them in arrays to fast array-range filtering.

The other methods partition the data space by hyperplanes. GH-tree [3,23] is the binary hyper-plane tree that was later generalized to recursive Voronoi tree, call GNAT [6]. The dynamic version is EGNAT [15], which bulk-loads the tree and then allows minor updates. Since metric balls provide a simple yet efficient way of filtering tree branches, they were incorporated into Voronoi diagrams in NOBH-tree [16]. A common disadvantage of Voronoi-diagram-based indexes is the difficulty of redefining the partitioning at reasonable costs when the tree becomes unbalanced. This is tackled in VD-tree [13]. The objects are swapped between Voronoi cells to reduce overlaps, which is analogous to the slim-down algorithm [22].

The concept of metric ball regions is widely used and proved advantageous when combined with Voronoi partitioning or pivot-based filtering (Linear AESA). This paper exploits the brand new proposal of metric hulls that can bound a set of objects tightly by outliers. Metric hulls are an alternative to selecting external pivots and consequential definition of constraints on distances for each individual tree nodes.

4 MH-Tree – The Proposed Method

This section describes the proposed Metric Hull Tree (MH-tree) that represents data partitions by metric hulls. The hulls are constructed bottom-up by following the grounds of *Incremental Hull Algorithm* [9]. Literally, it gradually merges hulls until only one final hull representation is obtained. However, the original merging procedure needs to be generalized to support larger arity than two and any capacity of leaf buckets.

4.1 Structure and Bulk Loading

The MH-tree is a hierarchical tree structure composed of two node types, as depicted in Fig. 1. Each *Leaf node* encapsulates a bucket – a storage of $[c, 2c]$ objects. Each leaf node is rooted for an internal node and represented there by a metric hull. The *Internal node* manages up to a pairs of a hull representation ($\mathcal{H}_i$) and a pointer (ptr_i) to a leaf node. Each hull is constructed by calling the *Optimized Hull Algorithm*, see Sect. 2.1 and [2].

We construct the MH-tree by a bulk-loading procedure. Firstly, we group the database objects into leaf nodes containing c objects. Secondly, a closest leaf nodes are merged, thus obtaining a level of internal nodes. This merging is repeated until one node is obtained, becoming the root of the MH-tree. This procedure creates a balanced a-ary tree. We present it in pseudo-code in Algorithm 1. If there are too few objects (incapable of forming at least two leaves), we create just one leaf node that forms the root. In the following, we detail the sub-algorithms.

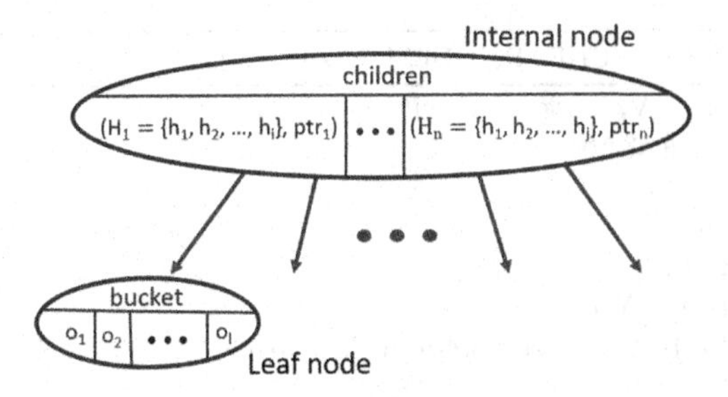

Fig. 1. A schema of MH-tree structure.

Algorithm 1: BulkLoad(X, c, a)

Input: a database X, bucket capacity c, arity a
Output: a root node of MH-tree

```
1  if |X| < 2c then
2  |   root ← create a new leaf node;
3  |   root.H ← compute Optimized Hull representation of X;
4  |   root.bucket ← X;
5  |_  return root;
6  nodes ← CreateLeafNodes(X, c);
7  while |nodes| ≠ 1 do
8  |_  nodes ← MergeNodes(nodes, a);
9  root ← nodes[0];
10 return root;
```

The database X is clustered by forming compact clusters of c objects. Thus, $\lfloor \frac{|X|}{c} \rfloor$ leaf nodes are created. Algorithm 2 presents the pseudo-code of CreateLeafNodes. The procedure starts with selecting the furthest object o_f from a random object in X (Lines 7–8), a new cluster's nucleus. The second object in the cluster is the nearest neighbor of o_f. To add the next object, we execute 1NN queries for each object already assigned to the cluster and choose the neighbor that minimizes the sum of distances to objects already in the cluster. This is repeated until the cluster contains c objects (Lines 9–12). The next cluster is formed by analogy but omitting the already clustered objects. If there are fewer than c unprocessed objects, they get assigned to closest clusters directly, i.e., some clusters can contain more than c objects. Finally, the leaf nodes storing the clusters' objects in buckets are returned.

The motivation of Algorithm 2 is to create compact clusters of up to $2c$ objects also for data with outliers and/or overlapping clusters. Here, agglomerative clustering linking closest objects/clusters would create much more overlaps among hulls. In particular, whenever a cluster exceeds c objects, it is taken out and forms a leaf node. Consequently, the remaining objects would very likely be outliers. They would group to a hull that would span over all the other nodes.

The next bulk-loading stage is merging leaf nodes to create a balanced structure of internal nodes. It is specified in Algorithm 3. By analogy we start with

Algorithm 2: CREATELEAFNODES(X, c)

```
   Input: a database X, bucket capacity c
   Output: a set of leaf nodes
1  leafNodes ← ∅;
2  while X ≠ ∅ do
3      if |X| < c then
4          foreach o ∈ X do
5              add o to the closest node in leafNodes;
6          break
7      o_f ← the furthest object in X for a randomly picked object from X;
8      X ← X \ o_f; cluster ← { o_f };
9      while |cluster| < c do
10         NNs ← { o_n | ∃o_c ∈ cluster : o_n ∈ 1NN(o_c) };
11         o ← argmin_{o_n ∈ NNs} Σ_{o_c ∈ cluster} d(o_n, o_c);
12         X ← X \ o; cluster ← cluster ∪ { o };
13     leaf ← create a new leaf node;
14     leaf.bucket ← cluster;
15     leafNodes ← leafNodes ∪ { leaf };
16 return leafNodes;
```

the furthest leaf node (n_f) and execute the aNN(n_f) query to get a cluster of a near leaf nodes, n_f inclusively. These nodes form an internal node for the next tree level. We repeat this procedure until all nodes are processed. The identification of furthest and close nodes is based on a comparison of nodes' hulls. We consider the distance between hulls $\mathcal{H}_1$ and $\mathcal{H}_2$ to be defined as:

$$d(\mathcal{H}_1, \mathcal{H}_2) = \min_{\forall h_1 \in \mathcal{H}_1, \forall h_2 \in \mathcal{H}_2} d(h_1, h_2). \quad (2)$$

So, the furthest node is thus the node whose hull is furthest from the hull of a randomly picked node (out of not-yet-processed ones). The outcome of Algorithm 3 is a list of nodes constituting the next level of MH-tree. We apply it until only one node is returned – the root node.

To create the hull representations we utilize the Optimized Hull Algorithm (called from Algorithm 3). When merging leaf nodes, the Optimized Hull Algorithm is invoked on the objects of the leaf node's bucket to obtain a hull. In this course, we would collect all objects from the previously merged nodes to create a hull. But the computational requirements would grow steeply then. Instead, in the next generations we gather the hull objects from all hulls in the internal node and compute the new hull on them solely. This practice introduces imprecision of hulls – some objects stored in the sub-tree may not be covered. We address this issue on the kNN search algorithm.

Algorithm 3: MERGENODES(N, a)

```
   Input: a set of nodes N (at the same level of the tree), arity a
   Output: a set of internal nodes of the upper level
 1 level ← ∅; notProc ← N;
 2 while notProc ≠ ∅ do
 3     if |notProc| ≤ a then
 4         create a new internal node nNode;
 5         foreach node ∈ notProc do
 6             H ← call Optimized Hull Algo on node;
 7             ptr ← pointer to node;
 8             nNode.HullChildPairs ← nNode.HullChildPairs ∪ { (H, ptr) };
 9         level ← level ∪ { nNode };
10         break
11     n_f ← extract the furthest node from notProc;
12     C ← execute (a − 1)NN(n_f) query on notProc;
13     notProc ← notProc \ C;
14     create a new internal node nextNode;
15     foreach node ∈ C ∪ { n_f } do
16         H ← call Optimized Hull Algo on node;
17         ptr ← pointer to node;
18         nextNode.HullChildPairs ← nextNode.HullChildPairs ∪ { (H, ptr) };
19     level ← level ∪ { nextNode };
20 return level;
```

4.2 Searching in the MH Tree

We outline the kNN search algorithm in Algorithm 4. We assume a limit on the number of visited data objects is passed, so the STOP function can terminate the search early. Such limitation together with the imprecisions introduced during the building procedure result in acquiring an approximate result. The algorithm starts from the root node and maintains a queue of nodes to be inspected. This queue is ordered by the "likelihood" of the node to contain relevant data. It can be defined as a lower bound or upper bound distance for the query object q to the nearest/furthest object in a node's hull $\mathcal{H}$. We define it using the hull objects exclusively, so the lower and upper bound are defined as

$$d_l(q, \mathcal{H}) = \min_{\forall h \in \mathcal{H}} d(q, h); \tag{3}$$

$$d_u(q, \mathcal{H}) = \max_{\forall h \in \mathcal{H}} d(q, h). \tag{4}$$

The actual definition of RANK function is investigated in the experiments in Sect. 5.

The exact evaluation of the kNN query can be obtained by setting the approximate limit to 100%. Even though this being a straightforward solution, it leads to scanning the whole database. Rather, the check comparing the distance to

Algorithm 4: ApproximateKNNSearch(q, k, Rank, $limit$)

```
Input: a query object q, number of nearest neighbors k, rank function Rank,
       approximation parameter limit
Output: a set of nearest neighbors found
1  answer ← ∅; // ordered set by objects' distances from q
2  PQ ← create a priority queue with the priority determined by Rank;
3  insert the root node into PQ with zero priority;
4  while PQ is not empty do
5      // early termination after a certain percentage of visited objects
6      if Stop(limit) then
7          break
8      node ← extract the node with highest priority from PQ;
9      if node is a leaf node then
10         foreach o ∈ node.bucket do
11             if |answer| < k then
12                 add o into answer;
13                 continue;
14             o_k ← the k^th object from query object q in answer;
15             if d(q, o) < d(q, o_k) then
16                 insert o in answer and remove o_k from answer;
17     else
18         foreach pair ∈ node.HullChildPairs do
19             insert pair.ptr into PQ with the priority RANK(q, pair.H, k);
20 return answer;
```

the kth nearest-neighbor candidate and the distance to the hull of PQ's head element must be defined. Since the current bulk-loading algorithm does ensure coverage such a test may not be ensuring result correctness. Our primary aim in this paper is to show the viability of the application of metric hulls to indexing, and we do not study the exact evaluation. A promising direction to define a more efficient exact-search algorithm is presented in [8]. The author exploits transformation of a metric space by multiple pivots, and defines a constraint on distance of an object in such a pivot space.

4.3 Dynamicity

After bulk-loading the MH-tree, we can insert new objects to corresponding leaf nodes. Firstly, we locate the most suitable leaf $\mathcal{N}_{bl}$ by executing kNN search algorithm with the newly inserted object as the query object (Algorithm 4). When a leaf node is extracted from the search queue (Line 8), we stop the search and insert the new object o into the leaf node's bucket. Next, if o is not covered by the node's hull $\mathcal{H} = \{h_1, \ldots, h_l\}$, it is updated with linear costs: $\forall i$, try to replace h_i with o and test whether h_i is covered by such updated hull. If so, this hull is stored in the leaf's parent. Otherwise, o is added as a new hull object, i.e., the new hull $\mathcal{H} \cup \{o\}$ is stored. The hulls of parent nodes need not be updated, since the closeness of hulls to an object is addressed in the tree traversal (by the RANK function).

If the capacity of the leaf $\mathcal{N}_{bl}$ is double exceeded, we split the leaf into two new nodes $\mathcal{N}_1$ and $\mathcal{N}_2$ by these steps: (i) we identify an outlier in $\mathcal{N}_{bl}$'s bucket (the furthest object from a random one), and halve the bucket's objects (incl. o) according to the distance from the outlier – $c+1$ objects closest to the outlier from the bucket of $\mathcal{N}_1$ and the remaining objects are stored in the bucket of $\mathcal{N}_2$. The new nodes are linked to $\mathcal{N}_{bl}$'s parent if there is room for them. Otherwise, a new internal node is created and roots the new leaf nodes. So, the MH-tree becomes unbalanced then. If the structure becomes highly unbalanced, it should be rebuilt from scratch.

5 Experimental Evaluation

This section provides an experimental study of the proposed Metric Hull Tree and compares its performance with M-tree with Slim-down.

We used the Profimedia [14] dataset in the experiments. It consists of a series of 4096-dimensional vectors extracted from Photo-stock images using a convolutional neural network. To measure the similarity between the data, we utilize the Euclidean distance. Our experiments were executed upon two sizes of Profimedia dataset – 10,000 and 100,000 objects.

To compare the performance, we execute kNN queries and measure the acquired recall. Specifically, we employ the approximate kNN search with an early termination strategy of visiting a certain percentage of database objects. The evaluation starts with the approximation parameter set to 5%, continued by a gradual increase of it in five percentage-point steps up to 100%. To quantify the trade-off between the accuracy and the efficiency of the approximate search, we compute $recall = \frac{|S \cap S_a|}{S}$, where the set S corresponds to the result of (precise) kNN query and set S_a to the result of approximate kNN query.

The pivotal part of effective search in MH-tree is the selection of the ranking function in priority queues. Naturally, we exploit just hull objects to define the ranking function. In the following sections, we study the influence of various ranking functions on recall in a shallow tree configuration, where we examine the ranking of the leaf nodes solely, and in deep tree configuration, where we take into consideration the ranking of both internal and leaf nodes. Lastly, we compare the best definition of ranking function in MH-tree with M-tree.

5.1 Ordering Leaf Nodes

We analyse the way of ordering leaf nodes by comparing their hulls to an object, e.g., a query object. The most encouraging way uses the distance from the query object to the nearest hull object. Formally, the rank of a leaf node is defined as follows

$$\text{RANK}_{LEAF}(q, \mathcal{H}) = \min_{\forall h \in \mathcal{H}} d(q, h). \tag{5}$$

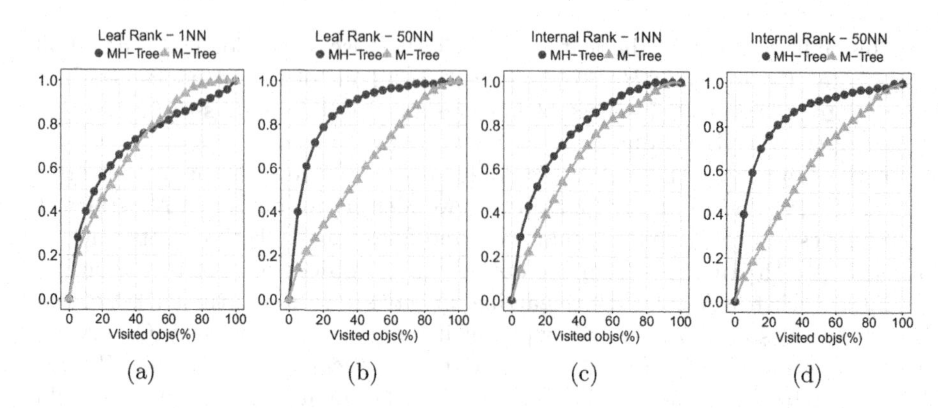

Fig. 2. Average recall of MH-tree and M-tree for 1NN and 50NN queries: (a, b) ranking leaf nodes in shallow structures, and (c, d) final ranking of both types of nodes in hierarchical structures. All results are on 10,000 Profimedia images.

To determine the efficiency of the $rank_{LEAF}$ function, we evaluate the shallow configurations of MH-tree as well as M-tree, i.e., the root node referencing all leaf nodes. In Fig. 2 ((a) and (b)), we provide the comparison for 1NN and 50NN queries varying the approximation limit. Regarding the 1NN query (Fig. 2a), the MH-tree surpasses the performance of the M-tree in the first 50% of visited objects. After that, the M-tree manages to gather faster growth of recall. However, the performance of both approaches is more or less the same.

On the contrary, the average recall of the 50NN query (Fig. 2b) reveals a much better performance of the MH-tree. It reaches 80% recall while visiting 20% of the database only. The M-tree manifests almost linear growth.

We also tested another variant of leaf-node ranking functions, e.g., the distance to the furthest hull object and average distance to all hull objects, but this nearest variant performed the best. All results are available in the bachelor's thesis [17].

5.2 Ordering Internal Nodes

To efficiently traverse also deep MH-tree structures, i.e., the multi-level ones, we need to determine the best-ranking strategy with respect to both leaf and internal nodes. Firstly, we examined ordering based solely on the distance between the query and the node without taking into consideration whether or not the query is covered by the node's hull representation. Such ranking roughly corresponds to RANK_{LEAF}. However, we experienced only a linear growth of recall to the number of visited nodes for the 50NN query. Thus, to improve navigation, the rank used in the priority queue needs to be more sophisticated.

The $\text{RANK}(q, \mathcal{H}, k)$ function, defined in Table 1, computes the rank depending on whether or not the query object is covered by the hull and also on the number of neighbors k to be retrieved. We included the ranking of leaves there for completeness. For 1NN queries, it is more convenient to prefer hulls that

Table 1. The definition of the most efficient rank function $\text{RANK}(q, \mathcal{H}, k)$.

Conditions			Value
Node type	k	*covered*	$\text{RANK}(q, \mathcal{H}, k)$
Internal	1	YES	$-\max_{\forall h \in \mathcal{H}} d(q,h)$
		NO	$\max_{\forall h \in \mathcal{H}} d(q,h)$
	>1	YES	$-\min_{\forall h \in \mathcal{H}} d(q,h)$
		NO	$\min_{\forall h \in \mathcal{H}} d(q,h)$
Leaf			$\min_{\forall h \in \mathcal{H}} d(q,h)$

are closer overall, so the furthest hull object is used. Whereas the nearest one is the best performing for any $k > 1$, since we do not know how many objects are present there in advance. Notably, the covered condition provides a better ordering of hulls that contain the query object, so hulls with q more to its center are preferred. More details are in the bachelor's thesis [17].

Figure 2c presents the average recall of the MH-tree and M-tree in the 1NN query. Notice that the recall has almost doubled per the same amount of visited nodes compared to Fig. 2a. Therefore, the *rank* is able to reflect the performance of RANK_{LEAF} while proving that it is also able to navigate the tree effectively. We observe similar behavior when comparing Figs. 2d and 2b on 50NN queries. The MH-tree achieves significantly better recall than M-tree. The difference in MH-tree's performance on 50NN when being shallow or deep is marginal, proving the node navigation by the rank function is robust.

5.3 Comparison

We compare the MH-tree with M-tree on 100,000 objects from Profimedia. The M-tree was built with the slim-down algorithm [22] to make it as efficient and compact as possible. To validate the quality and performance of the structures, we set the structure parameters to obtain similar trees. They are summarized in Table 2. We report the values of fat-factor [22] that quantifies overlaps of covering regions representing tree nodes. The fat factor is a relative quantity computed as an average performance of zero-radius range queries for each database object. If the search for an object visits exactly one node per level, the fat factor is zero. In the worst case, all tree nodes are visited. The fat factor then grades the tree with one.

In Figs. 3a and 3b, we summarize the average recall of MH-tree compared to M-tree in 1NN and 50NN queries. The recall of MH-tree rises much steeper than M-tree's up to visiting 15% of the database. For example, MH-tree provides 88% recall for 50NN queries compared to 29% of M-tree. Figures 3c and 3d showcase the details on performance of various kNN queries when approximation is fixed to 10% and 20% of dataset. The recall of M-tree deteriorates with increasing k. MH-tree's recall is more steady. The results manifest that the MH-tree is able to outperform the M-tree even on large datasets significantly.

Table 2. Features of MH-tree and M-tree build on 100,000 Profimedia images.

	Params		Building statistics					
	Arity	Leaf cap.	Height	Internal nodes	Leaf nodes	Routing objects	Fat factor	Building time (s)
MH-tree	100	100	2	11	1000	3262	0.03	1548
M-tree	100	200	3	51	2546	2596	0.56	67

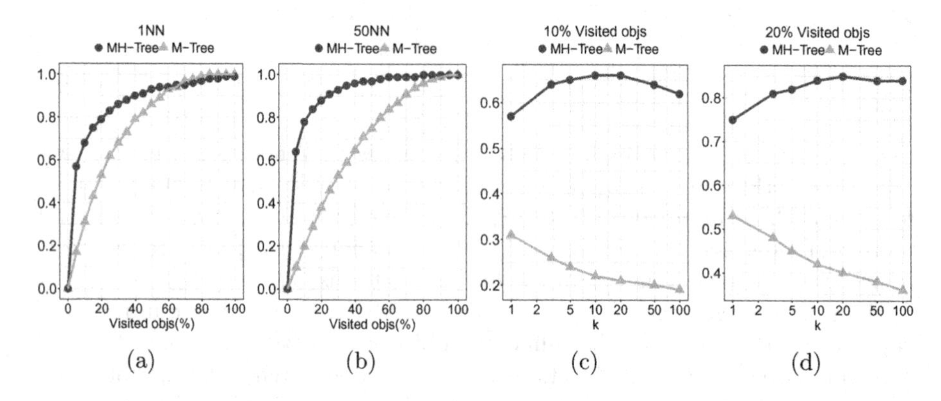

Fig. 3. Average recall of MH-tree and M-tree in 1NN and 50NN queries in Deep tree structure (a), (b). Average recall of MH-tree, M-tree for varying k in the kNN queries (c), (d). All results are on 100,000 Profimedia images.

The negative point is the construction costs that are quite high and can only be amortized when managing mostly static data. We did not focus on optimizing the building routine, but the concept of hulls can eliminate node overlaps to a large extent.

6 Conclusions

MH-tree is an index structure build upon the novel concept of metric hulls. We proposed algorithms for building a hierarchy of metric hulls that organizes data objects into leaf nodes, which are gradually merged into internal nodes constrained by metric hulls. In addition to such a bulk-loading procedure, we outlined the dynamic insertion of new objects. The fat factor of MH-tree is by one order of magnitude smaller than of M-tree with slimming-down. This proves the compactness of metric hull representation. Admittedly, this can also be accounted to the building process of MH-tree that groups close objects primarily.

We proposed and analysed a node-ranking function that orders nodes by their closeness to a query object. The bases of leaf and internal nodes' ranking differ – the distance to the closest hull object is taken as the measure for leaf nodes. In contrast, the distance to the furthest hull object is the means for internal nodes. We also showed that coverage of the query object by a hull needs to be employed in order to navigate deeper tree structures effectively. In addition, we achieved the highest recall when distinguishing between retrieval of one neighbor and multiple nearest neighbors.

Finally, we compared the best-performing setup of MH-tree with the M-tree built by the slim-down algorithm. The results showcase that MH-tree outperforms M-tree significantly – fewer nodes are visited for the same recall or vise versa. Specifically, the performance of MH-tree was higher by 20–30% on average compared to M-tree per the same amount of visited objects on smaller datasets. The differences were even more pronounced on larger data, 30–40% higher average recall of MH-tree depending on the number of extracted neighbors.

The future work would focus on generating representations with more hull objects, thus keeping the hulls even more compact. This could result in a significant improvement in approximate search. In addition, formulating mature ranking strategies could lead to a finer-grained tree traversal. Lastly, we will compare MH-tree with techniques using external pivots, e.g., Pivoting M-tree.

References

1. Amato, G., Gennaro, C., Savino, P.: MI-File: using inverted files for scalable approximate similarity search. Multimedia Tools Appl. **71**(3), 1333–1362 (2012). https://doi.org/10.1007/s11042-012-1271-1
2. Antol, M., Janosova, M., Dohnal, V.: Metric hull as similarity-aware operator for representing unstructured data. Pattern Recognit. Lett. 1–8 (2021). https://doi.org/10.1016/j.patrec.2021.05.011
3. Batko, M.: Distributed and scalable similarity searching in metric spaces. In: Lindner, W., Mesiti, M., Türker, C., Tzitzikas, Y., Vakali, A.I. (eds.) EDBT 2004. LNCS, vol. 3268, pp. 44–53. Springer, Heidelberg (2004). https://doi.org/10.1007/978-3-540-30192-9_5
4. Batko, M., Dohnal, V., Zezula, P.: M-grid: similarity searching in grid. In: P2PIR 2006: International Workshop on Information Retrieval in Peer-to-Peer Networks (2006). https://doi.org/10.1145/1183579.1183583
5. Böhm, C., Berchtold, S., Keim, D.A.: Searching in high-dimensional spaces: index structures for improving the performance of multimedia databases. ACM Comput. Surv. **33**(3), 322–373 (2001). https://doi.org/10.1145/502807.502809
6. Brin, S.: Near neighbor search in large metric spaces. In: Proceedings of the International Conference on Very Large Data Bases (1995)
7. Ciaccia, P., Patella, M., Zezula, P.: M-tree: an efficient access method for similarity search in metric spaces. In: Proceedings of the 23rd International Conference on Very Large Data Bases (VLDB), pp. 426–435. Morgan Kaufmann (1997)
8. Hetland, M.L.: Comparison-based indexing from first principles. arXiv preprint arXiv:1908.06318 (2019)
9. Jánošová, M.: Representing sets of unstructured data. Master thesis, Masaryk University, Faculty of Informatics (2020). https://is.muni.cz/th/vqton/
10. Krizhevsky, A., Sutskever, I., Hinton, G.E.: ImageNet classification with deep convolutional neural networks. Commun. ACM (2017). https://doi.org/10.1145/3065386

11. Laverde, N.A., Cazzolato, M.T., Traina, A.J., Traina, C.: Semantic similarity group by operators for metric data. In: Beecks, C., Borutta, F., Kröger, P., Seidl, T. (eds.) Similarity Search and Applications. LNCS, vol. 10609, pp. 247–261. Springer, Cham (2017). https://doi.org/10.1007/978-3-319-68474-1_17
12. Mic, V., Novak, D., Zezula, P.: Binary sketches for secondary filtering. ACM Trans. Inf. Syst. **37**(1), 1:1–1:28 (2019). https://doi.org/10.1145/3231936
13. Moriyama, A., Rodrigues, L.S., Scabora, L.C., Cazzolato, M.T., Traina, A.J.M., Traina, C.: VD-Tree: how to build an efficient and fit metric access method using voronoi diagrams. In: Proceedings of the 36th Annual ACM Symposium on Applied Computing (SAC), pp. 327–335. ACM, New York (2021)
14. Novak, D., Batko, M., Zezula, P.: Large-scale image retrieval using neural net descriptors. In: Proceedings of the 38th International ACM SIGIR Conference on Research and Development in Information Retrieval, pp. 1039–1040. ACM (2015)
15. Paredes, R.U., Navarro, G.: EGNAT: a fully dynamic metric access method for secondary memory. In: 2nd International Workshop on Similarity Search and Applications, SISAP 2009 (2009). https://doi.org/10.1109/SISAP.2009.20
16. Pola, I.R.V., Traina, C., Traina, A.J.M.: The NOBH-tree: improving in-memory metric access methods by using metric hyperplanes with non-overlapping nodes. Data Knowl. Eng. (2014). https://doi.org/10.1016/j.datak.2014.09.001
17. Procházka, D.: Indexing structure based on metric hulls. Bachelor thesis, Masaryk University, Faculty of Informatics (2021). https://is.muni.cz/th/jk21s/
18. Samet, H.: Foundations of Multidimensional and Metric Data Structures. The Morgan Kaufmann Series in Data Management Systems. Morgan Kaufmann (2006)
19. Skopal, T., Pokorný, J., Snasel, V.: PM-tree: Pivoting Metric Tree for Similarity Search in Multimedia Databases. ADBIS, Computer and Automation Research Institute Hungarian Academy of Science (2004)
20. Skopal, T., Pokorný, J., Snášel, V.: Nearest neighbours search using the PM-tree. In: Zhou, L., Ooi, B.C., Meng, X. (eds.) DASFAA 2005. LNCS, vol. 3453, pp. 803–815. Springer, Heidelberg (2005). https://doi.org/10.1007/11408079_73
21. Smith, J.R.: MPEG7 standard for multimedia databases. SIGMOD Record (2001). https://doi.org/10.1145/376284.375814
22. Traina, C., Traina, A., Faloutsos, C., Seeger, B.: Fast indexing and visualization of metric data sets using Slim-trees. IEEE Trans. Knowl. Data Eng. (2002). https://doi.org/10.1109/69.991715
23. Uhlmann, J.K.: Satisfying general proximity/similarity queries with metric trees. Inf. Process. Lett. **40**(4), 175–179 (1991)
24. Vilar, J.M.: Reducing the overhead of the AESA metric-space nearest neighbour searching algorithm. Inf. Process. Lett. **56**(5), 265–271 (1995)
25. Zhou, X., Wang, G., Yu, J.X., Yu, G.: M+-tree: a new dynamical multidimensional index for metric spaces. In: Proceedings of the 14th Australasian Database Conference, pp. 161–168 (2003)
26. Zhou, X., Wang, G., Zhou, X., Yu, G.: BM$^+$-tree: a hyperplane-based index method for high-dimensional metric spaces. In: Zhou, L., Ooi, B.C., Meng, X. (eds.) DASFAA 2005. LNCS, vol. 3453, pp. 398–409. Springer, Heidelberg (2005). https://doi.org/10.1007/11408079_36

Scaling Up Set Similarity Joins Using a Cost-Based Distributed-Parallel Framework

Fabian Fier(✉) and Johann-Christoph Freytag

Humboldt-Universität zu Berlin, Berlin, Germany
{fier,freytag}@informatik.hu-berlin.de

Abstract. The set similarity join (SSJ) is an important operation in data science. For example, the SSJ operation relates data from different sources or finds plagiarism. Common SSJ approaches are based on the filter-and-verification framework. Existing approaches are sequential (single-core), use multi-threading, or Map-Reduce-based distributed parallelization. The amount of data to be processed today is large and keeps growing. On the other hand, the SSJ is a compute-intensive operation. None of the existing SSJ methods scales to large datasets. Single- and multi-core-based methods are limited in terms of hardware. MapReduce-based methods do not scale due to too high and/or skewed data replication. We propose a novel, highly scalable distributed SSJ approach. It overcomes the limits and bottlenecks of existing parallel SSJ approaches. With a cost-based heuristic and a data-independent scaling mechanism we avoid intra-node data replication and recomputation. A heuristic assigns similar shares of compute costs to each node. A RAM usage estimation prevents swapping, which is critical for the runtime. Our approach significantly scales up the SSJ execution and processes much larger datasets than all parallel approaches designed so far.

1 Introduction

A major challenge in data science today is to compare and relate data of similar nature. One important operation to relate data is the *join* operation known from relational databases. The join operation finds all record pairs from two tables, which fulfill a given predicate. For basic predicates, such as equality, there exist efficient methods to compute the join. However, for many real-world problems the predicate is more complex: it involves similarity. If we assume that records are represented by sets, we could use existing set similarity measures to compare them pairwise. Given a collection of *records* (sets) R, formed over the universe U of tokens (set elements), and a similarity function between two records, $sim : \mathscr{P}(U) \times \mathscr{P}(U) \rightarrow [0, 1]$; the *set similarity self-join (SSJ)* of R computes all pairs of sets $(r, s) \in R \times R$ whose similarity exceeds a user-defined threshold θ, $0 < \theta \leq 1$, i. e., all pairs (r, s) with $sim(r, s) \geq \theta$. Without loss of generality, we focus on the Jaccard similarity function $sim(r, s) = \frac{|r \cap s|}{|r \cup s|}$ and the self-join.

N. Reyes et al. (Eds.): SISAP 2021, LNCS 13058, pp. 17–31, 2021.
https://doi.org/10.1007/978-3-030-89657-7_2

A naive approach to compute the SSJ compares all possible pairs. Since the complexity of such an approach is quadratic, it is not feasible even for small datasets. The most prominent approaches in the literature to compute the SSJ more efficiently are based on the filter-and-verification framework. Filter-and-verification-based approaches do not reduce the worst-case complexity (which is quadratic), but reduce the practical compute effort when favorable input data characteristics are present. The framework first generates candidate pairs by creating and probing an inverted index [1] and verifies the candidates in a second step. Sophisticated filters such as the prefix filter keep the number of candidate pairs low [2]. This method is efficient on single cores [6]. However, it does not scale easily to large datasets.

We proposed a novel data-parallel filter-and-verification approach using multi-threading [5]. It significantly scales up the SSJ computation. However, the number of available CPU cores limits scalability. The maximum amount of input we could process with this method on our hardware was roughly 25 GB. To compute the SSJ on larger datasets, various MapReduce-based distributed approaches evolved. The MapReduce programming paradigm requires independently computable work shares. The approaches use existing filters from the filter-and-verification framework to replicate and group data into such independent shares. We showed that the amount of data these approaches can process is limited [3]. In our experimental setup, the maximum possible input was roughly 12 GB, which is even smaller than what the multi-threaded approach could process. Users cannot shift the limit by adding more compute nodes due to high and skewed data replication.

The input dataset size and scalability limitations of the previously mentioned approaches motivate our novel distributed-parallel SSJ approach, which pushes these limits significantly. We experimentally show that our new approach scales to hundreds of gigabytes and that it is robust against unfavorable data characteristics[1]. We use existing filter-and-verification techniques as a basis and leverage intra-node multicore parallelization by default. The major advances compared to existing distributed approaches are as as follows. First, our approach *avoids intra-node replication* since replication is the main bottleneck of the MapReduce approaches due to our previous analysis. It assures that each record is present only once in the main memory of each node. Each node runs only one single multi-threaded SSJ instance in order to efficiently share commonly used data, such as the inverted index. Second, it *avoids recomputation*, i. e., the repeated validation of the same candidate record pair. Third, it *removes algorithmic data dependencies* that lead to a skewed execution load as observed in MapReduce approaches using prefix filtering [3].

Our approach solely requires a standard shared nothing architecture for a distributed execution. Our approach is generic, thus it is independent of a specific distributed system. The quadratic nature of the SSJ problem implies that scaling up to larger input dataset sizes may require adding a quadratic number of nodes in the worst case. To avoid the worst case, our distributed-parallel approach

[1] Our implementation is available at https://github.com/fabiyon/dist-ssj-sisap.

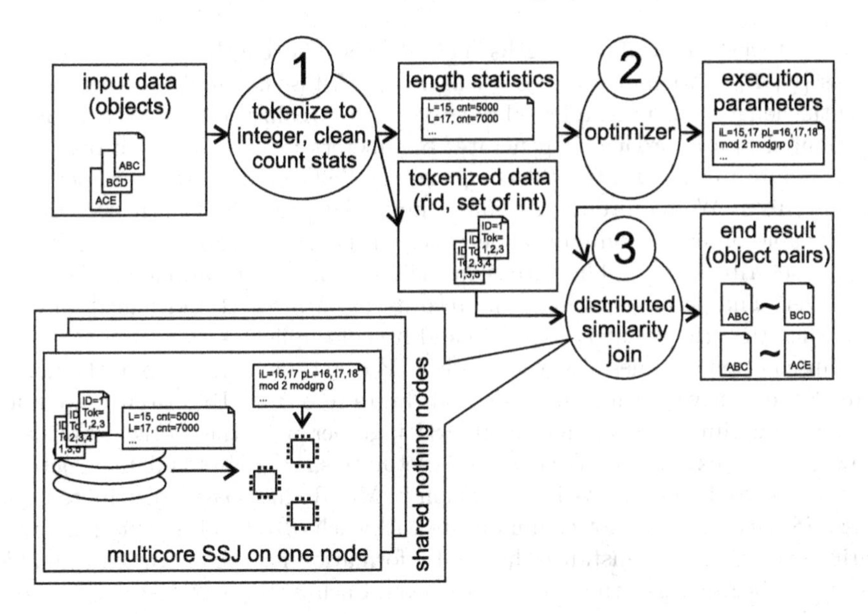

Fig. 1. Schematic dataflow of our distributed-parallel SSJ approach.

uses techniques to distribute the compute load evenly among nodes. However, depending on the dataset size, token distribution, and similarity threshold, the demand for compute nodes might still be high. Modern cloud computing allows to obtain a high number of compute nodes for a limited timeframe. Thus, we may safely assume that it is realistic today to have hundreds or even thousands of compute nodes available for just one operation. The main contributions of this paper are as follows:

- We introduce a cost-based heuristic to break down the SSJ computation into units that are computed independently in parallel.
- We additionally provide a data-independent scaling mechanism that allows to further subdivide each unit if necessary and a RAM usage estimation to avoid swapping.
- We experimentally verify that our distributed SSJ approach scales to hundreds of gigabytes of input data.

In the following section, we introduce our solution in detail. Section 3 experimentally shows its behavior on large datasets and large numbers of compute nodes. Section 4 concludes this paper. We provide an extended version of this paper with an additional description of experimental datasets as well as comprehensive tables and figures of our experimental results [4].

2 Distributing Filter-and-Verification-Based SSJ

Figure 1 provides an overview on our distributed-parallel SSJ approach. Step (1) preprocesses and tokenizes the raw input data. In addition, we require this step to compute a statistic of the lengths of all records. The length statistic consists of tuples $\{(l, |R_l|)\}$ where l is a record length and $|R_l|$ is the number of the records with this length. Step (2), referred to as optimizer, realizes the major part of our distributed SSJ approach. It generates parameters for each node to distribute the compute workload. Step (3) computes the SSJ based on the parameters of the optimizer. We require the tokenized input data and the length statistics to be available on every compute node. The join is an extension of our multicore SSJ as described in [5]. The extension includes the set of parameters from the optimizer. The parameters limit the records to be indexed and joined on each node such that the result is complete and free of duplicates.

Our solution assumes that each node runs exactly one instance of the multicore SSJ, exclusively using the nodes' hardware resources. By instance, we refer to the main thread of our multicore SSJ together with the worker threads it spawns during execution. We choose this setup to share common data structures such as inverted indexes. As it is common in MapReduce-based distributed systems, SSJ instances cannot communicate with each other and do not share data during execution. The instances have all information for the execution available before the beginning of the join computation. Each instance indexes and probes only subsets of the input dataset to independently compute a partial join result.

In the following, we introduce the optimizer. It runs before the actual join computation and divides the SSJ computation into independently computable units. The optimizer consists of a data-dependent cost-based heuristic and a data-independent scaling mechanism. Furthermore, we provide estimations of RAM demand and cost distribution and a heuristic to find suitable optimizer parameters. We first describe our cost-based heuristic.

2.1 Data-Dependent Cost-Based Heuristic

One goal of our cost-based heuristic is to avoid the cross product by only regarding record pairs with *matching* lengths. Regarding lengths to filter out hopeless pairs is a common technique, which most filter-and-verification approaches use [1]. This filter is effective on datasets with varying lengths and cheap to apply by using the length statistic computed beforehand. As discussed in the introduction, we focus on the Jaccard similarity function and the self-join.

Regarding Jaccard similarity and a record r, the length of a similar record s has to be in the interval $[\lceil \theta \cdot |r| \rceil; \lfloor \frac{|r|}{\theta} \rfloor]$. In the self-join case, the probe record set is equal to the index record set. To avoid duplicates and unnecessary recomputation, we subsequently consider only probe records larger than the length of an index record r: $[|r|; \lfloor \frac{|r|}{\theta} \rfloor]$. Figure 2 shows this length relationship for a similarity threshold of $\theta = 0.7$. For each record length on the y axis, it shows on the x axis, which record lengths have to be considered as join candidates. Now consider that we index the lengths on the y axis and probe the lengths on the x axis. Then each square in the figure represents a pair of index and probe lengths (i, p), which has to be joined for a complete result without duplicates. Each square can potentially be joined independently. However, for our heuristic, we choose to group squares with the same index lengths together and refer to them as *slices*.

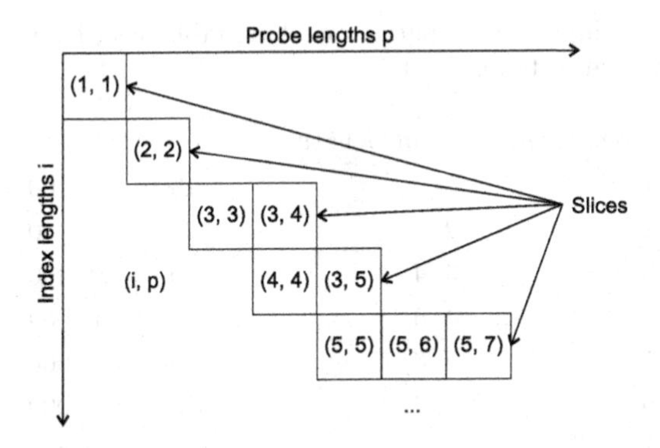

Fig. 2. Example join matrix for $\theta = 0.7$. Squares with the same index length compose one slice.

Table 1. Symbol reference.

R	Input dataset
θ	Similarity threshold
$\|r\|$	Number of tokens in r
$\|R_l\|$	Number of records with length l
$\mathcal{P}(l)$	Prefix length of length l: $\mathcal{P}(l) = l - \lceil \theta \cdot l \rceil + 1$
i	Index prefix length
p	Probe prefix length
rid	Record ID
n	Node parameter for cost-based heuristic
m	Modulo: data-independent scaling parameter
$modgroup$	Group parameter to check if a record is in a sub slice
$indexLengths$	Set of index lengths for one SSJ instance
$probeLengths$	Set of probe lengths for one SSJ instance

For each slice i, we estimate the probe costs $\mathcal{C}(i)$ as follows:

$$\mathcal{C}(i) = \mathcal{P}(i) \cdot |R_i| \cdot \sum_{p=i}^{\lfloor \frac{i}{\theta} \rfloor} \mathcal{P}(p) \cdot |R_p| \tag{1}$$

Table 1 serves as a symbol reference for the symbols we use in the equation and throughout this paper. For the cost estimation we assume that each probe of the inverted index causes a cost of the length of the postings list. We do not know the exact sizes of the postings lists a priori, because they are dependent on the token distribution. Instead, we assume the worst case, where all index records of the probed length are contained in the postings list. With regard to an index length i, the possible probe lengths p are in $[i, \lfloor \frac{i}{\theta} \rfloor]$. The total number of probes of one slice is the sum over the prefix of p (denoted as $\mathcal{P}(p)$) multiplied by the number of records with this length $|R_p|$ for all probe lengths. The number of index tokens of the slice is computed the same way and multiplied.

Table 2. Example of input data lengths, matching probe lengths, number of records, and corresponding slice costs for $\theta = 0.7$.

Index length i	Probe lengths $\{p\}$	$\|R_i\|$	$\mathcal{C}(i)$
1	1	10	100
2	2	30	900
3	3, 4	80	86 400
4	4, 5	500	1 800 000
5	5, 6, 7	400	1 416 000
6	6, 7, 8	200	568 000
7	7, 8	190	581 400
8	8	150	202 500

Example 1. Table 2 shows the cost computation for a hypothetical dataset. The dataset has eight length values as shown in the first column. The second column shows matching probe lengths for each index length. $|R_i|$ shows the hypothetical length count per index length. Column $\mathcal{C}(i)$ shows the resulting slice costs. □

Example 1 highlights that slices can exhibit uneven costs. Thus, we assign *sets* of slices to compute nodes with the intention to distribute the costs evenly. We use a greedy heuristic to achieve an even cost distribution. We assume that the user chooses a seed number of compute nodes n (the total number of compute nodes for the SSJ computation can be higher depending on further parameters). We sort the slice costs $\mathcal{C}(i)$ in ascending order. Then we assign each slice to each node in a round robin fashion. Thus, the first node receives the slice with the largest cost, the second node receives the second-largest, and when the last node is reached, the first node obtains the next slice again. The following example shows our greedy cost distribution heuristic:

Example 2. Consider again Table 2 and $n = 2$. The highest cost appears for $i = 4$. Thus, we assign this slice to the first node. The next highest cost appears for $i = 5$. Thus, we assign it to the second node. The third one is $i = 7$, assigned to node 1, and so on. This approach generates the following index and probe lengths:

Node 1: index lengths 2, 4, 7, 8, probe lengths 2, 4, 5, 7, 8, total costs 2 584 800 and

Node 2: index lengths 1, 3, 5, 6, probe lengths 1, 3, 4, 5, 6, 7, 8, total costs 2 070 500. □

As discussed before, our cost estimation cannot consider the specific sizes of the postings lists. The estimation assumes that all records with matching lengths are present in the postings lists, which is only the worst case and pessimistic. On the other hand, the heuristic ignores the costs for the verification. The verification is dependent on the number of candidates, which we cannot estimate a priori without actually computing the join. Thus, our heuristic potentially underestimates the costs if a dataset has many candidates. In our experiments, we show the strengths and limits of our approach. Next, we introduce the second part of the optimizer, the data-independent scaling mechanism.

2.2 Data-Independent Scaling Mechanism

The scaling mechanism subdivides each slice (cf. Subsect. 2.1) by partitioning its probe records. Our join computation assigns subsequent integer record IDs (*rids*) to each input record. We use the modulo function to assign a probe record to one partition as shown in the following equation:

$$isRecordInProbeSubset(rid, m, modgroup) = (rid \bmod m \stackrel{?}{=} modgroup) \quad (2)$$

The user-defined parameter m sets the number of sub slices to generate. The *modgroup* is in the interval $[0, m-1]$ and determines the sub slice a record is assigned to. The following example illustrates how our scaling approach assigns records to sub slices:

Example 3. Assume $m = 2$. One sub slice receives all records where the function returns true for $modgroup = 0$ and another sub slice obtains the ones for $modgroup = 1$. We ordered the records in our input datasets by ascending record lengths. Thus, we expect this approach to be robust against length skew in the input data. It assigns records of all probe lengths to each sub slice round robin. □

The scaling mechanism together with the cost-based heuristic form the main building blocks of the optimizer of our SSJ approach. To find suitable parameter values for m and n, we next discuss how to evaluate the quality of concrete instances of these parameters. We start with an estimation of RAM demand.

2.3 RAM Demand

Our heuristic and the scaling mechanism do not guarantee that the computation of one (sub) slice stays within the RAM size of a given compute node. If the SSJ computation allocates more memory than the system physically provides, swapping occurs. Swapping leads to severe runtime penalties, which we must avoid. The main idea to avoid RAM overutilization is to find optimization parameters m and n such that the RAM usage stays within system limits. With the heuristic from Sect. 2.1, a concrete value for n, a similarity threshold θ, and the length statistics of a concrete dataset $\{(r, |R_l|)\}$ we compute sets of lengths *indexLengths* and *probeLengths* for each node. We use these length sets for RAM demand estimations subsequently.

We use an extension of our multicore SSJ on each compute node [5]. The extension includes the parameters *indexLengths*, *probeLengths*, m, and *modgroup* to limit the index and probe records. Considering the extended multicore SSJ, the *inverted index*, *probe records*, and *candidates* demand the largest parts of main memory. Without loss of generality, we estimate the demands for all three categories for our concrete SSJ implementation. The estimation is applicable to possible other join implementations by adjusting the size factors of the employed data structures.

First, we focus on the inverted index. Our implementation of the inverted index holds the postings list entries in a struct of 12 Bytes. The number of postings list entries is the prefix length times the number of records $\mathcal{P}(l) \cdot |R_l|$ for each index length l. We can estimate the size of the inverted index (in Bytes) as follows:

$$indexRamDemand(indexLengths) = \sum_{l \in indexLengths} \mathcal{P}(l) \cdot |R_l| \cdot 12 \quad (3)$$

Similarly, we estimate the RAM demand for the probe records. One record in our implementation uses 60 Bytes plus each token stored as 4 Byte integer. We estimate the space requirement for the probe records (in Bytes) as follows:

$$probeRamDemand(probeLengths, m) = \sum_{l \in probeLengths} \frac{|R_l| \cdot (60 + l \cdot 4)}{m} \quad (4)$$

Lastly, we focus on the candidate size. Our SSJ uses 12 Bytes to store each candidate record in main memory until verification. Each thread keeps a local list of candidates for its subset of probe records. In the worst case, all indexed records are candidates. However, it is pessimistic to assume that all threads hold all index records as candidates at the same time. In our experiments, we found that it is safe to assume $\frac{1}{3}$ to $\frac{2}{3}$ of the index records to be present on each thread at a time on our datasets. Thus, we include a candidate factor $candFact$ in our estimation. We estimate the candidate RAM demand (in Bytes) as follows:

$$candidateRamDemand(indexLengths, numberThreads, candFact) = \sum_{l \in indexLengths} |R_l| \cdot 12 \cdot numberThreads \cdot candFact \quad (5)$$

To avoid swapping, the sum of all demands must stay below the system limit of a compute node leaving space for other storage needs and the operating system. We found the static space demand to be below 4 GB on the system we run our experiments on and thus consider this value in the following.

Example 4. Consider the dataset ORKU with scaling factor 100, $\theta = 0.6$, $m = 64$, $n = 8$, and $numberThreads = 24$. Over all slices, we can compute a maximum index RAM demand of 21 GB, 2 GB for the probe records, and up to 10 GB for candidates. We estimate the total demand including the static demand to be 37 GB. In fact, on our system with 32 GB RAM, this parameter combination leads to heavy swapping. The runtime of each slice is above 12 h. When we changed the parameters to $m = 16$ and $n = 32$ (which equals the total number of nodes in the previous configuration, 512) the total estimated RAM demand decreases to 24 GB. The maximum runtime per slice in this configuration is 300 s and no swapping occurs. The example motivates that it is crucial to find a suitable parameter configuration, which keeps the memory demand below the system limit to achieve an acceptable runtime for the join operation. □

Note that our data-independent scaling approach focuses only on probe records. In case the set of *indexLengths* contains solely one length and the corresponding *indexRamDemand* exceeds the available main memory, our approach does not provide a means to further reduce the index size. However, if an index exceeds available main memory it is possible to partition the index records, i.e., with a modulo function in the same way as we applied it to the probe records. We do not elaborate on further reducing the index size, because we cannot observe such an extreme index skew within our experiments even on highly enlarged datasets. Next, we discuss the cost distribution among the compute nodes.

2.4 Cost Distribution Quality

Even without swapping, the choice of parameter n might be crucial for the runtime depending on the length distribution of the input dataset. Example 5 illustrates and motivates the need for an appropriate parameter choice.

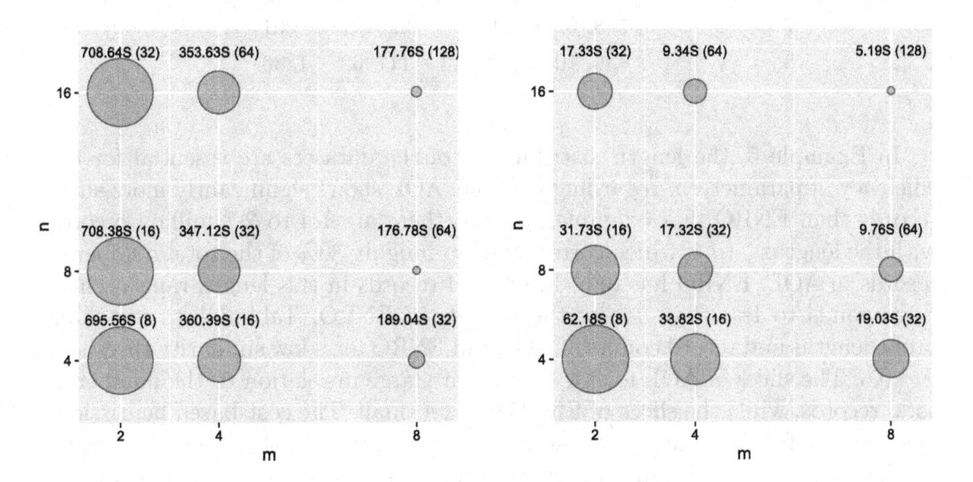

Fig. 3. AOL×10 runtimes.

Fig. 4. ENRO×10 runtimes.

Example 5. Figures 3 and 4 visualize the runtimes of AOL and ENRO, both increased with scaling factor 10, for $\theta = 0.6$ varying both parameters m and n.

The circle sizes represent the runtime. The same color marks combinations of parameters with the same total number of nodes. For example, the parameter combination $m = 8$ and $n = 4$ uses 32 nodes in total. Parameter combination $m = 4$ and $n = 8$ also uses 32 nodes and therefore has the same color assigned. The numbers above the circles are the maximum runtimes over all slices in seconds followed by the total number of nodes in brackets. For ENRO×10 a higher n is beneficial for an improved runtime. That is, the runtime with parameters $m = 2$ and $n = 16$ is lower than with parameters $m = 8$ and $n = 4$ for the same total amount of nodes of 32. On the other hand, for AOL×10, a higher value of n does not lead to improved runtimes. A higher m parameter is effective for both datasets. The effectiveness of parameter m on both datasets is expected, because it linearly scales the number of probe records. □

Table 3. Example for input data length skew. Columns show hypothetical input data lengths, matching probe lengths, and the number of records for AOL and ENRO for $\theta = 0.6$.

Index length i	Probe lengths $\{p\}$	AOL $\|R_i\|$	ENRO $\|R_i\|$
1	1	2705785	149
2	2, 3	2026952	361
3	3, 4, 5	2051010	594
4	4, 5, 6	1457075	814
5	5, 6, 7, 8	849944	1029
6	6, 7, 8, 9, 10	445489	1141
7	7, 8, 9, 10, 11	225401	1301
8	8, 9, 10, 11, 12, 13	117962	1386

In Example 5, the length distributions of the datasets are essential for the efficiency of parameter n regarding runtime. AOL shows significantly more short records than ENRO. For example, in AOL there are 1.4 to 2.7 million records with the lengths 1 to 4, which corresponds to roughly 80% of the total number of records in AOL. ENRO has only 149 to 814 records in this length range, which corresponds to less than 1% of the records in ENRO. Table 3 lists matching probe lengths and record counts of AOL and ENRO for a low similarity threshold $\theta = 0.6$. The slices of AOL for $i \in 1, 2, 3, 4$ are large in relation to the number of total records, while the slices of ENRO remain small. The cost-based heuristic is less effective for AOL due to its skewed record lengths. Furthermore, depending on the choice of n, this length skew results in cost skew over the slices. In this example, the costs for AOL are less skewed for $n = 4$ compared to higher values of n.

To evenly distribute the compute costs over the nodes, we aim to find the best n out of a given value range regarding a distribution quality function. Given one n, we can compute the maximum cost deviation over all slices with

$\max\{\mathcal{C}(i)\} \div \min\{\mathcal{C}(j)\}$ for $i, j \in [0; n-1]$. Given a *valueRange* for n, we can then minimize this deviation as follows:

$$\min_{n \in valueRange} = \left\{ \max_{i \in [0;n-1]} \{\mathcal{C}(i)\} \div \min_{j \in [0;n-1]} \{\mathcal{C}(j)\} \right\} \tag{6}$$

Example 6. Consider AOL×10, $\theta = 0.6$, and $n \in \{4, 8, 16, 32\}$. Using Eq. 6, $n = 4$ has the lowest maximum cost deviation of 4.16. For higher values of n the deviation varies between 200 and 230 000. For ENRO×10 and the same parameters, the lowest deviation is 1.02 for $n = 4$, followed by 1.05 for $n = 8$, 1.09 for $n = 16$, and 1.21 for $n = 32$. For both datasets, our cost distribution quality estimation chooses a good value for parameter n. Our estimation might not necessarily lead to the optimal parameter value regarding runtime, but it avoids unfavorable values. □

In the following subsection, we discuss how to use these cost distribution considerations together with the RAM estimation to find suitable parameter values m and n.

2.5 Finding Suitable Parameter Values

Our approach uses the two parameters m and n. Based on the previous discussion about RAM demand and cost distribution we propose the following strategy to determine parameter values, which avoid RAM overutilization and cost skew. We assume that the user chooses a total number of compute nodes t as a seed, which should preferably be a power of two for practical reasons. For each possible m and n (such that $m \cdot n = t$) we compute the estimated demand for RAM (cf. Sect. 2.3) and the minimum and maximum cost over all slices (cf. Sect. 2.4). We can prune all parameter combinations with a RAM demand above the system limit. We then choose the parameter combination (m, n) with the lowest cost deviation. In case all parameter combinations are pruned, we set the total number of nodes $t = t \cdot 2$ and re-run the previous computation until a suitable combination is found. If the resulting t is above the number of available compute nodes, the computation should be split into subsequent phases. The described strategy finds only the minimum m parameter value with respect to t. Users may increase m to achieve lower runtimes. In our experiments, we show the applicability of our approach to find suitable parameters.

3 Experiments

This section presents our experimental analysis. We focus on scalability, varying the parameters m and n, the input dataset sizes, and the similarity threshold θ. Based on the shortcomings of manually choosing parameter values, we subsequently discuss our strategy to find suitable parameter values m and n.

To compute the join on one slice we use a multicore C++ SSJ implementation running it on each compute node by extending our previous multicore SSJ with the parameters *indexLengths*, *probeLengths*, m, and *modgroup*. By default, we run the multicore SSJ with the optimal parameters [5]. We enable the position filter and set the number of threads to 24, which is optimal on our hardware: Each node is equipped with two Xeon E5-2620 2 GHz of 6 cores each (with hyper-threading enabled, i. e., 24 logical cores per node), 24 GBs of RAM, and two 1 TB hard disks. Whenever we report runtimes, we refer to the maximum runtime over all slices since the maximum runtime determines the overall runtime.

As input datasets, we use the 10 real-world and two synthetic datasets (cf. extended paper [4]). Since we focus on larger datasets, we use only increased datasets with the scaling factors 10, 25, 50, and 100. We start our experiments with a scaling factor of 10, because these are the largest datasets joinable with both the MapReduce and the multicore approaches so far. Our novel distributed approach is able to compute the join on much larger datasets as we show subsequently.

3.1 Impact of Cost-Based Heuristic

In this experiment, we show how the runtime develops varying parameter n. We do not set parameter m. Thus, the probe records per slice remain complete with regard to the *probeLengths* computed with the heuristic from Sect. 2.1. We use all datasets increased by factor 10, $\theta \in \{0.6, 0.75, 0.9\}$, $n \in \{4, 8, 16, 32\}$ and compare it to the non-distributed multicore SSJ (cf. Fig. 5).

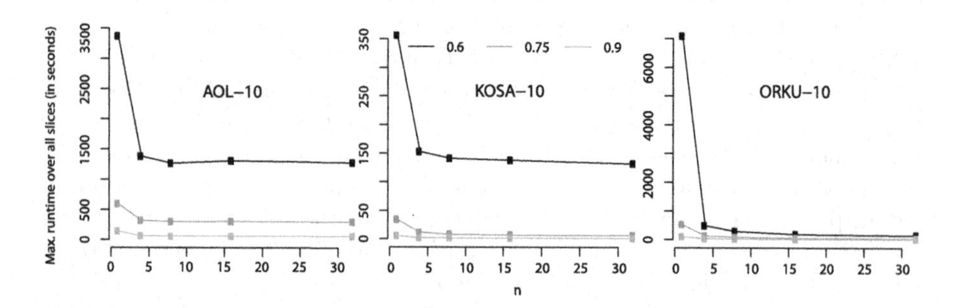

Fig. 5. Maximum runtimes over all slices for $n \in \{4, 8, 16, 32\}$ for three exemplary datasets AOL, KOSA, and ORKU. $n = 1$ represents the multicore SSJ without distributed parallelization. Thresholds $\theta \in \{0.6, 0.75, 0.9\}$.

For all datasets and all thresholds, $n = 4$ significantly reduces all runtimes compared to $n = 1$. The speedups vary between 1.8 (AOL×10, $\theta = 0.75$) and 13.9 (ORKU×10, $\theta = 0.6$). The average speedup over all datasets and thresholds is 3.7. For higher values of n the speedups decrease. Adding more than 8 or 16 nodes leads to only small runtime decreases for most datasets and thresholds. This effect is due to the nature of our heuristic. Recall that one slice consists of an index length and all its possible probe lengths. The length skew of the input datasets and the similarity threshold determine the largest and potentially slowest slice, which cannot be further partitioned with the heuristic. AOL×10 is exemplary for this circumstance. As we discussed in Sect. 2.4, AOL has roughly 80% of its records within the length range 1 to 4. n values higher than 4 are not beneficial for this dataset. Other datasets show different length distributions, which lead to optimal n values higher than 4.

KOSA×10 also shows a limited scalability for $\theta = 0.6$, but for a different reason than length skew. We observe that amongst all slices for each n there exists one slice with a runtime between 130 and 150 s, while all other slices have lower runtimes. The reason for the outlier slices in KOSA×10 are their high number of candidates compared to all other slices. The runtimes of KOSA×10 show a limitation of our heuristic. It optimizes the runtime based on length information and is thus not robust against candidate skew by design.

3.2 Impact of Data-Independent Scaling Mechanism

In this experiment, we study how the scaling parameter m influences the runtimes. We continue to use the datasets using scaling factor 10 and fix parameter n to 8, since this parameter setting showed good runtimes in the previous experiment. We again use $\theta \in \{0.6, 0.75, 0.9\}$ and vary $m \in \{2, 4, 8\}$. The results indicate that $m \geq 2$ is beneficial to achieve a lower runtime for all datasets and thresholds, including AOL×10 and KOSA×10, which showed scalability boundaries for $n \geq 4$ in the previous experiment (cf. Fig. 6).

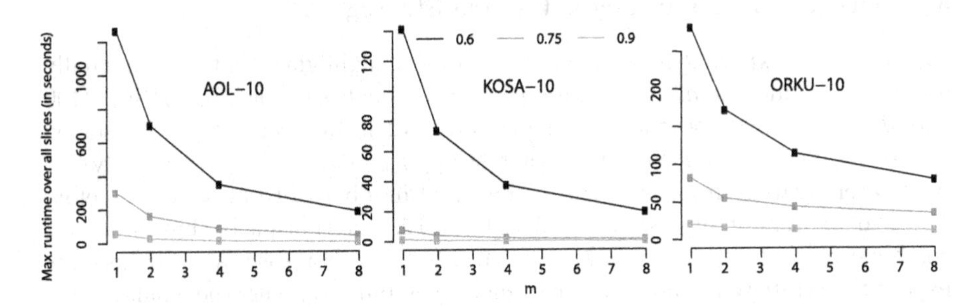

Fig. 6. Maximum runtimes for exemplary datasets over all slices for $n = 8$, $\theta \in \{0.6, 0.75, 0.9\}$, $m \in \{2, 4, 8\}$. $m = 1$ indicates runtimes without the scaling mechanism.

Since the modulo function evenly distributes different probe lengths among sub slices we expect the runtimes to scale linearly with m, which experimental results partially confirm. Regarding the minimum, maximum, and average speedups for $m \in \{2, 4, 8\}$ in relation to $m = 1$, grouped by θ, there is a maximum speedup close to the optimum m for each threshold group. The averages over all thresholds for $m = 2$ are close to the optimum 2. The average speedups for larger values for m decrease.

3.3 Impact of Dataset Size

In this subsection, we investigate how the runtimes evolve when increasing the dataset size by scaling factors $s \in \{10, 25, 50, 100\}$. We statically set $n = 8$ and $m = 64$. We compare the maximum runtimes per slice for $s \in \{25, 50, 100\}$ relative to maximum runtime for $s = 10$.

In many cases, the runtime does not increase linearly with the dataset size. A non-linear runtime increase is expected, because the SSJ has a quadratic complexity. A perfectly linear runtime relative to $s = 10$ would be $\frac{s}{10}$ for $s \in \{25, 50, 100\}$. Only few combinations of datasets, θ, and s fall in this category. For ENRO and $\theta = 0.9$, ORKU and $\theta = 0.9$, and SPOT (all thresholds) the relative runtimes for $s \in \{25, 50, 100\}$ are better than linear. ENRO and $\theta = 0.75$, FLIC and $\theta \in \{0.75, 0.9\}$, LIVE and $\theta = 0.9$, ZIPF and $\theta \in \{0.75, 0.9\}$ are close to linear. We can observe that the runtimes of higher thresholds increase more linearly than the ones of lower thresholds relative to s. This runtime behavior can be explained by the prefix filter, which is more effective for higher thresholds.

With our approach, it is possible to compute the SSJ on all datasets of all sizes in our evaluation and all thresholds except ENRO-100 and $\theta = 0.6$. We manually stopped the computation after 12 h. In Sect. 2.3, we discussed that for ORKU×100 the parameter combination $n = 8$ and $m = 64$ is not optimal, because it causes swapping. We next discuss our proposed parameter finding strategy.

3.4 Discussion of Parameter Finding Strategy

The previous experiment on enlarged datasets highlights that the manually assigned parameters $m = 64$ and $n = 8$ are not suitable for ORKU×100 and $\theta = 0.6$, because the runtime exceeds 12 h. In Sect. 2.3, we discussed the same example and concluded that swapping occurs. When we apply the parameter strategy from Sect. 2.5 to the equal number of total nodes as before ($t = 8 \cdot 64 = 512$), it suggests $m = 32$ and $n = 16$. The runtime of this parameter combination is 1314 s, so the strategy avoids the worst case. We furthermore expect the strategy to choose the parameter combination with the smallest cost deviation. In the example in Sect. 2.4, we discussed that for AOL×10 $\theta = 0.6$ $n = 4$ is better than a larger n. Running the parameter finding strategy for $t = 16$, it indeed suggests the parameter value $n = 4$.

4 Conclusion

In this paper we introduced our novel distributed SSJ approach. We showed experimentally that it scales the computation to potentially hundreds of compute nodes if needed. Our method computes the SSJ on our hardware on datasets up to roughly 240 GB, which is much larger than the ones which could be computed with existing parallel methods so far. We discussed how to a priori estimate limits of parameter values from which we cannot expect an efficient execution, especially regarding main memory usage. We proposed a parameter finding strategy, which avoids poor parameter values leading to either RAM overutilization or a skewed cost distribution. One remaining challenge is to better estimate or manipulate the maximum number of candidates of each slice, which occur at one instance of time.

Acknowledgements. This work was supported by a research grant from LexisNexis Risk Solutions.

References

1. Bayardo, R.J., Ma, Y., Srikant, R.: Scaling up all pairs similarity search. In: Proceedings of the International Conference on World Wide Web (2007)
2. Chaudhuri, S., Ganti, V., Kaushik, R.: A primitive operator for similarity joins in data cleaning. In: International Conference on Data Engineering (ICDE) (2006)
3. Fier, F., Augsten, N., Bouros, P., Leser, U., Freytag, J.C.: Set similarity joins on MapReduce: an experimental survey. In: Proceedings of the International Conference on Very Large Data Bases (PVLDB) (2018)
4. Fier, F., Freytag, J.C.: Scaling up set similarity joins using a cost-based distributed-parallel framework [extended paper] (2021). https://doi.org/10.18452/23209
5. Fier, F., Wang, T., Zhu, E., Freytag, J.-C.: Parallelizing filter-verification based exact set similarity joins on multicores. In: Satoh, S., et al. (eds.) SISAP 2020. LNCS, vol. 12440, pp. 62–75. Springer, Cham (2020). https://doi.org/10.1007/978-3-030-60936-8_5
6. Mann, W., Augsten, N., Bouros, P.: An empirical evaluation of set similarity join techniques. In: Proceedings of the International Conference on Very Large Data Bases (PVLDB) (2016)

A Triangle Inequality for Cosine Similarity

Erich Schubert(✉)

TU Dortmund University, Dortmund, Germany
erich.schubert@tu-dortmund.de

Abstract. Similarity search is a fundamental problem for many data analysis techniques. Many efficient search techniques rely on the triangle inequality of metrics, which allows pruning parts of the search space based on transitive bounds on distances. Recently, cosine similarity has become a popular alternative choice to the standard Euclidean metric, in particular in the context of textual data and neural network embeddings. Unfortunately, cosine similarity is not metric and does not satisfy the standard triangle inequality. Instead, many search techniques for cosine rely on approximation techniques such as locality sensitive hashing. In this paper, we derive a triangle inequality for cosine similarity that is suitable for efficient similarity search with many standard search structures (such as the VP-tree, Cover-tree, and M-tree); show that this bound is tight and discuss fast approximations for it. We hope that this spurs new research on accelerating exact similarity search for cosine similarity, and possible other similarity measures beyond the existing work for distance metrics.

Keywords: Cosine similarity · Triangle inequality · Similarity search

1 Introduction

Similarity search is a fundamental problem in data science and is used as a building block in many tasks and applications, such as nearest-neighbor classification, clustering, anomaly detection, and of course information retrieval. A wide class of search algorithms requires a metric distance function, i.e., a dissimilarity measure $d(x, y)$ that satisfies the triangle inequality $d(x, y) \leq d(x, z) + d(z, y)$ for any z. Intuitively, this is the requirement that the direct path from x to y is the shortest, and any detour over another point z is at least as long. Many dissimilarity measures such as the popular Euclidean distance and Manhattan distance satisfy this property, but not all do: for example the *squared* Euclidean distance (minimized, e.g., by the popular k-means clustering algorithm) does not, even on univariate data: $d^2_{\text{Euclid}}(0, 2) = 2^2 = 4$ but $d^2_{\text{Euclid}}(0, 1) + d^2_{\text{Euclid}}(1, 2) = 1^2 + 1^2 = 2$.

Part of the work on this paper has been supported by Deutsche Forschungsgemeinschaft (DFG), project number 124020371, within the Collaborative Research Center SFB 876 "Providing Information by Resource-Constrained Analysis", project A2.

N. Reyes et al. (Eds.): SISAP 2021, LNCS 13058, pp. 32–44, 2021.
https://doi.org/10.1007/978-3-030-89657-7_3

The triangle inequality $d(x, y) \leq d(x, z) + d(z, y)$ is a central technique in accelerating similarity search because it allows us to compute a bound on a distance $d(x, y)$ without having to compute it exactly if we know $d(x, z)$ and $d(z, y)$ for some object z. With trivial rearrangement and relabeling of y and z, we can also obtain a lower bound: $d(x, y) \geq d(x, z) - d(z, y)$. Given a maximum search radius ε, if $d(x, z) - d(z, y) > \varepsilon$, we can then infer that y cannot be part of the search result. This technique is often combined with a search tree, where each subtree Z is associated with a routing object z, and stores the maximum distance $d_{\max}(z) := \max_{y \in Z} d(z, y)$ for all y in the subtree. If $d(x, z) - d_{\max}(z) > \varepsilon$, none of the objects in the subtree can be part of the search result, and we can hence skip many candidates at once. This technique can also be extended to k-nearest neighbor and priority search, where we can use the minimum possible distance $d(x, z) - d_{\max}(z)$ to prioritize or prune candidates.

Metric similarity search indexes using this approach include the ball-tree [15], the metric tree [20] aka. the vantage-point tree [21], the LAESA index [11,17], the Geometric Near-neighbor Access Tree (GNAT) [3] aka. multi-vantage-point-tree [2], the M-tree [5], the SA-tree [13] and Distal SAT [4], the iDistance index [7], the cover tree [1], the M-index [14], and many more. (Neither the k-d-tree, quad-tree, nor the R-tree belong to this family, these indexes are coordinate-based, and require lower-bounds based on hyperplanes and bounding boxes, respectively). While they differ in the way they organize the data (e.g., with nested balls in the ball tree, M-tree, and cover tree, by splitting into ball-and-remainder in the VP-tree, or by storing the distances to reference points in LAESA and iDistance), all of these examples rely on the triangle inequality for pruning candidates as central search technique, and should not be used with a distance that does not satisfy this condition, as the search results may otherwise be incomplete (this may, however, be acceptable for certain applications).

In this paper, we introduce a triangle inequality for cosine similarity that allows lifting most of these techniques from metric distances to cosine similarity, and we hope that future research will allow extending this to other popular similarity functions.

2 Cosine Distance and Euclidean Distance

Cosine similarity (which we will simply denote as "sim" in the following) is commonly defined as the cosine of the angle θ between two vectors $\mathbf{x}$ and $\mathbf{y}$:

$$\operatorname{sim}(\mathbf{x}, \mathbf{y}) := \operatorname{sim}_{\text{Cosine}}(\mathbf{x}, \mathbf{y}) := \frac{\langle \mathbf{x}, \mathbf{y} \rangle}{\|\mathbf{x}\|_2 \cdot \|\mathbf{y}\|_2} = \frac{\sum_i x_i y_i}{\sqrt{\sum_i x_i^2} \cdot \sqrt{\sum_i y_i^2}} = \cos\theta$$

Cosine similarity has some interesting properties that make it a popular choice in certain applications, in particular in text analysis. First of all, it is easy to see that $\operatorname{sim}(\mathbf{x}, \mathbf{y}) = \operatorname{sim}(\alpha\mathbf{x}, \mathbf{y}) = \operatorname{sim}(\mathbf{x}, \alpha\mathbf{y})$ for any $\alpha > 0$, i.e., the similarity is invariant to scaling vectors with a positive scalar. In text analysis, this often is a desirable property as repeating the contents of a document multiple times does not change the information of the document substantially. Formally, cosine similarity can be seen as being the dot product of L_2 normalized vectors. Secondly, the computation of cosine similarity is fairly efficient for sparse vectors: rather than storing the vectors as a long array of values, most of which are zero, they can be encoded for example as pairs (i, v) of an index i and a value v, where only the non-zero pairs are stored and kept in sorted order. The dot product of two such vectors can then be efficiently computed by a *merge* operation, where only those indexes i need to be considered that are in both vectors because in $\langle \mathbf{x}, \mathbf{y} \rangle = \sum_i x_i y_i$ only those terms matter where both x_i and y_i are not zero.

In popular literature, you will often find the claim that cosine similarity is more suited for high-dimensional data. As we will see below, it cannot be superior to Euclidean distance because of the close relationship of the two, hence this must be considered a myth. Research on intrinsic dimensionality has shown that cosine similarity is also affected by the distance concentration effect [12] as well as the hubness phenomenon [16], two key aspects of the "curse of dimensionality" [22]. The main difference is that we are usually using the cosine similarity on sparse data, which has a much lower intrinsic dimensionality than the vector space dimensionality suggests.

Consider the Euclidean distance of two *normalized* vectors $\mathbf{x}$ and $\mathbf{y}$. By expanding the binomials, we obtain:

$$
\begin{aligned}
d_{\text{Euclidean}}(\mathbf{x}, \mathbf{y}) :=& \sqrt{\sum\nolimits_i (x_i - y_i)^2} = \sqrt{\sum\nolimits_i (x_i^2 + y_i^2 - 2x_i y_i)} \\
=& \sqrt{\|\mathbf{x}\|^2 + \|\mathbf{y}\|^2 - 2\langle \mathbf{x}, \mathbf{y} \rangle} = \sqrt{\langle \mathbf{x}, \mathbf{x} \rangle + \langle \mathbf{y}, \mathbf{y} \rangle - 2\langle \mathbf{x}, \mathbf{y} \rangle} \quad (1) \\
\text{if } \|\mathbf{x}\| = \|\mathbf{y}\| =& 1 := \sqrt{2 - 2 \cdot \text{sim}(\mathbf{x}, \mathbf{y})} \quad (2)
\end{aligned}
$$

where the last step relies on the vectors being normalized to unit length. Hence we have an extremely close relationship between cosine similarity and (squared) Euclidean distance of the normalized vectors:

$$
\text{sim}(\mathbf{x}, \mathbf{y}) = 1 - \tfrac{1}{2} d^2_{\text{Euclidean}}\left(\tfrac{\mathbf{x}}{\|\mathbf{x}\|}, \tfrac{\mathbf{y}}{\|\mathbf{y}\|}\right) . \quad (3)
$$

While we can also compute Euclidean distance more efficiently for sparse vectors using the scalar product form of Eq. 1, this computation is prone to a numerical problem called "catastrophic cancellation" for small distances (when $\langle \mathbf{x}, \mathbf{x} \rangle \approx \langle \mathbf{x}, \mathbf{y} \rangle \approx \langle \mathbf{y}, \mathbf{y} \rangle$ that can be problematic in clustering (see, e.g., [9,18]). Hence, working with cosines directly is preferable when possible, and additional motivation for this work was to work directly with a triangle inequality on the similarities, to avoid this numerical problem (as we will see below, we cannot completely avoid this, unless we can afford to compute many trigonometric functions).

In common literature, the term "cosine distance" usually refers to a dissimilarity function defined as

$$
d_{\text{Cosine}}(\mathbf{x}, \mathbf{y}) := 1 - \text{sim}(\mathbf{x}, \mathbf{y}) , \quad (4)
$$

which unfortunately is *not* a metric, i.e., it does not satisfy the triangle inequality.

There are two less common alternatives, namely:

$$d_{\text{SqrtCosine}}(\mathbf{x},\mathbf{y}) := \sqrt{2-2\,\text{sim}(\mathbf{x},\mathbf{y})} \quad \left(= d_{\text{Euclidean}}\left(\tfrac{\mathbf{x}}{\|\mathbf{x}\|}, \tfrac{\mathbf{y}}{\|\mathbf{y}\|}\right) \right) \tag{5}$$

$$d_{\text{arccos}}(\mathbf{x},\mathbf{y}) := \arccos(\text{sim}(\mathbf{x},\mathbf{y})) \ . \tag{6}$$

which are less common (but, e.g., available in ELKI [19]) and which are metric. Equation 5 directly follows from Eq. 3, while the second one is the angle between the vectors itself (the arc length, not the cosine of the angle), for which we easily obtain the triangle inequality by looking at the arc through x, y, z. We will use these metrics below to obtain a triangle inequality for cosines.

3 Constructing a Triangle Inequality for Cosine Similarity

Because the triangle inequality is the central rule to avoiding distance computations in many metric search indexes (as well as in many other algorithms), we would like to obtain a triangle inequality for cosine similarity. Given the close relationship to squared Euclidean distance outlined in the previous section, one obvious approach would be to just use Euclidean distance instead of cosine. If we know that our data is normalized (which is a best practice when using cosine similarities), we can make the computation slightly more efficient using Eq. 5, but we wanted to avoid this because (i) computing the square root takes 10–50 CPU cycles (depending on the exact CPU, precision, and input value) and (ii) the subtraction in this equation is prone to catastrophic cancellation when the two vectors are similar, i.e., we may have precision issues when finding the nearest neighbors. Hence, we would like to develop techniques that primarily rely on similarity instead of distance, yet allow a similar pruning to the (very successful) metric search acceleration techniques.

Using Eq. 5 and the triangle inequality of Euclidean distance, we obtain

$$\begin{aligned}
\sqrt{1-\text{sim}(\mathbf{x},\mathbf{y})} &\le \sqrt{1-\text{sim}(\mathbf{x},\mathbf{z})} + \sqrt{1-\text{sim}(\mathbf{z},\mathbf{y})} \\
\text{sim}(\mathbf{x},\mathbf{y}) &\ge 1 - \left(\sqrt{1-\text{sim}(\mathbf{x},\mathbf{z})} + \sqrt{1-\text{sim}(\mathbf{z},\mathbf{y})}\right)^2 \\
\text{sim}(\mathbf{x},\mathbf{y}) &\ge \text{sim}(\mathbf{x},\mathbf{z}) + \text{sim}(\mathbf{z},\mathbf{y}) - 1 \\
&\quad - 2\sqrt{(1-\text{sim}(\mathbf{x},\mathbf{z}))(1-\text{sim}(\mathbf{z},\mathbf{y}))}
\end{aligned} \tag{7}$$

which, unfortunately, does not appear to allow much further simplification. In order to remove the square root, we can approximate it using the smaller of the two similarities $\text{sim}_{\perp}(\mathbf{x},\mathbf{y},\mathbf{z}) := \min\{\text{sim}(\mathbf{x},\mathbf{z}), \text{sim}(\mathbf{z},\mathbf{y})\}$:

$$\begin{aligned}
\text{sim}(\mathbf{x},\mathbf{y}) &\ge \text{sim}(\mathbf{x},\mathbf{z}) + \text{sim}(\mathbf{z},\mathbf{y}) - 1 - 2(1-\text{sim}_{\perp}(\mathbf{x},\mathbf{y},\mathbf{z})) \\
\text{sim}(\mathbf{x},\mathbf{y}) &\ge \text{sim}(\mathbf{x},\mathbf{z}) + \text{sim}(\mathbf{z},\mathbf{y}) + 2\,\text{sim}_{\perp}(\mathbf{x},\mathbf{y},\mathbf{z}) - 3
\end{aligned} \tag{8}$$

This is highly efficient to compute, a strict bound to Eq. 7, but unfortunately also a rather loose bound if one of the similarities is high, but the other is not.

Besides the relationship to squared Euclidean distance, there is another way to obtain a metric from cosine similarity, namely by using the arc length as in Eq. 6 (i.e., using the angle θ itself, rather than the cosine of the angle):

$$d_{\text{arccos}}(\mathbf{x}, \mathbf{y}) := \arccos(\text{sim}(\mathbf{x}, \mathbf{y}))$$

This also yields a metric on the sphere that satisfies the triangle inequality:

$$\arccos(\text{sim}(\mathbf{x}, \mathbf{y})) \leq \arccos(\text{sim}(\mathbf{x}, \mathbf{z})) + \arccos(\text{sim}(\mathbf{z}, \mathbf{y}))$$

and, hence,

$$\text{sim}(\mathbf{x}, \mathbf{y}) \geq \cos(\arccos(\text{sim}(\mathbf{x}, \mathbf{z})) + \arccos(\text{sim}(\mathbf{z}, \mathbf{y}))) \tag{9}$$

Computationally, the trigonometric functions involved here are even more expensive (60–100 CPU cycles each), hence using this variant directly is not for free. However, this can be further transformed (c.f., angle addition theorems) to the following equivalent triangle inequality for cosine similarity:

$$\begin{aligned} \text{sim}(\mathbf{x}, \mathbf{y}) \geq\ & \text{sim}(\mathbf{x}, \mathbf{z}) \cdot \text{sim}(\mathbf{z}, \mathbf{y}) \\ & - \sqrt{(1 - \text{sim}(\mathbf{x}, \mathbf{z})^2) \cdot (1 - \text{sim}(\mathbf{z}, \mathbf{y})^2)}\ . \end{aligned} \tag{10}$$

This triangle inequality is *tighter* than the one based on Euclidean distance, and hence we can expect better pruning power than using an index for Euclidean distance or $d_{\text{SqrtCosine}}$ (Eq. 5) in a metric index; while the computational cost has been reduced to the low "overhead" of Euclidean distances. Equation 9 suggests that it is the tightest possible bound we can obtain because it is directly using the angles, rather than the chord length as used by Euclidean distance. This bound yields a very interesting insight: while the triangle inequality for Euclidean distances – and in the arc lengths – was additive, the main term of this equation in the cosine domain is multiplicative.

We also investigated approximations to further reduce the computation overhead. By approximating the last term using the smaller similarity only, we get

$$\text{sim}(\mathbf{x}, \mathbf{y}) \geq \text{sim}(\mathbf{x}, \mathbf{z}) \cdot \text{sim}(\mathbf{z}, \mathbf{y}) + \min\{\text{sim}(\mathbf{x}, \mathbf{z})^2, \text{sim}(\mathbf{z}, \mathbf{y})^2\} - 1 \tag{11}$$

which is a cheap bound, tighter than Eq. 8, but still too loose.

We can also expand and approximate the last term using both the smaller and the larger value $\text{sim}_\top(\mathbf{x}, \mathbf{y}, \mathbf{z}) := \max\{\text{sim}(\mathbf{x}, \mathbf{z}), \text{sim}(\mathbf{z}, \mathbf{y})\}$:

$$\begin{aligned} & \sqrt{(1 - \text{sim}(\mathbf{x}, \mathbf{z})^2) \cdot (1 - \text{sim}(\mathbf{z}, \mathbf{y})^2)} \\ = & \sqrt{(1 - \text{sim}(\mathbf{x}, \mathbf{z})) \cdot (1 + \text{sim}(\mathbf{x}, \mathbf{z})) \cdot (1 - \text{sim}(\mathbf{z}, \mathbf{y})) \cdot (1 + \text{sim}(\mathbf{z}, \mathbf{y}))} \\ \leq & \sqrt{(1 + \text{sim}_\top(\mathbf{x}, \mathbf{y}, \mathbf{z}))^2 \cdot (1 - \text{sim}_\perp(\mathbf{x}, \mathbf{y}, \mathbf{z}))^2} \\ = & (1 + \text{sim}_\top(\mathbf{x}, \mathbf{y}, \mathbf{z})) \cdot (1 - \text{sim}_\perp(\mathbf{x}, \mathbf{y}, \mathbf{z})) \\ = & 1 + \text{sim}_\top(\mathbf{x}, \mathbf{y}, \mathbf{z}) - \text{sim}_\perp(\mathbf{x}, \mathbf{y}, \mathbf{z}) - \text{sim}(\mathbf{x}, \mathbf{z}) \cdot \text{sim}(\mathbf{z}, \mathbf{y}) \end{aligned}$$

and hence obtain the inequality

$$\text{sim}(\mathbf{x}, \mathbf{y}) \geq 2 \cdot \text{sim}(\mathbf{x}, \mathbf{z}) \cdot \text{sim}(\mathbf{z}, \mathbf{y}) - 1 - |\,\text{sim}(\mathbf{x}, \mathbf{z}) - \text{sim}(\mathbf{z}, \mathbf{y})| \tag{12}$$

but this approximation is strictly inferior to Eq. 11.

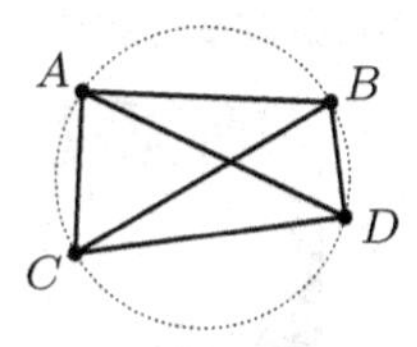

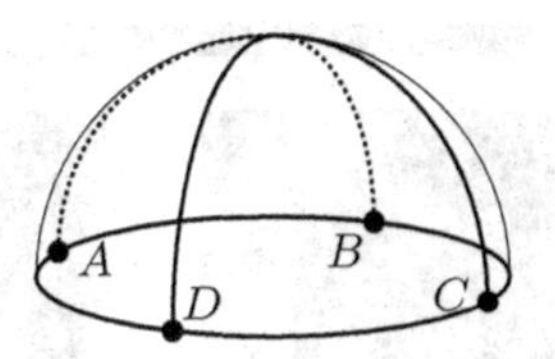

Fig. 1. Illustration of Ptolemy's (in-) equality in Euclidean space, and a simple counterexample for it in angular space.

3.1 Opposite Direction

The opposite direction of the triangle inequality is often as important as the first direction. For distances and the angles, it is simply obtained by moving one term to the other side and renaming. It can then be simplified as before

$$\begin{aligned} \arccos(\operatorname{sim}(\mathbf{x},\mathbf{y})) &\geq \arccos(\operatorname{sim}(\mathbf{x},\mathbf{z})) - \arccos(\operatorname{sim}(\mathbf{z},\mathbf{y})) \\ \operatorname{sim}(\mathbf{x},\mathbf{y}) &\leq \cos(\arccos(\operatorname{sim}(\mathbf{x},\mathbf{z})) - \arccos(\operatorname{sim}(\mathbf{z},\mathbf{y}))) \\ \operatorname{sim}(\mathbf{x},\mathbf{y}) &\leq \operatorname{sim}(\mathbf{x},\mathbf{z}) \cdot \operatorname{sim}(\mathbf{z},\mathbf{y}) + \sqrt{(1-\operatorname{sim}(\mathbf{x},\mathbf{z})^2)\cdot(1-\operatorname{sim}(\mathbf{z},\mathbf{y})^2)} \end{aligned} \tag{13}$$

It is interesting to see that Eqs. 10 and 13 together imply that

$$|\operatorname{sim}(\mathbf{x},\mathbf{y}) - \operatorname{sim}(\mathbf{x},\mathbf{z}) \cdot \operatorname{sim}(\mathbf{z},\mathbf{y})| \leq \sqrt{(1-\operatorname{sim}(\mathbf{x},\mathbf{z})^2)\cdot(1-\operatorname{sim}(\mathbf{z},\mathbf{y})^2)}$$

i.e., a symmetric error bound for $\operatorname{sim}(\mathbf{x},\mathbf{y}) \approx \operatorname{sim}(\mathbf{x},\mathbf{z}) \cdot \operatorname{sim}(\mathbf{z},\mathbf{y})$.

3.2 Angles are Not Ptolematic

Recently, people have also looked into using the Ptolemaic inequality for similarity search [6,10]. Ptolemy's inequality states that for a rectangle of ordered points A, B, C, D with diagonals AC and BD:

$$d(A,B)\cdot d(C,D) + d(B,C)\cdot d(D,A) \geq d(A,C)\cdot d(B,D)$$

i.e., the two products of opposing sides sum to more than the product of diagonals. If the four points are concyclic (as illustrated in Fig. 1), this becomes an equality. While this has some interesting properties for data indexing, it does not appear to be suitable for angular space, as shown by the counter example illustrated in Fig. 1, despite the similarity of the setting. The key difference is that we are interested in the arc lengths, whereas Ptolemy's inequality uses the chord lengths. For a simple counter example, we place the four points equally spaced on the equator (or any other great circle) of a 3d sphere for illustrative purposes (the example will also work in 2 dimensional data) The rectangle (on the sphere) becomes a great circle, and as we spaced them equally the angle from one to the next is $\frac{\pi}{2}$. The diagonals connect antipodal points, and hence have angle π. It is easy to see that $\frac{\pi}{2}\cdot\frac{\pi}{2}+\frac{\pi}{2}\cdot\frac{\pi}{2} = \frac{\pi^2}{2} \not\geq \pi^2$, and hence angles are not Ptolemaic. Using the cosines of the angles and Eq. 4, we get $1-\cos\frac{\pi}{2} = 1$

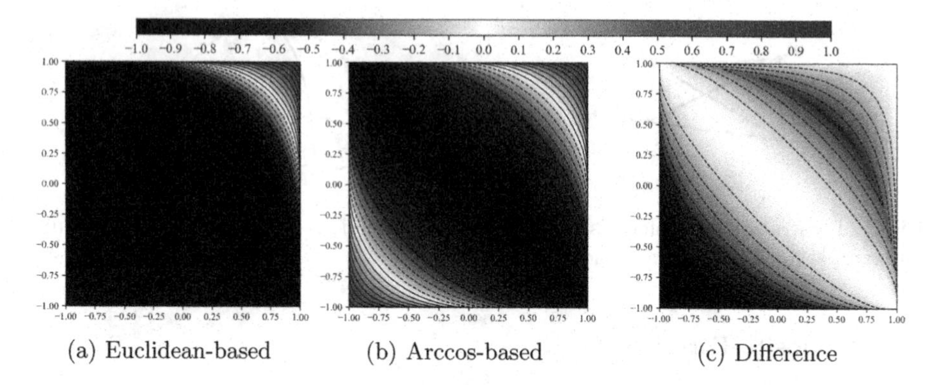

(a) Euclidean-based (b) Arccos-based (c) Difference

Fig. 2. Euclidean vs. Arccos-based triangular inequalities

respectively $1 - \cos\pi = 2$ and $1 + 1 \not\geq 4$. Only when using the Euclidean equivalency via Eq. 5 we obtain $\sqrt{2}$ respectively 2 we have $2 + 2 \geq 4$. The recent results in Ptolematic indexing hence will supposedly not easily transfer to angular space (i.e., to arc length not chord length), but remain usable for such data via Euclidean distance at a potential loss in numerical accuracy. It may also be possible to find a similar equation for the angular case.

4 Experiments

Table 1 summarizes the six bounds that we compare concerning their suitability for metric indexing. Note that we will not investigate the actual performance in a similarity index here, but plan to do this in future work. Instead, we want to focus on the bounds themselves concerning three properties:

1. how tight the bounds are, i.e., how much pruning power we lose
2. whether we can observe numerical instabilities
3. the differences in the computational effort necessary

Table 1. Triangle inequalities/bounds compared

Name	Eq.	Equation
Euclidean	(7)	$\text{sim}(\mathbf{x},\mathbf{z}) + \text{sim}(\mathbf{z},\mathbf{y}) - 1 - 2\sqrt{(1-\text{sim}(\mathbf{x},\mathbf{z}))(1-\text{sim}(\mathbf{z},\mathbf{y}))}$
Eucl-LB	(8)	$\text{sim}(\mathbf{x},\mathbf{z}) + \text{sim}(\mathbf{z},\mathbf{y}) + 2\cdot\min\{\text{sim}(\mathbf{x},\mathbf{z}),\text{sim}(\mathbf{y},\mathbf{z})\} - 3$
Arccos	(9)	$\cos(\arccos(\text{sim}(\mathbf{x},\mathbf{z})) + \arccos(\text{sim}(\mathbf{z},\mathbf{y})))$
Mult	(10)	$\text{sim}(\mathbf{x},\mathbf{z})\cdot\text{sim}(\mathbf{z},\mathbf{y}) - \sqrt{(1-\text{sim}(\mathbf{x},\mathbf{z})^2)\cdot(1-\text{sim}(\mathbf{z},\mathbf{y})^2)}$
Mult-LB1	(11)	$\text{sim}(\mathbf{x},\mathbf{z})\cdot\text{sim}(\mathbf{z},\mathbf{y}) + \min\{\text{sim}(\mathbf{x},\mathbf{z})^2,\text{sim}(\mathbf{y},\mathbf{z})^2\} - 1$
Mult-LB2	(12)	$2\cdot\text{sim}(\mathbf{x},\mathbf{z})\cdot\text{sim}(\mathbf{z},\mathbf{y}) - \lvert\text{sim}(\mathbf{x},\mathbf{z}) - \text{sim}(\mathbf{z},\mathbf{y})\rvert - 1$

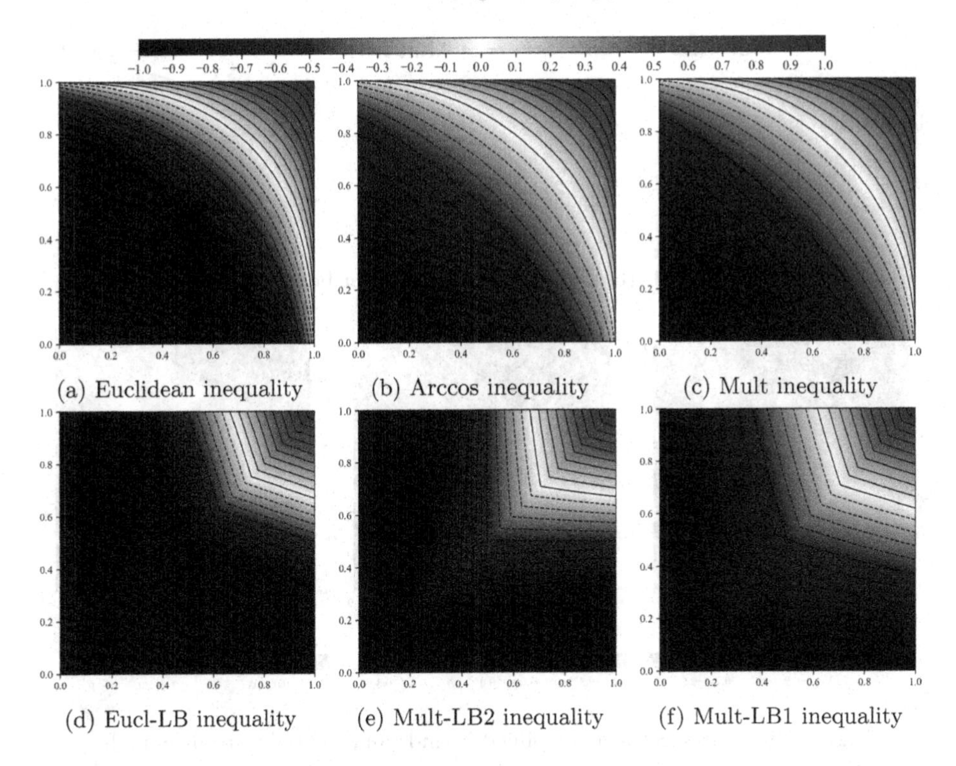

(a) Euclidean inequality (b) Arccos inequality (c) Mult inequality

(d) Eucl-LB inequality (e) Mult-LB2 inequality (f) Mult-LB1 inequality

Fig. 3. Lower bounds for the similarity $\mathrm{sim}(\mathbf{x}, \mathbf{y})$ given $\mathrm{sim}(\mathbf{x}, \mathbf{z})$ and $\mathrm{sim}(\mathbf{z}, \mathbf{y})$ using different inequalities from Table 1.

4.1 Approximation Quality

In Fig. 2, we plot the resulting lower bound for the similarity $\mathrm{sim}(\mathbf{x}, \mathbf{y})$ given $\mathrm{sim}(\mathbf{x}, \mathbf{z})$ and $\mathrm{sim}(\mathbf{z}, \mathbf{y})$ using the Euclidean-based bound (Eq. 7) in Fig. 2a and the Arccos-based bound (Eq. 9) in Fig. 2b. A striking difference is visible in the negative domain: if $\mathbf{x}$ and $\mathbf{z}$ are opposite directions, and $\mathbf{z}$ and $\mathbf{y}$ are also opposite, then $\mathbf{x}$ and $\mathbf{y}$ must in turn be similar. The Arccos-based bound produces positive bounds here, while the Euclidean-based bound can go down to -7. But in many cases where we employ cosine similarity, our data will be restricted to the non-negative domain, so this is likely not an issue, and could maybe be solved with a simple sign check. Upon closer inspection, we can observe that the bounds found by the Arccos-based approach tend to be substantially higher in particular for input similarities around 0.5. Figure 2c visualizes the difference between the two bounds. We can see that the Euclidean bounds are never higher than the Arccos bound, which is unsurprising as the latter is tight, and the first is a proper lower bound. But we can also see that the difference between the two (and hence the pruning power) can be as big as 0.5. This maximum is attained when the input cosine similarities are 0.5 (i.e., the known angles are 60°): The Euclidean bound is -1 then, while the Arccos-based bound is 0. In the typical use case of cosine on non-negative values, both bounds are effectively trivial. However, there still

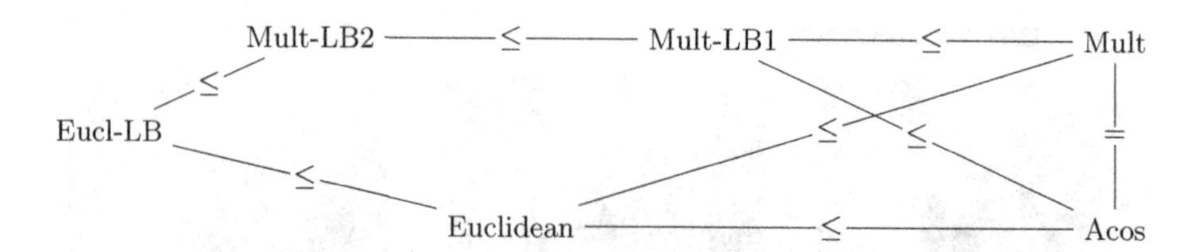

Fig. 4. Relationships between lower bounds

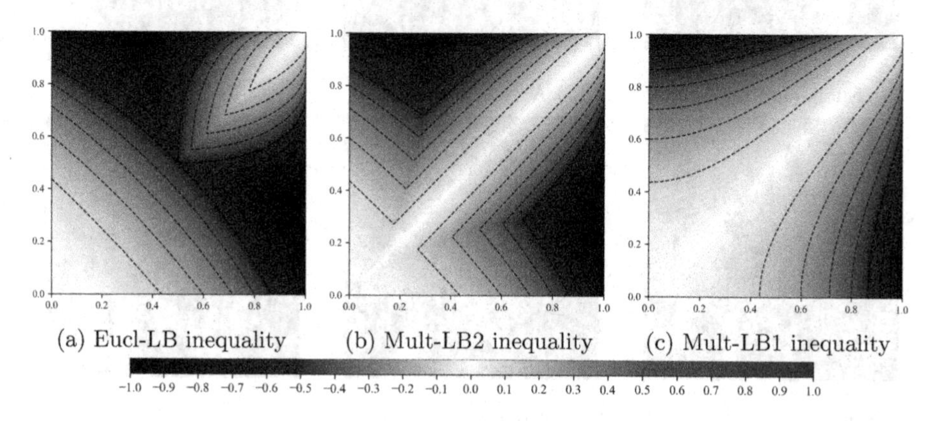

Fig. 5. Differences between simplified bounds and the tight arccos bound.

is a substantial difference for larger input similarities. Averaging over a uniform sampled grid of input values, considering only those where both bounds are non-negative, the average Euclidean bound is 0.2447, while the average Arccos-based bound is 0.3121, about 27.5% higher. Hence, using the Arccos-based bound is likely to yield better performance.

In the following, we focus on the non-negative domain, to improve the readability of the figures. In Fig. 3, we show all six bounds from Table 1. Figure 3a is the Euclidean bound, Fig. 3b is the Arccos bound we just saw. Figure 3c is the multiplicative version (Eq. 10), which yields no noticeable difference to the Arccos bound (mathematically, they are equivalent). Figure 3d is the simplified bound derived from cosine, whereas Fig. 3e and Fig. 3f are the two bounds derived from the multiplicative version of the Arccos bound. We observe that Mult-LB1 (Fig. 3f) is the best of the lower bounds, but also that none of the simplified bounds is a very close approximation to the optimal bounds in the first row. We obtain the following relationship of the presented lower bounds (c.f., Fig. 4):

$$\text{Eucl-LB} \leq \text{Euclidean} \leq \text{Arccos} = \text{mult}$$

$$\text{Eucl-LB} \leq \text{Mult-LB2} \leq \text{Mult-LB1} \leq \text{mult} = \text{Arccos}$$

In Fig. 5 we compare the three simplified bounds (we already compared the Euclidean bound to the tight Arccos bound in Fig. 2c). While the Mult-LB1

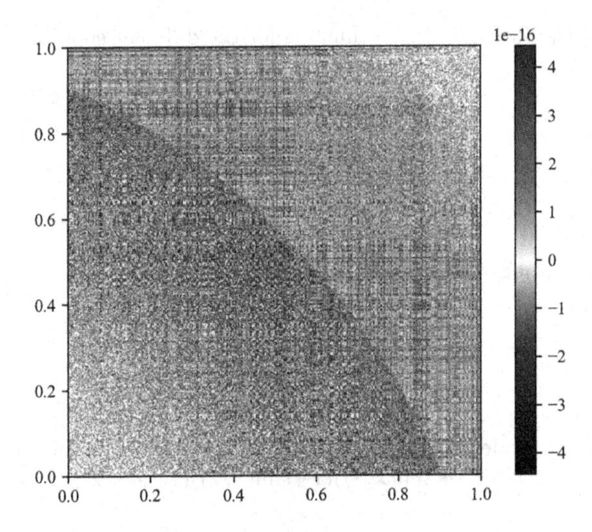

Fig. 6. Mult inequality

bound is the best of the simplified bounds, the divergence from the arccos bound can be quite substantial, at least when the two input similarities are not very close. As it can be seen from the isolines in the figures (at steps of 0.1), even if we would consider a bound that is worse by 0.1 or 0.2 acceptable, there remains a fairly large region of relevant inputs (e.g., where one similarity is close to 1.0, the other close to 0.8), where the loss in pruning performance may offset the slightly larger computational cost of using the Mult bound instead.

4.2 Numerical Stability

Mathematically, the Mult bound (Eq. 10) is equivalent to the arccos bound, but more efficient to compute. Given the prior experience with the numerical problem of catastrophic cancellation, we were concerned that this equation might be problematic because of the $(1 - \text{sim}^2)$ terms. Fortunately, if $\text{sim}^2 \to 1$, when the problem occurs, the entire square root will become negligible. We have experimented with some alternatives (such as expanding the square root to $\sqrt{(1+\text{sim}(\mathbf{x},\mathbf{z}))\cdot(1-\text{sim}(\mathbf{x},\mathbf{z}))\cdot(1+\text{sim}(\mathbf{z},\mathbf{y}))\cdot(1-\text{sim}(\mathbf{z},\mathbf{y}))}$), but could not find any benefits. We also compared Mult with the Arccos bound in Fig. 6. While the result appears largely chaotic, the values in this plot are all in the magnitude of 10^{-16}, i.e., they are at the expected limit of floating-point precision. Hence, there does not appear to be a numerical instability in this inequality.

4.3 Runtime Experiments

We benchmarked the different equations using Java 11 with double precision floats and the Java Microbenchmarking Harness JMH 1.32.[1] The experiments

[1] https://openjdk.java.net/projects/code-tools/jmh/.

Table 2. Runtime benchmarks for the different equations

Name	Eq.	Duration	Std. dev.	Accuracy
Euclidean	(7)	10.361 ns	±0.139 ns	○
Eucl-LB	(8)	10.171 ns	±0.132 ns	−−
Arccos	(9)	610.329 ns	±3.267 ns	++
Arccos (JaFaMa)	(9)	58.989 ns	±0.630 ns	++
Mult (recommended)	(10)	9.749 ns	±0.096 ns	++
Mult-variant	[a]	10.485 ns	±0.022 ns	++
Mult-LB1	(11)	10.313 ns	±0.025 ns	−
Mult-LB2	(12)	8.553 ns	±0.334 ns	−−
Baseline (sum)		8.186 ns	±0.146 ns	n/a

[a]Equation 10 expanded using $(1-x^2) = (1+x)(1-x)$ to obtain the variant $\text{sim}(\mathbf{x},\mathbf{z}){\cdot}\text{sim}(\mathbf{z},\mathbf{y}) - \sqrt{(1+\text{sim}(\mathbf{x},\mathbf{z}))(1-\text{sim}(\mathbf{x},\mathbf{z}))(1+\text{sim}(\mathbf{z},\mathbf{y}))(1-\text{sim}(\mathbf{z},\mathbf{y}))}$

were performed on an Intel i7-8650U using a single thread, and with the CPU's turbo-boost disabled such that the clock rate is stable at 1.9 GHz to reduce measurement noise as well as heat effects. As a baseline, we include a simple add operation to measure the cost of memory access to a pre-generated array of 2 million random numbers. Because trigonometric functions are fairly expensive, we also evaluate the JaFaMa library for fast math as an alternative to the JDK built-ins. JMH is set to perform 5 warmup iterations and 10 measurement iterations of 10 s each, to improve the accuracy of our measurements. We try to follow best practices in Java benchmarking (JMH is a well-suited tool for micro-benchmarking in Java), but nevertheless, the results with different programming languages (such as C) can be different due to different compiler optimization, and the usual pitfalls with runtime benchmarks remain [8]. Table 2 gives the results of our experiments. In these experiments, the runtime benefits of the simplified equations are minuscule. Apparently, the CPU can alleviate the latency of the square root to a large extend (e.g., via pipelining), and compared to the memory access cost of the baseline operation, the additional 1.6 nanoseconds will likely not matter for most applications. The benchmark, however, clearly shows the benefit of the "Mult" version over the "Arccos" version, which mathematically is equivalent but differs considerably in run time. While the use of JaFaMa as replacement reduces the runtime considerably, the much simpler "Mult" version still wins hands-down and hence is the version we ultimately recommend using. While "Mult-LB2" is marginally faster, it is also much less accurate and hence useful, as seen in Sect. 4.1.

5 Conclusions

In this article, we introduce a triangle inequality for cosine similarity. We study different ways of obtaining a triangle inequality, as well as different attempts at finding an even faster bound. The experiments show that a mathematically equivalent version of the Arccos-based bound is the best trade-off of accuracy (as it has optimal accuracy in our experiments) as well as run-time, where it is only marginally slower than the less accurate alternatives.

Hence, the recommended triangle inequalities for cosine similarity are:

$$\mathrm{sim}(\mathbf{x},\mathbf{y}) \geq \mathrm{sim}(\mathbf{x},\mathbf{z}) \cdot \mathrm{sim}(\mathbf{z},\mathbf{y}) - \sqrt{(1-\mathrm{sim}(\mathbf{x},\mathbf{z})^2)\cdot(1-\mathrm{sim}(\mathbf{z},\mathbf{y})^2)}$$
$$\mathrm{sim}(\mathbf{x},\mathbf{y}) \leq \mathrm{sim}(\mathbf{x},\mathbf{z}) \cdot \mathrm{sim}(\mathbf{z},\mathbf{y}) + \sqrt{(1-\mathrm{sim}(\mathbf{x},\mathbf{z})^2)\cdot(1-\mathrm{sim}(\mathbf{z},\mathbf{y})^2)}$$

We can not, however, rule out that there exists a more efficient equation that could be used instead. As this paper shows, there can be more than one version of the same bound that performs very differently due to the functions involved.

We hope to spur new research in the domain of accelerating similarity search with metric indexes, as this equation allows many existing indexes (such as M-trees, VP-trees, cover trees, LAESA, and many more) to be transformed into an efficient index for cosine similarity. Integrating this equation into algorithms will enable the acceleration of data mining algorithms in various domains, and the use of cosine similarity directly (without having to transform the similarities into distances first) may both allow simplification as well as optimization of algorithms. Furthermore, we hope that this research can eventually be transferred to other similarity functions besides cosine similarity. We believe it is a valuable insight that the triangle inequality for cosine distance contains the product of the existing similarities (but also a non-negligible correction term), whereas the triangle inequality for distance metrics is additive. We wonder if there exists a similarity equivalent of the definition of a metric (i.e., a "simetric"), with similar axioms but for the dual case of similarity functions, but the results above indicate that we will likely *not* be able to obtain a much more elegant general formulation of a triangle inequality for similarities.

References

1. Beygelzimer, A., Kakade, S.M., Langford, J.: Cover trees for nearest neighbor. In: International Conference on Machine Learning, ICML, pp. 97–104 (2006). https://doi.org/10.1145/1143844.1143857
2. Bozkaya, T., Özsoyoglu, Z.M.: Indexing large metric spaces for similarity search queries. ACM Trans. Database Syst. **24**(3), 361–404 (1999). https://doi.org/10.1145/328939.328959
3. Brin, S.: Near neighbor search in large metric spaces. In: Dayal, U., Gray, P.M.D., Nishio, S. (eds.) International Conference on Very Large Data Bases, VLDB, pp. 574–584. Morgan Kaufmann (1995)
4. Chávez, E., Ludueña, V., Reyes, N., Roggero, P.: Faster proximity searching with the distal SAT. In: International Conference on Similarity Search and Applications, SISAP, pp. 58–69 (2014). https://doi.org/10.1007/978-3-319-11988-5_6
5. Ciaccia, P., Patella, M., Zezula, P.: M-tree: an efficient access method for similarity search in metric spaces. In: International Conference on Very Large Data Bases, VLDB, pp. 426–435 (1997)
6. Hetland, M.L., Skopal, T., Lokoc, J., Beecks, C.: Ptolemaic access methods: challenging the reign of the metric space model. Inf. Syst. **38**(7), 989–1006 (2013). https://doi.org/10.1016/j.is.2012.05.011

7. Jagadish, H.V., Ooi, B.C., Tan, K., Yu, C., Zhang, R.: iDistance: an adaptive B^+-tree based indexing method for nearest neighbor search. ACM Trans. Database Syst. **30**(2), 364–397 (2005). https://doi.org/10.1145/1071610.1071612
8. Kriegel, H.-P., Schubert, E., Zimek, A.: The (black) art of runtime evaluation: are we comparing algorithms or implementations? Knowl. Inf. Syst. **52**(2), 341–378 (2016). https://doi.org/10.1007/s10115-016-1004-2
9. Lang, A., Schubert, E.: BETULA: numerically stable CF-trees for BIRCH clustering. In: International Conference on Similarity Search and Applications, SISAP, pp. 281–296 (2020). https://doi.org/10.1007/978-3-030-60936-8_22
10. Lokoc, J., Hetland, M.L., Skopal, T., Beecks, C.: Ptolemaic indexing of the signature quadratic form distance. In: International Conference on Similarity Search and Applications, pp. 9–16 (2011). https://doi.org/10.1145/1995412.1995417
11. Micó, L., Oncina, J., Vidal, E.: A new version of the nearest-neighbour approximating and eliminating search algorithm (AESA) with linear preprocessing time and memory requirements. Pattern Recognit. Lett. **15**(1), 9–17 (1994). https://doi.org/10.1016/0167-8655(94)90095-7
12. Nanopoulos, A., Radovanovic, M., Ivanovic, M.: How does high dimensionality affect collaborative filtering? In: ACM Conference on Recommender Systems, RecSys, pp. 293–296 (2009). https://doi.org/10.1145/1639714.1639771
13. Navarro, G.: Searching in metric spaces by spatial approximation. VLDB J. **11**(1), 28–46 (2002). https://doi.org/10.1007/s007780200060
14. Novak, D., Batko, M., Zezula, P.: Metric index: an efficient and scalable solution for precise and approximate similarity search. Inf. Syst. **36**(4), 721–733 (2011). https://doi.org/10.1016/j.is.2010.10.002
15. Omohundro, S.M.: Five balltree construction algorithms. Technical report. TR-89-063, International Computer Science Institute (ICSI) (1989)
16. Radovanovic, M., Nanopoulos, A., Ivanovic, M.: Nearest neighbors in high-dimensional data: the emergence and influence of hubs. In: International Conference on Machine Learning, ICML, pp. 865–872 (2009). https://doi.org/10.1145/1553374.1553485
17. Ruiz, G., Santoyo, F., Chávez, E., Figueroa, K., Tellez, E.S.: Extreme pivots for faster metric indexes. In: International Conference on Similarity Search and Applications, SISAP, pp. 115–126 (2013). https://doi.org/10.1007/978-3-642-41062-8_12
18. Schubert, E., Gertz, M.: Numerically stable parallel computation of (co-)variance. In: International Conference on Scientific and Statistical Database Management, SSDBM, pp. 10:1–10:12 (2018). https://doi.org/10.1145/3221269.3223036
19. Schubert, E., Zimek, A.: ELKI: a large open-source library for data analysis - ELKI release 0.7.5, Heidelberg. CoRR abs/1902.03616 (2019). http://arxiv.org/abs/1902.03616
20. Uhlmann, J.K.: Satisfying general proximity/similarity queries with metric trees. Inf. Process. Lett. **40**(4), 175–179 (1991). https://doi.org/10.1016/0020-0190(91)90074-R
21. Yianilos, P.N.: Data structures and algorithms for nearest neighbor search in general metric spaces. In: ACM/SIGACT-SIAM Symposium on Discrete Algorithms, SODA, pp. 311–321 (1993)
22. Zimek, A., Schubert, E., Kriegel, H.: A survey on unsupervised outlier detection in high-dimensional numerical data. Stat. Anal. Data Min. **5**(5), 363–387 (2012). https://doi.org/10.1002/sam.11161

A Cost Model for Reverse Nearest Neighbor Query Processing on R-Trees Using Self Pruning

Felix Borutta[1], Peer Kröger[2(✉)], and Matthias Renz[2]

[1] Institute for Computer Science, Ludwig-Maximilians-Universität München, München, Germany
borutta@dbs.ifi.lmu.de

[2] Institute for Computer Science, Christian-Albrechts-Universität zu Kiel, München, Germany
{pkr,mr}@isdm.informatik.uni-kiel.de

Abstract. In this short paper, we propose the first cost model for a class of index structures designed for reverse nearest neighbor (RNN) search, so-called self pruning approaches. These approaches use estimations of the nearest neighbor distances of database objects for pruning. We will particularly detail our cost model for R-Trees but our concepts can easily applied to any tree-like index structure that implements a self pruning strategy. Our cost model estimates the number of disk accesses of a given RNN query and, thus, allows to predict the required I/O costs in any hardware environment. We further explore three variants regarding the trade-off between estimation accuracy and model efficiency/storage overhead. Preliminary experiments on synthetic data confirm that the estimations are accurate compared to the exact query costs.

1 Introduction

Reverse nearest neighbor (RNN) queries are prevalent in many practical applications since they determine the set of data objects influenced by the query. Specifically, an RNN query retrieves those objects from the database having the query as one of their nearest neighbors (NNs). Variants of this basic query introduce the parameter k specifying the number of NNs that are considered (i.e., the query must be among the kNNs of a true hit), and/or a distinction between query set and answering set (so-called bi-chromatic scenario compared to the "normal" case that is referred to as monochromatic).

Beside a plethora of index structures and algorithms especially designed to optimize RNN query processing, to the best of our knowledge, no work has been done for estimating the costs of these approaches so far. Predicting the costs for processing a given query is mandatory for (relational) query optimizers. A cost model allows to generate efficient query plans and enables effective scheduling. Thus, this work is a first step towards the practical use of the existing query processing algorithms and their respective data structures in real database systems.

In this short paper, we sketch the idea for a general cost model for RNN queries for a rather general class of query processing algorithms. Existing algorithms for RNN query processing can be classified according to the applied strategy of pruning objects from

N. Reyes et al. (Eds.): SISAP 2021, LNCS 13058, pp. 45–53, 2021.
https://doi.org/10.1007/978-3-030-89657-7_4

the search space. *Self pruning* approaches typically rely on special index structures that usually offer a higher selectivity and, hence, are more efficient. In turn, these methods are usually less flexible than *mutual pruning* approaches in terms of query parametrization and database updates. Since both pruning strategies are rather different paradigms, we focus on self pruning methods only. We explore a general way to estimate the number of index pages that need to be accessed for a given query which allows to predict the I/O costs on any hardware environment. The cost model will be explained using a concrete instance of self pruning methods, the RdNN tree [1] which basically uses a member of the R-Tree family. However, we want to emphasize that our general idea is independent of the used index and can be adapted to any tree-like index implementing a self pruning method very easily. The reason for this is that the only basic assumption of our cost model is that information on the size of the page regions, e.g. minimum bounding rectangles (MBRs) in case of an R-Tree, of index nodes is know. This information typically contains only a very few numbers and can easily be materialized in the cache.

The remainder is organized as follows. Section 2 gives an overview of related work. In Sect. 3, we explain our new cost model for RNN query processing using self pruning methods. Some preliminary experiments are presented in Sect. 4. Finally, Sect. 5 concludes the paper.

2 Related Work

Self pruning approaches like [1,2] are usually designed on top of a tree-like index structure. They are based on the observation that if the distance between any database object o and the query q is smaller than the kNN distance of o, o is part of the result set, i.e., a RNN of q. Otherwise, o can be pruned. Consequently, self pruning approaches try to exactly compute or (conservatively) approximate the kNN distance of each index entry e. If this estimate is smaller than the distance of e to the query q, then e can be pruned. Often, self pruning approaches simply pre-compute kNN distances of database points and propagate maxima of these distances to higher level index nodes for pruning. The major limitation of these approaches is that the pre-computation is time and memory consuming and less flexible to database updates. These methods are usually limited to one specific or very few values of k. Approaches like [3–9] try to overcome these limitations by using approximations of kNN distances (for any k) but this yields an additional refinement overhead – or only approximate results. *Mutual pruning approaches* such as [10,11] use other points to prune a given index entry e. For instance, [11] iteratively constructs Voronoi hyper-planes around the query q from a nearest neighbor ranking w.r.t. q. Points and index entries that are beyond k Voronoi hyper-planes w.r.t. q can be pruned. Mutual pruning approaches need an additional refinement of candidates (i.e., a kNN query for not pruned objects) to compute the final results.

As mentioned above, to the best of our knowledge, there are no cost models for RNN algorithms proposed so far. However, there are a lot of cost models for other query types on R-Trees, including NN queries (e.g. [12–14]) and spatial join queries (e.g. [15,16]). Closest to our work is the method for range queries independently proposed by [17] and [18]. Both approaches assume that the MBR of each node in the underlying R-Tree is given and estimate the disc accesses using the concept of the Minkowski sum. We will revisit details on this model later.

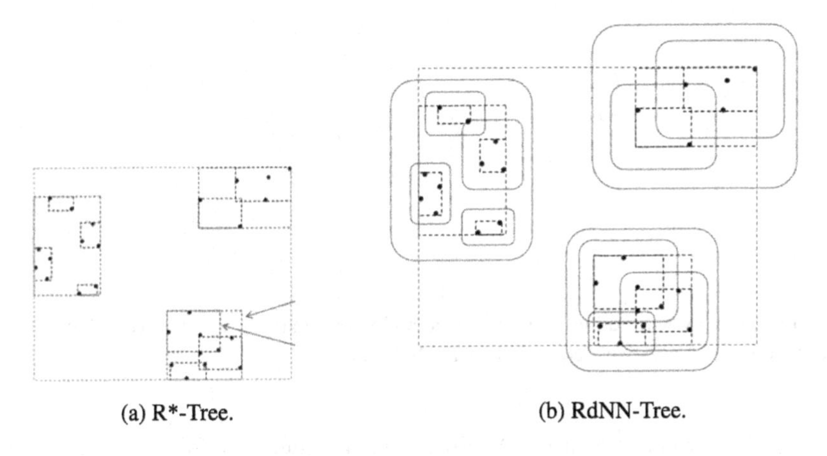

(a) R*-Tree. (b) RdNN-Tree.

Fig. 1. Visualization of an RdNN-Tree (b) as an extension of the R*-Tree (a).

3 A Cost Model for Self Pruning Approaches

For a positive integer k and a query object q, a k-NN query retrieves the set $NN(q,k)$ including those k points having the smallest distance $dist(.,.)$ to q. In case of ties, this set may have more than k elements. The distance between p and its kNN is called kNN-distance (denoted by $kNNdist(p)$) of p. A RkNN query with query object q can be defined as those objects having q as one of their kNN, i.e., $RNN(q,k) = \{o \in DB | o \in NN(q,k)\}$. If k is clear from context, we omit it and use NN, $NNdist(.)$, and RNN instead of kNN, $kNNdist(.)$, and RkNN.

The basic observation behind self-pruning approaches is that an object p qualifies for a given RNN query if and only if $dist(p,q) \leq NNdist(p)$. Thus, materializing (exact or approximate) NN-distances of all database objects provides a powerful and very selective pruning possibility. Self-pruning approaches use any conventional index, e.g. an R*-tree (as in the RdNN-Tree [1]), to organize the data objects but additionally stores the pre-computed NN-distances. For a data page of the index containing a set of database objects, an approximation of the NN-distances of all points of this page needs to be derived and this approximation needs to be conservative for producing exact results. This can easily be done by aggregating the maximum of all NN-distances in that page. For directory pages of a tree-based index containing child pages each associated with a NN-distance estimate of its corresponding sub-tree, the procedure is similar: the maximum NN-distance of all child nodes need to be aggregated. Thus, each node N of the index aggregates the maximum NN-distance of all objects represented by N. The extension of an R*-Tree to an RdNN-Tree is visualized in Fig. 1. The aggregated maximum NN-distance of each node N, denoted by $NNdist(N)$, is visualized as box with rounded corners around the corresponding page regions (PRs). These distances can be used during query processing: node N may contain a true hit if for the minimum distance $MINDIST$ between an object q and the PR of N it holds: $MINDIST(q,N) \leq NNdist(N)$. In this case, the subtree of N needs to be traversed. Else, node N can be pruned.

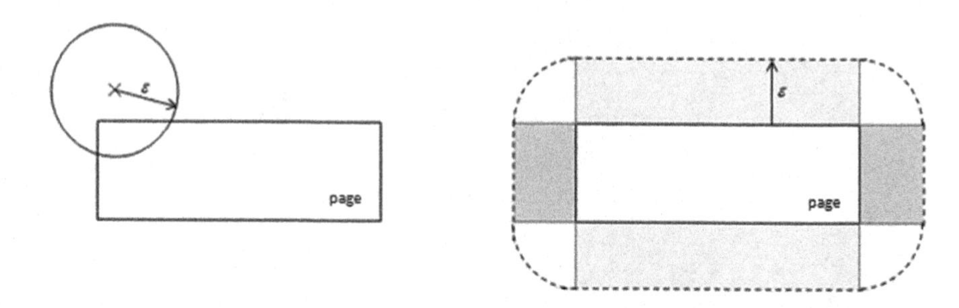

Fig. 2. Range query with radius ε and the page region of an arbitrary index page (left) and the corresponding spatially extended region a.k.a. Minkowski sum (right).

We will show our ideas using the RdNN, assuming that the PRs of the underlying index are minimum bounding rectangles (MBRs), the distance approximations of directory nodes are conservative, and the index is tree-based in the following but our ideas can be extended to any other shapes, approximations, and indexes. The basic observation of our approach is that the situation depicted in Fig. 1(b) is related to the cost model for (aka ε-)range queries [17,18] which estimates the probability of an intersection between the query (circle around q with radius ε) and the PRs of each level of the tree, i.e., the probability that the corresponding subtree needs to be traversed. For this purpose, the PRs are spatially extended by the radius ε as it is done in self-pruning approaches. However, for ε-range queries the spatial extend of all PRs is fixed to ε, while for the RdNN-Tree each PR has its own aggregated maximum NN-distance specifying the spatial extend.

Our cost model basically estimates for each node in the index tree the probability of being traversed when a given query is launched in order to determine the average amount of nodes that need to be accessed. It is based on the same assumptions claimed in [17,18]; the most basic assumption is that we have information on the PRs of each node in the index. We first start with revisiting the cost model of range queries.

For a range query with radius ε, the access probability of a node N with page region $N.Reg$[1] is given by

$$PR(Access(N)) = \frac{Vol(N.Reg\, spatially\, extended\, by\, \varepsilon)}{Vol(data\ space)},$$

where $Vol(.)$ computes the volume of a region. The page region $N.Reg$ spatially extended by the query radius ε correspond to a bounding box with rounded corners such that the edges have distance ε to the original page region $N.Reg$. This region is known as the Minkowski sum. The idea is visualized in Fig. 2.

In order to be able to compute the probability of accessing a given node N, we need to compute the volume of the Minkowski sum of the page region $N.Reg$ of N and ε (corresponding to the numerator in the above formular). Let $N.e$ be the edge length of the page region of N, then the volume of the Minkowski sum of $N.Reg$ and ε is

[1] As mentioned above, our method is not restricted to the exact geometry of page regions.

$$Vol(N.Reg\ spatially\ extended\ by\ \varepsilon) = Vol_{Minkowski}(N.Reg, \varepsilon)$$

$$= \sum_{0 \le i \le d} \binom{d}{i} \cdot 2^{d-i} \cdot N.e^i \cdot \frac{V(Sphere_{(d-i)}(\varepsilon))}{2^{d-i}} = \sum_{0 \le i \le d} \binom{d}{i} \cdot N.e^i \cdot V(Sphere_{(d-i)}, \varepsilon).$$

where $Sphere_{(d-i)}(\varepsilon)$ is a $(d-i)$-dimensional sphere of radius ε. The volume of this sphere, $Vol(Sphere_{(d-i)}(\varepsilon))$, can be computed using the Gamma function:

$$Vol(Sphere_{(d-i)}(\varepsilon)) = \frac{\pi^{\frac{d-i}{2}} \cdot \varepsilon^{d-i}}{\Gamma(\frac{d-i}{2} + 1)}.$$

The Minkowski volume can be used to calculate the volume of any node N of the index and can be used to compute the probability that N needs to be accessed. In order to estimate the costs for the entire index, we need to determine the access probabilities for all index nodes. For that purpose, we need the edge lengths $N.e$ of all index nodes N. One way to get this is to materialize these values which is typically not a significant overhead and can often even be hold in main memory. If the overhead of storing and updating this information is too large, [17] and [18] offer a way to estimate these values. This estimation is done level-wise: the number of nodes N_i on level i of the tree, denoted by $Card(N_i)$, can recursively be obtained from the average storage utilization. Under the assumption that the MBR of each node N_i is a hyper-cube with equal edge length $N_i.e$ and that its expected volume is $Vol(N_i) = Vol(data\ space)\,/\,Card(N_i)$ we can estimate the average edge length of N_i as

$$N_i.e = \sqrt[d]{\frac{Vol(data\ space)}{Card(N_i)}}.$$

Thus, the total number of index nodes (i.e., pages) accessed while processing an ε-range query can be approximated as:

$$\#\,page\,accesses = \sum_{i=1}^{\text{indexheight}} Card(N_i) \cdot Vol_{Minkowski}\Big(\sqrt[d]{\frac{Vol(data\ space)}{Card(N_i)}}, \varepsilon\Big).$$

For the transformation of this model from range queries to RNN queries we first explore the relationship between these two query types. Intuitively, range queries retrieve those objects o that are enclosed in a sphere centered at the query object q having the query range ε as radius, i.e., $dist(q, o) \le \varepsilon$. When a self pruning approach is implemented using pre-computed NN-distances, RNN queries retrieve those objects o that are the center of a sphere which has the NN-distance of o as radius and in which q is enclosed, i.e., $dist(q, o) \le NNdist(o)$. It should be mentioned, however, that this relationship cannot be used to process RNN queries like range queries in general. Only the aggregation and materialization of the NN-distances in the index as proposed in the literature enables to build this relationship.

During the processing of a range query, a page must be accessed if its page region (e.g. MBR) intersects with the query range, i.e., the sphere with radius ε centered at q. For a RNN query, a page must be accessed if the Minkowski sum of its page region and its maximum NN-distance includes the query. Thus, for any node N in the index, we need to use its maximum NN-distance, $NNdist(N)$ to compute the Minkowski volume:

$$Vol_{Minkowski}(N.e, NNdist(N)) = \sum_{0 \leq i \leq d} \binom{d}{i} \cdot N.e^i \cdot Vol(Sphere_{(d-i)}(NNdist(N))).$$

The edge length can be approximated as described above. The remaining challenge now is that while for range queries, the radius ε is fixed in all Minkowski volumes, the aggregated maximum NN-distances of the nodes on level index i can be rather different. In the following, we propose three variants to solve this.

Variant 1: The first variant accounts for the variation of NN-distances and sums up all Minkowski volumes of all index nodes. Note that this requires to have access to all NN-distance values of all nodes N_i on all levels i of the index which could be materialized (for small data sets even in the cache). The number of page accesses is

$$\#page\,accesses = \sum_{i=1}^{\text{indexheight}} \sum_{n \in N_i} Vol_{Minkowski}\left(\sqrt[d]{\frac{Vol(data\ space)}{Card(n)}}, NNdist(n)\right).$$

Variant 2: If the index is large and pre-computing/materialization of NN-distances for all index entries is not an option, the necessary information needs to be fetched from disc involving a huge overhead of I/O accesses (all nodes of the tree need to be accessed). We can circumvent this by taking the NN-distance of the root *Root* of the index which is the maximum NN-distances of all data objects. Obviously, this comes to the cost of decreasing the accuracy of the estimation. If M is the number of all nodes of the index, then, we can estimate the number of page accesses by

$$\#page\,accesses = M \cdot Vol_{Minkowski}\left(\sqrt[d]{\frac{Vol(data\ space)}{Card(N_i)}}, NNdist(Root)\right).$$

Variant 3: The variants discussed above basically trade-off the accuracy of the estimation and the costs for obtaining the estimation (in terms of storage overhead or, if the required information needs to be fetched from disc, in terms of time). As a compromise we propose to aggregate the average NN-distances for each index level which causes much less overhead to maintain and materialize than in Variant 1 but should give better estimates than Variant 2. The average NN-distance of all nodes N_i of level i is

$$NNdist_i^{avg} = \frac{\sum_{N_i} NNdist(N_i)}{Card(N_i)}.$$

Then the number of page accesses can be calculated as

$$\#page\,accesses = \sum_{i=1}^{\text{indexheight}} Card(N_i) \cdot Vol_{Minkowski}\left(\sqrt[d]{\frac{Vol(data\ space)}{Card(N_i)}}, NNdist_i^{avg}\right).$$

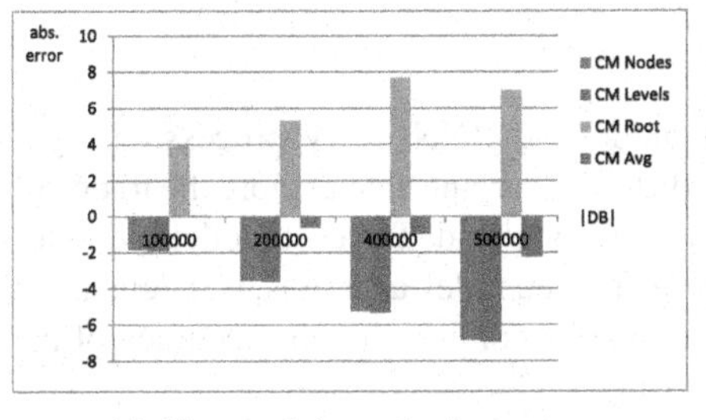

(a) Clustered data: absolute error.

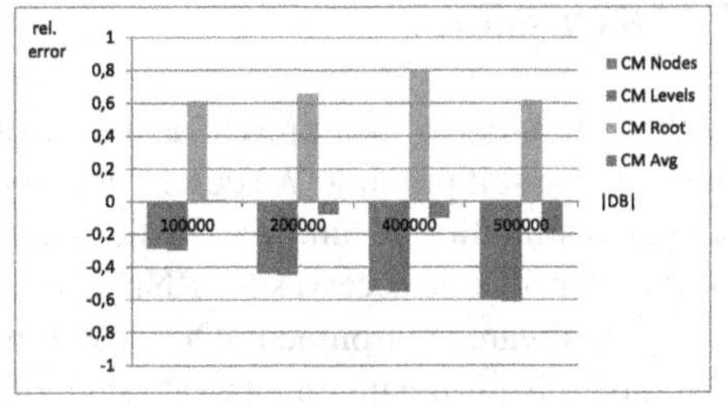

(b) Clustered data: relative error.

Fig. 3. Accuracy of the estimation w.r.t. varying data size.

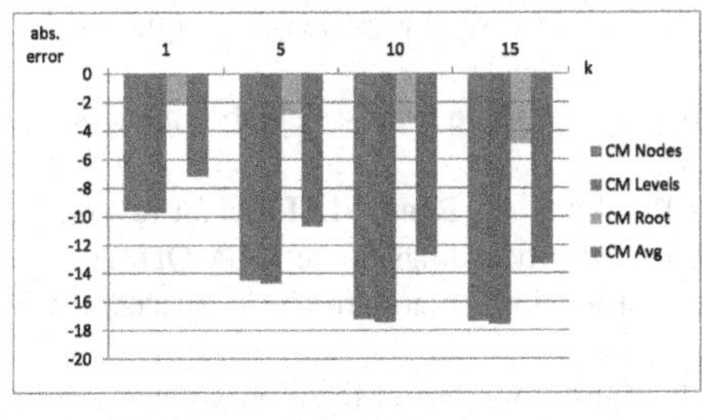

(a) Clustered data: absolute error.

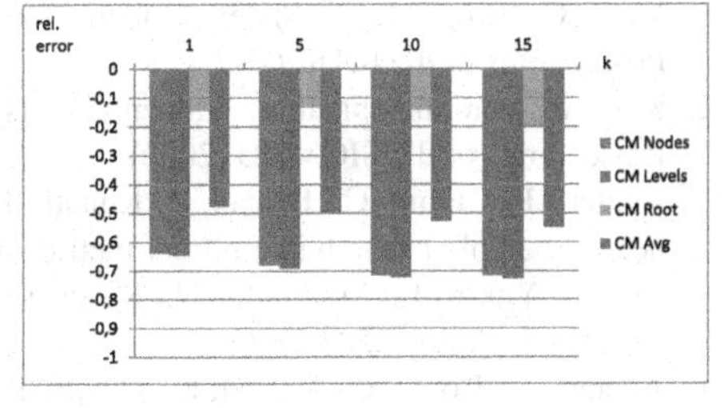

(b) Clustered data: relative error.

Fig. 4. Accuracy of the estimation w.r.t. varying k.

4 Preliminary Empirical Study

We evaluate how accurate our model can estimate the real page accesses for a given query and report absolute and relative estimation errors for all three variants discussed above for the RdNN-tree implementation of ELKI [19] with a page size of 8K. We used a 3D synthetic data sets with 10 clusters of equal size each following a Gaussian distribution with random mean and standard deviation and an additional 10% uniformly distributed noise. In all runs, we used 50% of the database points and another 50% of randomly generated points as query objects and averaged the results.

Figure 3 displays the accuracy w.r.t. the database size ($k = 1$). Both the absolute and relative error is considerably small and stable and only grows slowly with increasing database size. Variant 2 that only considers the root node overestimates the costs while all other estimations are conservative. Figure 4 depicts the accuracy of the model variants w.r.t. the query parameter k. The database size is fixed at 200,000 points. Here, all estimates are conservative. With increasing k, the error increases most likely because the kNN distance exponentially contributes to the Minkowski volume.

We also conducted first experiments on the impact of the data dimensionality (omitted due to space limitations). The results show low effects of the data dimensionality as long as it is moderate (>20), but we assume that a potential break-down of the index may not be accommodated adequately in the cost model.

5 Discussion

In this short paper, we present a first cost model for RNN query processing algorithms using self pruning. We described three different variants that explore the trade-off between estimation accuracy and efficiency/storage overhead. The cost model estimates the number of page accesses for RNN queries on a given index and, thus, is independent of any hardware environment. Our preliminary results confirm that the accuracy of the cost model is promising in a broad range of settings.

References

1. Yang, C., Lin, K.I.: An index structure for efficient reverse nearest neighbor queries. In: Proceedings of the ICDE (2001)
2. Korn, F., Muthukrishnan, S.: Influenced sets based on reverse nearest neighbor queries. In: Proceedings of the SIGMOD (2000)
3. Achtert, E., Böhm, C., Kröger, P., Kunath, P., Pryakhin, A., Renz, M.: Efficient reverse k-nearest neighbor search in arbitrary metric spaces. In: Proceedings of the SIGMOD (2006)
4. Tao, Y., Yiu, M.L., Mamoulis, N.: Reverse nearest neighbor search in metric spaces. IEEE TKDE **18**, 1239–1252 (2006)
5. Achtert, E., Böhm, C., Kröger, P., Kunath, P., Pryakhin, A., Renz, M.: Approximate reverse k-nearest neighbor search in general metric spaces. In: Proceedings of the CIKM (2006)
6. Achtert, E., Böhm, C., Kröger, P., Kunath, P., Pryakhin, A., Renz, M.: Efficient reverse k-nearest neighbor estimation. In: Proceedings of the BTW (2007)
7. Figueroa, K., Paredes, R.: Approximate direct and reverse nearest neighbor queries, and the k-nearest neighbor graph. In: Proceedings of the SISAP (2009)
8. Casanova, G., et al.: Dimensional testing for reverse k-nearest neighbor search. In: Proceedings of the VLDB (2017)
9. Berrendorf, M., Borutta, F., Kröger, P.: k-distance approximation for memory-efficient RkNN retrieval. In: Amato, G., Gennaro, C., Oria, V., Radovanović, M. (eds.) SISAP 2019. LNCS, vol. 11807, pp. 57–71. Springer, Cham (2019). https://doi.org/10.1007/978-3-030-32047-8_6
10. Singh, A., Ferhatosmanoglu, H., Tosun, A.S.: High dimensional reverse nearest neighbor queries. In: Proceedings of the CIKM (2003)
11. Tao, Y., Papadias, D., Lian, X.: Reverse kNN search in arbitrary dimensionality. In: Proceedings of the VLDB (2004)
12. Papadopoulos, A., Manolopoulos, Y.: Performance of nearest neighbor queries in R-trees. In: Afrati, F., Kolaitis, P. (eds.) ICDT 1997. LNCS, vol. 1186, pp. 394–408. Springer, Heidelberg (1997). https://doi.org/10.1007/3-540-62222-5_59
13. Berchtold, S., Böhm, C., Keim, D.A., Kriegel, H.P.: A cost model for nearest neighbor search in high-dimensional data space. In:1997 Proceedings of the 16th ACM SIGACT-SIGMOD-SIGART Symposium on Principles of Database Systems, Tucson, Arizona (1997)
14. Korn, F., Pagel, B., Faloutsos, C.: On the dimensionality curse and the self-similarity blessing. IEEE TKDE **13**, 96–111 (2001)
15. Huang, Y., Jing, N., Rundensteiner, E.: A cost model for estimating the performance of spatial joins using R-trees. In: Proceedings of the SSDBM (1997)
16. Böhm, C., Kriegel, H.P.: A cost model and index architecture for the similarity join. In: Proceedings of the ICDE (2001)
17. Kamel, I., Faloutsos, C.: On packing R-trees. In: Proceedings of the CIKM (1993)

18. Pagel, B.U., Six, H., Toben, H., Widmayer, P.: Towards an analysis of range query performance. In: PODS (1993)
19. Achtert, E., Kriegel, H.-P., Zimek, A.: ELKI: a software system for evaluation of subspace clustering algorithms. In: Ludäscher, B., Mamoulis, N. (eds.) SSDBM 2008. LNCS, vol. 5069, pp. 580–585. Springer, Heidelberg (2008). https://doi.org/10.1007/978-3-540-69497-7_41

How Many Neighbours for Known-Item Search?

Jakub Lokoč(✉) and Tomáš Souček

SIRET Research Group, Department of Software Engineering, Faculty of Mathematics and Physics, Charles University, Prague, Czech Republic
lokoc@ksi.mff.cuni.cz

Abstract. In the ongoing multimedia age, search needs become more variable and challenging to aid. In the area of content-based similarity search, asking search engines for one or just a few nearest neighbours to a query does not have to be sufficient to accomplish a challenging search task. In this work, we investigate a task type where users search for one particular multimedia object in a large database. Complexity of the task is empirically demonstrated with a set of experiments and the need for a larger number of nearest neighbours is discussed. A baseline approach for finding a larger number of approximate nearest neighbours is tested, showing potential speed-up with respect to a naive sequential scan. Last but not least, an open efficiency challenge for metric access methods is discussed for datasets used in the experiments.

Keywords: Similarity search · Known-item search · Data indexing

1 Introduction

Deep learning here, deep learning there, deep learning everywhere! Words that have come to mind of a multimedia retrieval researcher since 2012. Besides other retrieval challenges, similarity search [5,6,29] has also been significantly affected by the impressive deep learning paradigm [9]. The cornerstone of the general similarity search approach, similarity space (U, σ) consisting of a descriptor universe U and a similarity measure σ, started to be narrowed to "just" vector spaces with a cheap bin-to-bin similarity measure used during a deep model training process. In other words, a similarity of any multimedia data objects x, y is now often modeled with a cheap similarity function (usually linear time complexity) evaluated for their vector representations $v_x, v_y \in R^n$ obtained from a deep model[1]. Regardless of deep learning trends, there still exists a need for querying a large database for similar objects to a query object, assuming database objects are mapped to descriptors $S \subset U$. For a query q, applications usually require a set of most similar objects from a multimedia database. Assuming a popular approach to model similarity with a distance function δ, two popular similarity queries are $range(v_q, \theta) = \{v_o \in S | \delta(v_q, v_o) < \theta\}$, and k nearest neighbours query $kNN(v_q, S) = \{X \subset S : |X| = k, \forall v_x \in X, \forall v_y \in S - X : \delta(v_q, v_x) \leq \delta(v_q, v_y)\}$.

[1] In the following text, we follow the notation x, v_x to formally distinguish objects and their descriptors, where descriptors are means of object similarity evaluations.

N. Reyes et al. (Eds.): SISAP 2021, LNCS 13058, pp. 54–65, 2021.
https://doi.org/10.1007/978-3-030-89657-7_5

The aforementioned similarity queries are useful for search needs initiated with a query object addressing the contents of multimedia objects[2]. Usually, users provide either a text query for a text-multimedia cross modal search approach [14,22], or an example multimedia object. The ultimate problem is whether the provided query is good enough to ensure desired objects in the result set, i.e., in the set of the most similar objects to the query. The problem can be divided to two sub-problems – whether the user can provide a sufficient query object (or detailed text description), and whether the system implements a similarity model consistent with user expectation of similarity between two objects. In this paper, we further consider user need aspects [28], but we keep general formal specification of search needs. Let C be a subset of database objects representing some target class/topic. The search problem complexity differs if users want to find just an arbitrary item $x \in C$ (i.e., high precision is sufficient), or all items from C are required (i.e., high precision and also recall are necessary). Due to potentially high variability of objects in C, it is way more challenging to find all dataset instances of the class.

A special variant of the all-instance search task is for $|C| \rightarrow 1$, which corresponds to search need for a very narrow class of objects. In extreme case, only one multimedia object (e.g., image or shot) is required, which is referred to as *known-item search* (KIS). Although unique properties of a single searched object might seem as an advantage for the search engine, users often do not actively remember all the specific details for query formulation. On the other hand, there is an assumption that users can rely on (limited) passive knowledge of the known item when refining and browsing candidate result sets. The passive knowledge can include also a temporal context of the item in the case of video sequences. The complexity of a KIS task depends also on the number of similar dataset objects matching provided (potentially imperfect) query description. For example, searching for some specific scene of a surfing person would be way more easier if there are no other scenes of people surfing in the database. In case there exist near-duplicates (e.g., some small audio-visual transformations of the target object), multiple instances could be considered as the correct result. From this perspective, known-item search can be generalized from $|C| = 1$ to $|C| \geq 1$, but the set consists just of near-duplicate objects satisfying the need for one searched multimedia object. This is the main difference from an ad-hoc search task with a specific narrow search focus, where different objects can fulfill the specification.

In this paper, we argument that known-item search is often very challenging even with a state-of-the-art text-image search model (demonstrated in Sect. 3). In order to find a searched known item, an interactive search approach [25,27] is therefore a preferred option as reported by respected evaluation campaigns [10,16]. In the last decade, several interactive search systems were designed and tested [1,11–13,19,24]. To deal with a known-item search task, users can either iteratively reformulate queries after unsuccessful inspection of top ranked items (kNN queries with low k), or, use advanced visualization [4], relevance feedback

[2] We consider challenging content-based search cases, where users do not know unique structured attributes (e.g., filename or ID) of searched multimedia objects.

[7] or other exploration methods when top ranked result set inspection fails. With a single query available, a substantial portion of the database has to be considered for inspection to guarantee a higher chance for success. Therefore, finding a larger set of (approximate) nearest objects to a query represents a suitable search step. At the same time, larger numbers of nearest neighbours represent a challenge for query processing methods.

2 Known-Item Search

Imagine a large collection of funny videos, where a user wants to find one particular scene which made the user laugh for days. Definitely, the user might want to find this one particular scene again in the future, which would restrict the set of all funny scenes to one searched instance. Another example might be a memory of some experience, captured by a wearable camera to a personal lifelog database [10]. Again, the search need might focus just on the one specific memory. These examples illustrate that known-item search tasks are natural part of the set of possible search needs. The tasks are also well-suited for comparative evaluations [17] and benchmarking as the ground truth is determined by the one searched item (e.g., image or temporal segment), compared to partially unknown ground truth of more generally formulated Ad-hoc search tasks evaluated at TRECVID [3] (KIS tasks used to be evaluated at TRECVID in the past). We note that the discussed near-duplicates might be a missing part of ground truth for KIS tasks as well. Nevertheless, in an automatic evaluation of ranked lists the missing near-duplicates might achieve similar ranks as the available correct objects and also this approximation issue represents a consistent obstacle for all compared methods.

2.1 Problem Formulation

Known-item search corresponds to a search scenario, where a user has just a mental picture of an existing multimedia object from a given database. Either the known object has been seen before, or a specific enough description (potentially including hand-drawn sketches) of the object was provided to the user. In the context of this paper, a generalized KIS task can be formulated as:

Definition 1. *Let DB be a multimedia collection, the task is to find one* $t \in C_T \subset DB$*, where* C_T *contains one known target object and its near-duplicates differing from the target object by a small audio-visual transformation, negligible for the search need (e.g., different encoding or minor image enhancement).*

For automatic evaluations analyzing ranking of database objects with respect to a query, the top ranked $t \in C_T$ is considered, optimistically assuming that users do not overlook a correct item in a displayed ranked result set. For search needs targeting just a part of a multimedia object (e.g., segment of a video), the definition can be modified by using an appropriate data representation unit.

2.2 Ranking Model Evaluation

In order to measure known-item search effectiveness of a model (U, δ), a set of pairs $B = \{[q_i, C_{T_i}]\}_{i=1}^{n}$ can be created for a multimedia database DB, where q_i represents a user defined query addressing selected $C_{T_i} \subset DB$ presented in some convenient form to the user in advance. We remind that near-duplicates might be missing in ground truth, which limits objects evaluated as correct. For each query q_i, all database objects $o \in DB$ are ranked with respect to $\delta(v_{q_i}, v_o)$ and the rank r_i of the top ranked object $t \in C_{T_i}$ is stored. Either the average of all ranks r_i can be computed, or an empirical cumulative graph detailing effectiveness for growing rank is reported using

$$F_B(r) = \frac{|\{r_i : r_i \in Ranks, r_i \leq r\}|}{|Ranks|},$$

where $Ranks$ represents all obtained top ranks r_i for all benchmark pairs $[q_i, C_{T_i}]$ and a tested model (U, δ). For example, see the cumulative graph in Fig. 1 illustrating the percentage of findable known items when users browse a ranked list up to a rank r, provided that a correct item is not overlooked (which is generally not guaranteed [15]).

3 Experiments

This section presents an evaluation benchmark dataset and several experiments demonstrating challenges of effective and efficient known-item search.

3.1 Known-Item Search Benchmark Set

We analyze the performance of two respected text-image search approaches CLIP [22] and W2VV++ [14] (its BERT variant [15]) for a benchmark set comprising 327 pairs $[q_i, C_{T_i}]$, where all sets C_{T_i} are subsets of a 20K benchmark image dataset extracted from the V3C1 collection [23]. The search need (i.e., known item) was represented by one randomly selected image and no near-duplicates were considered during benchmark construction (i.e., $|C_{T_i}| = 1$). Free-form text descriptions (queries) for target images were provided by human annotators. Each annotator observed a target image for the whole annotation time (i.e., perfect memory was assumed). Although the size 20000 objects does not conform to the idea of big data, it might still represent for example a personal image database where known-item search can be expected.

Both CLIP and W2VV++ BERT text-image search approaches provide functions f_{visual}, f_{text} for joint image and text embedding to $\mathbb{R}^n$ (n = 2048 for BERT, n = 640 for CLIP). Using the functions, all database images (including known items) and text queries q_i were transformed to n-dimensional vectors. For ranking of the 20K images with respect to q_i, a similarity model based on $1 - \sigma_{cos}(f_{visual}(o), f_{text}(q_i))$ can be utilized to identify the rank of $t_i \in C_{T_i}$. For all 327 pairs, Fig. 1 shows the performance of both compared models, revealing more effective known-item search performance for the recently released CLIP model. Nevertheless, there were individual benchmark pairs (about 30%) where the CLIP model was outperformed by W2VV++ BERT.

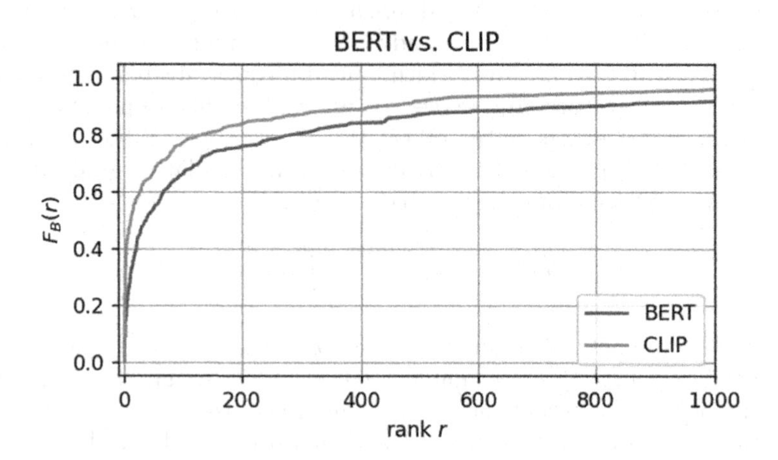

Fig. 1. Performance of text-image search models for the benchmark set, first 1000 ranks.

For both models, it is apparent that finding "just" one or ten nearest neighbours is not sufficient to solve all known-item search tasks even for the relatively small 20K dataset (though the performance of the new CLIP model is impressive!). With 100 nearest neighbours, more than 65% (75% for CLIP) of known items t_i searched by the query q_i would be directly findable in the result set. However, to provide a chance to solve 90% of all tasks by one query, hundreds of nearest neighbours are necessary for the 20K dataset. There also exist queries where even thousands of nearest neighbours are not enough. We emphasize that all the presented numbers are bound to the dataset size, for larger datasets the numbers of necessary nearest neighbours are significantly higher [15].

With the growing number of the nearest objects it becomes way more difficult to find the target with sequential result set browsing. Indeed, known-item search is a challenge that cannot be easily solved with just a single ranked list and scroll bar (at least yet). On the other hand, efficient construction of a larger candidate set is a promising first step that can be followed by a plethora of interactive search approaches. Assuming that the user cannot remember more details to extend/change the query, there are still options to inspect results for text query subsets, provide relevance feedback for displayed set of images, browse images in an exploratory structure, etc. However, these methods are beyond the scope of this paper.

3.2 Upper Performance Estimates for kNN Browsing

Before we proceed to a large candidate set selection study in the next section, we investigate kNN based browsing with small candidate sets to solve KIS tasks for the small 20K dataset. To analyze the search strategy, we run simulations for the 327 benchmark pairs and the W2VV++ BERT model.

Each simulated search session started with a text query q_i. From the result, top ranked k items were selected as a display D_j out of which one item $q_j \in D_j$ was selected as a new query object for the next display presenting $\text{kNN}(v_{q_j}, S)$. This process was repeated until the target item t_i was found or the maximal limit of iterations was reached. The automatic selection (i.e., simulation of user interaction [7,8]) of the new query considered two optimistic options based on $\text{kNN}(v_{t_i}, D'))$, where $t_i \in C_{T_i}$ is the searched target image and D' are descriptors of images on the current display. We consider an IDEAL user automatically selecting as the new query the most similar object from the display D to the target t_i. In addition, we consider also a randomized TOP user, where the new query object is selected randomly from $\text{kNN}(v_{t_i}, D'))$, $k = 8$. To prevent from cycles, once selected queries q_j were removed from the dataset in a given search session.

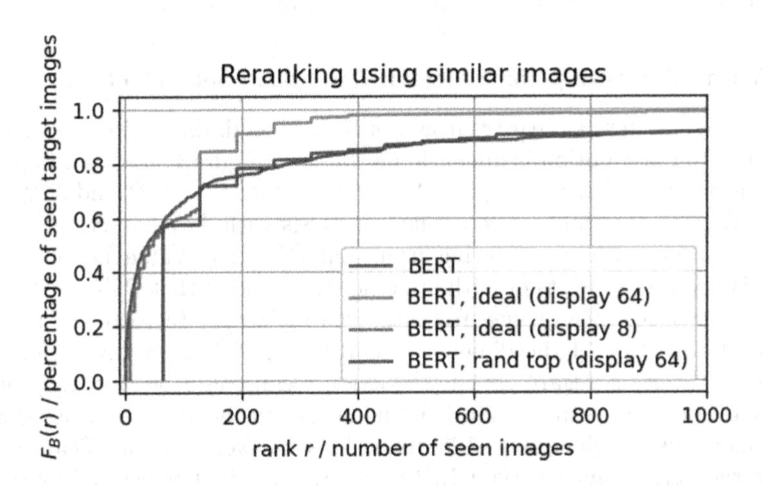

Fig. 2. Browsing simulations using 16 kNN displays, browsing performance ("stairs") related to performance of the W2VV++ BERT text search model (first 1000 ranks).

Figure 2 compares the W2VV++ BERT text search model fine-grained ranking (i.e., browsing the original ranked set) with iterative reformulations providing always 64 nearest objects for one selected query object (IDEAL or TOP) from the current display. For each iteration, the graph shows the increase of solved tasks for the whole display at once (therefore the staircase pattern). For the IDEAL user and $|D| = k = 64$ the kNN browsing would boost the performance compared to the original ranked set. However, the IDEAL user is too optimistic, real users are not 100% consistent with the similarity model. Furthermore, for smaller display size $|D| = 8$, even the IDEAL user performance is worse than the original ranking. For the randomized TOP user and display size 64, the performance of kNN browsing has a similar performance effect as sequential search of the original ranked list. However, even the randomized TOP user is still rather optimistic, real users may select from a display also less similar items to $t_i \in C_{T_i}$.

To sum up the simulations, kNN based browsing with IDEAL selections and small $k = 8$ is not effective enough, while, for $k = 64$, such browsing would be a competitive strategy with respect to the original ranking. However, to the best of our knowledge selections by real users are usually not ideal which decreases recall gains by the kNN browsing strategy. The effect of less optimal selections is illustrated by the performance drop between IDEAL and TOP users in Fig. 2. kNN browsing by the TOP user and $k = 64$ resulted in "just" similar effectiveness as sequential browsing of the original ranked list, where users do not have to select a good example query in each iteration. In other words, top ranked 1000 items for a text query could be browsed directly. For more effective browsing, advanced models based on relevance feedback were proposed [7], maintaining relevance scores for all objects in the database. In order to make the maintenance process more efficient, a larger candidate set can be selected for the models (e.g., 10% of top ranked items guaranteeing 90% of searched items).

3.3 A Baseline Study for Efficient Candidate Set Selection

In order to find top k nearest neighbours in a high-dimensional space efficiently, one popular option is dimension reduction. Figure 3 shows a comparison of dimension reduction techniques [21] for both models CLIP and W2VV++ BERT. We consider principal component analysis with data centering as a first step (PCA) and without centering using only Singular Value Decomposition (SVD). We compare effects of both approaches, provided that PCA might harm data by subtracting mean values to center (normalized) data vectors. The graph shows that reduction of the dimension to 128 does not affect the performance of the BERT variant regardless the reduction technique. However, the benefits of the CLIP model seem to vanish with the dimension reduction using SVD. Furthermore, PCA reduction to 128 dimensions (or even 256) significantly deteriorated the performance of the CLIP model which might be caused by specific properties of the text-image similarity space (see the next section).

Focusing just on the W2VV++ BERT model, Fig. 4 presents ranking performance for decreasing number of dimensions selected after SVD. We may observe that up to 64 dimensions, the performance of the model does not deteriorate. In other words, 32 times smaller dataset of descriptors and faster computation can be achieved with a standard pre-processing technique. Furthermore, even lower dimensional versions are useful for approximate search in a filter and refine mode. For the data, 50% of the database can be filtered with the 16 dimensional version of descriptors and the remaining part can be refined with the 64 dimensional version. This simple approach would reduce computation costs using a bin-to-bin measure like $-\sigma_{cos}(v_x, v_y)$ from $64 \cdot DBSize$ to $16 \cdot DBSize + 48 \cdot DBSize/2$

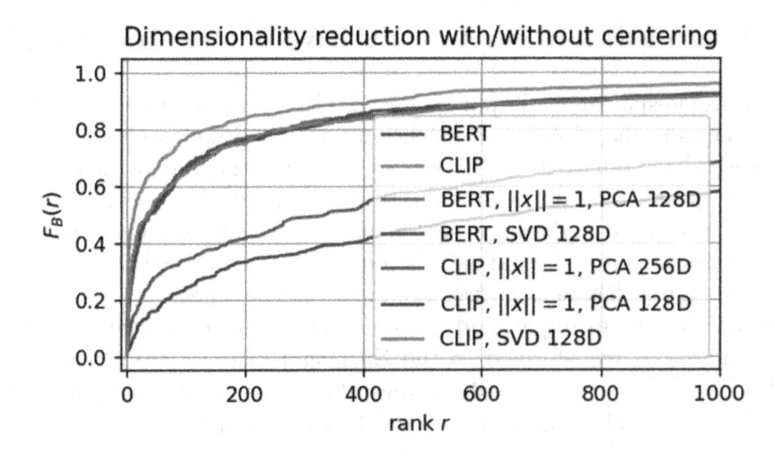

Fig. 3. Comparison of PCA and SVD for first 1000 ranks.

bin-to-bin operations. Please note that intermediate results for filtering can be re-used for refining and there emerge additional sorting costs for the refined half of the database. Allowing a small drop in recall, approximate 1000 most similar objects could be computed by refining just 10% of the 20K database filtered with 16 dimensional vectors, resulting in $16 \cdot DBSize + 4.8 \cdot DBSize$ bin-to-bin operations. At the same time, the approximate filtering approach still allows easy parallelization of the computation. We note that a bin-to-bin distance function for the first a dimensions of (normalized) data vectors can lower bound the distance for $b > a$ dimensions (e.g., for a similarity model based on squared Euclidean distance $\sum_{i=1}^{a}(v_{x_i} - v_{y_i})^2 \leq \sum_{i=1}^{b}(v_{x_i} - v_{y_i})^2$). Hence an optimal kNN query processing strategy [26] could be tested instead of a fixed hard filter of x% of the database.

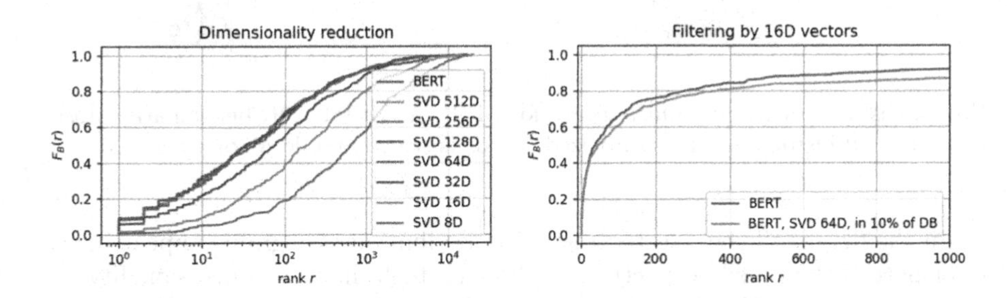

Fig. 4. Performance for decreasing dimensionality of descriptors after SVD. On the right, the effect of refining of 10% of database filtered with 16 dimensional vectors.

4 Is There a Room for Competitive Metric Indexing?

Using the single-space-for-all approach (e.g., $(\mathbb{R}^n, \sigma_{cos})$) for various application domains reminds the motivation of the metric space approach providing one access method for different metric spaces. The question raised in this section is, whether general distance-based metric indexing [6,29] can provide a competitive approach to methods presented in the previous section. Since metric indexing relies on lower bound estimation $LB(v_q, v_o) = |\delta(v_p, v_q) - \delta(v_p, v_o)|$ from precomputed distances between objects $v_o, v_p, v_q \in \mathbb{R}^n$, we show distance distributions for the example models from Sect. 2. For normalized vectors, the cosine similarity is transformed to the Euclidean distance using $L_2(v_x, v_y) = \sqrt{2 \cdot (1 - \sigma_{cos}(v_x, v_y))}$. Figure 5 shows several L_2 distance distributions for CLIP and W2VV++ BERT models for the 20K benchmark dataset:

- Image-image variant shows the distance distribution histogram for all pairs of images in the 20K dataset.
- Text-image variant shows the distance distribution histogram for pairs between all text query vector representations and vectors of all images.
- Text-target variant shows the distance distribution histogram for pairs between all text query vector representations and their corresponding target item.
- Distance at rank 2000 variant shows the histogram of distances at rank 2000 from all 327 result sets for all benchmark queries.

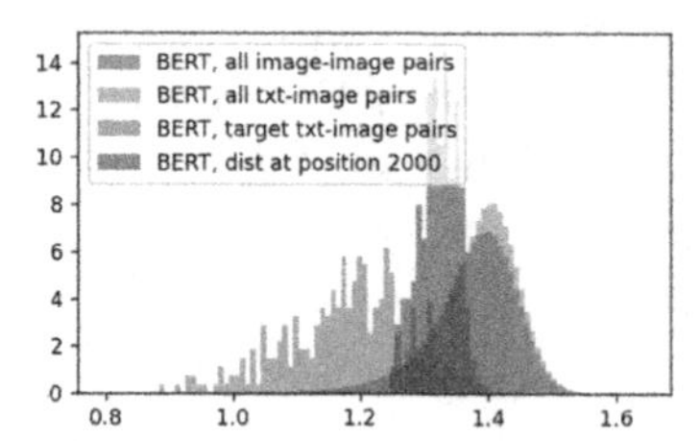

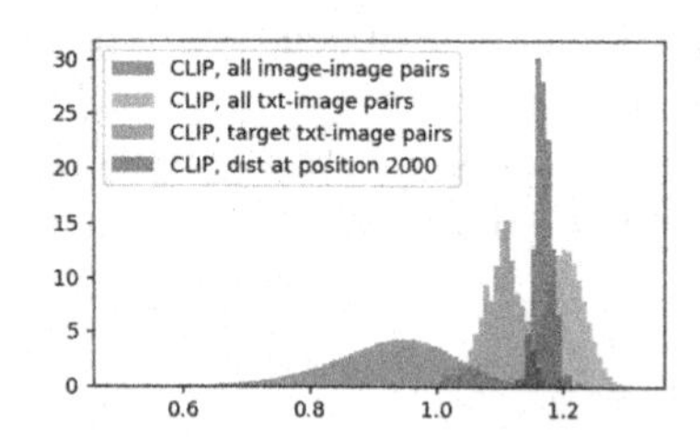

Fig. 5. Distance distribution histograms for CLIP and BERT, 20K benchmark dataset was used. All histograms are normalized, x axis is scaled and does not start at 0.

In the figure, all the selected distance distribution histograms appear in the right part of the possible spectrum, indicating high intrinsic dimensionality [6]. Whereas the W2VV++ BERT model has a similar distance distribution for image-image and text-image pairs, for the CLIP model the two histograms are significantly different. We hypothesise that this inconsistency is caused by different concepts used to design and train the CLIP model. Nevertheless, for both models the necessary distance from query to the searched item is high as well as distances between potentially indexed images. For a fixed high k, the distances at rank k are even higher. This questions filtering power of exact metric filtering rules and leads to the need for approximate search methods. Although there have been proposed and empirically tested efficient approximate search approaches for metric spaces (e.g., pivot tables [18], permutation approaches [2], or M-Index [20]), the question is whether metric search methods could outperform (for the discussed KIS problem and high k) the simple sequential SVD based filtering approach for W2VV++ BERT (see the previous section) or could deal with specifics of the CLIP based similarity space. We leave this open question as well as all the descriptors of the benchmark dataset for the metric indexing community.

5 Conclusions

In this paper, we focused on the known-item search problem where a larger number of nearest neighbours may be necessary to achieve a high recall. After a brief introduction of the problem, experimental evaluations with two state-of-the-art text-image search models were presented. The difficulty of the task was demonstrated with a benchmark dataset comprising hundreds of query-target pairs. An analysis of browsing performance with simulated user actions provided additional motivation for larger candidate sets. A baseline model for high-dimensional vectors was studied and an open challenge for metric indexing community was provided in the form of a new benchmark dataset accessible at github repository https://github.com/soCzech/KIS-Neighbours.

Acknowledgments. This paper has been supported by Czech Science Foundation (GAČR) project 19-22071Y.

References

1. Amato, G., et al.: The visione video search system: exploiting off-the-shelf text search engines for large-scale video retrieval. J. Imaging **7**(5), 76 (2021). https://doi.org/10.3390/jimaging7050076, https://www.mdpi.com/2313-433X/7/5/76
2. Amato, G., Savino, P.: Approximate similarity search in metric spaces using inverted files. In: Proceedings of the 3rd International Conference on Scalable Information Systems. InfoScale 2008. ICST (Institute for Computer Sciences, Social-Informatics and Telecommunications Engineering), Brussels, BEL (2008)
3. Awad, G., et al.: Trecvid 2019: an evaluation campaign to benchmark video activity detection, video captioning and matching, and video search & retrieval. In: TRECVID 2019. NIST, USA (2019)
4. Barthel, K.U., Hezel, N.: Visually exploring millions of images using image maps and graphs. In: Big Data Analytics for Large-Scale Multimedia Search. John Wiley and Sons Inc. (2018)
5. Böhm, C., Berchtold, S., Keim, D.A.: Searching in high-dimensional spaces: index structures for improving the performance of multimedia databases. ACM Comput. Surv. **33**(3), 322–373 (2001)
6. Chávez, E., Navarro, G., Baeza-Yates, R., Marroquín, J.L.: Searching in metric spaces. ACM Comput. Surv. **33**(3), 273–321 (2001)

7. Cox, I., Miller, M., Omohundro, S., Yianilos, P.: Pichunter: Bayesian relevance feedback for image retrieval. In: Proceedings of 13th International Conference on Pattern Recognition, vol. 3, pp. 361–369 (1996). https://doi.org/10.1109/ICPR.1996.546971
8. Cox, I.J., Miller, M.L., Minka, T.P., Papathomas, T.V., Yianilos, P.N.: The Bayesian image retrieval system, Pichunter: theory, implementation, and psychophysical experiments. IEEE Trans. Image Process. **9**(1), 20–37 (2000)
9. Goodfellow, I., Bengio, Y., Courville, A.: Deep Learning. MIT Press (2016). http://www.deeplearningbook.org
10. Gurrin, C., et al.: [Invited papers] Comparing approaches to interactive lifelog search at the lifelog search challenge (lsc2018). ITE Trans. Media Technol. Appl. **7**(2), 46–59 (2019). https://doi.org/10.3169/mta.7.46
11. Jónsson, B.Þ, Khan, O.S., Koelma, D.C., Rudinac, S., Worring, M., Zahálka, J.: Exquisitor at the video browser showdown 2020. In: Ro, Y.M., et al. (eds.) MMM 2020, Part II. LNCS, vol. 11962, pp. 796–802. Springer, Cham (2020). https://doi.org/10.1007/978-3-030-37734-2_72
12. Kratochvíl, M., Veselý, P., Mejzlík, F., Lokoč, J.: SOM-Hunter: video browsing with relevance-to-SOM feedback loop. In: Ro, Y.M., et al. (eds.) MMM 2020, Part II. LNCS, vol. 11962, pp. 790–795. Springer, Cham (2020). https://doi.org/10.1007/978-3-030-37734-2_71
13. Leibetseder, A., Münzer, B., Primus, J., Kletz, S., Schoeffmann, K.: diveXplore 4.0: the ITEC deep interactive video exploration system at VBS2020. In: Ro, Y.M., et al. (eds.) MMM 2020, Part II. LNCS, vol. 11962, pp. 753–759. Springer, Cham (2020). https://doi.org/10.1007/978-3-030-37734-2_65
14. Li, X., Xu, C., Yang, G., Chen, Z., Dong, J.: W2VV++: fully deep learning for ad-hoc video search. In: Proceedings of the 27th ACM International Conference on Multimedia, MM 2019, Nice, France, October 21–25, 2019, pp. 1786–1794 (2019). https://doi.org/10.1145/3343031.3350906
15. Lokoč, J., et al.: A w2vv++ case study with automated and interactive text-to-video retrieval. In: Proceedings of the 28th ACM International Conference on Multimedia, MM 2020. Association for Computing Machinery, New York (2020)
16. Lokoč, J., Bailer, W., Schoeffmann, K., Muenzer, B., Awad, G.: On influential trends in interactive video retrieval: video browser showdown 2015–2017. IEEE Trans. Multimed. **20**(12), 3361–3376 (2018)
17. Lokoč, J., et al.: Interactive search or sequential browsing? A detailed analysis of the video browser showdown 2018. ACM Trans. Multimed. Comput. Commun. Appl. **15**(1), 29:1–29:18 (2019). https://doi.org/10.1145/3295663
18. Micó, M.L., Oncina, J., Vidal, E.: A new version of the nearest-neighbour approximating and eliminating search algorithm (AESA) with linear preprocessing time and memory requirements. Pattern Recognit. Lett. **15**(1), 9–17 (1994)
19. Nguyen, P.A., Wu, J., Ngo, C.-W., Francis, D., Huet, B.: VIREO @ video browser showdown 2020. In: Ro, Y.M., et al. (eds.) MMM 2020, Part II. LNCS, vol. 11962, pp. 772–777. Springer, Cham (2020). https://doi.org/10.1007/978-3-030-37734-2_68
20. Novák, D., Batko, M., Zezula, P.: Metric index: an efficient and scalable solution for precise and approximate similarity search. Inf. Syst. **36**, 721–733 (2011). https://doi.org/10.1016/j.is.2010.10.002
21. Pearson, K.: On lines and planes of closest fit to systems of points in space. Philos. Mag. **2**, 559–572 (1901)
22. Radford, A., et al.: Learning transferable visual models from natural language supervision. CoRR abs/2103.00020 (2021). https://arxiv.org/abs/2103.00020

23. Rossetto, L., Schuldt, H., Awad, G., Butt, A.A.: V3C – a research video collection. In: Kompatsiaris, I., Huet, B., Mezaris, V., Gurrin, C., Cheng, W.-H., Vrochidis, S. (eds.) MMM 2019, Part I. LNCS, vol. 11295, pp. 349–360. Springer, Cham (2019). https://doi.org/10.1007/978-3-030-05710-7_29
24. Sauter, L., Amiri Parian, M., Gasser, R., Heller, S., Rossetto, L., Schuldt, H.: Combining Boolean and multimedia retrieval in vitrivr for large-scale video search. In: Ro, Y.M., et al. (eds.) MMM 2020, Part II. LNCS, vol. 11962, pp. 760–765. Springer, Cham (2020). https://doi.org/10.1007/978-3-030-37734-2_66
25. Schoeffmann, K., Hudelist, M.A., Huber, J.: Video interaction tools: a survey of recent work. ACM Comput. Surv. **48**(1), 14:1–14:34 (2015)
26. Seidl, T., Kriegel, H.P.: Optimal multi-step k-nearest neighbor search. SIGMOD Rec. **27**(2), 154–165 (1998)
27. Thomee, B., Lew, M.S.: Interactive search in image retrieval: a survey. Int. J. Multimed. Inf. Retr. **1**(2), 71–86 (2012). https://doi.org/10.1007/s13735-012-0014-4
28. Worring, M., Sajda, P., Santini, S., Shamma, D.A., Smeaton, A.F., Yang, Q.: Where is the user in multimedia retrieval? IEEE MultiMedia **19**(4), 6–10 (2012)
29. Zezula, P., Amato, G., Dohnal, V., Batko, M.: Similarity Search - The Metric Space Approach. Advances in Database Systems, vol. 32. Kluwer (2006)

On Generalizing Permutation-Based Representations for Approximate Search

Lucia Vadicamo(✉), Claudio Gennaro, and Giuseppe Amato

Institute of Information Science and Technologies (ISTI), Italian National Research Council (CNR), Via G. Moruzzi 1, 56124 Pisa, Italy
{lucia.vadicamo,claudio.gennaro,giuseppe.amato}@isti.cnr.it

Abstract. In the domain of approximate metric search, the Permutation-based Indexing (PBI) approaches have been proved to be particularly suitable for dealing with large data collections. These methods employ a permutation-based representation of the data, which can be efficiently indexed using data structures such as inverted files. In the literature, the definition of the permutation of a metric object was derived by reordering the distances of the object to a set of pivots. In this paper, we aim at generalizing this definition in order to enlarge the class of permutations that can be used by PBI approaches. As a practical outcome, we defined a new type of permutation that is calculated using distances from pairs of pivots. The proposed technique permits us to produce longer permutations than traditional ones for the same number of object-pivot distance calculations. The advantage is that the use of inverted files built on permutation prefixes leads to greater efficiency in the search phase when longer permutations are used.

Keywords: Permutation-based indexing · Metric space · Metric search · Similarity search · Approximate search · Planar projection

1 Introduction

Searching a database for objects that are most similar to a query object is a fundamental task in many application domains, like multimedia information retrieval, pattern recognition, data mining, and computational biology. In this context, the *Metric Search* framework [24] provides us with a wide class of indexing and searching techniques for similarity data management. A common factor in all these approaches is that they are applicable on generic metric spaces, i.e. these techniques are not specialised for a particular type of data. A *metric space* is a pair (D, d) formed by a domain D and a distance function $d : D \times D \to \mathbb{R}$ that satisfies the metric postulates of non-negativity, identity of indiscernibles, symmetry, and triangle inequality [24]. In a general metric space we cannot use any algebraic function, e.g. sum of two objects or product by scalars, but the only operation that can be exploited is calculating the distance between any two objects. Therefore any technique that aims mapping a metric object $o \in D$

N. Reyes et al. (Eds.): SISAP 2021, LNCS 13058, pp. 66–80, 2021.
https://doi.org/10.1007/978-3-030-89657-7_6

to another (more tractable) space, e.g. a vector space, must rely only on algorithms that use the distances of the object o to other metric objects, e.g. a set of reference objects selected within the space.

Many *approximate metric search* approaches employ transformations of the original metric space to overcome the *curse of dimensionality*, which affects exact metric search techniques whose performance may be not better than a sequential scan for spaces with high intrinsic dimensionality [18,23]. Successful examples of approximate methods are the *Permutation-based Indexing* (PBI) techniques that transform the metric data into permutations of a set of integers $\{1, \ldots, N\}$, which are then indexed and searched using data structures like prefix trees [13,19] and inverted files [3,20]. The original definition of permutation-based representation of a metric object was derived by computing the distances of the object to a set of *pivots* (reference objects) and then by reordering the pivot identifiers according these distances [4,9,10]. This characterization have been adopted in several research papers that further investigated the properties of this data representations and ways to efficiently index them, e.g. [2,13–15,17–19]. Moreover, some alternative permutation-based representations have been defined in the literature [1,21], but only for representing objects of specific metric spaces.

In this work, we aim at generalizing the definition of permutation associated with a metric object, by introducing the concept of permutation induced by a transformation $f : (D, d) \to \mathbb{R}^N$. The function f simply projects the metric objects of D into an N-dimensional vector space. This function typically relies only on some distance calculations to transform the objects, such as distances to a set of pivots as done in traditional permutations, but the way distances are combined and exploited to represent objects may be different from what is done in the traditional approach. We believe that this generalization can open up new lines of research, on the one hand, to understand theoretically what properties the function f should have in order to generate permutations that have good performance for the approximate search, and on the other hand, to define alternative permutation-based representations. In this paper, we have started investigating the latter aspect by defining permutations that rely on distances of objects to pairs of pivots. In this way, for a fixed set of n pivots, we can generate permutations with length $N > n$, while the length of traditional permutations is fixed equal to the number of pivots n. The advantage of having longer permutations (at the same cost in terms of original distance computations) is the more efficiency at searching time when using inverted index build upon permutation prefixes (e.g. MI-File [3]). In fact, the inverted index contains as many posting lists as the number N of permutants (i.e. the length of the full permutation) and so, for a fixed permutation prefix length λ, the higher N, the shorter the posting lists, and hence the smaller the fraction of the database accessed to answer a query.

The rest of the paper is structured as follows. Section 2 reviews and generalizes the concept of permutation-based representation of metric objects. Permutations built using distances to pivot-pairs are introduced in Sect. 3. Section 4 reports the experimental evaluation, and Sect. 5 draws the conclusions.

2 Permutation-Based Representation(s) of a Metric Object

Chavez et al. [9,10] and Amato et al. [4] originally defined the permutation-based representation of a metric object by ordering the identifiers of a fixed set of pivots according to their distances to the object to be represented, that is

Definition 1 (Permutation of a metric object given a set of pivots). *The permutation-based representation Π_o (briefly* permutation*) of an object $o \in D$ with respect to the pivot set $\{p_1, \ldots, p_n\} \subset D$ is the sequence $\Pi_o = [\pi_1, \ldots, \pi_n]$ that lists the pivot identifiers $\{1, \ldots, n\}$ (called* permutants*) in an order such that $\forall\, i \in \{1, \ldots, n-1\}$*

$$d(o, p_{\pi_i}) < d(o, p_{\pi_{i+1}}) \qquad or \qquad \left[d(o, p_{\pi_i}) = d(o, p_{\pi_{i+1}})\right] \wedge \left[\pi_i < \pi_{i+1}\right]. \quad (1)$$

This representation is also referred to as the *full-length* permutation to distinguish it from the *permutation prefix* adopted in several PBI methods [3,13,19]. In facts, based on the intuition that the most relevant information in the permutation is present in its very first elements, i.e. the identifiers of the closest pivots to an object, several researchers proposed to represent the data by using a fixed-length prefix of the permutation, i.e. $\Pi_{o,\lambda} = [\pi_1, \ldots, \pi_\lambda]$ with $\lambda < n$. The use of permutation prefixes may be dictated by either the employed data structure (e.g. prefix tree), efficiency issues (more compact data encoding and better performance when using inverted files) or even by effectiveness reasons (in some cases the use of prefixes gives better results than full-length permutations [2,3]).

Alternative permutation-based representations have been defined in literature, but only for specific metric spaces. For example, the *Deep Permutations* [1,5] were defined by reordering the dimensions of a vector according to the corresponding element values. This approach can only be used in vector spaces and has so far only been tested on Convolutional Neural Network features. The *SPLX-Perms* [21] use the n-Simplex projection [12] followed by a random rotation to transform a metric object into a Euclidean vector and then computes the permutation by reordering the components of the vector as done in the Deep Permutations. This method can be used on the large class of spaces meeting the n-point property [12] but it is not applicable on general metric spaces.

We now observe that all these approaches belong to the same family of transformations, as explained hereafter, and thus the traditional definition of permutation associated to a metric object (Def. 1) could be generalized to be more inclusive. In this context, the first trivial but useful observation to make is that any sorting function defined on a finite-dimensional Coordinate space implicitly produce a permutation representation of the data. Suppose $\sigma : \mathbb{R}^N \to \mathbb{R}^N$ is a function that sorts the coordinate elements of a N-dimensional real vector with respect to a predefined criterion (e.g. ascending order). For any $\boldsymbol{v} \in \mathbb{R}^N$, the sort function σ is described by the permutation Π_v^σ of the indices $\{1, \ldots, N\}$ that specifies the arrangement of the elements of $\boldsymbol{v}$ into $\boldsymbol{v}' = \sigma(\boldsymbol{v})$. Specifically, if $\boldsymbol{v} = [v_1, \ldots, v_N]$ and $\sigma(\boldsymbol{v}) = [v_{i_1}, \ldots, v_{i_N}]$ then $\Pi_v^\sigma = [i_1, \ldots, i_N]$. In other words, the j-th element of the permutation Π_v^σ is the index $i \in \{1, \ldots N\}$ such that the i-th element of $\boldsymbol{v}$ is equal to the j-th element of $\sigma(v)$. For example, if $\boldsymbol{v} = [8, 10, 6]$ and $\sigma(\boldsymbol{v}) = [6, 8, 10]$ then $\Pi_v^\sigma = [3, 1, 2]$. However, this characterization is not well defined if the vector $\boldsymbol{v}$ contains duplicate values, therefore we give the following definition.

Definition 2 (Permutation of a vector induced by a sort function σ). *A permutation representation of a vector $\boldsymbol{v} = [v_1, \ldots, v_N] \in \mathbb{R}^N$ associated to a given sort function $\sigma : \mathbb{R}^N \to \mathbb{R}^N$ is the permutation $\Pi_v^\sigma = [\pi_1, \ldots, \pi_N]$ of the index identifiers $\{1, \ldots, N\}$ such that for any $j = 1, \ldots, N$ the element π_j is the smallest index for which v_{π_j} equals the j-th element of $\sigma(\boldsymbol{v})$.*

The Deep Permutations can be formalized by using the above definition, but this of course cannot be used to describe the SPLX-perms or the traditional permutations. However, these approaches share a common idea, that is using the distance to a set of pivots to first transform the metric object into a Cartesian coordinate space and then obtaining the permutation by applying a sort function. Therefore, for any function $f : (D, d) \to \mathbb{R}^N$ and a given *sort* function we may define a permutation representation of a metric objects as follows:

Definition 3 (Permutation of a metric object induced by a space transformation f and a sort function σ). *Let $f : (D, d) \to \mathbb{R}^N$ a space transformation, and $\sigma : \mathbb{R}^N \to \mathbb{R}^N$ a function that sorts the components of a N-dimensional vector according to some predefined criteria. We define the permutation representation of a object $o \in D$ induced by the functions f and σ as the permutation $\Pi_o^{\sigma,f} = [\pi_1, \ldots, \pi_N]$ that lists the index identifiers $\{1, \ldots, N\}$ in an order such that for any $j = 1, \ldots, N$ the permutant π_j is the smallest index for which the π_j-th element of the vector $f(o)$ is equals to the j-th element of $\sigma(f(o))$.*

For the sake of simplicity, in the following we assume that the sort function σ is the sorting of the elements in *ascending* order and we omit the dependency of this function in the definition of the permutation. Please note that the effect of using a different sorting function in most cases could be reproduced by changing the function f. For example, for a given f and object o the permutation obtained by sorting $f(o)$ in descending order is equal to the permutation obtained by applying the function $-f$ to the object o and then sorting the elements in ascending order. Therefore, we use the following characterization:

Definition 4 (Permutation of a metric object induced by a space transformation f). *The permutation representation of a object $o \in (D, d)$ with respect to the transformation $f : (D, d) \to \mathbb{R}^N$ is the sequence $\Pi_o^f = [\pi_1, \ldots, \pi_N]$ that lists the permutants $\{1, \ldots, N\}$ in an order such that $\forall\, i \in \{1, \ldots, N-1\}$,*

$$f(o)_{\pi_i} < f(o)_{\pi_{i+1}} \qquad or \qquad \left[f(o)_{\pi_i} = f(o)_{\pi_{i+1}}\right] \wedge \left[\pi_i < \pi_{i+1}\right] \tag{2}$$

where $f(o)_j$ indicates the j-th value of the vector $f(o)$.

Note that, according to this definition, the traditional permutation is induced by the transformation $f(o) = [d(o, p_1), \ldots, d(o, p_N)]$, where $\{p_1, \ldots, p_N\}$ is a fixed set of pivots. The Deep Permutation is induced by the identity function.

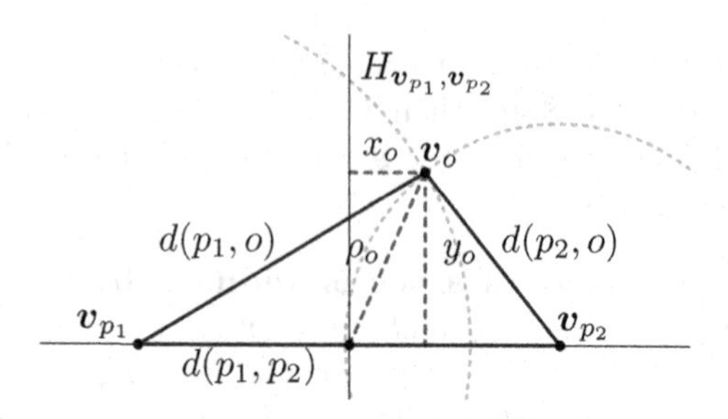

Fig. 1. Planar Projection of two pivots p_1, p_2 and a data point o.

The SPLX-Perm, instead, is induced by the composition of the n-Simplex projection and a random rotation. Moreover, this generalization suggests that new permutation representations of generic metric objects can be defined but assuming that we use a transformation $f : (D, d) \rightarrow \mathbb{R}^N$ that relies only on distance computations and metric postulates to transform the objects. The function f in some cases can also be generated using machine learning techniques, as in [8]. Nevertheless, for particular metric spaces, e.g. vector spaces, other operations or properties of the space can be employed when defining the transformation f. However, not all the transformations may produce permutations which are suitable for metric search, as we would like that similar objects are projected into similar permutations. An in-depth theoretical and experimental study on the properties that the function f should have to produce "good" permutations for approximate metric search is beyond the scope of this paper and we reserve it for future work. Here, as proof of concepts, we define a novel permutation-based representation that uses a transformation f that relies not only on the distance of the objects to a fixed set of pivots but also exploits information on the distances between pivot pairs.

3 Pivot Pairs Permutations

Thanks to the triangle inequality, we know that any three points of a metric space can be isometrically embedded in a two dimensional Euclidean space. Specifically, let $p_1, p_2 \in D$ two pivots and $o \in D$ an arbitrarily metric object. Without loss of generality, we could consider an isometric embedding that maps the points p_1, p_2, o to the vectors $v_{p_1}, v_{p_2}, v_o \in (\mathbb{R}^2, \ell_2)$, such that (i) v_{p_1} and v_{p_2} lies in the X-axis; (ii) v_o is above the X-axis and its coordinate are given by the intersection of the ball centered on p_1 with radius $d(p_1, o)$ and the ball centered on p_2 with radius $d(p_2, o)$. Figure 1 depicts this situation in a 2D coordinate space where the two pivots are projected in the X-axis symmetrically with respect to the origin and a single data object o is mapped to the point $\boldsymbol{v}_o = (x_o, y_o)$, where

$$x_o = \frac{d(o, p_1)^2 - d(o, p_2)^2}{2 \cdot d(p_1, p_2)}, \qquad y_o = \sqrt{d(o, p_1)^2 - \left(x_o + \frac{d(p_1, p_2)}{2}\right)^2} \tag{3}$$

Note that the only information used in the projection is the distances of the object o to the two pivots and the inter-pivot distance. Moreover, all the three distances between the points are preserved, i.e. $\ell_2(\boldsymbol{v}_{p_1}, \boldsymbol{v}_{p_2}) = d(p_1, p_2)$, and $\ell_2(\boldsymbol{v}_{p_i}, \boldsymbol{v}_o) = d(p_i, o)$ for $i = 1, 2$. This projection, called *planar projection* [11], could be repeated for all the data points $o \in D$ while fixing the two pivots p_1, p_2. So we have a projection $\phi_{p_1,p_2} : (D, d) \to (\mathbb{R}^2, \ell_2)$ that preserves the distances of data objects to the two pivots. Therefore, since the distance to the pivots is preserved for each data point, it can be easily proved that all the objects in the hyperplane separating the two pivots in the original space are projected in the hyperplane $H_{\boldsymbol{v}_{p_1},\boldsymbol{v}_{p_2}} = \{\boldsymbol{v} \in \mathbb{R}^2 \mid \ell_2(\boldsymbol{v}, \boldsymbol{v}_{p_1}) = \ell_2(\boldsymbol{v}, \boldsymbol{v}_{p_2})\}$ separating the pivots in the 2D projection.

The Euclidean norm of a projected object ($\rho_o = \|\boldsymbol{v}_o\|$) could be interpreted as the distance of the point o to a synthetic pivot that is equidistant to the two original pivots, i.e. a sort of midpoint which may not exist in the original metric space. Its calculation is immediate if we already know $d(p_1, p_2)$, $d(o, p_1)$, and $d(o, p_2)$ as it is equals to

$$\rho_o = \sqrt{(x_o)^2 + (y_o)^2} = \frac{1}{2}\sqrt{2\,d(o,p_1)^2 + 2\,d(o,p_2)^2 - d(p_1,p_2)^2} \tag{4}$$

We can repeat this procedure for several pairs of pivots to characterize a metric object based on the distribution of its distance from the synthetic midpoints between the original pivots. Formally, given a set $\{p_1, \ldots, p_n\} \subset D$ of n pivots, we select $m < \binom{n}{2}$ pivot pairs that we enumerate using an index i, so that (p_{i_1}, p_{i_2}) indicates the i-th pivot pair. For each object $o \in D$ and for each selected pair of pivots (p_{i_1}, p_{i_2}) we use Eq. 4 to compute the norm $\rho_o^{(i)}$ of the projected point $\phi_i(o) = (x_o^{(i)}, y_o^{(i)})$. Then we generate a permutation $\Pi_o^{f'}$ of length m by reordering the components of $f'(o) = \left(\rho_o^{(1)}, \ldots, \rho_o^{(m)}\right)$. Moreover, since we can interpreted the values ρ_i as the distance to some synthetic pivots, we may combine these information with the distances to the actual pivots by computing the permutations induced by the function

$$f'' : o \in D \to \left(d(o,p_1), \ldots, d(o,p_n), \rho_o^{(1)}, \ldots, \rho_o^{(m)}\right) \in \mathbb{R}^{n+m} \tag{5}$$

In the following we refer to the permutations $\Pi_o^{f'}$ and $\Pi_o^{f''}$ as *Pairs Permutation* (`P-Perms`), and *Pivot-Pairs Permutation* (`PP-Perms`), respectively.

4 Experiments

In this section, we compare the performance of `P-Perms`, `PP-Perms`, and the traditional permutations (`Perms`) for approximate k-nearest neighbors (k-NN) search. The experiments were conducted both on real-word and publicly available datasets (CoPhIR and ANN-SIFT) and on synthetic datasets, which are described below. In the following, we first introduce the measures used for the evaluation and then we present the experimental results.

Evaluation Protocol. For each dataset we build a ground-truth for the exact k-NN search related to $1,000$ randomly-selected queries. The ground-truths were used to evaluate the quality of the approximate results obtained either by performing a k-NN search in the permutation space or by using the actual distance to re-rank a candidate result set of size $k' \geq k$ that was selected using a k'-NN search in the permutation space. Please note that the latter is a filter-and-refine approach, which requires to store the original dataset and access to it at query time to refine the permutation-based candidate results. In the experiments we used $k = 10$. For the filter-and-refine approach we used $k' = 100$. The quality of the approximate results was evaluated using the *recall@k*, defined as $|\mathcal{R} \cap \mathcal{R}^A|/k$, where $\mathcal{R}$ is the result set of the exact k-NN search in the original metric space and $\mathcal{R}^A$ is the approximate result set.

As done by many PBI approaches [3,13,19], we index and search the data using fixed-length permutation prefixes instead of the full-length permutations. The permutation prefixes were compared using the *Spearman's rho with location parameter* ($S_{\rho,\lambda}$), defined as in [1, Sect. 3.5]), where the location parameter is the length λ of the permutation prefixes. If N is the length of the full permutations (i.e. we have N different permutants that may appear in a permutation prefix) and we index the permutation prefixes using inverted files [3], we have that

- the inverted index is composed of N posting lists (one for each permutant).
- each object is stored in exactly λ posting lists (corresponding to the permutants appearing in its permutation prefix). Thus, the i-th posting list contains t_i entries related only to the data objects whose permutations prefixes contain the permutant i.
- each entry of the i-th posting list is of the form $(ID_o, pos_o(i))$, where ID_o is the identifier of a data object, $pos_o(i)$ is the position of the permutant i in the permutation prefix associated to the object o.
- at query time, we access only to the λ posting lists corresponding to the permutants in the query permutation prefix. For each object o in those selected posting lists, we use the stored $pos_o(i)$ to compute the $S_{\rho,\lambda}$ distance to the query permutation prefix.

In this setting, *the size in bits of the inverted index* is a function of the number of permutants N, the prefix length λ, and the number of data objects $|X|$:

$$\text{Size(Inverted Index)} = \underbrace{N \lceil \log_2 N \rceil}_{\text{posting list identifiers}} + \underbrace{\lambda |X| \left(\lceil \log_2 |X| \rceil + \lceil \log_2 \lambda \rceil \right)}_{\text{posting list entries}} \quad (6)$$

The cost at query time includes 1) the cost of transforming the query into the permutation representation; 2) the search cost; 3) the cost of re-ranking the candidate set using the actual distance (only for the filter-and-refine approach).

The cost for computing the permutations (Table 1) varies with the employed permutation-based representation and the specific metric of the space. For a given set of n pivots the traditional permutation has $N = n$ permutants and requires the calculation of n object-pivot distances. For the same set of pivots and for m selected pivot pairs the `P-Perms` and the `PP-Perms` have $N = m$

Table 1. Distance computations needed for generating various permutation-based representations given the same set of n pivots. m is the number of pairs used in the Pairs and Pivot-Pair Permutations. N is the number of permutants.

<table>
<tr><th rowspan="2">Approach</th><th rowspan="2">N</th><th colspan="2">Number of distance calculations</th></tr>
<tr><th>Computed for each object/query</th><th>Computed once at indexing time</th></tr>
<tr><td><code>Perms</code></td><td>n</td><td>n actual distances</td><td></td></tr>
<tr><td><code>P-Perms</code></td><td>m</td><td rowspan="2">n actual distances + m 2D Euclidean distances</td><td rowspan="2">$\min(m, n(n-1)/2)$ actual distances</td></tr>
<tr><td><code>PP-Perms</code></td><td>$n+m$</td></tr>
</table>

and $N = n + m$ permutants, respectively, and require n object-pivot distance calculation plus m 2D Euclidean distances to calculate the ρ_o (Eq. 4), which in most cases is a negligible cost with respect to object-pivot distance calculations. The `P-Perms` and the `PP-Perms` also require $\min(m, n(n-1)/2)$ pivot-pivot distances, that can be computed and stored once at indexing time and then reused for calculating all the objects/query permutations. The *search cost* (SC), calculated as the *number of bits accessed per query*, is given by

$$SC = \left(\sum_{i=1}^{N} \delta_i t_i \right) C(\text{pEntry}) \tag{7}$$

where t_i is the number of objects stored in the i-th posting list, δ_i is the fraction of samples in the database having the permutant i in their permutation prefixes (i.e. $\delta_i = t_i/|X|$), and $C(\text{pEntry}) = \lceil \log_2 |X| \rceil + \lceil \log_2 \lambda \rceil$ is the size in bits of a single entry of a posting list. In facts, given a query q, we access the i-th posting list only if the index i is in the permutation prefix associated to the query, which is true with probability δ_i as query and database objects share the same distribution. Therefore, the number of elements accessed per query is $\sum_{i=1}^{N} \delta_i t_i$ since the i-th posting list contains t_i entries, and we access it with δ_i probability. Note that for a fixed N, the larger the prefix λ, the greater t_i, thus the higher the search cost. Moreover, for the filter-and-refine approach, the bytes accessed per query are those needed to select the k' candidate results (given by Eq. 7) plus those needed to re-rank the candidate results using the actual distance, i.e. $k' * C(Obj)$, where $C(Obj)$ is the size in bits of one original data object.

4.1 Experiments on Synthetic Data

The first question that may arise when considering the `P-Perms` representation as an alternative to the traditional permutation (`Perms`) is whether using the distances to the synthetic midpoint pivots instead of the actual pivots still helps in distinguishing similar data points from dissimilar ones in an approximate search scenario. Moreover, since the `P-Perms` and the `PP-Perms` allows producing permutations that are longer than the number of the employed pivots, it would be also interesting to analyze the performance of these permutations when the number m of pairs is increased while fixing the number n of pivots (i.e. fixing the number of actual distance computation needed to generate the permutations). To this scope, we performed experiments on two representative typologies of synthetic data, clustered and not clustered Euclidean vectors, because it was already proved in the literature [3] that the traditional permutations have different behaviors on these data when varying the number n of pivots and the prefix length λ. Specifically, we considered two datasets, each containing 100K vectors in a 30-dimensional Euclidean space. The first dataset, named *Gaussian Euclid30*, contains vectors whose coordinates are generated using a Gaussian distribution centered in the origin and with a standard deviation $\sigma = 0.1$. The second dataset, named *Clustered Euclid30*, contains vectors arranged in 20 clusters. The cluster centers were randomly selected in the hypercube $[0,1]^{30}$. For each cluster we generated $5K$ vectors using a Gaussian distribution with a small standard deviation ($\sigma = 0.01$).

Figures 2a and 2b show, for the two datasets, the *recall@10* achieved by the traditional permutation when varying the number n of pivots and the prefix length λ. Please note the different behaviors of the permutations on these two kind of data. On Gaussian Euclid30, the recall increases when increasing both n and λ, but for a fixed λ the recall is almost unchanged when increasing only n (Fig. 2a). For the clustered data, instead, the performance of the *full-length* permutations (i.e. the cases $\lambda = n$) is not improved when increasing the number of pivots more than $n = 500$. However, for a fixed n there exists an optimal prefix length $\lambda < n$ for which the recall achieves a maximum. Amato el al. [3] noted that this maximum is systematically achieved around the prefix length $\lambda = n/cl$, where *cl* is the number of clusters. Note that n/cl represents the average number of pivots taken from each cluster since we use n random pivots. This suggests that an object of a cluster is well represented by the pivots that belong to its same cluster, but when we increase the length of the permutation prefixes we also include pivots taken from other clusters which seems to introduce noisy information. In facts, when we fix n the recall begins to decrease sharply for $\lambda > n/cl$ (Fig. 2b).

Regarding our initial question, that is, whether `P-Perms` represent a valid alternative to classical permutations in distinguishing objects, for a preliminary analysis we selected a number m of random pivot pairs equal to n, so that the full permutations have the same length (i.e. $N = n = m$). For this settings, we discovered that on *Gaussian Euclid30* the `P-Perms` had similar behaviour and slightly lower effectiveness than the classical `Perms` when varying n and λ (Fig. 2c), thus confirming us that the synthetic pivots computed from the pivot pairs could be used as alternative pivots for generating permutations. However, on the clustered data, the `P-Perms` seems to be completely useless (Fig. 2d) with the exception of the case $\lambda = n$ for which the full-length `P-Perms` slightly outperforms the traditional full-length `Perms` (nevertheless both the approaches achieved very low recall when using their full-length representations). One possible reason for the low performance of the `P-Perms` on clustered data is that we are using as reference objects the synthetic midpoints of just $m = n$ random pivot pairs out of $n(n-1)/2$ possible pairs. In facts, since the data is uniformly

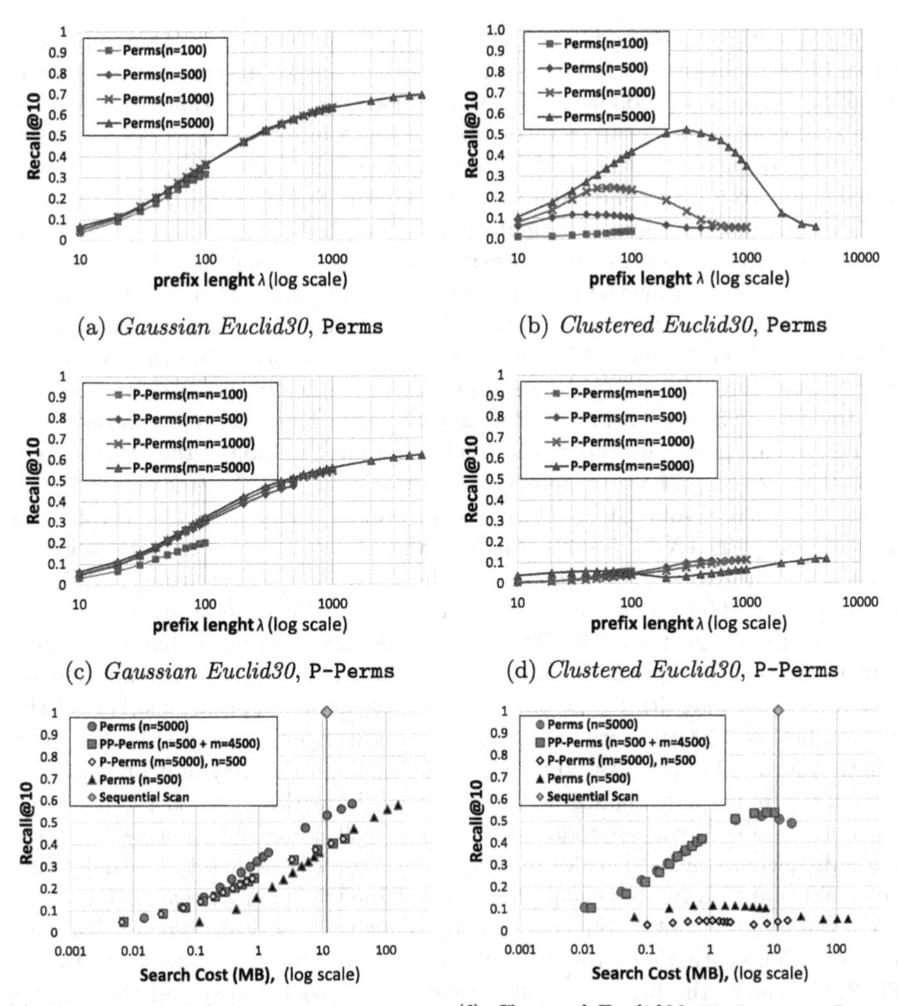

(a) *Gaussian Euclid30*, `Perms` (b) *Clustered Euclid30*, `Perms`

(c) *Gaussian Euclid30*, `P-Perms` (d) *Clustered Euclid30*, `P-Perms`

(e) *Gaussian Euclid30*, `PP-Perms` vs `Perms` (f) *Clustered Euclid30*, `PP-Perms` vs `Perms`

Fig. 2. *Synthetic Datasets*: Recall@10 for the traditional permutations (graphs (a) and (b)) and the `P-Perms` (graphs (c) and (d)) varying the number of pivots and the prefix lengths. Graphs (e) and (f) show the recall for increasing prefix lengths as function of the Search Cost for `Perms` and `PP-Perms`. For each method, the points in the graphs correspond to the prefix lengths $\lambda = 10, 20, 30, 40, 50, 60, 70, 80, 90, 100, 200, 300, 400, 500$

distributed over the cl clusters, the synthetic midpoint of a pair (p_{i_1}, p_{i_2}) is representative of a cluster $\mathcal{C}$ if both p_{i_1} and p_{i_2} belong to the same cluster $\mathcal{C}$, which happen with probability $(n - cl)/cl(n-1)$. Conversely, with probability $n(cl-1)/cl(n-1)$ the two pivots belong to different clusters. For example, if $cl = 20$ and $n = 5K$ the probability of picking a pair of pivots of different clusters is about 95%, so when we use only $m = n = 5K$ random pairs we have on average

4, 750 pairs of pivots form different clusters and just about 12–13 pairs representative of each cluster. To mitigate this inconvenience we may try to use $m \gg n$ or, as proposed in the following, use the PP-Perms representation that employs both traditional and synthetic pivots. The latter approach guarantees to have a percentage of pivots that are still representative of the original data distribution which seems to be fundamental for clustered data. Note that for n pivots and m pairs the PP-Perms produces permutations of length $N = m+n$. For some prefix lengths the effectiveness of the PP-Perms may decrease when considering $m \gg n$ as the percentage of synthetic pivots will be a way larger than the percentage of actual pivots, which may be a issue for clustered data. Anyway the loss in effectiveness is compensated by the more efficiency at searching time since the Search Cost (number of bits accessed per query) typically increases proportionally to λ^2/N. Therefore, since PP-Perms and Perms have different lengths N, in the rest of this paper, we report the *recall* values as function of the Search Cost.

In Figs. 2e and 2f we compared the performance of the traditional Perms using $n = 500$ pivots ($N = 500$), the PP-Perms using $n = 500$ pivots and $m = 4,500$ pairs ($N = 5,000$), the P-Perms using $m = 5,000$ pairs selected from $n = 500$ pivots ($N = 5,000$), and the traditional Perms using $n = 5,000$ pivots ($N = 5,000$). The latter approach is plotted for reference as it has the same length of the tested PP-Perms and P-Perms, but note that it requires 5, 000 actual object pivot distance computations, while the other approaches uses a order of magnitude less object-pivot distances computations. For all the approaches we plot the recall versus the search cost when increasing the prefix length λ form 10 to 500. As expected, the P-Perms, which rely only on synthetic pivots, has poor performance on the clustered data. However, on the clustered dataset, the PP-Perms approach, which uses both actual and synthetic pivots, not only outperforms the Perms techniques that use the same set of actual pivots ($n = 500$) but also reaches the performance of the traditional permutations built upon the larger set of pivots ($n = 5,000$). Thus we observed a great advantage in combining synthetic and real pivots to represent clustered data. In facts, the PP-Perms shows the best trade-off between recall, search cost and the cost for computing the permutations (i.e. the actual object-pivot distance computations). On Gaussian data, both the P-Perms and the PP-Perms still outperforms the traditional permutation built on the same pivot set (we are not interested in the recalls when the search costs is greater than the sequential scan). Moreover, for small search cost values, it achieves recalls in line with the more expensive traditional permutation built upon the larger pivot set. Given these outcomes, in the following we focus our attention only on the PP-Perms and Perms approaches.

4.2 Experiments on Real-World Data

For the experiments on real-world data we used two sets of 1M objects from the *CoPhIR* [7] and *ANN-SIFT* [16] datasets, for which we used different kinds of image features compared with distinct metrics. On the *CoPhIR* data we used as metric the linear combination of the five distance functions (Manhattan, Euclidean, and other special metrics) for the five MPEG-7 descriptors that

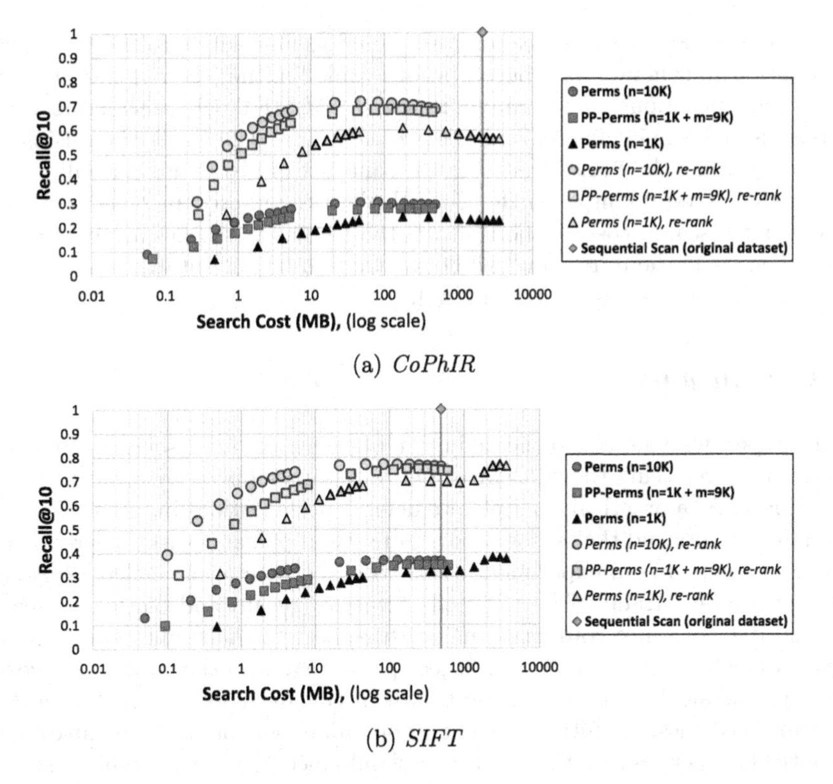

(a) *CoPhIR*

(b) *SIFT*

Fig. 3. Recall@10 as function of the Search Cost (with and without re-rank based on the actual distance), for increasing permutation prefix lengths. For each method, the points plotted in the graphs correspond to the prefix lengths $\lambda = 10, 20, 30, 40, 50, 60, 70, 80, 90, 100, 200, 300, 400, 500, 600, 700, 800, 900, 1000$

have been extracted from each image. We adopted the weights proposed in [6, Table 1]. The *ANN-SIFT* contains SIFT local features (128-dimensional vectors) compared with the Euclidean distance. Note that the SIFT data contains some clusters as the distance distribution is a mixture of Gaussians (see [22, Fig. 1]). On both the datasets, we tested the traditional `Perms` using $n = 1,000$ pivots ($N = 1,000$), the `PP-Perms` using $n = 1,000$ pivots and $m = 9,000$ pairs ($N = 10,000$), and the traditional `Perms` using $n = 10,000$ pivots ($N = 10,000$). For each approach we varied the prefix length λ from 10 to 1,000. The results are depicted in Figs. 3a and 3b for *CoPhIR* and *SIFT* data, respectively. For reference we also reported the cost of the sequential scan for searching the original data descriptors using the actual distance. Moreover, we also include the results when the actual distance is used to refine (*re-rank*) the candidate results selected in the permutation space. We observed that on both the datasets the `PP-Perms` performs better than the traditional permutation build upon the same set of pivots. Moreover, for $\lambda > 100$ it achieves recall values in line with that of the more expensive permutation built upon the 10 times larger set of pivots.

Therefore, the `PP-Perms` can be profitably used as alternative to the traditional permutation to generate long permutations while limiting the number of actual distance computations. For example, to search 1M SIFT data with a query cost of about 8 MB, the `PP-Perms` achieves a *recall@10* of 0.29 (0.69 when using the re-ranking) while the `Perms` that uses the same set of pivots has a recall of 0.24 (0.56 with the re-ranking). On the CoPhIR data the improvement is even more evident: for a search cost of about 4 MB the `PP-Perms` reaches a recall of 0.24 (0.61 when using the re-ranking) while the traditional permutations has a recall of 0.16 (0.47 when using the re-ranking).

5 Conclusions

In this paper, we generalized the definition of permutations associated to metric objects by introducing the concept of permutations induced by a metric transformation f. As a practical example, we defined permutations induced by a combination of pivots and the tensor product of several planar projections related to some pivot pairs. In our experiments, this novel representation, called `PP-Perms`, achieved the best trade-off between effectiveness (recall) and efficiency (search cost and data distance computations) with respect to the traditional permutations. In facts, for the same set of object-pivot distance calculations, `PP-Perms` allows producing longer permutations, which can be more efficiently searched using inverted files. As future work, on one hand, we would like to investigate theoretical properties that the function f should meet in order to induce effective permutation-based representations; on the other hand, we would like to exploit artificial intelligence techniques to automatically learn suitable functions f.

Acknowledgements. The work was partially supported by H2020 project AI4EU under GA 825619, by H2020 project AI4Media under GA 951911, and by NAUSICAA (CUP D44E20003410009).

References

1. Amato, G., Falchi, F., Gennaro, C., Vadicamo, L.: Deep permutations: deep convolutional neural networks and permutation-based indexing. In: Amsaleg, L., Houle, M.E., Schubert, E. (eds.) SISAP 2016. LNCS, vol. 9939, pp. 93–106. Springer, Cham (2016). https://doi.org/10.1007/978-3-319-46759-7_7
2. Amato, G., Falchi, F., Rabitti, F., Vadicamo, L.: Some theoretical and experimental observations on permutation spaces and similarity search. In: Traina, A.J.M., Traina, C., Cordeiro, R.L.F. (eds.) SISAP 2014. LNCS, vol. 8821, pp. 37–49. Springer, Cham (2014). https://doi.org/10.1007/978-3-319-11988-5_4
3. Amato, G., Gennaro, C., Savino, P.: MI-File: using inverted files for scalable approximate similarity search. Multimed. Tools. Appl. **3**, 1333–1362 (2014)
4. Amato, G., Savino, P.: Approximate similarity search in metric spaces using inverted files. In: Proceedings International ICST Conference on Scalable Information Systems, pp. 28:1–28:10. ICST/ACM (2008)

5. Amato, G., Carrara, F., Falchi, F., Gennaro, C., Vadicamo, L.: Large-scale instance-level image retrieval. Inf. Process. Manag. **57**(6), 102100 (2020)
6. Batko, M., et al.: Building a web-scale image similarity search system. Multimed. Tools. Appl. **47**(3), 599–629 (2010)
7. Bolettieri, P., et al.: Cophir: a test collection for content-based image retrieval (2009)
8. Carrara, F., Gennaro, C., Falchi, F., Amato, G.: Learning distance estimators from pivoted embeddings of metric objects. In: Satoh, S., et al. (eds.) SISAP 2020. LNCS, vol. 12440, pp. 361–368. Springer, Cham (2020). https://doi.org/10.1007/978-3-030-60936-8_28
9. Chávez, E., Figueroa, K., Navarro, G.: Effective proximity retrieval by ordering permutations. IEEE Trans. Pattern Anal. Mach. Intell. **30**(9), 1647–1658 (2008)
10. Chávez, E., Figueroa, K., Navarro, G.: Proximity searching in high dimensional spaces with a proximity preserving order. In: Gelbukh, A., de Albornoz, Á., Terashima-Marín, H. (eds.) MICAI 2005. LNCS (LNAI), vol. 3789, pp. 405–414. Springer, Heidelberg (2005). https://doi.org/10.1007/11579427_41
11. Connor, R., Vadicamo, L., Cardillo, F.A., Rabitti, F.: Supermetric search. Inf. Syst. **80**, 108–123 (2019)
12. Connor, R., Vadicamo, L., Rabitti, F.: High-dimensional simplexes for supermetric search. In: Similarity Search and Applications, pp. 96–109. Springer International Publishing (2017). https://doi.org/10.1007/978-3-319-68474-1_7
13. Esuli, A.: Use of permutation prefixes for efficient and scalable approximate similarity search. Inf. Process. Manage. **48**(5), 889–902 (2012)
14. Figueroa, K., Chavez, E., Navarro, G., Paredes, R.: Speeding up spatial approximation search in metric spaces. ACM J. Exp. Algorithmics **14**, 3–6 (2010)
15. Hetland, M.L., Skopal, T., Lokoč, J., Beecks, C.: Ptolemaic access methods: challenging the reign of the metric space model. Inf. Syst. **38**(7), 989–1006 (2013)
16. Jegou, H., Douze, M., Schmid, C.: Product quantization for nearest neighbor search. IEEE Trans. Pattern Anal. Mach. Intell. **33**(1), 117–128 (2010)
17. Kruliš, M., Osipyan, H., Marchand-Maillet, S.: Employing gpu architectures for permutation-based indexing. Multimed. Tools. Appl. **76**(9), 11859–11887 (2017)
18. Naidan, B., Boytsov, L., Nyberg, E.: Permutation search methods are efficient, yet faster search is possible. Proc. Int. Conf. Very Large Data Bases **8**(12), 1618–1629 (2015)
19. Novak, D., Zezula, P.: PPP-codes for large-scale similarity searching. In: Hameurlain, A., Küng, J., Wagner, R., Decker, H., Lhotska, L., Link, S. (eds.) Transactions on Large-Scale Data- and Knowledge-Centered Systems XXIV. LNCS, vol. 9510, pp. 61–87. Springer, Heidelberg (2016). https://doi.org/10.1007/978-3-662-49214-7_2
20. Tellez, E.S., Chávez, E., Navarro, G.: Succinct nearest neighbor search. Inf. Syst. **38**(7), 1019–1030 (2013)
21. Vadicamo, L., Connor, R., Falchi, F., Gennaro, C., Rabitti, F.: SPLX-Perm: a novel permutation-based representation for approximate metric search. In: Amato, G., Gennaro, C., Oria, V., Radovanović, M. (eds.) SISAP 2019. LNCS, vol. 11807, pp. 40–48. Springer, Cham (2019). https://doi.org/10.1007/978-3-030-32047-8_4
22. Vadicamo, L., Mic, V., Falchi, F., Zezula, P.: Metric embedding into the hamming space with the n-simplex projection. In: Amato, G., Gennaro, C., Oria, V., Radovanović, M. (eds.) SISAP 2019. LNCS, vol. 11807, pp. 265–272. Springer, Cham (2019). https://doi.org/10.1007/978-3-030-32047-8_23

23. Weber, R., Schek, H.J., Blott, S.: A quantitative analysis and performance study for similarity-search methods in high-dimensional spaces. In: Proceedings International Conference on Very Large Data Bases, vol. 98, pp. 194–205 (1998)
24. Zezula, P., Amato, G., Dohnal, V., Batko, M.: Similarity search: the metric space approach, vol. 32. Springer Science & Business Media (2006)

Data-Driven Learned Metric Index: An Unsupervised Approach

Terézia Slanináková[✉], Matej Antol, Jaroslav Oľha, Vojtěch Kaňa, and Vlastislav Dohnal

Faculty of Informatics, Masaryk University,
Botanická 68a, 602 00 Brno, Czech Republic
{xslanin,matejantol,olha,456598,dohnal}@mail.muni.cz

Abstract. Metric indexes are traditionally used for organizing unstructured or complex data to speed up similarity queries. The most widely-used indexes cluster data or divide space using hyper-planes. While searching, the mutual distances between objects and the metric properties allow for the pruning of branches with irrelevant data – this is usually implemented by utilizing selected anchor objects called pivots. Recently, we have introduced an alternative to this approach called Learned Metric Index. In this method, a series of machine learning models substitute decisions performed on pivots – the query evaluation is then determined by the predictions of these models. This technique relies upon a traditional metric index as a template for its own structure – this dependence on a pre-existing index and the related overhead is the main drawback of the approach.

In this paper, we propose a data-driven variant of the Learned Metric Index, which organizes the data using their descriptors directly, thus eliminating the need for a template. The proposed learned index shows significant gains in performance over its earlier version, as well as the established indexing structure M-index.

Keywords: Index structures · Learned index · Unstructured data · Content-based search · Metric space · Machine learning

1 Introduction

Searching within collections of unstructured or complex data (such as images, audio files or protein structures) is a challenging task. Whereas in structured data-sets, the order of the data objects is determined using a straightforward key (e.g., their alphabetical order) and the match to a search filter is objectively given (e.g., retrieve all records where `created_on` $\leq$ `2010-04-01`), in the realm of unstructured data, such properties do not exist. Since there is no intrinsic

This research has been supported by the Czech Science Foundation project No. GA19-02033S. Computational resources were supplied by the project "e-Infrastruktura CZ" (e-INFRA LM2018140) provided within the program Projects of Large Research, Development and Innovations Infrastructures.

N. Reyes et al. (Eds.): SISAP 2021, LNCS 13058, pp. 81–94, 2021.
https://doi.org/10.1007/978-3-030-89657-7_7

ordering to the data, there is no single agreed-upon response to a given search query.

The issue can be addressed using metric spaces, where the pairwise similarity of objects can be leveraged to organize the data and formulate search queries. If we can design a suitable distance function that meets certain criteria (such as symmetry and triangle inequality), any indexing structure or search algorithm designed for generic metric spaces can be applied to our data, and various pruning rules can be used to reduce the search space.

The search itself is usually performed using various similarity queries, wherein we specify a query object and choose the properties of the desired result in relation to this object (e.g., the k closest objects to the query object – kNN query, or all objects within a certain range from the query object – range query). Even after applying the metric spaces and similarity searching methods, a major challenge remains – since these complex data-sets tend to have a very high number of intrinsic dimensions [5], the distance computations needed for index construction and query evaluation are computationally expensive.

This problem can be addressed using an alternative approach – finding the similarity in large groups of data can be reformulated into a pattern searching task, which can be solved by machine learning. We have previously introduced such a solution, using supervised machine learning to imitate the structure of a pre-existing index, resulting in a hierarchy of several learned models that we call *Learned Metric Index (LMI)* [2]. While this approach achieves very good performance in the query evaluation phase by eliminating costly distance computations, its main downside is obvious – to train such an index, we first need to construct one of the traditional index structures as a template.

In our current work, we have evolved the LMI's approach beyond the need for a pre-existing index built using traditional methods. Instead, we can now construct the LMI from scratch, using nothing but the pattern recognition capability of the machine learning models to discern the natural distributions of the data in the metric space.

To the best of our knowledge, this is a completely novel method for tackling the problem of indexing unstructured data. This paper describes our approach and implementation in detail and evaluates its performance, comparing it to the traditional state-of-the-art indexes and our previous implementations of supervised learned indexing.

2 Related Work

More and more research work has recently addressed the possibilities of enhancing or even replacing standard database index models (B^+-trees) with machine learning [8,14,17]. The authors argue that machine learning models can be trained for the same purpose of answering queries (categorizing a query object to the most suitable class, which represents a child node) while presenting several performance benefits.

For instance, in inverted indexes, a hierarchical machine learning model is used to reduce index size at the expense of performance [29,34]. In cases involving multidimensional data, learned indexes attempt to approximate the search to be reasonably efficient. The density distribution of multidimensional data is approximated to create a new index structure in [31,33]. Another application of learned models [24] presents an index named Flood that creates not only a consistently performing index for multidimensional data, but also optimizes both index and data storage layout. In [18], a learned variation of Bloom filters for multidimensional data can save a significant portion of space. A wide study [16] of various algorithms for kNN queries over multidimensional Euclidean spaces concludes that it is still a research challenge to provide a solution of highly precise approximate kNN search due to the curse of dimensionality.

We carry on with the proposition of utilizing machine learning to index structured data and apply it to complex data and metric space model. In this paper specifically, we follow up on the Learned Metric Index method we introduced in [2]. Even though we believe that our research is original, the idea of learned models has been applied before in the domain of similarity searching in metric data. In [13], ANN-tree was introduced to solve the 1-NN problem for metric space scenarios. Authors of [22] consequently introduced the FLANN library to perform the 1-NN search significantly faster than a previous, nearly brute force implementation.

Recently, a new partitioning procedure focused on nearest neighbor search performance, called Neural Locality-Sensitive Hashing (Neural LSH) [7], has been shown to outperform traditional partitioning methods (k-means) consistently. A learned model that approximates bounds on k nearest neighbor distances and consequently allows precise and memory-efficient computation of reverse nearest neighbors has been introduced in [4]. The authors conducted experiments on up to 8-dimensional and low-volume data. Finally, Hünemörder et al. [11] explored the application of various predictive models to learn an index for approximate nearest-neighbor queries. Their evaluation on synthetic data as well as the MNIST data-set further demonstrates the research potential of this topic.

3 Indexing in Metric Spaces

A *metric space* $\mathcal{M} = (\mathcal{D}, d)$ is defined over a universe $\mathcal{D}$ of data objects and a distance function $d(\cdot, \cdot)$ that satisfies metric postulates. A database $X \subseteq \mathcal{D}$ of data objects forms a collection to be queried by a *k-nearest neighbors query* ($kNN(q)$ – k objects closest to the query object q), or the *range query* ($range(q, r)$ – all database objects closer to q than the distance r).

To avoid tedious sequential scanning, which is costly on large data-sets or with an expensive distance function, various indexing structures have been developed. Firstly, hierarchical structures include variations of the original M-tree [6], Spatial Approximation Trees [25], or Rank Cover Trees [10]. These structures divide data objects into groups or clusters, respecting their distribution in space.

They provide sub-linear search time $\mathcal{O}(n^{\alpha})$, where $\alpha \leq 1$ depends on data distribution. Next, permutations of preselected anchor objects (pivots) and their prefixes define (Voronoi-like) space cells at bounded costs, so M-index [26] and PPP-Codes [28] improve search efficiency substantially. Rearrangement of such cells is applied in [1,21]. Lastly, independent filtering techniques can be applied to further eliminate accessing excessive amounts of data objects, e.g. Binary Sketches for Secondary Filtering [19].

The properties of the metric function (namely symmetry and triangle inequality) are typically indispensable for constructing index structures and for the correctness of search. Learned indexes do not inherently depend on these properties, so the query evaluation based on predictions can be advantageous for non-metric distance functions as well.

4 Learned Metric Index

Learned Metric Index (LMI), as introduced in [2], is a hierarchical tree index structure of nodes containing machine learning models. These models are trained to search for (i.e., categorize) query objects, which emulates the behaviour of traditional index nodes. However, instead of determining the objects' positions according to their distances, a query is resolved by applying a series of predictions. This changes the standard paradigm of index building and query evaluation, resulting in very different performance characteristics and outperforming traditional similarity searching methods in many cases, both in terms of efficiency and effectiveness.

In general, the concept of LMI can be realized in two distinct ways. The first one involves using a pre-existing index and its data partitioning as labels for *supervised* training. In such a case, each data object has a label corresponding to its position in the original index, i.e., a concatenated list of integer values per index level. We have examined this variant in [2] and demonstrated that it can achieve more than competitive performance with state-of-the-art methods.

The other option is to assemble LMI "from scratch" by letting it create its own meaningful divisions of the data. Such approach exploits the information embedded in the descriptors of data objects to emulate the similarity function. This constitutes an *unsupervised* learning problem, which is the subject of this paper.

4.1 Training Unsupervised LMI

Training an unsupervised LMI requires: (i) digital fingerprint of objects to train on, and (ii) the number of clusters each model is expected to create, which defines the shape of the learned index structure. The training procedure of the whole LMI then starts with the root model, which is trained on the entirety of the given data-set, while its descendants are trained on smaller and smaller portions of the data as we dive deeper into the structure. The training is therefore sequential – the input of every model depends on the output of its parent.

Algorithm 1: Unsupervised Learned Metric Index training

```
Input: a data-set X, max. depth H (tree height),
       max. number of children per level A[]
Output: a tree of trained models T[][]
part[1][1] = X;
for lvl ← 1 to H do
    for chld ← 1 to A[lvl] do
        if part[lvl][chld] = ∅ then
            continue
        end
        M ← new model trained on part[lvl][chld] clustering the data into A[lvl]
        groups;
        if lvl < H then
            for obj ∈ part[lvl][chld] do
                p = M.predict(obj);
                part[lvl+1][p].add(obj);
            end
        end
        T[lvl][chld] = M;
    end
end
return T;
```

Algorithm 1 formally describes the entire training procedure. During the training, each model is presented with a clustering problem. The objective is to organize the data into a pre-specified number of groups according to their mutual similarity obtained from the descriptors. Each training epoch re-organizes the data to allow mutually similar objects to end up in the same cluster. A single instance of LMI is then created by connecting the parent models with their children, resulting in a tree structure.

4.2 Searching in LMI

We define the overall goal of LMI as finding as many of the query's k nearest neighbors as possible in the shortest time. The output of every learned model in the searching (inference) phase is a probability distribution, which can be viewed as the query's correspondence to each of the classes (i.e., child nodes). We expect LMI to be able to assign higher probabilities (and therefore higher search priorities) to categories where the query object and its nearest neighbors reside. The priority queue can then be formed in a naïve manner by sorting the child nodes based on the probabilities assigned by their parent model. This contrasts with traditional indexing methods, which need to calculate the distances to all of the child objects to form their priority queues.

The searching process of LMI is shown in Fig. 1. From the LMI's point of view, an answer to a query is gradually updated with objects from the visited leaf nodes. Note that the small sub-sections of the data-set contained within the

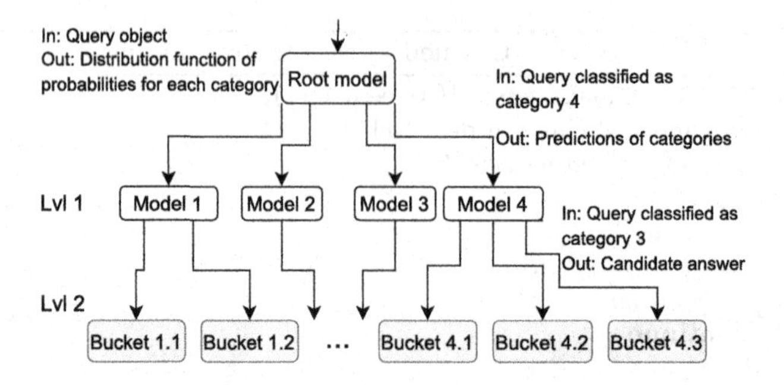

Fig. 1. Example of a few initial steps of searching within a two-level LMI with four models on Level 1. The search continues until a stop condition is met.

leaf nodes are searched linearly – once a leaf node gets to the top of the priority queue, all of its objects are evaluated (i.e., added to the answer or discarded based on their distance) and the leaf node is removed from the queue.

4.3 Machine Learning Models

In the previous sections, we introduced a basic version of unsupervised LMI wherein we can use the machine learning models to build the index, and then use the probability outputs of these models to search the resulting structure.

However, in practice, very few unsupervised algorithms can operate probabilistically. To use a non-probabilistic unsupervised algorithm, we need to modify the approach in one of two ways. The first option is to build the structure using Algorithm 1, and use distance calculations for searching in the case of distance-based algorithms. The second option is to substitute the distance function with a supervised machine learning model. However, this second approach requires a modification of the building phase described by Algorithm 1, splitting the training into two steps[1].

We selected two basic machine learning algorithms to implement unsupervised LMI – K-Means and Gaussian Mixture Models (GMM).

K-Means is a well-established distance-based algorithm, which requires the Euclidean space to suitably place cluster centers – so-called *centroids* – within the data. The algorithm runs until a local optimum is reached by iteratively recalculating the centroids' position to minimize the sum of squares within clusters. Logistic Regression was selected as the supervised algorithm for the two-step version of this process.

Gaussian Mixture Model (GMM) employs a more flexible approach to data modelling, using soft clustering instead of the hard cluster assignments

[1] This training procedure consists of two separate phases: one for clustering the data, and the second for their categorization. For every level, the data is firstly clustered in the same way as described above. Subsequently, a supervised categorization machine learning algorithm is trained on the relevant portion of the data and the clustered labels.

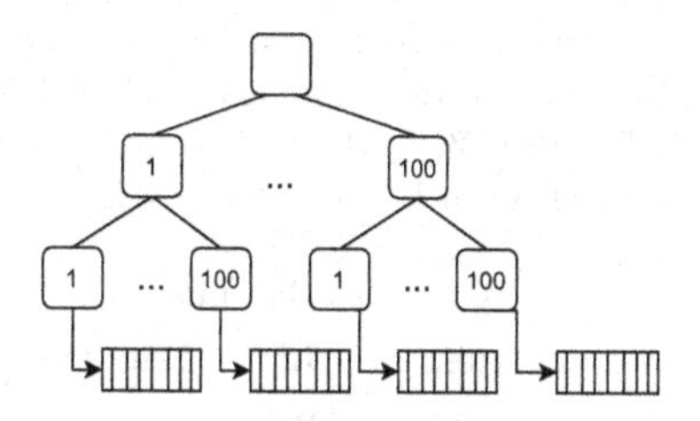

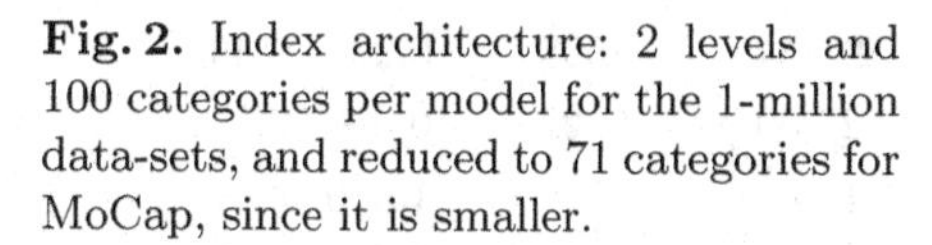
Fig. 2. Index architecture: 2 levels and 100 categories per model for the 1-million data-sets, and reduced to 71 categories for MoCap, since it is smaller.

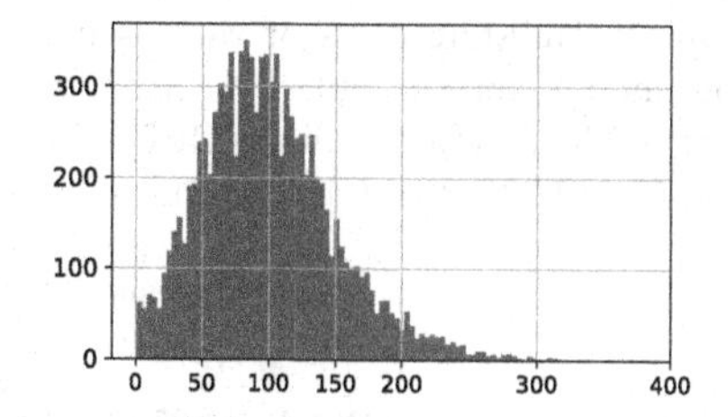

Fig. 3. Histogram of object distribution in the buckets (distance-based K-Means, Profiset).

made by K-Means. As its name suggests, GMM assumes that each data point could have been generated by any number of its k Multi-variate Gaussian distributions (k being the chosen number of clusters) with a given probability. To evaluate this probability, we must approximate the posterior probability of an object belonging to a cluster, given the observed data. *Bayesian GMM* is an extension of GMM, which estimates the object's cluster membership by Bayesian Variational Inference instead of calculating the marginal probabilities.

As a result, we evaluate four separate algorithms in the experimental phase – distance-based K-Means, K-Means with Logistic Regression, GMM and Bayesian GMM. The index-building and searching operations were implemented in Python, and algorithms used to prototype unsupervised LMI came from the scikit-learn library [30] with the exception of K-means with Logistic regression, where we employed an efficient GPU implementation of K-means [12].

5 Experiments

We have executed a wide range of experiments with three different multimedia data-sets: *CoPhIR*, *Profiset* and *MoCap*. CoPhIR [3] is a data collection of 282-dimensional vectors derived from five visual descriptors of images. Profiset [27] is a series of 4096-dimensional vectors extracted from Photo-stock images using a convolutional network. Finally, MoCap is HDM05 data-set [23] that consists of sequences of 3D skeleton poses, which were segmented to extract 4096-dimensional descriptors using AlexNet [15]. The data-set sizes were fixed at 1-million objects for CoPhIR and Profiset. MoCap contains 354,893 segments.

In contrast with the supervised version of LMI, unsupervised LMI has a unique architectural flexibility provided by the unsupervised mode of training, where one can specify the index architecture via the number of clusters per each model and thus optimize the performance. As a results, we chose to use a single architectural configuration throughout the experiments, consisting of two levels with a fixed number of nodes, as detailed in Fig. 2. As opposed to the traditional indexing structures, such as M-tree or M-index, LMI does not limit leaf node capacity. However, this fact does not cause the distribution of objects within buckets to be uncontrollably skewed, as Fig. 3 shows. The vast majority of the

Table 1. Building costs of various unsupervised setups and baselines. Unsupervised experiments were executed on a machine with 1 CPU – Intel Xeon E5-2650v2 2.60 GHz. K-Means (LR) utilized GPU - nVidia Tesla T4 16 GB. LMI baseline used Intel Xeon Gold 6230 2.10 GHz. M-index baseline used Intel Xeon E5-2620 2.00 GHz.

		Bayesian GMM	GMM	K-means (LR)	K-means (dist.)	Baselines	
						LMI	M-index
Build t. (h)	CoPhIR	0.305	0.351	1.926	0.639	2.670	0.330
	Profiset	1.419	1.553	9.554	3.698	0.230	0.490
	MoCap	0.351	0.467	2.200	0.627	0.390	0.170
Memory (gb)	CoPhIR	10.0	13.6	15.6	8.6	150.0	3.4
	Profiset	74.4	86.6	75.0	85.0	150.0	20.7
	MoCap	55.5	71.0	32.0	49.0	85.0	6.4

bucket occupancies is within the 75–125 interval, guaranteeing similar sequential search costs in the final part. This property allows us to skip searching of the leaf nodes in evaluation, and focus on the performance of the internal index navigation, where the various indexes truly differ.

In each of the experiments, we perform a 30-NN query for 1,000 randomly chosen query objects. The performance is measured in terms of recall – i.e., how many of the actual 30 nearest neighbors are returned when visiting a limited portion of data-set. We set such search limits (*stop-conditions*) as increasing thresholds spanning from 0.05% of the indexing structure searched (the lowest stop-condition) to 75% searched (the highest stop-condition).[2] As is the case with all indexes, we are primarily interested in optimizing the trade-off between *recall* and the searching time (i.e., the time needed to evaluate a query).

5.1 Building Costs

To provide a clear comparison of various indexes, we have to consider the costs of their construction. Table 1 documents the RAM usage and time required to build each of LMIs and M-index.

The table shows that the construction cost of a given setup is strongly influenced by the data-set dimensionality, which is consistent with the results observed in [2]. Specifically, the dimensionality of Profiset and MoCap descriptors is almost 15 times that of the CoPhIR data-set (282 vs 4096 features), which results in greater memory and building time requirements. The amount of the data present in the data-set influences the cost as well – in the case of MoCap, the number of objects the structure has to index is about one-third of the amount of Profiset, resulting in shorter building times and lower memory requirements.

[2] Full enumeration of stop-conditions used: 0.05%, 0.1%, 0.3%, 0.5%, 1%, 5%, 10%, 20%, 30%, 50% and 75% of the data-set size.

Table 2. The hyperparameters and their various settings for all four implemented algorithms. The highlighted values enabled the models to reach the best performance on majority of the data-sets. For further details, see documentation of scikit-learn [30].

	Covariance type	Initialization alg.	Prior type	No. init.	Max. iters
GMM	full, **spher.**, diag, tied	**K-Means, rand.**	–	–	1,**2**,**5**
Bayesian GMM	full, **spher.**, diag, tied	**K-Means, rand.**	**process, distr.**	–	1,**2**
K-Means (LR)	–	–	–	**5**,10,15,20	**5**,10,15,20
K-Means (d.)	–	–	–	1,5,**10**	5,**10**,25

The results show that the least time-consuming LMI models are GMM and Bayesian GMM. On the other hand, the most time-consuming model is K-Means with Logistic regression due to its two-step training design. In this case, the time expenditure can be attributed mainly to the second (supervised) part of the training (Logistic regression), which does not have the advantage of the time-efficient GPU-optimized K-Means implementation. In comparison with the building costs of the supervised LMI baseline, it appears that the unsupervised models exhibit lower RAM usage in all cases.

M-index requires the least time and memory out of all the examined indexes. Its performance in terms of building costs, compared to the LMI models, can be justified primarily by the fact that M-index is a mature index with many heuristics developed over the years to improve its baseline performance, which provides it with a considerable advantage over our newly-developed index.

5.2 Tuning of Learned Models

In all of the machine learning models, we identified several hyperparameters that influence the quality of the run in a major way – we list them in Table 2.

In Mixture models, i.e., GMM and Bayesian GMM, *Covariance type* influences the shape of the covariance matrix, and whether each cluster has its own covariance matrix, or all components share a common one. *The initialization algorithm* represents the pre-training initialization procedure. Bayesian GMM has one extra hyperparameter, *Prior type*, which influences the initial setting of the weight concentration prior. In the case of the K-Means algorithms, we considered different *Numbers of initializations*, where we let the algorithm run multiple times with different initialization seeds to avoid stoppage in local optimum.

We have conducted more than a hundred trials with different combinations of data-sets and hyperparameter values. The best performing parameter setups per model were selected for experimental evaluation. We have chosen the best-performing setups to be the ones that achieve 90% recall for the lowest possible stop-condition, in the shortest searching time.

5.3 Results

Four unsupervised machine learning algorithms were selected, as described in Sect. 4.3 to test the capabilities of an unsupervised approach experimentally.

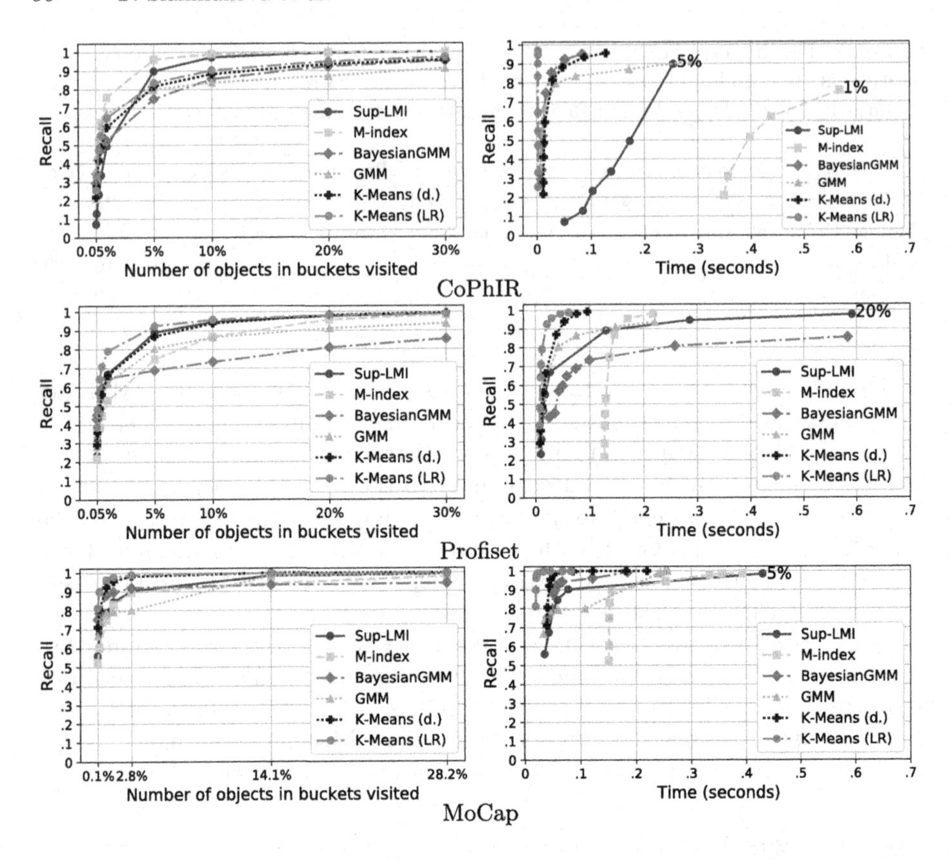

Fig. 4. Comparison between the recall of unsupervised LMI models, the best supervised setup from Antol et al. [2] (Sup-LMI), and M-index. The X-axis of the graphs on the left spans 30% of the total index size.

Two of them (*GMM*, *BayesianGMM*) represent a standard application of LMI unsupervised training and searching algorithm for data without labels. The third (*K-Means (d.)*) is constructed using unsupervised clustering combined with standard distance-based searching. The final one (*K-Means (LR)*) involves a training approach that combines unsupervised clustering and supervised learning using logistic regression. Figure 4 displays the achieved recall using two measures – the percentage of the structure searched and the time needed to evaluate one query. We compare the results of unsupervised LMI with two benchmarks: the best-performing M-index[3] and the supervised LMI[4] from our previous work [2].[5]

[3] The configurations of M-index selected as baselines for our three data-sets [2]: *M-index CoPhIR 200*, *M-index Profiset 2000* and *M-index MoCap 2000*.

[4] Best LMI setups in [2]: Multi-label trained on CoPhIR (M-index 200), Logistic Reg. trained on Profiset (M-tree 2000) and Neural net. trained on MoCap (M-index 2000).

[5] The best performing setup was the one achieving 90% recall in the lowest stop-condition and in the shortest time.

Our experiments show that in terms of navigation efficiency (i.e., recall per number of visited objects – left column), unsupervised indexes fall behind in the case of CoPhIR, but dominate both baselines in Profiset and MoCap. We attribute the poor performance seen in CoPhIR to two factors: the length of descriptors and their origin. CoPhIR's descriptors are composed of hand-picked features of the images, such as color histogram, whereas Profiset and MoCap's descriptors are extracted from machine learning models. Unsupervised LMI exhibits a better ability to traverse the indexing structure in case of more complex descriptors of machine-learning origins. In this type of descriptors, the average gain in recall over the CoPhIR data-set ranges from 4.5% to 13.5% (given the 5% stop-condition), depending on the algorithm used.

The performance difference is much more decisive when comparing time efficiency. Both of the baselines fall behind the unsupervised LMI setups significantly in all three data-sets (see the right column).[6] Specifically, the K-Means unsupervised methods reach 90% recall faster than M-index by a factor of approximately 70 (e.g., 0.02 s vs 1.55 s on MoCap) and faster than supervised LMI by a factor of approximately 8 (e.g., 0.02 s vs 0.123 s on Profiset).

K-Means with Logistic Regression is reaching the highest recall and the shortest search time. The index is able to find the relevant objects very quickly, achieving 90% recall in under 20 ms in every data-set. **Distance-based K-Means** also exhibits a favourable recall-to-speed trade-off on all of the data-sets, with performance similar to K-Means trained with the two-step approach. This setup also outperforms both of the baselines throughout all stop-conditions on Profiset and MoCap. **Mixture models** – GMM and BayesianGMM – generally show worse performance than K-Means-based indexes, and they are only competitive within the CoPhIR data-set. In most instances, mixture models only manage to outperform the baselines in the lower stop-conditions ($\leq$5%). However, they stay close behind in the higher stop-conditions in the case of Profiset and MoCap (except for BayesianGMM in Profiset).

5.4 Summary

We consider results of our experiments to be very encouraging. In the overwhelming majority of stop-conditions, both of the K-means-based setups were able to outperform M-index, as well as the best LMI setups from [2]. While this is true for both of our performance metrics – recall per number of objects searched and recall per time – the advantage of our unsupervised setups is much more prominent when considering the time-based metric.

The performance of distance-based K-Means demonstrates that the concept of LMI can be extended to work with distances instead of probabilities, with no degradation in performance. Out of all the indexes, the two-step training method achieved the most promising searching speeds and the highest recall per percentage of the structure searched in every stop-condition.

[6] For the sake of consistency of the environments across indexes, we used the Python 3.6 implementation of M-index from [2].

The GMM-based indexes perform better than the baselines on lower stop-conditions, but the performance gain disappears later in the search. As a result, these indexes still might be preferred in scenarios where one is limited by the time or the amount of the structure that can be searched, and tolerates lower recall, possibly in exchange for more favourable building costs.

6 Conclusion

In this paper, we extend the capabilities of the Learned Metric Index – a novel, machine-learning-based indexing paradigm introduced in [2]. We present a new means of LMI construction that builds the index from scratch – no pre-existing index is needed to guide the building process. Our experiments confirm that building an unsupervised LMI is a viable approach, and clustering algorithms within LMI create meaningful divisions of the data. In comparison to the formerly introduced supervised LMI, the building costs are significantly lower. By far the most significant benefit of unsupervised LMI is the overall search performance measured as recall in time – our new approach managed to beat both benchmarks (M-index and supervised LMI) by at least one order of magnitude in all cases. If we measured performance as recall per portion of the index structure visited (navigation), unsupervised LMI was superior to both benchmarks by approximately 10% in two out of the three tested data-sets. On the third data-set, the unsupervised methods fell behind when searching a larger portion of the structures. However, even in these cases, the computation speed of the unsupervised LMI outweighs the navigation deficit and reaches all accuracy thresholds in shorter time.

The performance of unsupervised LMI shown in our experiments invites for future research. This work has demonstrated the architecture of unsupervised LMI in a typical domain where similarity is obtained from vectors. These vectors are extracted directly from the objects' raw data, which is an ideal scenario for standard machine learning models. However, other types of complex data, e.g., protein structures, use different concepts of similarity – this means that their processing by LMI may not be so straightforward. In these domains, we need to employ more specialized machine learning models, such as LSTM [9], Transformer [32], or Word2vec [20] to produce vector data.

Furthermore, there is room for improvement in decreasing the construction costs by exploring different libraries and environments for the building of LMI. We also plan to inspect other machine learning models to improve LMI's pattern recognition potential even further. Finally, we plan to explore the topics of index dynamicity (i.e., the ability to locate objects outside of the indexed data-set), priority queue optimization, testing the LMI on different data-sets from different domains, and finding suitable hardware setups for LMI operations.

Overall, we view this work as an additional proof that the adoption of machine learning techniques in similarity searching is worth deep exploration, and that the concept of Learned Metric Index can provide significantly better results when it is built without a pre-existing traditional index as a template.

References

1. Antol, M., Dohnal, V.: BM-index: balanced metric space index based on weighted Voronoi partitioning. In: Welzer, T., Eder, J., Podgorelec, V., Kamišalić Latifić, A. (eds.) ADBIS 2019. LNCS, vol. 11695, pp. 337–353. Springer, Cham (2019). https://doi.org/10.1007/978-3-030-28730-6_21
2. Antol, M., Ol'ha, J., Slanináková, T., Dohnal, V.: Learned metric index — proposition of learned indexing for unstructured data. Inf. Syst. **100**, 101774 (2021)
3. Batko, M., et al.: Building a web-scale image similarity search system. Multimedia Tools Appl. **47**(3), 599–629 (2009)
4. Berrendorf, M., Borutta, F., Kröger, P.: k-distance approximation for memory-efficient RkNN retrieval. In: Amato, G., Gennaro, C., Oria, V., Radovanović, M. (eds.) SISAP 2019. LNCS, vol. 11807, pp. 57–71. Springer, Cham (2019). https://doi.org/10.1007/978-3-030-32047-8_6
5. Chávez, E., Navarro, G., Baeza-Yates, R.A., Marroquín, J.L.: Searching in metric spaces. ACM Comput. Surv. (CSUR 2001) **33**(3), 273–321 (2001)
6. Ciaccia, P., Patella, M., Zezula, P.: M-tree: an efficient access method for similarity search in metric spaces. In: Proceedings of the 23rd International Conference on Very Large Data Bases (VLDB 1997), Athens, Greece, 25–29 August 1997, pp. 426–435. Morgan Kaufmann (1997)
7. Dong, Y., Indyk, P., Razenshteyn, I.P., Wagner, T.: Learning space partitions for nearest neighbor search. In: 8th International Conference on Learning Representations, ICLR, Addis Ababa, Ethiopia, 26–30 April 2020 (2020)
8. Ferragina, P., Vinciguerra, G.: The PGM-index: a fully-dynamic compressed learned index with provable worst-case bounds. Proc. VLDB Endow. **13**(8), 1162–1175 (2020)
9. Hochreiter, S., Schmidhuber, J.: Long short-term memory. Neural Comput. **9**(8), 1735–1780 (1997)
10. Houle, M.E., Nett, M.: Rank cover trees for nearest neighbor search. In: Brisaboa, N., Pedreira, O., Zezula, P. (eds.) SISAP 2013. LNCS, vol. 8199, pp. 16–29. Springer, Heidelberg (2013). https://doi.org/10.1007/978-3-642-41062-8_3
11. Hünemörder, M., Kröger, P., Renz, M.: Towards a learned index structure for approximate nearest neighbor search query processing. In: Reyes, N., et al. (eds.) SISAP 2021. LNCS 13058, pp. 95–103 (2021)
12. Johnson, J., Douze, M., Jégou, H.: Billion-scale similarity search with GPUs. arXiv preprint arXiv:1702.08734 (2017)
13. Lin, K.-I., Yang, C.: The ANN-tree: an index for efficient approximate nearest neighbor search. In: Proceedings Seventh International Conference on Database Systems for Advanced Applications, DASFAA 2001, pp. 174–181, April 2001
14. Kraska, T., Beutel, A., Chi, E.H., Dean, J., Polyzotis, N.: The case for learned index structures. In: Proceedings of the 2018 International Conference on Management of Data, SIGMOD 2018, pp. 489–504. Association for Computing Machinery (2018)
15. Krizhevsky, A., Sutskever, I., Hinton, G.E.: ImageNet classification with deep convolutional neural networks. Adv. Neural Inf. Process. Syst. **25**, 1097–1105 (2012)
16. Li, W., et al.: Approximate nearest neighbor search on high dimensional data — experiments, analyses, and improvement. IEEE Trans. Knowl. Data Eng. **32**(8), 1475–1488 (2020)
17. Llaveshi, A., Sirin, U., Ailamaki, A., West, R.: Accelerating B+tree search by using simple machine learning techniques. In: AIDB — VLDB Workshop on Applied AI for Database Systems and Applications (2019)

18. Macke, S., et al.: Lifting the curse of multidimensional data with learned existence indexes. In: Workshop on ML for Systems at NeurIPS, pp. 1–6 (2018)
19. Mic, V., Novak, D., Zezula, P.: Binary sketches for secondary filtering. ACM Trans. Inf. Syst. **37**(1), 1:1–1:28 (2019). https://doi.org/10.1145/3231936
20. Mikolov, T., Chen, K., Corrado, G., Dean, J.: Efficient estimation of word representations in vector space. arXiv preprint arXiv:1301.3781 (2013)
21. Moriyama, A., Rodrigues, L.S., Scabora, L.C., Cazzolato, M.T., Traina, A.J.M., Traina, C.: VD-tree: how to build an efficient and fit metric access method using Voronoi diagrams. In: Proceedings of the 36th Annual ACM Symposium on Applied Computing (SAC), p. 327–335. ACM, New York (2021)
22. Muja, M., Lowe, D.G.: Fast approximate nearest neighbors with automatic algorithm configuration. In: International Conference on Computer Vision Theory and Applications (VISAPP), pp. 331–340 (2009)
23. Müller, M., Röder, T., Clausen, M., Eberhardt, B., Krüger, B., Weber, A.: Documentation Mocap database HDM05. Technical report, CG-2007-2, Universität Bonn (2007)
24. Nathan, V., Ding, J., Alizadeh, M., Kraska, T.: Learning multi-dimensional indexes. In: Proceedings of the 2020 International Conference on Management of Data (SIGMOD), pp. 985–1000. ACM (2020)
25. Navarro, G., Reyes, N.: Dynamic spatial approximation trees. J. Exp. Algorithmics **12** (2008). https://doi.org/10.1145/1227161.1322337
26. Novak, D., Batko, M., Zezula, P.: Metric index: an efficient and scalable solution for precise and approximate similarity search. Inf. Syst. **36**, 721–733 (2011)
27. Novak, D., Batko, M., Zezula, P.: Large-scale image retrieval using neural net descriptors. In: Proceedings of the 38th International ACM SIGIR Conference on Research and Development in Information Retrieval, pp. 1039–1040. ACM (2015)
28. Novak, D., Zezula, P.: Rank aggregation of candidate sets for efficient similarity search. In: Decker, H., Lhotská, L., Link, S., Spies, M., Wagner, R.R. (eds.) DEXA 2014. LNCS, vol. 8645, pp. 42–58. Springer, Cham (2014). https://doi.org/10.1007/978-3-319-10085-2_4
29. Oosterhuis, H., Culpepper, J.S., de Rijke, M.: The potential of learned index structures for index compression. In: Proceedings of the 23rd Australasian Document Computing Symposium (ADCS) (2018). https://doi.org/10.1145/3291992.3291993
30. Pedregosa, F., et al.: Scikit-learn: Machine learning in Python. J. Mach. Learn. Res. **12**, 2825–2830 (2011)
31. Sablayrolles, A., Douze, M., Schmid, C., Jégou, H.: Spreading vectors for similarity search. In: 7th International Conference on Learning Representations, ICLR 2019, New Orleans, LA, USA, 6–9 May 2019. OpenReview.net (2019)
32. Vaswani, A., et al.: Attention is all you need. In: Advances in Neural Information Processing Systems, pp. 5998–6008 (2017)
33. Wang, H., Fu, X., Xu, J., Lu, H.: Learned index for spatial queries. In: 20th IEEE International Conference on Mobile Data Management (MDM), pp. 569–574 (2019)
34. Xiang, W., Zhang, H., Cui, R., Chu, X., Li, K., Zhou, W.: Pavo: a RNN-based learned inverted index, supervised or unsupervised? IEEE Access **7**, 293–303 (2019)

Towards a Learned Index Structure for Approximate Nearest Neighbor Search Query Processing

Maximilian Hünemörder(✉), Peer Kröger, and Matthias Renz

Institute for Computer Science, Christian-Albrechts-Universität zu Kiel, Kiel, Germany
{mah,pkr,mr}@informatik.uni-kiel.de

Abstract. In this short paper, we outline the idea of applying the concept of a learned index structure to approximate nearest neighbor query processing. We discuss different data partitioning approaches and show how the task of identifying the disc pages of potential hits for a given query can be solved by a predictive machine learning model. In a preliminary experimental case study we evaluate and discuss the general applicability of different partitioning approaches as well as of different predictive models.

1 Introduction

Nearest neighbor (NN) search is prevalent in many applications such as image retrieval, recommender systems, and data mining. In order to process a NN query efficiently appropriate data structures (usually called index structures) that enable identifying the result of a query by examining only a sub set of the entire data set are typically used. Additional speed-up can be gained by approximate nearest neighbor (ANN) search that trades accuracy for query time which is acceptable in many applications.

In this short paper, we examine the applicability of a new emerging paradigm, so-called learned index structures (LIS), for ANN query processing. The idea of LIS has been coined in [1] where the authors show that an index for 1D search keys (e.g. a B+-tree) is essentially similar to a regression model: the index induces an ordering of the keys and stores the data objects according to this ordering on disc pages (blocks). The corresponding learning task is, given the keys (observations) as training data, to train a predictive model (function) that determines the physical page address for each key. Processing a query is then simply applying the predictive model to the query key, i.e., predicting the addresses of the blocks (pages) on disc where the results of the query are located. While this approach works pretty well for primary key search, such as exact match queries and range queries on 1D data, we present one of the first works towards extending LIS to multi-dimensional spatial queries such as (A)NN queries.

This work aims at exploring the general applicability of LIS for multi-dimensional indexing with a focus on ANN queries. We discuss the two basic challenges any index structure has to solve (see also Fig. 1). First, the database needs to be partitioned in order to store the objects in a clustered way on disc pages. We propose a new partitioning that adapts to the real data distribution and is based on a specific k-Means clustering here,

N. Reyes et al. (Eds.): SISAP 2021, LNCS 13058, pp. 95–103, 2021.
https://doi.org/10.1007/978-3-030-89657-7_8

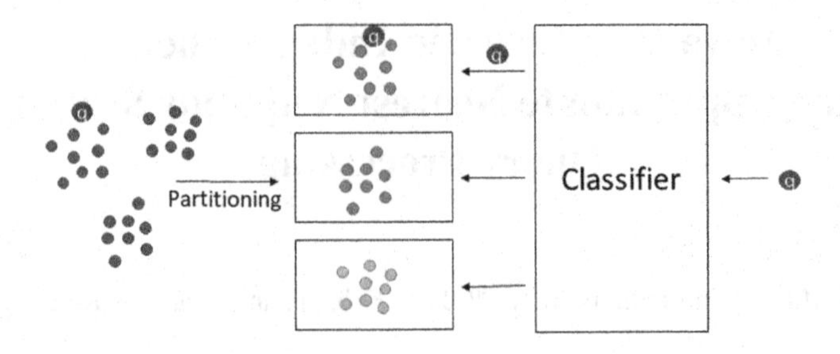

Fig. 1. A sketch of a spatial LIS: the data (left) is partitioned and these partitions are mapped onto disc pages. A predictive model (classifier) learns this mapping. Given a query object q, the model predicts the disc page containing the potential NN of q.

but any other partitioning scheme is possible, e.g. by simply taking the leaf nodes of any hierarchical index structure. Second, the relationship between observations (values of the data objects) and their corresponding disc page IDs are learned using a predictive model. An ANN query can be supported by applying the learned prediction function to the query object. Since the predictive model may be not 100% accurate, the predicted disc page may not contain the true nearest neighbor(s) and therefore only result in an approximation. We will discuss implications, potential extensions, etc. on this aspect in detail. This way, a LIS could offer a good compromise between existing indexing paradigms: it could combine

1. a data-centric partitioning which is usually done by hierarchical index structures such as search trees that typically suffer from higher query costs due to the traversal of the search tree,
2. a fast prediction of disc page IDs which can be generally achieved by hash functions that often suffer from data-agnostic partitioning which may lead to a large number of collisions (disc page overflows) and, as a consequence to higher query times.

The reminder is organized as follows. Section 2 discusses preliminaries and related work. We sketch an LIS for multi-dimensional ANN query processing in Sect. 3. A preliminary empirical evaluation is presented in Sects. 4, and 5 offers a summary and a discussion of directions for future research.

2 Background

2.1 ANN Query Processing: Preliminaries and Related Work

Given a query q, an number $k \in \mathbb{N}$ and a distance measure $dist$, a kNN query around q on a data set $\mathcal{D}$, $\text{NN}_k(q)$, retrieves the k objects having the smallest distance to q among all objects in $\mathcal{D}$ (ties need to be resolved). Without loss of generality, we set the query parameter $k = 1$ and omit it in the following. Sequentially scanning all data objects to retrieve the NNs involves loading all pages of the entire data file from disk. Since this is usually not acceptable performance-wise, many approaches for speeding up NN search using indexing techniques have been explored in recent years. A further way to achieve speed-ups is to trade performance for accuracy of the results using approximate algorithms that may report false hits. These ANN algorithms usually implement one of the following index paradigms:

Hierarchical indexes are typically based on balanced search trees [2–4] that recursively split the data space by some heuristics until a minimum number of objects remain in a partition. All nodes of the search tree are usually mapped to pages on disk. Searching theoretically requires $O(\log f)$ random page accesses on average for f data pages but the performance typically degrade with increasing data complexity.

Hashing such as locality sensitive hashing (LSH) and variants [5–8] applies one or more hash functions to map data objects into buckets (and store these buckets as pages on disc). If the number of objects in a bucket exceeds the maximum capacity of a page (e.g. due to an unbalanced partitioning), the objects are stored in any order on so-called "overflow pages" increasing the number of page accesses necessary to answer a given query. However, in the best case, query processing requires $O(1)$ page accesses.

Vector quantization and compression techniques (e.g. [9–11]) aim at reducing the data set size by encoding the data as a compact approximated representation such that (approximate) similarity among data objects is preserved.

A significant comparison of the different methods under varying realistic conditions is a generally challenging task. Thus, a benchmarking tool for ANN algorithms have been proposed in [12]. However, we do not aim for benchmarking LIS with other approaches here but rather explore the general applicability of LIS to ANN queries.

2.2 Learned Index Structures

The term LIS has been introduced by [1] where the authors show how to represent an index structures as a learning task. This pioneering work proposes a LIS for indexing 1D keys and supporting exact match and range queries. In recent years, the term LIS has been also used for methods that utilize machine learning techniques to support any aspect of query processing, e.g. [13] where kNN distance approximations are learned in order to support reverse NN queries, or [14] where the authors propose a new approach to generate permutations for permutation based indexing using deep neural networks. The most similar approach to ours can be found in [15] and [16] where the authors propose a learned metric index for ANN search. In contrast to our work they learn a whole tree of prediction models to index a metric space.

2.3 Contributions

LIS may offer the best of two worlds in spatial query processing, i.e., a data-centric, collision-free partitioning of the database and a search method that returns a result in constant time w.r.t. page accesses even in the worst-case. In this short paper, we explore the applicability of LIS to ANN query processing. In particular, we propose a general schema of a LIS for ANN query processing and implement this schema with existing techniques, e.g. k-means clustering for data partitioning. We present some first results on the performance of various predictive models from machine learning and derive implications for future work.

3 Towards a Learned Index for ANN Search

The data set $\mathcal{D}$ is stored on disk in blocks (pages) of a fixed capacity c. Thus, depending on c, $\mathcal{D}$ is distributed over a set $\mathcal{P}$ of p pages on disk. Processing an object $o \in \mathcal{D}$ in RAM requires to load the entire page $P_o \in \mathcal{P}$ on which o is stored.

The key to any search index is that the data objects are not randomly distributed over $\mathcal{P}$. Rather, objects that are similar to each other w.r.t. the distance $dist$ should be placed on the same page. There are many possible solutions for producing such a clustered partitioning, e.g. using the buckets of LSH, the leaf nodes of a search tree or use an unsupervised learning method. Here, we experimented with k-means clustering, which aims at partitioning the data into k disjoint clusters maximizing the compactness of these partitions. The idea is, to use k-means in such a way, that the number of points assigned to each cluster is constrained by a minimum capacity (for efficient storage usage) c_{min} and a maximum capacity C_{max} in order to map each cluster to one data page (C_{max} usually depends on c from above). Extensions such as Constraint k-means [17] are able to cope with these issues but are computationally very complex. Instead, in our study, we propose to just use traditional k-means clustering. The points assigned to a cluster $C_i (1 \leq i \leq k)$ are mapped to page $P_i \in \mathcal{P}$.

For query processing, we need to predict the page $P \in \mathcal{P}$, the query object q would have been placed on. This page likely contains the NN of q (depending on the partitioning, etc.). This prediction could be done by any machine learning model that can learn the mapping of an object to the corresponding disk page. Analogously to hashing, such a predictive model is a function

$$M : \mathbb{F} \rightarrow \mathcal{P}$$

from the feature space $\mathbb{F}$ of the data into the set of data pages that depends on some model-specific parameters θ_M. In general, we can learn (train) the corresponding parameters θ_M from $\mathcal{D}$ (and the corresponding partitioning $C_1, ..., C_k$). Given a query object $q \in \mathbb{F}$ and a predictive model M trained on $\mathcal{D}$, we can predict the disk page $P = M(q)$ by applying M on q. The page P can be loaded into main memory and the NN of q among all objects stored on P can be determined and returned as (approximate) result. Since our data partitioning does not produce overflow pages, we only need to access one page, i.e., $P = M(q)$. Thus, the time complexity is guaranteed to be in $O(1)$ in any cases (we can usually even assume that the model M fits into main memory). The accuracy of this procedure obviously depends on various aspects such as the accuracy of the prediction, the data partitioning, etc., some will be examined in Sect. 4. However, we consider the optimization of such aspects as an open challenge for future research, e.g. by aggregating more information from the partitions such as centrality measures, distance bounds, etc.

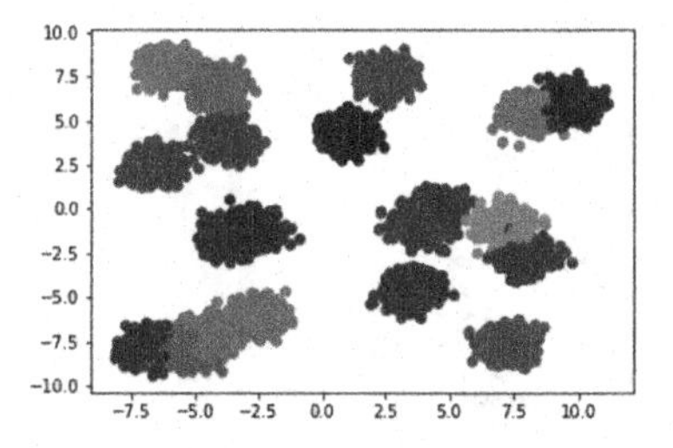

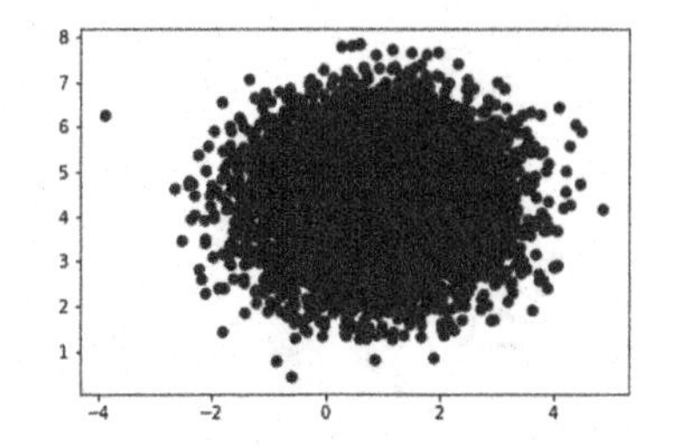

Fig. 2. Random 2D projections of sample clustered (left) and non-clustered (right) data.

4 Evaluation

4.1 Set-Up

In order to get a first impression of the proposed LIS for ANN query processing, we used synthetic data sets generated by the *make_blobs* function from sklearn[1]. In all experiments, we generated five different random datasets and report average results. We conducted two general runs w.r.t. the data distributions: clustered and non-clustered data. Figure 2 depicts arbitrary 2D projections of two sample data sets from both runs. We used 20-dimensional synthetic datasets consisting of 5000, 10000, 30000 and 50000 samples. The clustered datasets had 20 clusters with a cluster standart deviation of 0.5 and the non-clustered datasets have only a single Gaussian blob with a standard deviation of 1.0. Additionally, we used a low dimensional embedding of the popular MNIST data set generated by a fully connected Autoencoder (AE). Since this paper is a preliminary study of the general applicability of LIS to ANN search we did not yet compare to other ANN methods.

We used two different accuracy scores for evaluation. First, to explore the potential of the different predictive models to learn the mapping of objects to pages, we employed a classical train-validation split (called **validation accuracy**). Second, to measure the approximation accuracy of the query (called **test accuracy**), we used a withheld third sub-set of the data (not used in partitioning or training of the predictive model) as query objects, compared the results of these queries with the correct NN computed by a brute force search. The accuracy is determined by the ratio of the amount of zero distance hits and the amount of query objects. Additionally, we report the mean relative error for ANN search in our repository[2]).

For the partitioning step, we used the k-means implementation from sklearn. For comparison, we used the leaf nodes of a kd-tree (also from sklearn) as an alternative data partitioning. As predictive models, we used diverse classifiers from sklearn, including: Naïve Bayes, Decision Tree and Random Forest, Support Vector Machine (SVM) with a linear and an rbf kernel, and a simple dense multi-layer perceptron (MLP). For these preliminary experiments we did not perform hyper-parameter tuning but used reasonable default parameters. As a "Base Model", we assign each query object to its

[1] https://scikit-learn.org/stable/modules/generated/sklearn.datasets.make_blobs.html.
[2] https://github.com/huenemoerder/kmean-lis.git.

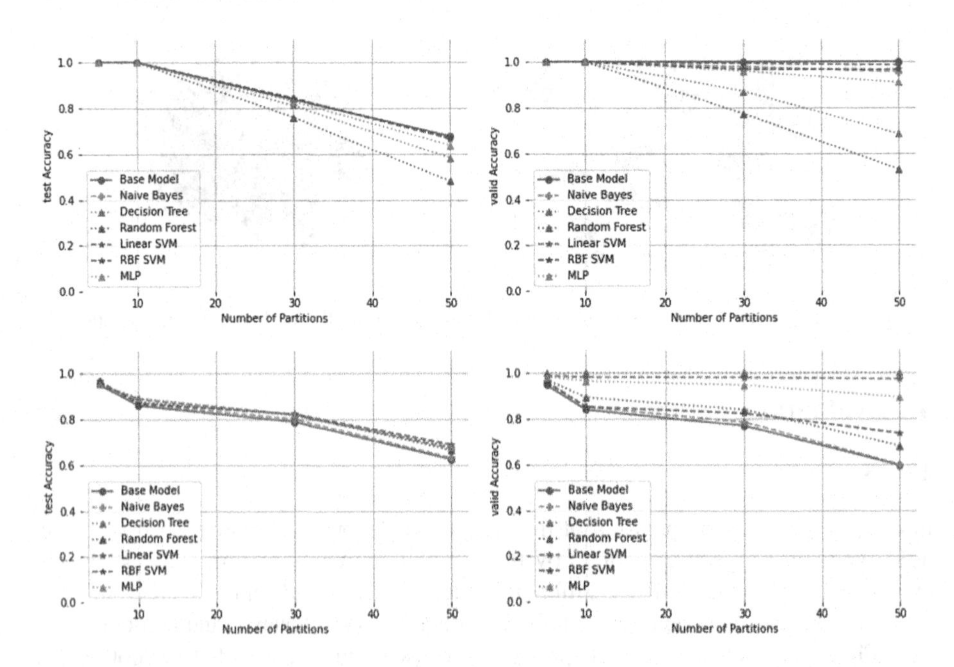

Fig. 3. Test accuracy (left charts) and validation accuracy (right charts) on **clustered** data sets (upper charts: k-means partitioning; lower charts: kdtree partitioning).

closest centroid of the corresponding partition (validation accuracy of 1.0 by design). The ''size'' of this model grows linearly with the number of partitions, i.e., database size, and is expected to not fit into the cache (requiring additional page accesses on application). The AE for the MNIST data set was implemented in pytorch[3] with only one single linear layer that maps the flattened images (784 dimensional array) to a latent space vector of 32 dimensions (usin Leaky ReLU as activation).

4.2 Results

We analysed the relationship between the test accuracy and the number of samples and number of partitions, i.e., data pages. In all runs, we kept the capacity of pages fixed but changed the number of data points n accordingly. Figure 3 displays this relationship on clustered data sets. In general, we can see that both the test accuracy and the validation accuracy drops with increasing number of partitions. This is somehow intuitive: with increasing number of partitions (and data points), the mapping that has to be learned by the predictive model becomes more and more complex. It is interesting to note that for most models the validation error (right charts) remains better than the test accuracy (left charts), i.e. even though, the mapping is learned well, the true NNs for the query objects are approximated not quite as well. In these cases, the partitioning model seems to not optimally fit the real data distribution and therefore even with a perfect predictive model some queries can be placed in an unsuitable data page. This is also reflected in

[3] https://pytorch.org/.

the fact that the kd-tree partitioning performs even worse in terms of test accuracy, since the clustered dataset was created in a way that favours k-means. We can also observe that the Decision Tree classifier shows perfect validation accuracy for the kd-tree partitioning, while showing the worst performance for k-means. This suggests that choosing a fitting pair of prediction and partitioning algorithm is vital to at least result in a high validation accuracy. These observations are further confirmed by the non-clustered data sets (the results can be found in our repository[4]). Additionally, this is further reflected in our results on MNIST in Table 1, where the test accuracies for the kdtree paritioning are significantly worse than the ones for k-means. Generally further experiments and benchmarking are obviously necessary to obtain more significant results.

Table 1. Results on MNIST data set ($k$menas partitioning)

	k-means		KDTree	
Classifier	Validation accuracy	Test accuracy	Validation accuracy	Test accuracy
Base Model	**1.000**	0.8808	0.5407	0.4974
Naïve Bayes	0.9140	0.8479	0.6140	0.5409
Decision Tree	0.8560	0.8121	**0.9997**	0.6160
Random Forest	0.7315	0.7089	0.4610	0.4066
Linear SVM	0.9973	0.8800	0.9630	0.6165
RBF SVM	0.9845	**0.8810**	0.8588	**0.6388**
MLP	0.9455	0.8736	0.8678	0.5994

5 Summary

In this short paper, we applied the idea of LIS to ANN query processing and examined its general applicability to this problem. We explored a new data partitioning based on k-means clustering and applied the standard predictive models from machine learning in a simple set up. The results are generally promising for synthetic (clustered/non-clustered) and real data such that we think it is worth putting more future focus on LIS. For example, exploring new ways for data partitioning including a more thorough evaluation of different existing partitioning schemes could be interesting. Also, understanding the relationship between data characteristics, properties of the partitioning, and the accuracy of different predictive models could be a promising research direction that may lead to approaches that better integrate partitioning and learning. Additionally, exploring postprocessing methods to increase accuracy, e.g. use additional information from training as well as from the partitioning like distance bounds would be helpful. Last not least, the application of LIS to other types of similarity queries is still an open research question.

[4] https://github.com/huenemoerder/kmean-lis.git.

References

1. Kraska, T., Beutel, A., Chi, E.H., Dean, J., Polyzotis, N.: The case for learned index structures. In: Proceedings of the International Conference on Management of Data (SIGMOD), Houston, TX, pp. 489–504 (2018)
2. Ciaccia, P., Patella, M., Zezula, P.: M-tree: an efficient access method for similarity search in metric spaces. In: Proceedings of the International Conference on Very Large Databases (VLDB) (1997)
3. Sakurai, Y., Yoshikawa, M., Uemura, S., Kojima, H., et al.: The A-tree: an index structure for high-dimensional spaces using relative approximation. In: Proceedings of the International Conference on Very Large Databases (VLDB), pp. 5–16 (2000)
4. Amsaleg, L., Jónsson, B.Þ, Lejsek, H.: Scalability of the NV-tree: three experiments. In: Marchand-Maillet, S., Silva, Y.N., Chávez, E. (eds.) SISAP 2018. LNCS, vol. 11223, pp. 59–72. Springer, Cham (2018). https://doi.org/10.1007/978-3-030-02224-2_5
5. Christiani, T.: Fast locality-sensitive hashing frameworks for approximate near neighbor search. In: Amato, G., Gennaro, C., Oria, V., Radovanović, M. (eds.) SISAP 2019. LNCS, vol. 11807, pp. 3–17. Springer, Cham (2019). https://doi.org/10.1007/978-3-030-32047-8_1
6. Jafari, O., Nagarkar, P., Montaño, J.: mmLSH: a practical and efficient technique for processing approximate nearest neighbor queries on multimedia data. In: Satoh, S., et al. (eds.) SISAP 2020. LNCS, vol. 12440, pp. 47–61. Springer, Cham (2020). https://doi.org/10.1007/978-3-030-60936-8_4
7. Jafari, O., Nagarkar, P., Montaño, J.: Improving locality sensitive hashing by efficiently finding projected nearest neighbors. In: Satoh, S., et al. (eds.) SISAP 2020. LNCS, vol. 12440, pp. 323–337. Springer, Cham (2020). https://doi.org/10.1007/978-3-030-60936-8_25
8. Ahle, T.D.: On the problem of p_1^{-1} in locality-sensitive hashing. In: Satoh, S., et al. (eds.) SISAP 2020. LNCS, vol. 12440, pp. 85–93. Springer, Cham (2020). https://doi.org/10.1007/978-3-030-60936-8_7
9. Weber, R., Schek, H.J., Blott, S.: A quantitative analysis and performance study for similarity-search methods in high-dimensional spaces. In: Proceedings of the International Conference on Very Large Databases (VLDB), pp. 194–205 (1998)
10. Ferhatosmanoglu, H., Tuncel, E., Agrawal, D., Abbadi, A.E.: Vector approximation based indexing for non-uniform high dimensional data sets. In: Proceedings of the ACM International Conference on Information and Knowledge Management (CIKM), McLean, VA, pp. 202–209 (2000)
11. Houle, M.E., Oria, V., Rohloff, K.R., Wali, A.M.: LID-fingerprint: a local intrinsic dimensionality-based fingerprinting method. In: Marchand-Maillet, S., Silva, Y.N., Chávez, E. (eds.) SISAP 2018. LNCS, vol. 11223, pp. 134–147. Springer, Cham (2018). https://doi.org/10.1007/978-3-030-02224-2_11
12. Aumüller, M., Bernhardsson, E., Faithfull, A.: ANN-benchmarks: a benchmarking tool for approximate nearest neighbor algorithms. In: Beecks, C., Borutta, F., Kröger, P., Seidl, T. (eds.) SISAP 2017. LNCS, vol. 10609, pp. 34–49. Springer, Cham (2017). https://doi.org/10.1007/978-3-319-68474-1_3
13. Berrendorf, M., Borutta, F., Kröger, P.: k-distance approximation for memory-efficient RkNN retrieval. In: Amato, G., Gennaro, C., Oria, V., Radovanović, M. (eds.) SISAP 2019. LNCS, vol. 11807, pp. 57–71. Springer, Cham (2019). https://doi.org/10.1007/978-3-030-32047-8_6
14. Amato, G., Falchi, F., Gennaro, C., Vadicamo, L.: Deep permutations: deep convolutional neural networks and permutation-based indexing. In: Amsaleg, L., Houle, M.E., Schubert, E. (eds.) SISAP 2016. LNCS, vol. 9939, pp. 93–106. Springer, Cham (2016). https://doi.org/10.1007/978-3-319-46759-7_7

15. Antol, M., Ol'ha, J., Slanináková, T., Dohnal, V.: Learned metric index-proposition of learned indexing for unstructured data. Inf. Syst. **100**, 101774 (2021)
16. Slanináková, T., Antol, M., Ol'ha, J., Vojtěch, K., Dohnal, V.: Data-driven learned metric index: an unsupervised approach. In: International Conference on Similarity Search and Applications, Springer (2021, to appear)
17. Bennett, K., Bradley, P., Demiriz, A.: Constrained k-means clustering. In: Technical Report MSR-TR-2000-65, Microsoft Research (2000)

Similarity vs. Relevance: From Simple Searches to Complex Discovery

Tomáš Skopal[1(✉)], David Bernhauer[1,2], Petr Škoda[1], Jakub Klímek[1], and Martin Nečaský[1]

[1] SIRET Research Group, Faculty of Mathematics and Physics, Charles University, Prague, Czech Republic
{tomas.skopal,david.bernhauer,petr.skoda,jakub.klimek, martin.necasky}@matfyz.cuni.cz

[2] Faculty of Information Technology, Czech Technical University in Prague, Prague, Czech Republic

Abstract. Similarity queries play the crucial role in content-based retrieval. The similarity function itself is regarded as the function of relevance between a query object and objects from database; the most similar objects are understood as the most relevant. However, such an automatic adoption of similarity as relevance leads to limited applicability of similarity search in domains like entity discovery, where relevant objects are not supposed to be similar in the traditional meaning. In this paper, we propose the meta-model of data-transitive similarity operating on top of a particular similarity model and a database. This meta-model enables to treat directly non-similar objects $\mathbf{x}$, $\mathbf{y}$ as similar if there exists a chain of objects $\mathbf{x}$, i_1, ..., i_n, $\mathbf{y}$ having the neighboring members similar enough. Hence, this approach places the similarity in the role of relevance, where objects do not need to be directly similar but still remain relevant to each other (transitively similar). The data-transitive similarity concept allows to use standard similarity-search methods (queries, joins, rankings, analytics) in more complex tasks, like the entity discovery, where relevant results are often complementary or orthogonal to the query, rather than directly similar. Moreover, we show the data-transitive similarity is inherently self-explainable and non-metric. We discuss the approach in the domain of open dataset discovery.

1 Introduction

When searching data, we can choose from a multitude of available models and paradigms. Some models assume exact data structure and semantics, such as the relational database model (and SQL) or graph database model (RDF+SPARQL, XML+XQuery). In such models, the relevance of a data entity to a particular query is binary (relevant/not relevant); specified by a binary predicate. The precision and recall in retrieval of structured data is always 100% as there is no uncertainty expected. Also, structured query languages offer high expressive power that allows the user to specify the relevance of data in many ways.

N. Reyes et al. (Eds.): SISAP 2021, LNCS 13058, pp. 104–117, 2021.
https://doi.org/10.1007/978-3-030-89657-7_9

On the other side of the data universe, when searching in unstructured or loosely structured data (like multimedia, text, time series), we do not have enough a-priori information on how to model the data features for exact search. In such situation the similarity search models could be used, representing a universal way of content-based retrieval in unstructured data. Instead of formulating a structured query aiming at binary relevance, in similarity search we use a ranking of the database objects determined by their similarity score to a query example (the query-by-example paradigm). Hence, the relevance is relaxed from binary to multiple-value. When compared to retrieval of structured data, the similarity search is more like an "emergency solution" for unstructured data. The expressive power of similarity queries is limited to a ranking induced by numeric aggregation of differences between the query example and the database objects; keeping it a black-box search for the user. The low expressive power of the query-by-example paradigm leads to a paradox – we search for what we already have. Specifically, we query for as good results as possible, having the best result already at hand – the query example. Of course, in practical applications the query-by-example paradigm makes sense, because the query example itself does not contain the whole information we search for. For instance, searching by the photo of Eiffel tower we not only get another Eiffel tower image, but also some context (the Wikipedia web page the result image was embedded in). Nevertheless, the context (external information attached to data) does not remove the essence of the paradox – based purely on the similarity of results, the query example itself is always the best result[1].

Historically, the low expressive power of similarity search has been accepted in the major application area – the multimedia retrieval. Here the semantics to be captured in multimedia objects (the descriptors) is rather vague, general and bound to human common knowledge. The similarity search is thus a perfect method for multimedia retrieval as the similarity concept itself is vague and general (and so is the human cognition – the inspiration for similarity search). When combined with descriptor models employing high-level "canonized" semantics, such as the bag of words using the vocabulary of deep features [11], then even the cosine similarity can perform well. Unfortunately, the domain experts are not always so lucky to work with nicely shaped semantic descriptors, while then the low expressive power of similarity search is fully revealed. A solution to this could be a proposal of similarity-aware relevance of data objects to an example object (query) that enables much more complex aggregation than just evaluating the direct similarity (the "exampleness" of the results). If we find a way of how to extend the concept of similarity into a relevance, we would be able to use the existing similarity search methods in more expressive retrieval scenarios. For example, consider a fashion e-shop where a user searches for a product by an example image, e.g., shoes. The result could not only consist of similar shoes, but it could also return related accessories (handbag, belt) sharing some design features with the shoes [14].

[1] Let's omit another problem; where to acquire such a "holy grail" example in real-world problems.

In the following, we continue the discussion in the specific domain of open datasets discovery. Unlike in multimedia retrieval, where direct audiovisual similarity to a query usually leads to good results, in open datasets with sparse descriptors we often do not find anything directly (non-trivially) similar. Here the similarity extended towards more general relevance could improve the retrieval effectiveness in a fundamental way.

1.1 Discovery of Open Datasets by Similarity

The similarity search models can utilize not only content features but also metadata (if available). The focus on metadata can be efficient and effective in domains where the content of the objects is too heterogeneous so that it is hard to extract features for measuring similarity (or relevance). On the other hand, such objects could be catalogued by a community to enable search of the objects by metadata.

This is the case of the domain of open datasets search and discovery [12]. There are various datasets published on the internet which are catalogued in open data catalogs [18]. They are extremely heterogeneous in structure and semantics so that modeling them by content is nearly impossible (consider tables and spreadsheets without schema, full-text reports, database dumps, geographical and map data, logs, etc.). Open data catalogs provide descriptive metadata about the datasets in a single place where potential consumers can search for datasets. However, the problem of metadata is that they are often sparse and poor. In the open data domain, dataset publishers usually limit their descriptive metadata to briefly describe the core semantics of their datasets (by title, keywords, text description). No broader context of a dataset including some description of its relationships to other datasets is specified in the metadata. Using such sparse metadata for similarity retrieval is therefore limited. We confirmed this in our previous work [26] where we showed that various similarity methods do not perform very well when applied to the descriptive metadata of open datasets.

In our experiments, we noticed situations where two datasets are relevant to each other but none of the similarity models is able to identify this relevance. Let us demonstrate this on a concrete example of open data published by public authorities in Czechia. The datasets are catalogued in the National Open Data Catalog (NODC)[2]. There are two datasets entitled *IDOL Integrated Transport System Tariff Zones* and *Traffic intensity on sections of motorways*. The similarity of both datasets based on their metadata descriptions is low according to various similarity models presented in [26]. However, when we reviewed the datasets manually we found out that they are very relevant to each other. The first one is related to public transport. The second one is related to transport on motorways. So when users find one of the datasets, they would like to get also the other dataset as well. What makes them relevant to each other is the background semantics which is not directly expressed in the descriptive metadata. Since it is not expressed in the metadata, no similarity model can work with

[2] https://data.gov.cz/english/.

this. However, there is a third dataset in NODC titled *BKOM transport yearbook*. The similarity models identify its similarity with the original two datasets on the base of available metadata. So using the third dataset we could say that the two original datasets are relevant to each other because they are both similar to the third one. In other words, they are *transitively similar* when using other datasets as a context. What is also interesting in the example is that metadata about the third dataset express explicitly the concept of transport. So, the third dataset is not just an intermediary dataset between the two. It explains why they are relevant, contributing thus to the discussion on *explainability* of similarity search.

2 Related Work

Before presenting the meta-model of data-transitive similarity, we discuss several related points.

2.1 Similarity Modeling

The research in the similarity search area had intensified some three decades ago by setting the metric space model as the golden standard [25]. The metric distances in place of (dis)similarity functions were introduced purely for database indexing reasons (i.e., for fast search). Though a good trade-off for many problems, the metric space model remains quite restrictive for modeling similarity. The restrictions are even more strict in follow-up models aiming at improving search efficiency, such as the ptolemaic [15] or supermetric [9] models. As mentioned in the previous section, this might not be a problem in case the descriptors are canonized and semantic (such as histograms referring to a vocabulary of deep features). However, for the lower-semantic cases there were alternative approaches to indexing similarity proposed in the past 15 years, ranging from dynamic combinations of multiple metrics [5] for multi-modal retrieval to completely unrestricted, non-metric approaches [23]. The rationale for their introduction was to increase the expressive power of similarity search (and effectiveness) and still provide an acceptable retrieval efficiency.

2.2 Retrieval Mechanisms

No matter if we choose metric or non-metric similarity, the expressive power of retrieval is also affected by the retrieval mechanism used. The query-by-example paradigm constitutes the basic functionality of similarity search in form of kNN or range queries. The similarity joins enable the use of similarity within the database JOIN operators [22]. The similarity queries could be also used with additional post-processing techniques for multi-modal retrieval and analytics, such as the late fusion [21] and content-based recommender systems [1]. Last but not least, there appear proposals and frameworks helping with the integration of similarity search constructs into query languages, such as SimilarQL [24], or MSQL [19]. The ultimate goal is to establish higher-level declarative query models for similarity search [3].

2.3 Dataset Discovery

Finding related datasets, also known as dataset discovery, is one of the important tasks in data integration [20]. Large companies such as Google have developed their own dataset search techniques and solutions [4]. New solutions for dataset search in specific domains started to appear recently. For example, *Datamed* [8] is an open source discovery index for finding biomedical datasets. The existing works emphasize the role of quality metadata for dataset findability while [6] points out that available metadata does not always describe what is actually in a dataset and whether a described dataset fits for a given task. Other studies [12,13,16] confirm that dataset discovery is highly contextual depending on the current user's task. The studies show that this contextual dependency must be reflected by the dataset search engines. This makes the task of dataset discovery harder as it may not be sufficient to search for datasets only by classical keyword-based search. More sophisticated approaches being able to search for similar or related datasets could be helpful in these scenarios. As shown by [6,20] many existing dataset discovery solutions are based on simple keyword search. Discovery of datasets by similarity is discussed in the recent survey [6]. Several papers propose dataset retrieval techniques based on metadata similarity. In [2] a method is described which enables to measure similarity between datasets on the base of papers citing the datasets and a citation network between datasets. In [10] four different metadata-based models are evaluated for searching spatially related datasets, i.e., datasets which are related because of the same or similar spatial area covered. To the best of our knowledge, none of the approaches does apply the following technique of data-transitive similarity in dataset discovery.

3 Data-Transitive Similarity

In this section, we introduce the meta-model of data-transitive similarity. The original inspiration was the omnipresent database operation JOIN, used in many data management use cases for interconnecting relevant pieces of information. In relational databases the join operations allow to connect data records by means of shared attribute(s). In an extensive interpretation, the mechanism in database joins has roots in an identification of relevant entities by partial matches (equality predicate) or by partial similarity (inequality predicate). Analogously, by introducing data-transitive similarity we aim at consecutively joining similar objects and evaluating the overall relevance as an aggregation over the partial similarity scores.

The basic assumption of data-transitive similarity is thus a chain of objects from the database that are similar to each other, but the beginning and end of the chain could be quite dissimilar (yet relevant). Remember the well-known example with the human and the horse, illustrating the violation of the triangle inequality [23]. These two creatures tend to be quite dissimilar, yet they can be relevant (transitively similar). The relevance here can be ensured by a connecting object in the middle of the chain – a horseman, or more poetically a centaur, creature that is half man and half horse. The data-transitive similarity itself, however, can be more complex; the connecting agent may not be a single object, but a whole chain of objects. This chain also serves as an explanation of why the two objects are relevant and in what context (addressing the explainability issue).

The connection itself can be formalized as an aggregation of several consecutive ground distances. The Eq. 1 defines general form of data-transitive distance function $\hat{d}$, where $\mathcal{D}$ is a set of objects (the database in practical applications), d is a ground distance (the direct similarity), n is the length of the chain. Operator $\bigodot$ is an outer aggregation over all permutations of length n over elements of database $\mathcal{D}$ (e.g., min, max, avg). Operator $\biguplus$ is an inner aggregation over the individual direct distances within a particular chain. Table 1 shows examples of various inner aggregation functions. They are also the aggregation functions we worked with in our preliminary experiments. A more complex alternative may be a combination of several kinds of aggregations or distances.

$$\hat{d}_{\uplus}^{\odot,n}(\mathbf{x},\mathbf{y}) = \bigodot_{(i_1,\ldots,i_n)\in\mathcal{D}^n} \biguplus \left(d(\mathbf{x},i_1), d(i_1,i_2), \ldots, d(i_n,\mathbf{y})\right) \quad (1)$$

Table 1. Examples of inner aggregation $\biguplus$.

$\mathrm{sum}(\delta_0,\delta_1,\ldots,\delta_n)$	=	$\sum_{j=0}^{n}\delta_j$
$\min(\delta_0,\delta_1,\ldots,\delta_n)$	=	$\min\{\delta_0,\delta_1,\ldots,\delta_n\}$
$\max(\delta_0,\delta_1,\ldots,\delta_n)$	=	$\max\{\delta_0,\delta_1,\ldots,\delta_n\}$
$\mathrm{prod}(\delta_0,\delta_1,\ldots,\delta_n)$	=	$\prod_{j=0}^{n}\delta_j$
$\mathrm{iprod}(\delta_0,\delta_1,\ldots,\delta_n)$	=	$1-\prod_{j=0}^{n}(1-\delta_j)$

To summarize, we define the data-transitive similarity $\hat{d}$ as a meta-model operating on top of a ground similarity model d and a particular database $\mathcal{D}$. The computation of a single data-transitive distance involves a series of similarity queries over the database. The computational complexity of the data-transitive similarity thus involves not just the complexity of d but also the size of the database $|\mathcal{D}|$. Depending on the implementation, the worst-case time complexity $O(\hat{d})$ can vary from $O(d)$ to $O(d)O(|\mathcal{D}|^n)$, assuming n as a constant or $n \ll |\mathcal{D}|$.

From the definitions above it immediately follows that data-transitive distances are not metric distances – not only due to the possibly non-linear combination of the particular ground distances, but mainly due to the database-dependent nature of the distance topology (non-uniform distribution of points in the data universe and its impact on the chain members).

One might say that such advanced relevance constructions should not be modeled at the level of similarity, as they are part of higher retrieval models closer to the application level (e.g., a part of content-based recommender system). However, we want to stress that we intentionally included the data-transitive similarity into the family of generic pair-wise non-metric similarities. As such, it can be plugged into any search engine that supports non-metric similarities. This would not be possible if designed as a proprietary late-fusion retrieval model.

3.1 Implementation

The fundamental problem we have addressed in the data-transitive similarity design was determining the number of intermediaries (the chain length n) to form a transitive similarity. Although our model assumes an arbitrary n, determining the specific value is not a straightforward problem itself. A significant issue may be that for some objects, there is no intermediary to form transitive similarity. In general, the number of intermediaries may not be constant, and for different objects this value needs to be chosen dynamically.

Thus, for our experiment, we have applied a simplification in this regard and assume that data-transitive similarity has at most one intermediary (i.e., $n = 1$). Therefore, we always have a triplet: a query, an intermediary, and a result. This decision reduces the number of hyperparameters with respect to longer chains (e.g., number of intermediaries, different aggregation functions). This approach also has the advantage of a higher level of explainability. For longer chains of intermediaries, we need to discuss whether each part of the sequence makes sense for given transitivity. Whereas in the case of a single intermediary, we can argue with a reasonable certainty whether the query and result are relevant from the perspective of the intermediary explanation.

The second problem is the transitivity involving duplicates or near-duplicates in the chain – intermediaries very d-close to the query or to the result. Such duplicate intermediaries usually do not add any value. Therefore, small distances d (the first 5% of distance distribution) are not considered (in fact, all such distances are set to infinity to become disqualified in $\hat{d}$).

Third, all ground distances are required to be normalized to 0–1 because some aggregations ($\uplus$ = prod, $\uplus$ = iprod) require a bounded distance. In our implementation, we do not implement any optimizations, while to compute the data-transitive similarity we need to iterate over all database objects in the role of an intermediary. At the moment, optimizations for reduction of the set of intermediaries are beyond the subject of our research.

3.2 Open Dataset Testbed

For the open dataset testbed presented in Sect. 1.1, we considered title, description, and keywords metadata. Since the original data provided by the National Open Data Catalog are in the Czech language, we used the automatic English translation [17], followed by the words lemmatization and filtering non-meaningful words (we consider only nouns, adjectives, verbs, and adverbs). In addition, we ignored several experimentally detected stop-words (data, dial, export, etc.). The metadata descriptors were represented in the bag of words model (BoW) with tf-idf weights.

Over these descriptors, the ground cosine distance was computed as $d_{\text{cos}}(\mathbf{x}, \mathbf{y}) = 1 - s_{\text{cos}}(\mathbf{x}, \mathbf{y})$ (where s_{cos} is cosine similarity) for all pairs of objects (all pairs of datasets in our case). Figure 1 shows the distribution of distances d_{cos} over this testbed. We can see that most of the datasets are not d_{cos}-similar, and the testbed exhibits high intrinsic dimensionality [7]. This is due to the relatively sparse metadata (average about 20 words). For some datasets, some parts, such as description or keywords are empty; there is only the title description.

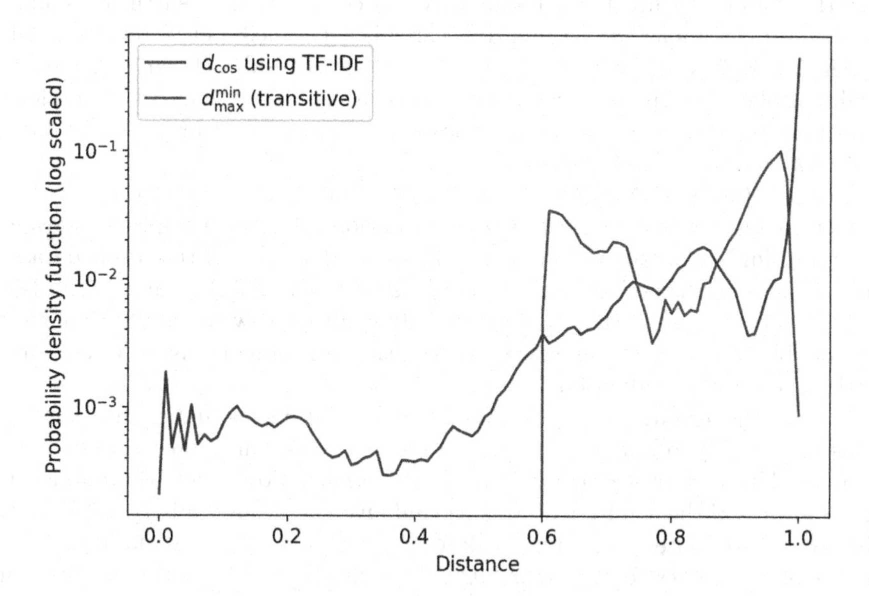

Fig. 1. Distance distribution of d_{cos} and $\hat{d}_{\text{max}}^{\text{min}}$ transitive similarity

In our experiment, we took only one intermediary, while $\hat{d}_{\text{max}}^{\text{min}}$ (Formula 2) was chosen as the data-transitive similarity function, since it exhibited the most robust aggregation in our preliminary experiments. Figure 1 shows how the distribution of $\hat{d}_{\text{max}}^{\text{min}}$-distances is different when compared to d_{cos}. Smaller distances (below approx. 0.6) are eliminated due to the removal of near-duplicate dataset pairs (set to 5% closest datasets), as mentioned in the previous subsection. The rest of the $\hat{d}_{\text{max}}^{\text{min}}$-distance domain is split into two categories representing relevance (more relevant around 0.7, less relevant around 0.9), with many d_{cos}-dissimilar datasets moving into the category of more $\hat{d}_{\text{max}}^{\text{min}}$-relevant datasets.

$$\hat{d}_{\text{max}}^{\text{min}}(\mathbf{x}, \mathbf{y}) = \min_{\forall i \in \mathcal{D}} \max \{d(\mathbf{x}, i), d(i, \mathbf{y})\} \tag{2}$$

4 Evaluation

As we have already discussed in [26], the findability evaluation in the open dataset discovery is complicated from several points of view. The database contains a relatively large number of datasets, but there is no sufficient ground truth for dataset similarity. To overcome the lack of ground truth, in this paper we evaluate the concept of relevance which is closer to dataset discovery, rather than direct context-independent similarity of datasets.

4.1 Methodology

Our evaluation targets the additional value of data-transitive similarity search over the standard (direct d_{cos}) similarity search. First, the search for similar datasets using standard d_{cos}-similarity search is performed. Let us represent this search as a k_d NN query, where k_d is the number of results. Then, there are k_t results displayed to the user using data-transitive similarity based k_t NN query, while filtering out results of the previous k_d NN query. For our experiment, we assume $k_d = 100$ and $k_t = 20$.

The user (evaluator) is given a list of triplets (query, intermediary, result) and then evaluates each such triplet as relevant or non-relevant. A triplet is relevant if the user finds a possible use case for the query dataset and the result dataset and, at the same time, the intermediary dataset reasonably connects the two datasets. Let us repeat that the user is only confronted with results that were not findable by standard (direct) similarity search. A total of 5 users (evaluators) participated in the evaluation.

During the evaluation, we encountered the problem that some pairs of datasets are only relevant if we ignore specific fine-grained attributes of the datasets. The first observed attribute is the information about the publisher, e.g., contracts of the Ministry of Finance and invoices of the Ministry of Finance. The second attribute is the time or date of repeatedly published datasets, e.g., the list of companies for the year 2020. The third attribute is the localization specified in the datasets, e.g., hospitals in Prague vs. hospitals in Brno. For the evaluation, we decided to ignore these attributes as they only contribute to fragmentation of the datasets that are otherwise relevant to each other. However, this problem might disappear if we consider more than just one intermediary in the data-transitivity model (subject of future evaluations).

As part of the experiment, we evaluated the relevance of the results for a set of prepared queries. This set was created based on previous experiments presented in [26]. A total of 64 transitive results were found for 11 different queries.

4.2 Results

During the evaluation, we looked at two main criteria: consistency and effectiveness. For every triplet, we have computed its score as sum of 0 (non-relevant) and 1 (relevant) ratings of all evaluators. In our case, the score ranges from 0 (all evaluators claim the triplet is non-relevant) to 5 (all evaluators claim the triplet is relevant). Figure 2 (left) shows the number of triplets with particular score, Fig. 2 (right) shows the number of triplets per data-transitive distance ranges and distribution of scores inside these ranges.

The consistency is validated based on the evaluators' agreement on the relevance of the evaluated triplets. Figure 2 (left) shows that in almost 78.13% of the cases, majority of evaluators (scores 0–1 and 4–5) agreed on the triplets' relevances. This observation confirms that the overall evaluation results are not just random noise.

Effectiveness is measured as the ratio of relevant datasets to all returned results. This gives us a measure of how much data-transitive similarity can improve the standard search. At Fig. 2 (left), we see that in 57.81% of the cases, the triplet was marked as relevant by a majority of evaluators (score 4–5).

Although the overall effectiveness may not seem significant, we must stress that all the relevant results found were not achievable by the direct similarity search (as already mentioned in Sect. 4.1). For 65.63% of the datasets, d_{cos} distances to query are maximal. We can also notice in Fig. 2 (right) that the data-transitive similarity model complies with the general thesis of similarity search (more distant datasets are less relevant and vice versa).

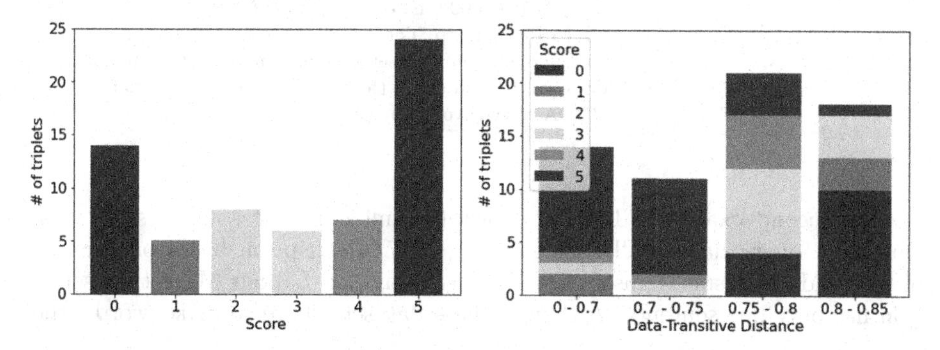

Fig. 2. The left figure shows the distribution of triplet ratings (how many triplets were rated by a particular relevancy score). For example, the score = 3 means that 3 evaluators thought the triplet was relevant (they rated it 1) and 2 evaluators thought the triplet was not relevant (they rated it 0). The right figure shows the distribution of ratings according to each data-transitive distance interval.

4.3 Qualitative Analysis

In Table 2 we see an example of triplet (Q, I, R) that was evaluated as relevant in our experiment (small data-transitive distance $\hat{d}_{\max}^{\min}(Q, R)$ through I). If we analyze the distance structure, the query dataset (Q) "Floods in the 19th century" does not have the "water" keyword in the metadata. However, thanks to the intermediary "5-year water" dataset (I), we have both "water" and "flood" in metadata and so the query dataset is transitively similar to the result dataset (R) "Water reservoirs". In the original similarity (the direct ground distance d_{cos}), the query Q and the result R datasets have maximum distance; they have nothing in common. In the data-transitive similarity search, however, the dataset R is within the first 20 results thanks to the connection with I. The relevance here can be explained by the fact that reservoirs can affect flooding and so the dataset R might be useful in flood prevention planning.

Table 2. Example of Query, Intermediary, Result triplet: floods vs water. Title, keywords and description metadata are provided for each dataset.

	Title	
	Keywords	Description
	Floods in the 19th century	
Q	Floods, Environment, GIS	Flooded areas in a 19th century flood in the Pilsen region
	5-year water	
I	GIS, Floods, Environment	Flooding areas of n-year water in the Pilsen region
	Water reservoirs under the management of the river basin and the forest of the Czech Republic under the territorial jurisdiction of the river Vltava	
R	water tanks, water management	The shp file contains points representing water reservoirs whose permitted volume of buoyant or accumulated water exceeds 1 000 000 m3 or to which the Forests of the Czech Republic, p. The registers are updated continuously, the dataset only once a year. The current data can be viewed on the water information portal VODA – www.voda.gov.cz

The second example (Table 3) shows the imbalance of some descriptions, where the query dataset "Housing Young 2017" description has 3 paragraphs of text and the result dataset "BUG[3] - Economy and Labour Market" description has only one sentence. Although these datasets share some keywords, the

Table 3. Example of Query, Intermediary, Result triplet: housing vs labour. Title, keywords and description metadata are provided for each dataset.

	Title	
	Keywords	Description
	Housing Young 2017	
Q	sociology, housing research, housing young, housing, Brno	The main objective of the Youth Housing survey conducted in 2017 was to identify and describe the housing needs of young people living in Brno, as well as their preferences in this area. ... *3 paragraphs of text here* ...
	BUG - people and housing	
I	Brno urban Grid, housing, people, BUG	Datasets from the Brno Urban Grid - theme people and housing
	BUG - Economy and Labour Market	
R	BUG, labour market, economy, Brno Urban Grid	Datasets from the Brno Urban Grid application - theme of economy and labour market

[3] BUG = Brno Urban Grid.

resulting position in ranking is too far when using the direct distance d_{cos}, so that the user cannot find the dataset. With the data-transitive similarity using the intermediary "BUG - people and housing" dataset the problem is mitigated. In this case, we are able to explain the relevance between the housing of young people and the state of the labour market.

5 Conclusion and Future Work

We proposed an extended concept of similarity search by introducing the meta-model of data-transitive similarity operating on top of a particular similarity model. In the evaluation focused on the open data domain, we have demonstrated that the user is able to find relevant datasets that were not findable using standard (direct) similarity search. Moreover, as the data-transitive similarity is a variant of pair-wise non-metric similarity, it can be plugged into any search engine that supports non-metric similarities. It also confirms the necessity of non-metric approaches in complex retrieval tasks, such as the entity discovery.

In the future we plan to investigate more general chains of intermediaries, as well as internal indexing techniques for the data-transitive similarity computation itself. We also plan to experiment with other domains that require more complex explainable similarity approaches.

Acknowledgments. This work was supported by the Czech Science Foundation (GAČR), grant number 19-01641S.

References

1. Aggarwal, C.C.: Recommender Systems. Springer, Cham (2016). https://doi.org/10.1007/978-3-319-29659-3
2. Altaf, B., Akujuobi, U., Yu, L., Zhang, X.: Dataset recommendation via variational graph autoencoder. In: 2019 IEEE International Conference on Data Mining (ICDM), pp. 11–20 (2019). https://doi.org/10.1109/ICDM.2019.00011
3. Augsten, N.: A roadmap towards declarative similarity queries. In: Bohlen, M., Pichler, R., May, N., Rahm, E., Wu, S.H., Hose, K. (eds.) Advances in Database Technology - EDBT 2018, pp. 509–512. Advances in Database Technology - EDBT, OpenProceedings.org, January 2018. https://doi.org/10.5441/002/edbt.2018.59
4. Brickley, D., Burgess, M., Noy, N.F.: Google dataset search: building a search engine for datasets in an open web ecosystem. In: The World Wide Web Conference, WWW 2019, San Francisco, CA, USA, 13–17 May 2019, pp. 1365–1375. ACM (2019). https://doi.org/10.1145/3308558.3313685
5. Bustos, B., Kreft, S., Skopal, T.: Adapting metric indexes for searching in multi-metric spaces. Multimedia Tools Appl. **58**, 1–30 (2012). https://doi.org/10.1007/s11042-011-0731-3
6. Chapman, A., et al.: Dataset search: a survey. VLDB J. **29**(1), 251–272 (2020)
7. Chávez, E., Navarro, G., Baeza-Yates, R., Marroquín, J.L.: Searching in metric spaces. ACM Comput. Surv. **33**(3), 273–321 (2001)

8. Chen, X., et al.: DataMed - an open source discovery index for finding biomedical datasets. J. Am. Med. Inform. Assoc. **25**(3), 300–308 (2018). https://doi.org/10.1093/jamia/ocx121
9. Connor, R., Vadicamo, L., Cardillo, F.A., Rabitti, F.: Supermetric search. Inf. Syst. **80**, 108–123 (2019)
10. Degbelo, A., Teka, B.B.: Spatial search strategies for open government data: a systematic comparison. CoRR abs/1911.01097 (2019). https://arxiv.org/abs/1911.01097
11. Gkelios, S., Sophokleous, A., Plakias, S., Boutalis, Y., Chatzichristofis, S.A.: Deep convolutional features for image retrieval. Exp. Syst. Appl. **177**, 114940 (2021)
12. Gregory, K., Groth, P., Scharnhorst, A., Wyatt, S.: Lost or found? Discovering data needed for research. Harvard Data Sci. Rev. **2**(2) (2020). https://doi.org/10.1162/99608f92.e38165eb. https://hdsr.mitpress.mit.edu/pub/gw3r97ht
13. Gregory, K.M., Cousijn, H., Groth, P., Scharnhorst, A., Wyatt, S.: Understanding data search as a socio-technical practice. J. Inf. Sci. **46**(4), 459–475 (2020)
14. Grosup, T., Peska, L., Skopal, T.: Towards augmented database schemes by discovery of latent visual attributes. In: Herschel, M., Galhardas, H., Reinwald, B., Fundulaki, I., Binnig, C., Kaoudi, Z. (eds.) Advances in Database Technology - 22nd International Conference on Extending Database Technology, EDBT 2019, Lisbon, Portugal, 26–29 March 2019, pp. 670–673. OpenProceedings.org (2019). https://doi.org/10.5441/002/edbt.2019.83
15. Hetland, M.L., Skopal, T., Lokoc, J., Beecks, C.: Ptolemaic access methods: challenging the reign of the metric space model. Inf. Syst. **38**(7), 989–1006 (2013)
16. Koesten, L.: A user centred perspective on structured data discovery. In: Companion Proceedings of the The Web Conference 2018, WWW 2018, International World Wide Web Conferences Steering Committee, Republic and Canton of Geneva, CHE, pp. 849–853 (2018). https://doi.org/10.1145/3184558.3186574
17. Košarko, O., Variš, D., Popel, M.: LINDAT translation service (2019). http://hdl.handle.net/11234/1-2922, LINDAT/CLARIAH-CZ digital library at the Institute of Formal and Applied Linguistics (ÚFAL), Faculty of Mathematics and Physics, Charles University
18. Kučera, J., Chlapek, D., Nečaský, M.: Open government data catalogs: current approaches and quality perspective. In: Kő, A., Leitner, C., Leitold, H., Prosser, A. (eds.) EGOVIS/EDEM 2013. LNCS, vol. 8061, pp. 152–166. Springer, Heidelberg (2013). https://doi.org/10.1007/978-3-642-40160-2_13
19. Lu, W., Hou, J., Yan, Y., Zhang, M., Du, X., Moscibroda, T.: MSQL: efficient similarity search in metric spaces using SQL. VLDB J., 3–26 (2017). https://www.microsoft.com/en-us/research/publication/msql-efficient-similarity-search-metric-spaces-using-sql/
20. Miller, R.J., Nargesian, F., Zhu, E., Christodoulakis, C., Pu, K.Q., Andritsos, P.: Making open data transparent: Data discovery on open data. IEEE Data Eng. Bull. **41**(2), 59–70 (2018). http://sites.computer.org/debull/A18june/p59.pdf
21. Novak, D., Zezula, P., Budikova, P., Batko, M.: Inherent fusion: towards scalable multi-modal similarity search. J. Database Manage. **27**(4), 1–23 (2016)
22. Silva, Y.N., Pearson, S.S., Chon, J., Roberts, R.: Similarity joins: their implementation and interactions with other database operators. Inf. Syst. **52**, 149–162 (2015). https://doi.org/10.1016/j.is.2015.01.008. Special Issue on Selected Papers from SISAP 2013
23. Skopal, T., Bustos, B.: On nonmetric similarity search problems in complex domains. ACM Comput. Surv. **43**(4) (2011). https://doi.org/10.1145/1978802.1978813

24. Traina, C., Moriyama, A., da Rocha, G.M., Cordeiro, R.L.F., de Aguiar Ciferri, C.D., Traina, A.J.M.: The SimilarQL framework: similarity queries in plain SQL. In: Proceedings of the 34th ACM/SIGAPP Symposium on Applied Computing, SAC 2019, Limassol, Cyprus, 8–12 April 2019, pp. 468–471. ACM (2019). https://doi.org/10.1145/3297280.3299736
25. Zezula, P., Amato, G., Dohnal, V., Batko, M.: Similarity Search: The Metric Space Approach, Advances in Database Systems, vol. 32. Springer, Boston (2006). https://doi.org/10.1007/0-387-29151-2
26. Škoda, P., Bernhauer, D., Nečaský, M., Klímek, J., Skopal, T.: Evaluation framework for search methods focused on dataset findability in open data catalogs. In: Proceedings of the 22nd International Conference on Information Integration and Web-based Applications & Services, pp. 200–209 (2020)

Non-parametric Semi-supervised Learning by Bayesian Label Distribution Propagation

Jonatan Møller Nuutinen Gøttcke[1(✉)], Arthur Zimek[1(✉)], and Ricardo J. G. B. Campello[2(✉)]

[1] Institute of Mathematics and Computer Science, University of Southern Denmark, Odense, Denmark
{goettcke,zimek}@imada.sdu.dk

[2] School of Mathematical and Physical Sciences, University of Newcastle, Callaghan, Australia
Ricardo.Campello@newcastle.edu.au

Abstract. Semi-supervised classification methods are specialized to use a very limited amount of labelled data for training and ultimately for assigning labels to the vast majority of unlabelled data. Label propagation is such a technique that assigns labels to those parts of unlabelled data that are in some sense close to labelled examples and then uses these predicted labels in turn to predict labels of more remote data. Here we propose to not propagate an immediate label decision to neighbors but to propagate the label probability distribution. This way we keep more information and take into account the remaining uncertainty of the classifier. We employ a Bayesian schema that is simpler and more straightforward than existing methods. As a consequence we avoid to propagate errors by decisions taken too early. A crisp decision can be derived from the propagated label distributions at will. We implement and test this strategy with a probabilistic k-nearest neighbor classifier, proving competitive with several state-of-the-art competitors in quality and more efficient in terms of computational resources.

Keywords: Semi-supervised classification · k-Nearest neighbor classification · Transductive learning · Label propagation

1 Introduction

While easily collectable unlabelled data become more abundant, and labelled data continue to be a scarce resource, semi-supervised learning remains relevant. Semi-supervised learning falls between supervised learning (learning from labelled data) and unsupervised learning (learning from unlabelled data) [5]. It is used for improving unsupervised learning by taking advantage of information traditionally used in supervised learning and for improving the performance of supervised learning methods by using unlabelled instances.

N. Reyes et al. (Eds.): SISAP 2021, LNCS 13058, pp. 118–132, 2021.
https://doi.org/10.1007/978-3-030-89657-7_10

In semi-supervised classification we have to distinguish between assigning labels to unlabelled data during training and the application of the resulting semi-supervised classifier to new unseen data during testing, where the classifier is built on the complete training data consisting of labelled and unlabelled instances. The labelling of unlabelled training instances is known as 'transduction' [20,21] or as 'label propagation' since it is often done by propagating labels from the labelled training instances to the unlabelled training instances. After label propagation, all training data can be used for inducing labels of new unseen query instances as it is conventionally done in supervised learning. This is therefore also known as 'induction'. Applying a learner for transduction therefore yields a training set with then all instances labelled that could be used by any other classifier for induction, i.e., conventional classification [27]. The classification model used for induction could therefore be different from the semi-supervised classifier used for transduction, but it could also be the same method that is used beyond transduction on the training data also for induction on new, unseen data. Besides describing different phases, tasks, or scenarios in the context of semi-supervised learning, the exact relationship between 'transduction', 'induction', and 'semi-supervised learning' remains debatable [4]. In this paper we evaluate methods in a transductive setting as it is common practice [8,9,12,23–26].

We argue that it might be important to account for uncertainties during transduction and to keep information on uncertain decisions possibly also beyond the transduction phase, if induction is treated separately and the classification algorithm employed for induction can make use of uncertain label information or label probability distributions. For a simple illustration, consider the one-dimensional distribution of classes in one attribute (sepal length) of the well-known Iris data, plotted in Fig. 1. Some classification model might be based on the estimated probability density distribution and decide for the maximum likelihood class at any given point in the data space. Considering the example of the figure, if we have a sepal length, say, between 5 and 6 cm, we can decide on a clear decision boundary but a high probability would remain to have chosen the wrong class. If just the resulting label is propagated and used for ensuing decisions, these later decisions are necessarily oblivious of a possibly considerable level of uncertainty that actually affects also these later decisions. To account for this uncertainty, we suggest that, instead of propagating a label, the label *probability distribution* should be propagated and would thus also be available for later decisions. We could see this as an attempt to keep as much information as possible as long as potentially useful for the classification of new instances.

This idea has been employed in some specific graph-based methods, as we will survey below. Here we propose a more general probabilistic, non-parametric semi-supervised classification schema and demonstrate its benefits by implementing it with a probabilistic k nearest neighbor classifier that is conceptually simpler and yet compares favorably against state-of-the-art methods for semi-supervised classification on a large collection of datasets being considerably more efficient.

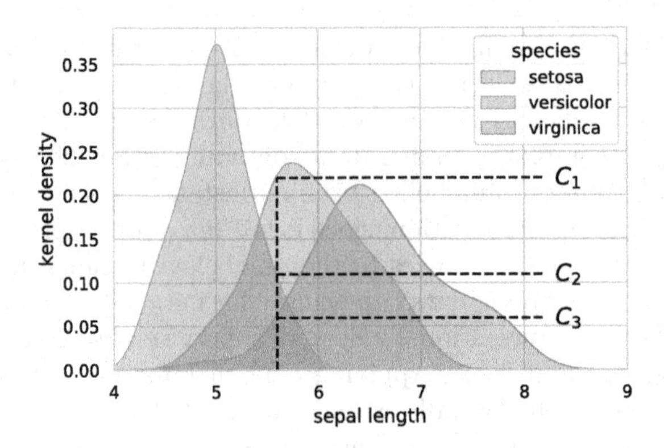

Fig. 1. Some classifier's model (estimated class-conditional probability density distributions). Deciding at any point according to the maximum class (posterior) probability and using only that label later renders further propagation or later induction oblivious of the evidence for other classes. Example data from the Iris dataset.

In the remainder, we give an overview of related work on semi-supervised classification (Sect. 2), introduce the general concept and a concrete implementation of our method (Sect. 3), study its performance on a large collection of datasets and compare against several state-of-the-art methods in terms of effectiveness and efficiency (Sect. 4), and conclude with a short discussion of some properties of the compared methods (Sect. 5).

2 Related Work

There are many variations of semi-supervised learning such as *self-training* [16] and *co-training* [2]. Transductive learning [20] is a part of the foundation of semi-supervised learning and relates to an approach that uses both labelled and unlabelled data as training data, $TR = L \cup U$, to predict labels for U. Some of the most popular methods in this field have been surveyed in books broadly discussing the area [5,27]. Well known methods such as graph classification [12, 23] and support vector machines have been adapted to the transductive setting [9], and have been used in combination with Laplacian regularization as for Gaussian Field Harmonic Function (GFHF) [25,26]. In the following we discuss some methods that are more closely related to our approach.

The basic idea of GFHF [25,26] is to model a transition probability in the graph representing the dataset, typically using the RBF kernel. All nodes are associated with a class label distribution which is updated following the transition probabilities until convergence. GFHF also was seminal for Laplacian support vector machines and Laplacian regularized least squares [1]. GFHF propagates the transition probabilities estimated by the RBF kernel on the complete graph. In each iteration the propagation of transition probabilities is normalized to maintain a probability interpretation, and the method iterates until convergence.

Learning with Local and Global Consistency (LGC) [23] is inspired by GFHF and mainly differs in the propagation process, including a parameter α that determines how the information from the previous and the current iteration are weighted. LGC also uses the RBF kernel to generate a weight matrix for the complete graph and normalizes the weights to maintain label probability distributions for propagation in each iteration.

Szummer and Jaakola [18] proposed one of the first semi-supervised classification algorithms, using Markov random walks from a number of random points in the dataset, and the RBF kernel (similarly as Zhu et al. [26]) for determining the edge weights to estimate the class conditional probability. They use *k*NNs for finding the local manifold structure. Substantial differences between the method presented in this paper and Szummer and Jaakola's is the choice of the kernel used to define the label distributions, their use of a symmetrized *k*NN graph, and their probability estimation procedure. Furthermore their method requires several additional parameters such as the time parameter t for the number of steps in the Markov process to govern smoothness and the σ parameter of the RBF kernel for edge weights, and the number of random starting points.

Liu and Chang [12] introduced the RMGT algorithm, a graph based algorithm that utilizes the combinatorial Laplacian matrix to describe the local manifold structure. They introduce a new underlying graph topology referred to as the symmetry favored *k*NN graph, which adds weights to bidirectional edges in the directed *k*NN graph. De Sousa and Batista [17] extended this algorithm further (RMGTHOR) by modifying the regularization framework to use a normalized Laplacian, or a Laplacian with a degree higher than 1, instead of the combinatorial Laplacian used in the earlier methods.

3 Label Probability Distribution Propagation

3.1 Motivation

In semi-supervised classification, a core assumption is that the labelled subset L of the training set is much smaller both in relative and absolute terms than the unlabelled subset U, i.e., $|L| \ll |U|$. Therefore a learner should be extra careful when propagating labels because of the high probability of propagating errors when decisions are based on insufficient information. With further propagation of potentially erroneous labels such errors can spread and have a severe impact on the quality of the transduction.

This is the core motivation for our proposal to not propagate class *labels* but instead to propagate the *distribution* of class labels (or class label *probabilities*) from instances in L to instances in U. Such class probabilities can be determined in principle using any probabilistic classifier.

3.2 General Schema

In Fig. 1 we see three probability density functions for the three classes over the sepal length attribute in the Iris dataset. We propose to propagate label probability distributions to unlabelled instances in a semi-supervised manner using the (estimated) probability density for each class starting from a straightforward Bayesian schema:

With the prior class probability $\Pr(c)$ for each class $c \in C$ and Bayes theorem, the probability for instance x to belong to a class c can be estimated by:

$$\Pr(c|x) \propto \frac{\hat{f}(x|c)\Pr(c)}{\sum_{c_i \in C} \hat{f}(x|c_i)\Pr(c_i)} \tag{1}$$

where $\hat{f}$ is some estimate of the probability density (which could be a direct probability estimate, if some classifier delivers that).

Such estimated label probability distributions are assigned to all instances $x \in U$, such that the label y_i of instance x_i is in fact a label distribution:

$$y_i = (\Pr(c_1|x_i), \Pr(c_2|x_i), \cdots, \Pr(c_n|x_i))^\top \tag{2}$$

Consider an unlabelled instance x in Fig. 1. The probability density for each class is given by c_1, c_2, and c_3, and the probability for belonging to each class can then be calculated by Eq. (1). The standard maximum likelihood prediction would predict class c_3 and immediately lose the information that the other two classes, albeit less likely, still carry a non-negligible probability that could be helpful to decide close cases downstream, when using x transductively (or even for induction later on). This is because, after the assignment of a label probability distribution to instance x, x is moved from the set of unlabelled data U to the set of labelled data L and is used for the transductive labelling process that continues until $U = \emptyset$ and all training instances are labelled.

3.3 *k*NN-Label Distribution

To test this concept in semi-supervised classification we employ a probabilistic *k*NN classifier to estimate label probability distributions in a non-parametric way and describe an algorithm to propagate label probability density distributions using *k*NN, resulting in an algorithm *k*NN *Label Distribution Propagation* (*k*NN-LDP). Propagating label probability distributions allows data instances to have a soft labelling and the transduction to account for this label distribution when calculating the label probability distributions of unlabelled neighbors.

While we keep all information of the label probability distributions as long as possible, we can at any point derive a crisp labelling of a query object if needed, taking the maximum of the assigned class probabilities. Also note that, although we focus on the transduction here, we could also apply the same algorithm for induction beyond the training data to predict the class (or the label probability distribution) for any unseen query object.

In the supervised scenario of using the *k*NN classifier, the label probability distribution for some instance x is given by the class-conditional density estimates based on the k nearest neighbors of x taken over the labeled training data L [7,22]:

$$\hat{f}(x|c_j) = \frac{|\{x_\ell \in k\text{NN}(x) \cap c_j\}|}{|\{x_\ell \in L \cap c_j\}| \cdot Vol_{k\text{NN}(x)}} \tag{3}$$

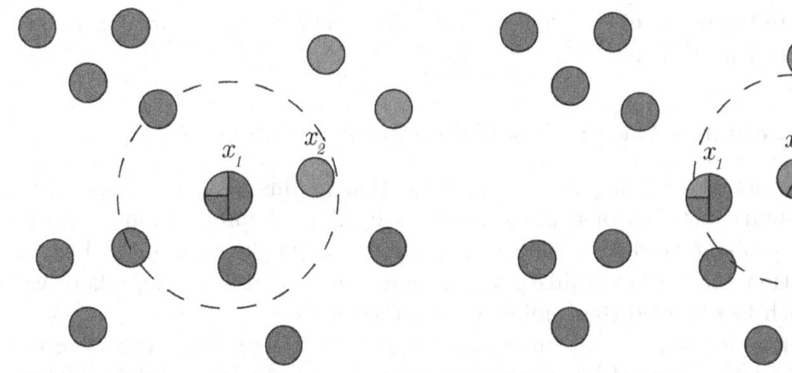

(a) An unlabelled instance in the training set receives a label distribution from labelled neighbors (note that we have a tie in the neighborhood and therefore effectively use $k + 1$ neighbors).

(b) An unlabelled instance in the training set receives the label distributions from neighbors, including examples labeled by label distribution propagation (no tie here).

Fig. 2. *k*NN label distribution propagation, using $k = 3$, without and with partially labelled examples.

where $|\cdot|$ denotes the cardinality of a set and $Vol_{k\mathrm{NN}(x)}$ denotes the volume needed to cover k nearest neighbors of x, centered at x. The shape of this volume will depend on the employed distance function. Note, however, that the volume cancels nicely out when putting this into Eq. (1).

In the semi-supervised scenario tackled here, an instance among the nearest neighbors might not have a crisp label but a label probability distribution itself, or no label for instances $\in U$. For getting a well-defined probability distribution we can treat the "unknown" case as a special class. Accounting for partial labels in Eq. 3 thus yields

$$\hat{f}(x|c_j) = \frac{\sum_{x_\ell \in k\mathrm{NN}(x)} \Pr(c_j|x_\ell)}{\sum_{x_\ell \in L} \Pr(c_j|x_\ell) \cdot Vol_{k\mathrm{NN}(x)}} \tag{4}$$

Using this in Eq. (1), the probability for each class c in the label distribution, depending on the label probability distributions of the k nearest neighbors, is therefore given by

$$\Pr(c|x) = \frac{\sum_{x_\ell \in k\mathrm{NN}(x)} \Pr(c|x_\ell)}{|\, k\mathrm{NN}(x)|} \tag{5}$$

We illustrate the method in Fig. 2. The example $x_1 \in U$ receives label information from its k nearest neighbors. One of the nearest neighbors, $x_2 \in U$, is unlabelled. As a result, x_1 would now carry partial label information which we can interpret as a label probability distribution (Fig. 2(a)). Next, example x_2 is processed and receives label information from its neighbors, including x_1, thus not just counting labels of classes but considering the label probability distribution over the k nearest neighbors (Fig. 2(b)).

3.4 Abstention in Case of Insufficient Information

We have to account for a potential complication in this process of assigning label probability distributions. There might be insufficient information to assign some label probability distribution to some given unlabelled instance. This is a complication that is not unlikely in the semi-supervised scenario, where we assume much less labeled than unlabeled training data.

If we encounter such an instance that cannot get assigned any class probabilities, we assign a special label signaling the fact that the label distribution is unknown, thus employing the concept of abstaining classifiers [15], although we do not propose here to optimize the classifier w.r.t. abstention.

3.5 Propagation Algorithm

To use as much information as possible for the label assignment in just two passes over the data we start with the instances where most information is available, that is where the sum of the class probabilities over the neighbors (except for the class "unknown") is maximal, and continue to process instances with decreasing order w.r.t. this available information (which might change over time). This requires checking all neighborhoods in advance. For the sake of efficiency, the forward and reverse k nearest neighbors should be indexed in this first pass.

The information that can be used for label distribution assignments can be captured in weights:

$$w(x) = \sum_{c \in C \setminus \{\text{“unknown”}\}} \Pr(c|x) \tag{6}$$

These weights w are used to keep the instances sorted in decreasing order w.r.t. the available information in some priority queue. This way, as much information as possible is used in one sweep over the data, updating the label probability distributions and the weights of the reverse k nearest neighbors (RkNN) of updated instances (i.e., those that are affected by an update of the current instance). This might make instances climb up in the priority queue if their weight changed because their neighbors got label distribution information assigned. In the example of Fig. 2, this would be the case for $x_2 \in \text{R}k\text{NN}(x_1)$: after having assigned a label probability distribution to x_1, $w(x_2)$ will increase. Then we can assign an estimate of the label distribution for each instance, as defined in Eqs. (2) and (5). Note that the definitions of probability distributions include the class "unknown", such that probabilities sum up to one. A sketch of the procedure is provided in Algorithm 1.

Algorithm 1. kNN-LDP

1: **for all** $x \in U$ **do**
2: index forward and reverse k nearest neighbors (kNN, RkNN)
3: $x.w \leftarrow \sum_{c \in C \setminus \{\text{unknown}\}} \Pr(c|x)$ {Eq. (6)}
4: **end for**
5: $PQ_U \leftarrow$ priority-queue(U) {decreasing order w.r.t. $x.w$}
6: **while** PQ_U.size > 0 **do**
7: $x \leftarrow PQ_U$.getMax()
8: **if** $x.w > 0$ **then**
9: $x.y \leftarrow (\Pr(c|x))_{c \in C}$ {Eqs. (2) and (5)}
10: **for all** $p \in$ RkNN(x) **do**
11: $p.w \leftarrow \sum_{c \in C \setminus \{\text{unknown}\}} \Pr(c|p)$ {Eq. (6)}
12: PQ_U.update(p)
13: **end for**
14: $L \leftarrow L \cup \{x\}$
15: **else**
16: $x.y \leftarrow$ unknown
17: **for all** $p \in PQ_U$ **do**
18: $p.y \leftarrow$ unknown
19: $L \leftarrow L \cup \{p\}$
20: **end for**
21: $PQ_U \leftarrow \emptyset$
22: **end if**
23: **end while**

3.6 Advantages and Disadvantages of kNN-LDP

The label distribution propagation algorithm proposed here also has an underlying graph interpretation when implemented with a k nearest neighbor classifier. It has some advantages over other graph-based methods. The asymptotic runtime of the kNN-based label distribution propagation algorithm is identical to that of finding the kNN, i.e., the operation of identifying the nearest neighbors is the computational bottleneck as in many applications and could naturally benefit from employing efficient neighborhood search methods [10,11]. Yet, due to the heuristic order of processing instances, our method, as opposed to many competitors, does not require any iterations.

The runtime for graph-based algorithms depends on the topology of the graph but is typically higher than the runtime of kNN-LDP which only makes neighborhood queries for the unlabelled data. A mutual-kNN graph requires the computation of the nearest neighbors of all labelled and unlabelled instances which takes $\mathcal{O}(n^2)$ for a dataset of size n. If Ozaki's graph connection method [13] is used, computing the complete similarity graph takes $\mathcal{O}(n^2)$, which cannot be improved. Finding the minimum spanning tree (or rather maximum spanning tree, as it is based on similarities, not distances) takes $\mathcal{O}((V+E)\log V)$, where V is the number of vertices, and E is the number of edges in the graph. For the complete graph this takes $n + \frac{n(n-1)}{2} \log \frac{n(n-1)}{2}$ using Prim's algorithm.

Table 1. Competitors and their implementations

Method	Impl. Name	Impl. Source
GFHF [26]	Label Propagation	Scikit-Learn [14]
LGC [23]	Label Spreading	
LapRLS [1]	LapRls	https://github.com/HugoooPerrin/semi-supervised-learning
LapSVM [1]	LapSvm	
RMGT [12]	RMGT	de Sousa & Batista [17]
RMGTHOR [17]	RMGTHOR	
This paper	kNN LDP	https://github.com/Goettcke/kNN_LDP

Although the kNN Label Distribution Propagation is different from the most common graph-based algorithms it also comes with some of the same disadvantages. When constructing an adjacency matrix in nearest-neighbor graph-based semi-supervised learning algorithms, the number of components plays an essential role in the success of the label propagation. A similar problem is present in kNN-LDP, if an unlabelled instance cannot be reached by the propagation through neighborhoods, i.e., if an unlabelled instance resides in a graph component without any labelled instances. This problem tends to occur more with a smaller ratio of $\frac{L}{U}$ and a smaller value of k, and tends to affect the label propagation late in the process.

There are different strategies to tackle this problem. One could be to increase k until at least one neighbor carries label information. However, sparse graphs are observed empirically to perform better than dense or complete graphs [28], as they have a higher sensitivity to detecting the local manifold which the data points lie on. Another solution, as seen in related work, is to assign the majority class if no other information is available.

The solution we have chosen for kNN-LDP is to make the classifier abstaining from making a decision where it does not have the information and giving it the class label "unknown". It should be noted that this is a clear disadvantage in the comparison, as such a label will always count as an error. We will see in the evaluation that trading these errors in the evaluation metric for not propagating potential errors to further decisions seems to pay off.

4 Experimental Evaluation

4.1 Competitors

As competitors we selected the more closely related methods GFHF [26], LGC [23], RMGT [12], and the more recent RMGTHOR [17] method. The Laplacian regression method, Laplacian Regularized Least Squares (LapRLS) and Laplacian Support Vector Machines (LapSVM) [1] were also evaluated. We used publicly available implementations, an overview is provided in Table 1.

Table 2. Datasets used for comparative evaluation.

Dataset	Classes	Attributes	Instances	Dataset	Classes	Attributes	Instances
australian	2	14	690	page-blocks	5	10	5472
banana	2	2	5300	phoneme	2	5	5404
breast	2	30	569	segment	7	19	2310
bupa	2	6	345	spambase	2	57	4597
cleveland	5	13	297	spectfheart	2	44	267
contraceptive	3	9	1473	tae	3	5	151
dermatology	6	34	358	vowel	11	13	990
glass	6	9	214	wine	3	13	178
hayes roth	3	4	160	wine-red	6	11	1599
heart	2	13	270	wine-white	7	11	4898
iris	3	4	150	wisconsin	2	9	683
led7digit	10	7	500	COIL	2	50	1500
mammo. mass	2	5	830	digit-1	2	50	1500
monk-2	2	6	432	G-241C	2	50	1500
movement libras	15	90	360	G-241N	2	50	1500
new thyroid	3	5	215	USPS	2	50	1500

4.2 Parameters

In the experiments all distances are Euclidean. LapSVM and LapRLS were tested in two settings, using the Linear kernel and the RBF kernel. GFHF and LGC use the RBF kernel. For LGC we used $\alpha = 0.2$ as it is the default value in the Scikit-Learn implementation. RMGT uses combinatorial Laplacian regularization. RMGTHOR uses the normalized Laplacian. Both methods use local linear embedding for building the weight matrices. RMGTHOR uses a Laplacian degree of 1. For all algorithms based on k nearest neighbors we invariably set $k = 10$, following findings that some small value is typically a good choice for the local density estimation and for determining the local manifold [28].

For the methods taking an RBF kernel, $\sigma_M \in \{0.1, 0.5, 1\}$ was tested. For the parameters γ_L and γ_M for LapSVM and LapRLS a grid search was performed for all combinations of the values $\gamma_L \in \{0.1, 0.5, 1\}$ and $\gamma_M \in \{0.1, 0.5, 1\}$, that is using the same range as the original publication [1]. The grid search tries all combinations of the parameters in the parameter sets, and for each dataset the best result achieved is extracted and used for comparison.

Parameter optimization is necessary for these methods to avoid poor performance. It should be noted, though, that selecting the best results for these methods gives them an advantage in the comparison.

4.3 Datasets

We have evaluated our method and state-of-the-art competitors on the 5 datasets commonly used in semi-supervised classification benchmarks [3] as well as on 27 datasets used by Triguero et al. [19]. The datasets have been selected such that

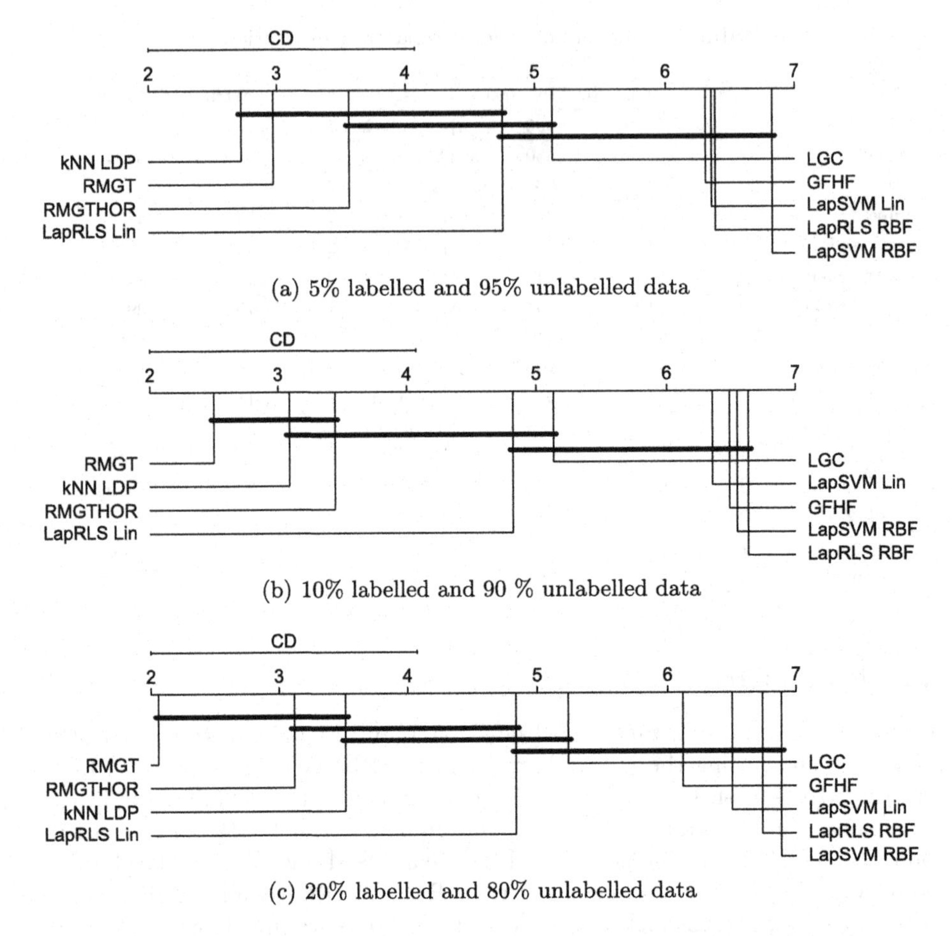

(a) 5% labelled and 95% unlabelled data

(b) 10% labelled and 90 % unlabelled data

(c) 20% labelled and 80% unlabelled data

Fig. 3. Critical difference plots showing the relation between the tested algorithms transductive performance on different proportions of available label information.

all algorithms compared to in this study could be applied without adjustments. An overview on the datasets is provided in Table 2.

4.4 Evaluation Setup

Each dataset has been split into labelled training data and unlabelled training data, in three different split proportions (labelled, unlabelled): $[0.05, 0.95]$, $[0.1, 0.9]$, and $[0.2, 0.8]$. In each setting, we average the test results over 64 random samples. We measure accuracy on crisp decisions derived from the label probability distributions post hoc in case of *k*NN-LDP, and the built-in predict functions for other methods. We count as error when *k*NN-LDP is abstaining from classifying an instance.

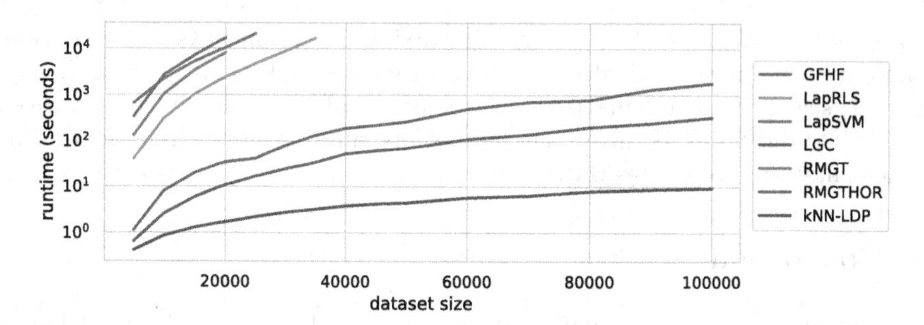

Fig. 4. Runtime in seconds scaling with dataset size.

4.5 Results

For assessing the performance of the different methods, we show "critical difference plots" following the methodology described by Demšar [6] for assessing the statistical significance in the ranking of the compared methods in terms of accuracy in Fig. 3. These plots visualize the mean rank over all datasets and the critical difference given the number of datasets and algorithms used in the statistic. If the mean rank for a method is connected to the mean rank of another method by a horizontal bar, the two methods are within the critical difference and their performance is not significantly different.

kNN-LDP often achieves the highest accuracy score and is significantly better than several competitors, and not different from the other methods with statistical significance. The advantage of kNN-LDP tends to be more prominent with a smaller proportion of label information. It performs best for the smallest fraction of labelled data, which is the most important scenario for semi-supervised learning. In all scenarios, kNN-LDP forms a top group together with RMGT and RMGTHOR where the differences are not statistically significant. We will see next, however, that kNN-LDP is much more efficient than all the other methods, in particular kNN-LDP beats RMGT and RMGTHOR by a large margin in terms of scalability.

4.6 Scalability

For testing the scalability of the algorithms we generated 2-dimensional datasets with two-class problems using the Scikit-Learn *make_classification* function, increasing the dataset size. The classification problems all have uniform class distributions, and the algorithms were given 10% labelled training data. The test hardware used consists of an AMD EPYC 7501 32-Core Processor, and 256 GB available memory. All algorithms were run with default parameters in these tests. We depict the results of the scalability experiment in Fig. 4.

The rapid increase in runtime for LapRLS, LapSVM, RMGT, and RMGTHOR and their also rapidly increasing demands on system memory prevented tests on larger dataset sizes. However, the disadvantage of these methods in terms of runtime and scalability behavior is already quite clear at this point (note the logarithmic scale of the runtime axis). GFHF and LGC remain more competitive to our method, although they are considerably slower. Given the logarithmic scale, the scalability performance of kNN-LDP is clearly superior even if we would disregard its advantage in absolute runtime accounting for possible implementation advantages [11].

5 Conclusion

We studied an elegant non-parametric method with a clear interpretation in terms of density estimation and Bayesian reasoning here that performs as good as or better than state-of-the-art methods on a large collection of datasets even though it was put on a disadvantage compared to other methods in two aspects:

First, it is a fundamental requirement in graph-based algorithms that each instance (i.e., a vertex in some kNN graph) must belong to a component in which at least one other vertex is labelled. While other methods use undirected, symmetrized, or even complete graphs to adhere to this assumption, in the case of kNN-LDP the assumption is more likely to be violated because the kNN graph is inherently directed. As a consequence each unlabelled vertex should not only be in a component with at least one other labelled vertex but also be directly connected to it. While other methods use various heuristics to solve cases where this assumption is violated and can be sometimes correct with that, we simply abstain from a decision which as such will always count as an error. However, this abstention fits and contributes to the fundamental motivation and strategy of our method: to avoid the propagation of errors, which will be of utmost importance in the induction on test or new data.

Second, we performed grid search optimization of several parameters for the competing Laplacian methods and the methods using an RBF kernel. Without such parameter tuning, these methods would not be able to achieve reasonable performance. No such parameter optimization was done for the k-value used in the kNN-LDP method, and the other methods using nearest neighbor information, where some small value is typically a good choice for the local density estimation and for determining the local manifold [28].

In terms of efficiency and scalability, our method is clearly outperforming the competitors.

References

1. Belkin, M., Niyogi, P., Sindhwani, V.: Manifold regularization: a geometric framework for learning from labeled and unlabeled examples. J. Mach. Learn. Res. **7**, 2399–2434 (2006)
2. Blum, A., Mitchell, T.M.: Combining labeled and unlabeled data with co-training. In: COLT, pp. 92–100 (1998)
3. Chapelle, O., Schölkopf, B., Zien, A.: Analysis of benchmarks. In: Semi-Supervised Learning [5], pp. 376–393

4. Chapelle, O., Schölkopf, B., Zien, A.: A discussion of semi-supervised learning and transduction. In: Semi-Supervised Learning [5], pp. 473–478
5. Chapelle, O., Schölkopf, B., Zien, A. (eds.): Semi-Supervised Learning. The MIT Press, Cambridge (2006)
6. Demšar, J.: Statistical comparisons of classifiers over multiple data sets. J. Mach. Learn. Res. **7**, 1–30 (2006)
7. Duda, R.O., Hart, P.E., Stork, D.G.: Pattern Classification, 2nd edn. Wiley, Hoboken (2001)
8. Castro Gertrudes, J., Zimek, A., Sander, J., Campello, R.J.G.B.: A unified view of density-based methods for semi-supervised clustering and classification. Data Min. Knowl. Discov. **33**(6), 1894–1952 (2019). https://doi.org/10.1007/s10618-019-00651-1
9. Joachims, T.: Transductive inference for text classification using support vector machines. In: ICML, pp. 200–209 (1999)
10. Kirner, E., Schubert, E., Zimek, A.: Good and bad neighborhood approximations for outlier detection ensembles. In: SISAP, pp. 173–187 (2017). Springer, Cham. https://doi.org/10.1007/978-3-319-68474-1_12
11. Kriegel, H.-P., Schubert, E., Zimek, A.: The (black) art of runtime evaluation: are we comparing algorithms or implementations? Knowl. Inf. Syst. **52**(2), 341–378 (2016). https://doi.org/10.1007/s10115-016-1004-2
12. Liu, W., Chang, S.: Robust multi-class transductive learning with graphs. In: CVPR, pp. 381–388 (2009)
13. Ozaki, K., Shimbo, M., Komachi, M., Matsumoto, Y.: Using the mutual k-nearest neighbor graphs for semi-supervised classification on natural language data. In: CoNLL, pp. 154–162. ACL (2011)
14. Pedregosa, F., et al.: Scikit-learn: machine learning in Python. J. Mach. Learn. Res. **12**, 2825–2830 (2011)
15. Pietraszek, T.: On the use of ROC analysis for the optimization of abstaining classifiers. Mach. Learn. **68**(2), 137–169 (2007)
16. Scudder, H.J., III.: Probability of error of some adaptive pattern-recognition machines. IEEE Trans. Inf. Theory **11**(3), 363–371 (1965)
17. de Sousa, A.R., Batista, G.E.A.P.A.: Robust multi-class graph transduction with higher order regularization. In: IJCNN, pp. 1–8 (2015)
18. Szummer, M., Jaakkola, T.S.: Partially labeled classification with Markov random walks. In: NIPS, pp. 945–952 (2001)
19. Triguero, I., Sáez, J.A., Luengo, J., García, S., Herrera, F.: On the characterization of noise filters for self-training semi-supervised in nearest neighbor classification. Neurocomputing **132**, 30–41 (2014)
20. Vapnik, V.: Statistical Learning Theory. Wiley, Hoboken (1998)
21. Vapnik, V.: Transductive inference and semi-supervised learning. In: Chapelle et al. [5], pp. 452–472
22. Zaki, M.J., Meira, W., Jr.: Data Mining and Analysis: Fundamental Concepts and Algorithms. Cambridge University Press, Cambridge (2014)
23. Zhou, D., Bousquet, O., Lal, T.N., Weston, J., Schölkopf, B.: Learning with local and global consistency. In: NIPS, pp. 321–328 (2003)
24. Zhou, D., Schölkopf, B.: Discrete regularization. In: Chapelle et al. [5], pp. 236–249
25. Zhu, X., Ghahramani, Z.: Learning from labeled and unlabeled data with label propagation. Technical Report CMU-CALD-02-107. School of Computer Science, Carnegie Mellon University (2002)
26. Zhu, X., Ghahramani, Z., Lafferty, J.D.: Semi-supervised learning using Gaussian fields and harmonic functions. In: ICML, pp. 912–919 (2003)

27. Zhu, X., Goldberg, A.B.: Introduction to Semi-Supervised Learning. Morgan & Claypool Publishers, San Rafael (2009)
28. Zhu, X.J.: Semi-supervised learning literature survey. Technical Report. University of Wisconsin-Madison Department of Computer Sciences (2005)

Optimizing Fair Approximate Nearest Neighbor Searches Using Threaded B+-Trees

Omid Jafari(✉), Preeti Maurya, Khandker Mushfiqul Islam, and Parth Nagarkar

New Mexico State University, Las Cruces, NM, USA
{ojafari,preema,mushfiq,nagarkar}@nmsu.edu

Abstract. Similarity search in high-dimensional spaces is an important primitive operation in many diverse application domains. *Locality Sensitive Hashing* (LSH) is a popular technique for solving the Approximate Nearest Neighbor (ANN) problem in high-dimensional spaces. Along with creating fair machine learning models, there is also a need for creating data structures that target different types of fairness. In this paper, we propose a fair variant of the ANN problem that targets *Equal opportunity* in group fairness in the ANN domain. We formally introduce the notion of fair ANN for *Equal opportunity* in group fairness. Additionally, we present an efficient disk-based index structure for finding Fair approximate nearest neighbors using Locality Sensitive Hashing (*FairLSH*). Moreover, we present an advanced version of *FairLSH* that uses cost models to further balance the trade-off between I/O cost and processing time. Finally, we experimentally show that *FairLSH* returns fair results with a very low I/O cost and processing time when compared with the state-of-the-art LSH techniques.

Keywords: Approximate nearest neighbor search · Similarity search · Locality sensitive hashing · Fairness · Equal opportunity

1 Introduction

In recent years, many real-world applications use machine learning algorithms for their decision making systems (e.g. job interviews, credit card offers, etc.). Often, these algorithms make discriminative and biased decisions towards specific individuals or group of individuals. There have been several works [7,14,18] that have studied different types of fairness and biases in these decision making systems. Moreover, even if algorithms may not be biased, they could amplify the latent bias that exists in the data. As a result, researchers have proposed new methods to deal with the algorithmic and data biases in classification [1], clustering [4,19], optimization [6], risk management [9], resource allocation [10], and many other domains.

Bias in the data used for training machine learning algorithms is a major challenge in developing fair algorithms. Here, in a rather different problem, we

N. Reyes et al. (Eds.): SISAP 2021, LNCS 13058, pp. 133–147, 2021.
https://doi.org/10.1007/978-3-030-89657-7_11

are interested in handling the bias imposed by the data structures used by such algorithms. In particular, data structures, regardless of how the data is handled and how it is collected, involve bias in the way they respond to searches.

In general, fairness can be divided into two categories: 1) individual fairness and 2) group fairness. The goal of individual fairness is to treat similar individuals similarly and the goal of group fairness is to treat similar groups of individuals similarly. Both categories can be further divided into sub-categories [22]. One such sub-category of group fairness is *Equal Opportunity*, which states that if individuals in multiple groups of people qualify for an outcome, those groups of people should receive the outcome at equal rates [22].

Finding nearest neighbors of a given query is an important problem in many domains. For high-dimensional datasets, the traditional index structures suffer from the well-known problem of *Curse of Dimensionality* [8]. It is shown that even linear searches are faster than using these traditional index structures for high-dimensional datasets [5]. A solution to this problem is to search for approximate nearest neighbors instead of exact neighbors that results in much better running times. Locality Sensitive Hashing (LSH) [12] is a popular technique for solving the Approximate Nearest Neighbor (ANN) problem in high-dimensional spaces that takes a sub-linear time (with respect to the dataset size) to find the approximate nearest neighbors of a given query. LSH maps points in the high-dimensional space to a lower-dimensional space by using random projections. The intuition behind LSH is that close points in the high-dimensional space will map to the same hash buckets in the lower-dimensional space with a high probability and vice-versa.

1.1 Motivation

An important benefit of LSH is that it provides theoretical guarantees on the accuracy of the results. Moreover, LSH is a data-independent method (i.e. the index structure is not affected by data properties such as data distribution). Therefore, when the distribution of data changes, data-dependent methods (such as deep hashing approaches) need to re-generate the indexes. Additionally, LSH is known for its ease of disk-based implementations, making it very scalable as the dataset size grows [20]. Often, in various applications, there is a need to run approximate nearest neighbor searches in order to find fair neighbors of a given query. In these applications, there is a growing need to remove discrimination and bias towards specific individuals or a group of individuals. Here, the goal of fairness is to remove arbitrariness of the search strategy and base it upon predefined conditions such that neighbors of a given query that belong to different groups would have the same probability of being chosen in the final results.

There is no existing work that studies the *equal opportunity* of *group fairness* in the domain of ANN search. Existing state-of-the-art LSH approaches lead to wasted I/O while tackling the *equal opportunity* notion in group fairness (because they are not designed to efficiently search for fair nearest neighbors). Therefore, in this paper, our goal is to design a fair, yet efficient, disk-based LSH index structure, called *FairLSH*, that can reduce the I/O costs and processing times for finding the fair nearest neighbors.

1.2 Contributions

In this paper, we propose an efficient, disk-based index structure for finding Fair approximate nearest neighbors using Locality Sensitive Hashing, called *FairLSH*. The following are the primary contributions of this paper:

- We formally introduce the notion of fair approximate near neighbors for equal opportunity in group fairness.
- We present a tree-based and disk-based index structure, called *FairLSH-Basic*, that reduces disk I/O costs and processing times for finding fair approximate nearest neighbors.
- We further improve the efficiency by proposing a cost model-based variant of our index structure, called *FairLSH-Advanced*, that uses a user-input threshold to tune the trade-off between I/O costs and processing times, and hence further improve performance.
- Lastly, we experimentally evaluate the both variants of *FairLSH* on several datasets for different fairness scenarios and show that *FairLSH* outperforms the state-of-the-art techniques in terms of performance efficiency.

To the best of our knowledge, we are the first work that tackles the group fairness notion of *equal opportunity* in the ANN domain.

2 Related Work

2.1 LSH and Its Variants

Locality Sensitive Hashing (LSH) is one the most popular techniques for solving the Approximate Nearest Neighbor (ANN) problem in high-dimensional spaces [17]. LSH was first proposed in [12] for the Hamming distance and was later extended for the Euclidean distance in [8]. Then, the concepts of *Collision Counting* and *Virtual Rehashing* were introduced in [11] that solved the two main drawbacks of E2LSH [8], which were large index sizes and a large search radius. The idea of using query-aware hash functions where the indexes are created such that the query is an anchor of a bucket was proposed in QALSH [15] to solve the issue when close points to the query were mapped to different buckets. Moreover, QALSH uses B+-trees as its index structure for efficient lookups and range queries on the hash functions.

I-LSH [20] was recently proposed to improve the I/O cost of QALSH by incrementally increasing the search radius in the projected space instead of using exponential radius increases. However, as shown in [16], I-LSH achieves this I/O cost optimization at the expense of a costly processing time spent on finding closest points in the projected space. Recently, PM-LSH [26] was proposed to utilize a confidence interval value and estimate the Euclidean distance with the goal of reducing the overall query processing time. Moreover, a method, called R2LSH [21], was proposed that uses two-dimensional projected spaces (instead of one-dimensional spaces) to improve the I/O cost of the query processing.

2.2 Fairness in ANN Search

Definitions of fairness are commonly categorized as 1) group fairness, in which the aim is to treat different groups equally, and 2) individual fairness, in which the purpose is to treat individual people of the same profile similarly [22]. So far, only two works have studied the idea of fairness in the ANN domain.

[13] proposes to remove the bias from exact neighborhood and approximate neighborhood (E2LSH) searches by using sampling techniques with the goal of providing individual fairness. Additionally, another work [3] has been proposed that considers equal opportunity in individual fairness in the sense that all points near a query should have the same probability to be returned. In [3], authors first use uniform sampling techniques and then build a data structure for fair similarity search under inner product. Very recently, a joint work [2] containing [13] and [3] has been proposed that connects ideas from both works (i.e. equal opportunity and independent range sampling). Our proposed work is different from the prior works [2,3,13] in two main aspects: 1) Unlike these prior works that focus on individual fairness, our work focuses on equal opportunity in group fairness, and 2) these prior works are designed specifically for the original LSH design (E2LSH). Particularly, their proposed data structures are applicable only for LSH designs that use the multiple hash functions in multiple hash tables (Compound Hash Keys). State-of-the-art LSH designs, such as [11,15,20], use advanced techniques such as *Collision Counting* that makes having multiple hash tables unnecessary (thus saving on space and time). Our proposed work is designed specifically for these state-of-the-art LSH designs.

3 Background and Key Concepts

In this section, we describe the key concepts behind LSH. Given a dataset $\mathcal{D}$ with n points in a d-dimensional Euclidean space $\mathcal{R}^d$ and a query point q in the same space, the goal of c-ANN search (for an approximation ratio $c > 1$) is to return points $o \in \mathcal{D}$ such that $\|o-q\| \leq c\times \|o^* - q\|$, where o^* is the true nearest neighbor of q in $\mathcal{D}$ and $\|\|$ is the Euclidean distance between two points. Similarly, c-k-ANN search aims at returning top-k points such that $\|o_i - q\| \leq c \times \|o_i^* - q\|$ where $1 \leq i \leq k$.

Definition 1 (LSH Family). *A hash function family $\mathcal{H}$ is called (r, c, p_1, p_2)-sensitive if it satisfies the following conditions for any two points x and y in a d-dimensional dataset $\mathcal{D} \subset \mathcal{R}^d$:*

- *if $\|x - y\| \leq R$, then $Pr[h(x) = h(y)] \geq p_1$, and*
- *if $\|x - y\| > cR$, then $Pr[h(x) = h(y)] \leq p_2$*

Here, p_1 and p_2 are probabilities, R is the distance between two points (commonly referred to as the radius), and c is an approximation ratio. LSH requires $c > 1$ and $p_1 > p_2$. The conditions show that the probability of mapping two points to a same hash value decreases as their distance increases.

Definition 2 (Collision Counting). *In [11], it is theoretically shown that only those points that collide (are mapped to the same bucket) with the query in at least l projections (out of m) are chosen as candidates. Here, l is the collision count threshold and calculated as $l = \lceil \alpha \times m \rceil$, where α is the collision threshold percentage calculated using $\alpha = \frac{zp_1+p_2}{1+z}$ and m is the total number of projections calculated as $m = \lceil \frac{\ln(\frac{1}{\delta})}{2(p_1-p_2)^2}(1+z)^2 \rceil$. Here, $z = \sqrt{\ln(\frac{2}{\beta})/\ln(\frac{1}{\delta})}$, where β is the allowed false positive percentage (i.e. the allowed number of points whose distance with a query point is greater than cR). [11] sets $\beta = \frac{100}{n}$, where n is the cardinality of the dataset.*

Definition 3 (Virtual Rehashing). *[11] starts query processing with a very small radius, and then, exponentially increases the radius in the following sequence: $R = 1, c, c^2, c^3, ...$, where c is an approximation ratio. If at level-R, enough candidates are not found, the radius is increased until found.*

4 Problem Specification

Definition 4 (Fair ANN). *The definition of Fair ANN in this paper is focused on the equal opportunity problem in group fairness in the ANN domain. Given a dataset $\mathcal{D}$ with n points in a d-dimensional Euclidean space $\mathcal{R}^d$, a query point q, and two groups of points in $\mathcal{D}$ labeled o^A and o^B, the goal of Fair ANN is to find top-$\lfloor k/2 \rfloor$ points $o^A \in \mathcal{D}$ and top-$\lfloor k/2 \rfloor$ points $o^B \in \mathcal{D}$, such that $\|o_i^A - q\| \leq c \times \|o_i^{A,*} - q\|$ and $\|o_i^B - q\| \leq c \times \|o_i^{B,*} - q\|$, where $1 \leq i \leq \lfloor k/2 \rfloor$, and $o_i^{A,*}$ and $o_i^{B,*}$ are the true nearest neighbors of q in $\mathcal{D}$ from each group.*

In this paper, our goal is to return Fair ANN for a given query q while reducing the overall I/O costs and processing times while maintaining the accuracy of the result. Note that, in this work, we only focus on two distinct groups in the dataset. We leave the problem of Fair ANN for multiple groups as future work. In Sect. 5, we present the design of our index structure, *FairLSH.*

5 FairLSH

In this section, we first describe the naive approaches for solving the Fair ANN problem using the existing LSH methods. We then present the design of our proposed index structure, *FairLSH,* which consists of two variants: *FairLSH-Basic* and *FairLSH-Advanced.* Given a query point, our goal is to efficiently return top-$\lfloor k/2 \rfloor$ NN from each of the two groups of points in the dataset.

5.1 Naive Approaches

In Sect. 3, we explained how LSH families are used to map high-dimensional points into a lower-dimensional space while preserving locality. In order to retrieve fair results, we make the following changes to existing LSH methods:

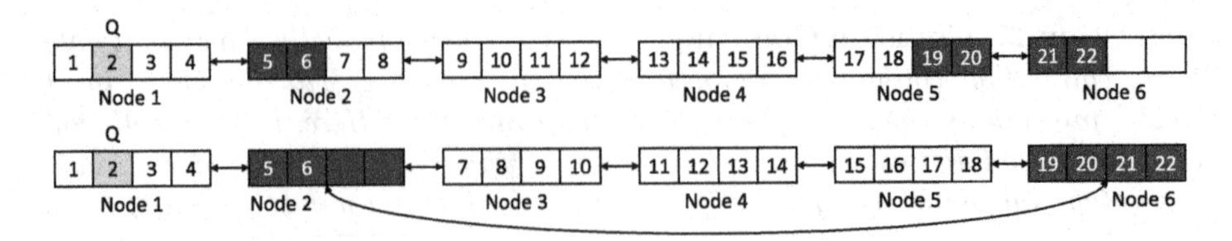

Fig. 1. (a) QALSH leaf nodes (top) compared to (b) FairLSH leaf nodes (bottom)

Naive Strategy 1: A naive strategy to find top-$\lfloor k/2 \rfloor$ nearest points from each group is to simply divide the dataset into two separate datasets (based on their labels), and run LSH individually on the two separate datasets. The drawbacks of this strategy are: 1) it is not space efficient since two sets of indexes need to be maintained, and 2) redundant processing needs to be performed on the two sets of indexes which results in an increase in the overall query processing time.

Naive Strategy 2: All LSH-based methods have several stopping conditions that make the search algorithm stop once enough points are found. In the second naive strategy, we change these stopping conditions to continue the search algorithm until enough points are found *from each group of points.*
For example, Fig. 1(a) shows the leaf nodes of a B+-tree from QALSH [15]. In this example, the query is point ID 2 which is located in *Node 1*, white points are from group A, and red points are from group B. Assuming that $k = 8$, then our goal is to find four nearest points from each of the groups A and B.

In this example, the original QALSH method will start reading *Node 1* and fetching the first four points and it will continue reading *Node 2*. After that, since the stopping conditions are met, the algorithm will stop. However, we only find two nearest points (5 and 6) and the results are not fair. Naive strategy 2 changes stopping conditions such that the algorithm continues reading *Node 3* and *Node 4* as well. By doing this, the algorithm finds enough points from each group and if more than enough points are found, they can be pruned at another step using Euclidean distance calculations.

The main drawback of this naive strategy (as it can also be seen in the given example) is that extra and unnecessary nodes are read which results in an increase in the I/O cost. As a result, we present two variants of a novel index structure in the next section that use threaded B+-trees and cost models to optimize the I/O cost and processing time.

5.2 Design of FairLSH-Basic

The main intuition behind *FairLSH-Basic* is that when enough nearest neighbors of a group are found, we should avoid reading the points of that group from the disk. Hence, our goal is to skip those points that will lead to unnecessary I/O in order to improve processing time.

Skipping the points in the current LSH index structures has several challenges that include: 1) Since hashed points are ordered without considering their groups, when reading a "page-size" of data from the disk, we might get points from

Algorithm 1: Query phase of FairLSH-Basic and FairLSH-Advanced

Input: $\mathcal{D}$ is the dataset; $\boldsymbol{q}$ is the query point; $\mathcal{L}$ is the label of dataset points; $\mathcal{I}$ is m index structures created in indexing phase; k is the number of nearest neighbors to find; l is the collision threshold, $\boldsymbol{a}$ is the random vector generated in the indexing phase.
Output: $\lfloor k/2 \rfloor$ nearest points to $\boldsymbol{q}$ in $\mathcal{D}$ from each label
Variable: cc is the collision count of points; $\mathcal{C}$ is the candidates list; R is the number of nodes to read

```
1  h(q) := a.q;
2  Find query leaf node in m index structures;
3  Let R = 1;
4  while TRUE do
5      for i = 1 to m do
6          Read R nodes around the query node;
7          if ⌊k/2⌋ label A nearest neighbors are found then
8              Set next round nodes to only label B nodes;
9          else
10             Set next round nodes to adjacent nodes;
11         foreach point o in leaf node do
12             cc[o] := cc[o] + 1 if cc[o] ≥ l then
13                 Add o to C;
14     if (|{o|o ∈ C ∧ L[o] = 1 ∧ ||o − q|| ≤ c × R}| ≥ ⌊k/2⌋) and
       (|{o|o ∈ C ∧ L[o] = 2 ∧ ||o − q|| ≤ c × R}| ≥ ⌊k/2⌋) then
15         break;
16     R := R × c;
17 return {o|o ∈ C ∧ L[o] = 1 ∧ ||o − q|| ≤ c × R} and
   {o|o ∈ C ∧ L[o] = 2 ∧ ||o − q|| ≤ c × R}
```

different groups (e.g. node 2 of Fig. 1(a)), and 2) The current index structures only have pointers to the sibling nodes and there is no possibility to avoid certain nodes of hash functions (that contain unnecessary data) in the index structure.

FairLSH-Basic uses a *group-aware* strategy when creating the leaf nodes to only allow points belonging to the same group to be added to a single leaf node and to create a new leaf node when a new point belongs to a different group. Furthermore, *FairLSH-Basic* uses threaded B+-tree structures that allow arbitrary pointers between different nodes of the tree. In the current version of *FairLSH-Basic*, these arbitrary pointers are created between leaf nodes belonging to group B and the idea of using smart pointers where the algorithm can detect which nodes require pointers between them is left for future work.

In the query phase of *FairLSH-Basic*, the goal is to find $\lfloor k/2 \rfloor$ nearest points from each group to a given query point. Similar to other tree-based LSH methods, the leaf nodes are searched in an exponential search radius manner (starting from the query node) and the collision counting process (explained in Sect. 3) is carried away. However, when enough nearest neighbors are found from a group

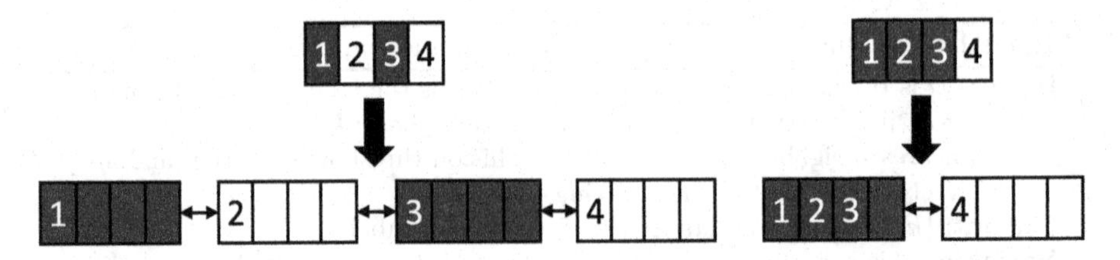

Fig. 2. Breakdown of node containing (a) sparse and (b) dense groups

and we only need to get nearest neighbors from the other group, *FairLSH* uses its arbitrary pointers to skip unnecessary nodes instead of doing a range query (line 8 of Algorithm 1).

An example of how the search process of *FairLSH-Basic* works is shown in Fig. 1 (b). In this example, the query resides in Node 1, white points are from group A, and red points are from group B. With an assumption of $k = 8$, our goal is to find four nearest neighbors from each group. *FairLSH-Basic* starts by reading Node 1 and fetching the first four points, and now, we have enough neighbors from group one (assuming that the query is also a point from group one). After that, *FairLSH-Basic* reads Node 2 and fetches two points from group two. From this moment, since we only need to read group two points, the algorithm jumps to Node 6 and reads the remaining points from that node. This way, *FairLSH-Basic* has saved the I/O cost of reading three unnecessary nodes (3, 4, and 5). Note that the indexing phase prevents points from multiple groups to be in the same node (e.g. points 17, 18, 19, and 20 in one node); therefore, *FairLSH-Basic* can always skip unnecessary nodes.

It is worth mentioning that Fig. 1 is only showcasing one projection as a simple example; in the real scenario, we have several hash projections with more complex distributions. In Sect. 6, we show that *FairLSH-Basic* performs much better than state-of-the-art techniques.

5.3 Drawbacks of FairLSH-Basic

The main benefit of *FairLSH-Basic* is that it is effective in reducing disk I/O costs when points from different groups are not sparsely distributed in the nodes. Although the indexes are created offline, *FairLSH-Basic* has a processing overhead in the query phase which is related to the extra pointers between the nodes (compared to only performing a range query search). However, this processing overhead is negligible compared to the savings in disk I/O cost, especially when we have a dense distribution of group points. In this scenario, as shown in Fig. 2 (a), *FairLSH-Basic* breaks down the nodes containing a mixed group of points (i.e. points from group A and group B) and eliminates the need to read unnecessary points from the indexes.

On the other hand, the processing overhead of *FairLSH-Basic* increases when points from different groups are sparsely distributed in the nodes. Figure 2(b) shows an example of this scenario. In this scenario, *FairLSH-Basic* breaks down the nodes which results in multiple nodes containing only one point. As a result, the indexes will contain more pointers and more "book-keeping" is required during the query processing phase. We also observed that in reality, both of the mentioned scenarios (i.e. sparse and dense distribution) happen in the indexes. In other words, different hash functions in the same index structure can have nodes containing a sparse distribution of group points, and also nodes containing a dense distribution of group points.

5.4 Design of FairLSH-Advanced

To remedy the drawbacks of *FairLSH-Basic*, we present a cost-based strategy, *FairLSH-Advanced*, that can smartly detect if breaking down a node is going to positively or negatively affect the overall performance.

There are three costs associated with reading a node into the main memory: 1) C_s: cost of disk seeks, 2) C_h: cost of reading the header (i.e. pointers), and 3) C_p: cost of reading the payload (i.e. points). C_s is defined as $IO_{calls} \times seek_{speed}$, where $seek_{speed}$ is the time it takes to perform a disk seek and can be obtained from disk manufacturer or by benchmarking the disk. C_h is defined as $node_{count} \times node_{headersize} \times IO_{speed}$, where $node_{count}$ is the number of nodes we are going to read, $node_{headersize}$ is the header size of each node, and IO_{speed} is the time it takes to read data from the disk and can be obtained from disk manufacturer. Finally, C_p is defined as $points_{count} \times IO_{speed}$, where $points_{count}$ is the number of points that we want to read from the disk.

Given a node in the indexing phase, the goal of *FairLSH-Advanced* is to decide whether breaking down a node is going to be beneficial or not. Therefore, we introduce a cost to represent each scenario (i.e. C_{Before} and C_{After} for the cost before and after breaking down the node respectively) and the difference of these two costs can be used to make the decision. Thus, we have:

$$C_{Before} = C_{s,B} + C_{h,B} + C_{p,B} \tag{1}$$

$$C_{After} = C_{s,A} + C_{h,A} + C_{p,A} \tag{2}$$

Note that $C_{s,B}$, $C_{h,B}$, and $C_{p,B}$ are the cost of disk seeks, cost of reading the header, and cost of reading the payload respectively before we break down a node, and $C_{s,A}$, $C_{h,A}$, and $C_{p,A}$ are the costs after we break down a node.

As an example, we consider the nodes in Fig. 2 and assume that $seek_{speed} = 6$, $node_{headersize} = 3$, and $IO_{speed} = 13$. In Fig. 2(a), before breakdown, we have $IO_{calls} = 1$, $node_{count} = 1$, and $points_{count} = 4$. When we break the node down, we have $IO_{calls} = 4$, $node_{count} = 4$, and $points_{count} = 4$. Therefore, we have $C_{Before} = (1 \times 6) + (1 \times 3 \times 13) + (4 \times 13) = 97$, $C_{After} = (4 \times 6) + (4 \times 3 \times 13) + (4 \times 13) = 232$, and $C_{After} - C_{Before} = 232 - 97 = 135$. Similarly, in Fig. 2(b), we have $C_{Before} = (1 \times 6) + (1 \times 3 \times 13) + (4 \times 13) = 97$, $C_{After} = (2 \times 6) + (2 \times 3 \times 13) + (4 \times 13) = 142$, and $C_{After} - C_{Before} = 142 - 97 = 45$.

$$\begin{cases} \textit{break down}, & \textit{if } C_{After} - C_{Before} \leq \theta \\ \textit{do not break down}, & \textit{if } C_{After} - C_{Before} > \theta \end{cases} \tag{3}$$

FairLSH-Advanced utilizes a user-input parameter, called θ, in the indexing phase. As shown in Eq. 3, if the $C_{After} - C_{Before}$ of a node is lower than θ, the node will be broken down and vice versa. Since nodes have different costs, it is crucial to find a good θ value such that the index will be efficient enough in the query processing phase. Note that the query processing phase of *FairLSH-Advanced* is similar to *FairLSH-Basic* (Algorithm 1). In Sect. 6, we show how using the cost-based strategy and the user-input parameter can improve the performance of *FairLSH-Basic.*

6 Experimental Evaluation

In this section, we evaluate the effectiveness and fairness of our two proposed methods, *FairLSH-Basic* and *FairLSH-Advanced.* All experiments were run on a machine with the following specifications: Intel Core i7-6700, 16 GB RAM, 2TB HDD, and Ubuntu 20.04 OS. All codes were written in C++ and compiled with gcc v9.3.0 with the -O3 optimization flag. Since the code of PM-LSH was not released when writing this paper, we compare our two strategies with the following state-of-the-art disk-based alternatives:

- **C2LSH:** Fair top-k results are found using C2LSH [11].
- **QALSH:** Fair top-k results are found using QALSH [15].

We modified existing state-of-the-art algorithms (C2LSH and QALSH) to output fair nearest neighbors by using the naive strategy explained in Sect. 5.1.

6.1 Datasets

We ran our experiments on two real datasets Mnist [23] and Sift [25] where the group labels are randomly assigned. In addition, in order to cover different scenarios that might happen in different applications, we construct seven synthetic datasets. There are two groups in each dataset and the goal of this paper is to give both of these groups the same opportunity (i.e. equal opportunity) to appear in the final results. We randomly assign a binary label (A or B in our explanation) to each dataset point to represent these groups. Each one of the groups contain 50% of the dataset. In this work, we experiment with scenarios where A and B have different distributions. Label A data points are generated using a Beta distribution with $\alpha = 2$ and $\beta = 8$, and label B data points are generated using a Beta distribution with $\alpha = 8$ and $\beta = 2$. Table 1 summarizes the characteristics of our datasets. We choose 100 random points as our queries and report the average as the final result. Due to space limitations and since we observed similar results for all synthetic datasets, we only include two synthetic datasets in this paper.

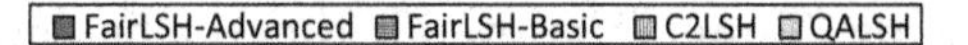

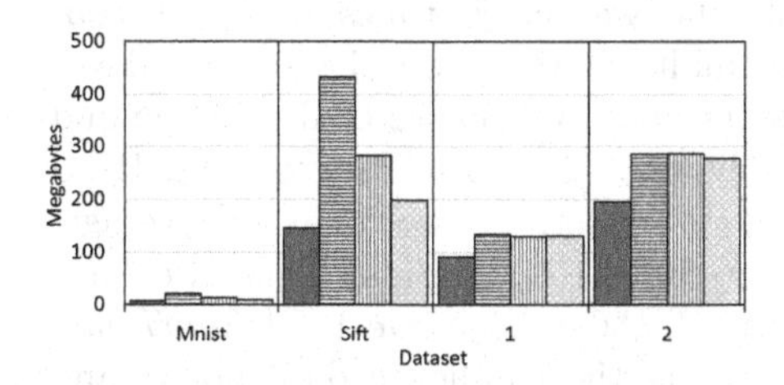

Fig. 3. Amount of data read

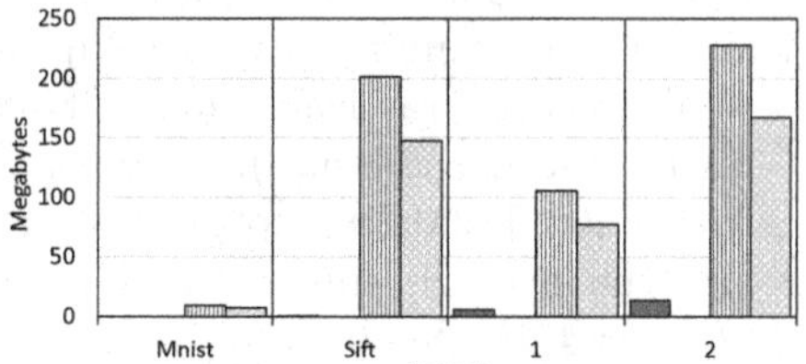

Fig. 4. Amount of Wasted IO

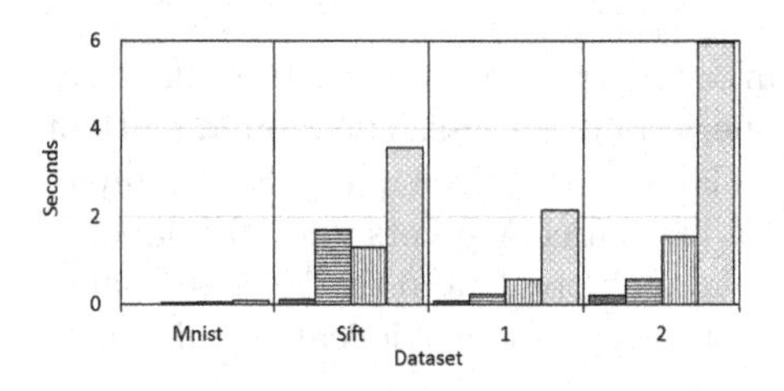

Fig. 5. Algorithm time

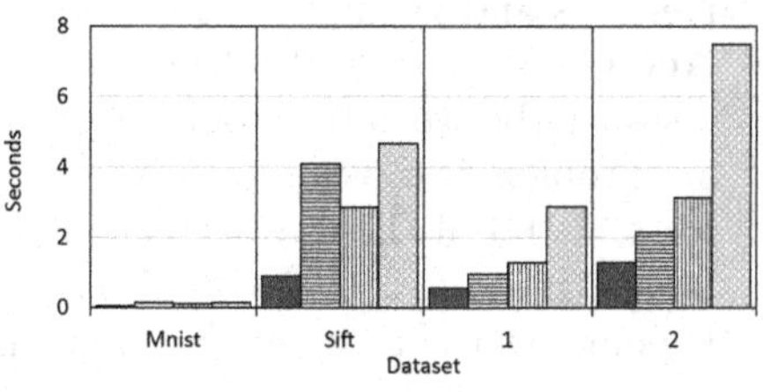

Fig. 6. Query processing time

Table 1. Characteristics of the datasets

Name	# of Points	# of Dim.
Mnist	60,000	50
Sift	1,000,000	128
D1	500,000	1,000
D2	1,000,000	1,000

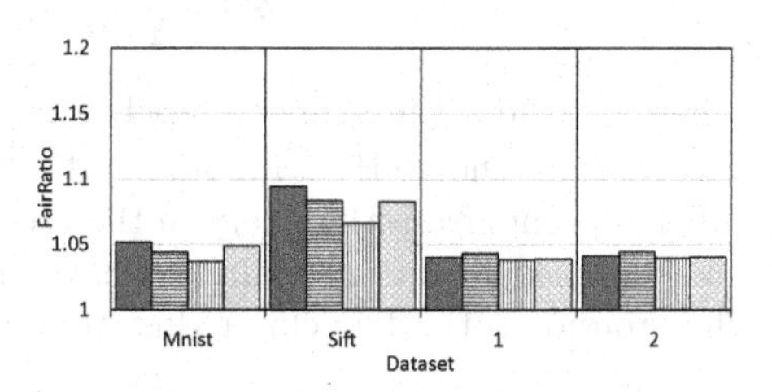

Fig. 7. Fairness accuracy ratio

6.2 Evaluation Criteria

The performance and fairness of the compared techniques are evaluated using the following criteria:

- **Index IO Size:** The amount of total data read from index files.
- **Wasted IO Size:** The amount of unnecessary data read from index files (e.g. reading label A points while we have enough label A candidates and should only look for label B candidates).
- **Algorithm Time:** The processing time of index files once they are read into the main memory. The algorithm time consists of operations such as Collision Counting which are explained in Sect. 3.

- **Query Processing Time:** The overall time of finding fair approximate nearest neighbors. We observed that the wall-clock times were not consistent (i.e. running the same query multiple times on the same indexes would return drastically different results, mainly because of disk cache and instruction cache issues). Therefore, following [24], for a Seagate 1TB HDD with 7200 RPM, we consider a random seek to cost 8.5 ms on average, and the average time to read data to be 0.156 MB/ms. Thus, we have $QPT = \#DiskSeeks \times 8.5 + dataRead \times 0.156 + AlgTime + FPRemTime$, where $FPRemTime$ is the cost of reading the candidate data points and computing their exact Euclidean distance for removing false positives. We do not report individual $FPRemTime$ results since they are similar for different methods and negligible (less than one millisecond).
- **Accuracy:** In order to define an accuracy metric, we consider the Euclidean distance (between the candidate and query) and fairness. The ground truth of our problem is the closest points of each label to the given query. For example, for $k = 100$, if the dataset is split into 50% label A points and 50% label B points, the goal is to find the 50 closest label A points and the 50 closest label B points to the given query. We define our accuracy metric, called FairRatio, as following:

$$FairRatio = \frac{1}{k}\left(\sum_{i=1}^{\lfloor\frac{k}{2}\rfloor}\frac{\left\|o_i^A - q\right\|}{\left\|o_i^{A,*} - q\right\|} + \sum_{i=1}^{\lfloor\frac{k}{2}\rfloor}\frac{\left\|o_i^B - q\right\|}{\left\|o_i^{B,*} - q\right\|}\right) \tag{4}$$

where o_i^A and o_i^B are the ith label A and label B points respectively that are returned by the LSH technique, and $o_i^{A,*}$ and $o_i^{B,*}$ are the ith label A and label B points from the query in the ground truth. FairRatio of 1 means that the returned results are fair and have the same distance from the query as the ground truth. The closer this value is to 1, the higher the accuracy.

6.3 Parameter Settings

For the state-of-the-art methods, we used the same parameters suggested in their papers. For QALSH, *FairLSH-Basic*, and *FairLSH-Advanced* (since we use the same hashing formula as QALSH as explained in Sect. 5), we used $w = 2.781$, $\delta = 0.1$, and $c = 2$. For C2LSH, we used $w = 2.184$, $\delta = 0.1$, and $c = 2$.

In this work, we focus on only two data point labels (A and B) in the dataset, $k = 100$, and the goal of finding 50 nearest points from label A and 50 nearest points from label B. We leave experimenting on other parameter settings for future work. For *FairLSH-Advanced*, we tried different values of θ and observed improvements over compared methods for all experimented values. Due to space limitations, we only show results for $\theta = 100$.

6.4 Discussion of the Results

In this section, we compare the performance, accuracy of *FairLSH-Basic* and *FairLSH-Advanced* against the state-of-the-art methods.

- **Index IO Size:** Figure 3 shows the total amount of data read from the index files. Since C2LSH has more index files compared to other methods, it has a higher I/O cost. For the real datasets, *FairLSH-Basic* has a higher amount of data read because of the sparse distribution of the groups (Sect. 5.3). *FairLSH-Advanced* has the lowest amount of index I/O since its index structure is further optimized to skip reading unnecessary node headers.
- **Wasted IO Size:** Figure 4 shows the amount of unnecessary data read from the index files. Wasted I/O happens when, in the query processing phase, enough nearest neighbors from group A are already found; but the algorithm keeps reading data related to group A. The wasted I/O size of *FairLSH-Basic* and *FairLSH-Advanced* are several orders of magnitude smaller compared to C2LSH and QALSH. *FairLSH-Advanced* has a slightly higher wasted I/O size than *FairLSH-Basic* since it is sacrificing wasted I/O over algorithm time.
- **Algorithm Time:** Figure 5 shows the time needed to find the candidates (excluding the time taken to read index files). QALSH has the highest algorithm time since it uses non-optimized B+-trees that become significantly larger as the dataset size grows. It is interesting to note that although *FairLSH-Basic* and *FairLSH-Advanced* use more complex tree structures (due to more number of pointers), their algorithm time is lower than C2LSH and QALSH. This is due to avoiding processing of unnecessary nodes which offsets the overhead of using additional tree pointers. However, because of the sparse distribution of the groups (Sect. 5.3) in the real datasets, the overhead of *FairLSH-Basic* is higher for real datasets. *FairLSH-Advanced* has a lower algorithm time than *FairLSH-Basic* since it is using cost models to optimize the index structures and balance the trade-off between wasted I/O cost and algorithm time.
- **Query Processing Time:** Figure 6 shows the overall time required to solve a given k-NN query and retrieve fair neighbors. *FairLSH-Advanced* is the fastest method compared to the others since its index structures are optimized to significantly reduce algorithm time while increasing wasted I/O cost slightly.
- **Accuracy:** Figure 7 shows the accuracy of all techniques. C2LSH and QALSH have a similar accuracy for all datasets, and *FairLSH-Basic* and *FairLSH-Advanced* have a slightly lower accuracy (i.e. higher FairRatio). The reason of this difference is because when *FairLSH-Basic* and *FairLSH-Advanced* find enough nearest neighbors of a group, they stop reading and processing points belonging to that group. However, C2LSH and QALSH continue this process and get more neighbors and return the closest k neighbors at the end. We should mention that a difference of 0.05 in FairRatio is very small compared to the savings in I/O costs and algorithm time. In addition, we analyzed the returned results and observed that similar nearest neighbors are returned by all experimented methods (i.e. 0.1 difference in terms of average precision).

7 Conclusion and Future Work

In this paper, we define the group fairness notion of *Equal Opportunity* in the context of Approximate Nearest Neighbor domain. We proposed a novel index structure for efficiently finding fair top-k approximate nearest neighbors using Locality Sensitive Hashing, called *FairLSH*. Existing LSH-based techniques are not capable of efficiently finding fair nearest neighbors to a given query. We proposed two novel strategies, *FairLSH-Basic* and *FairLSH-Advanced*, which uses threaded B+-trees and advanced cost models to optimize the overall query processing cost. Experimental results show the benefit of our proposed structures over state-of-the-art techniques. In the future, we plan to introduce a user-defined parameter to adjust the trade-off between fairness, accuracy, and processing time. We also plan to provide theoretical guarantees for the results of *FairLSH*.

References

1. Agarwal, A., et al.: A reductions approach to fair classification. arXiv (2018)
2. Aumüller, M., et al.: Fair near neighbor search via sampling. SIGMOD Rec. **50**(1), 42–49 (2021)
3. Aumüller, M., et al.: Fair near neighbor search: Independent range sampling in high dimensions. In: SIGMOD (2020)
4. Bera, S., et al.: Fair algorithms for clustering. In: NIPS (2019)
5. Chávez, E., et al.: Searching in metric spaces. CSUR **33**(3), 273–321 (2001)
6. Chierichetti, F., et al.: Matroids, matchings, and fairness. In: AISTATS (2019)
7. Chouldechova, A.: Fair prediction with disparate impact: a study of bias in recidivism prediction instruments. Big Data **5**(2), 153–163 (2017)
8. Datar, M., et al.: Locality-sensitive hashing scheme based on p-stable distributions. In: SOCG (2004)
9. Donini, M., et al.: Empirical risk minimization under fairness constraints. In: NIPS (2018)
10. Elzayn, H., et al.: Fair algorithms for learning in allocation problems. In: FAccT (2019)
11. Gan, J., et al.: Locality-sensitive hashing scheme based on dynamic collision counting. In: SIGMOD (2012)
12. Gionis, A., et al.: Similarity search in high dimensions via hashing. In: VLDB (1999)
13. Har-Peled, S., et al.: Near neighbor: who is the fairest of them all? In: NIPS (2019)
14. Hardt, M., et al.: Equality of opportunity in supervised learning. In: NIPS (2016)
15. Huang, Q., et al.: Query-aware locality-sensitive hashing for approximate nearest neighbor search. VLDB **9**(1), 1–12 (2015)
16. Jafari, O., Nagarkar, P.: Experimental analysis of locality sensitive hashing techniques for high-dimensional approximate nearest neighbor searches. In: Qiao, M., Vossen, G., Wang, S., Li, L. (eds.) ADC 2021. LNCS, vol. 12610, pp. 62–73. Springer, Cham (2021). https://doi.org/10.1007/978-3-030-69377-0_6
17. Jafari, O., et al.: A survey on locality sensitive hashing algorithms and their applications. arXiv (2021)
18. Kleinberg, J., et al.: Human decisions and machine predictions. QJE **133**(1), 237–293 (2018)
19. Kleindessner, M., et al.: Guarantees for spectral clustering with fairness constraints. arXiv (2019)
20. Liu, W., et al.: I-LSH: I/O efficient c-approximate nearest neighbor search in high-dimensional space. In: ICDE (2019)

21. Lu, K., Kudo, M.: R2LSH: a nearest neighbor search scheme based on two-dimensional projected spaces. In: ICDE (2020)
22. Mehrabi, N., et al.: A survey on bias and fairness in machine learning. arXiv (2019)
23. MNIST (1998). http://yann.lecun.com/exdb/mnist
24. Seagate ST2000DM001 Manual (2011). https://www.seagate.com/files/staticfiles/docs/pdf/datasheet/disc/barracuda-ds1737-1-1111us.pdf
25. SIFT (2004). http://corpus-texmex.irisa.fr
26. Zheng, B., et al.: PM-LSH: a fast and accurate LSH framework for high-dimensional approximate NN search. VLDB **13**(5), 643–655 (2020)

Fairest Neighbors

Tradeoffs Between Metric Queries

Magnus Lie Hetland(✉) and Halvard Hummel

Norwegian University of Science and Technology, Trondheim, Norway
{mlh,halvard.hummel}@ntnu.no

Abstract. Metric search commonly involves finding objects similar to a given sample object. We explore a generalization, where the desired result is a fair tradeoff between multiple query objects. This builds on previous results on complex queries, such as linear combinations. We instead use measures of inequality, like ordered weighted averages, and query existing index structures to find objects that minimize these. We compare our method empirically to linear scan and a post hoc combination of individual queries, and demonstrate a considerable speedup.

Keywords: Metric indexing · Multicriteria decisions · Fairness

1 Introduction

From the early days, indexing metric spaces has mainly been in service of straightforward similarity search: Given some query object q, find other objects o for which the distance $d(q, o)$ is low—either all points within some search radius, or a certain number of the nearest neighbors. Alternative forms of search have been explored, certainly. Of particular interest to us is using multiple query objects q_i, without restricting the indexing methods used. That is, we wish to take any existing metric index, already constructed, and execute a combination query on it. Such a query may be specified directly by the user, or it may be a form of interactive refinement. A user first performs a query using a single object. Then she indicates which of the returned objects are most relevant (possibly to varying degrees), and these are then used as a second, combined query. The result should ideally be a tradeoff between the query objects. In particular, we wish to ensure that it is a *fair* tradeoff, borrowing measures of fairness from the field of multicriteria decision making.

Our Contributions. In this short paper, we introduce the idea of *fairest neighbors* (kFN), i.e., items that are close to multiple query objects at once, as measured by some kind of fairness measure. For example, if we are looking for a centaur, using a human and a horse, a simple linear combination will not do, as the best results are then just as likely to be similar to just the human or just the horse; only a fair combination would give us what we want. We formulate such queries in the context of the complex queries of [4], but extend the formalism by applying linear ambit overlap [7] to ordered weighted averages (OWA) and

N. Reyes et al. (Eds.): SISAP 2021, LNCS 13058, pp. 148–156, 2021.
https://doi.org/10.1007/978-3-030-89657-7_12

weighted OWA, for improved bounds. The resulting queries may be resolved using existing metric index structures *without modification*. We perform preliminary experimental feasibility tests, showing that such combined kFN queries outperform both linear scan and using multiple separate kNN queries.

Related Work. Others have studied metric search with multiple simultaneous criteria. As discussed by [4], Fagin's $\mathcal{A}_0$ algorithm also resolves complex queries, but makes additional assumptions about synchronized independent subsystems. Bustos and Skopal [1] study a superficially similar problem that involves a linear combination of multiple metrics, while still using a single query object. More closely related are *metric skylines*, which are essentially Pareto frontiers in pivot space [3]. These result sets will be diverse, and will tend to include both fair and unfair solutions. Our approach moves beyond non-Pareto-dominated solutions to non-Lorenz-dominated solutions [cf. 6].

2 Complex Queries as Multicriteria Decisions

In 1998, Ciaccia et al. introduced a formalism for dealing with what they called *complex queries* in metric indexes—queries involving multiple query objects, along with some domain-specific query *language*, specifying which objects are relevant and which aren't [4]. Part of their formalism involves mapping distances to similarity measures, which are then constrained by some query predicate; however, the core ideas apply equally well to distances directly. A central insight is that *monotone* predicates may be used not just to detect whether an *object* is relevant, but also whether certain *regions* might contain relevant objects.

Let $x = [d(q_i, o)]_{i=1}^m$ be a vector of distances between query objects q_i and some potentially relevant object o. Relevance is then defined by some predicate on these distances, $\mathrm{P}(x)$. This predicate is *monotone* if for all $x \leq y$ we have that $\mathrm{P}(y)$ implies $\mathrm{P}(x)$. That is, if we start with the distance vector of a relevant object, and we reduce one or more of the distances, the resulting vector should *also* be judged as relevant. In this case, using *lower bounds* for the individual distances is safe (i.e., it will not cause false negatives). So, for example, if we know that o is in a ball with center c and radius r, we can safely replace $[d(q_i, o)]_i$ with $[d(q_i, c) - r]_i$ and apply P to determine whether or not to examine the ball. Using this approach, one can find the k best objects by maintaining a steadily shrinking search radius encompassing the k best candidates found so far, just as one would for kNN. The idea is illustrated in Fig. 1: The vector $x - r$ of lower bounds corresponds to the lower left corner of the square enveloping the region in pivot space (indicated by a '+' in the right-hand subfigure). A monotone query and a ball region may overlap only if this lower left point is inside the query definition in this space [cf. 7]. Similarly, we may conclude that the region is *entirely inside* the query (and thus return all its objects without further examination) if the upper right corner $(x + r)$ satisfies the query predicate.

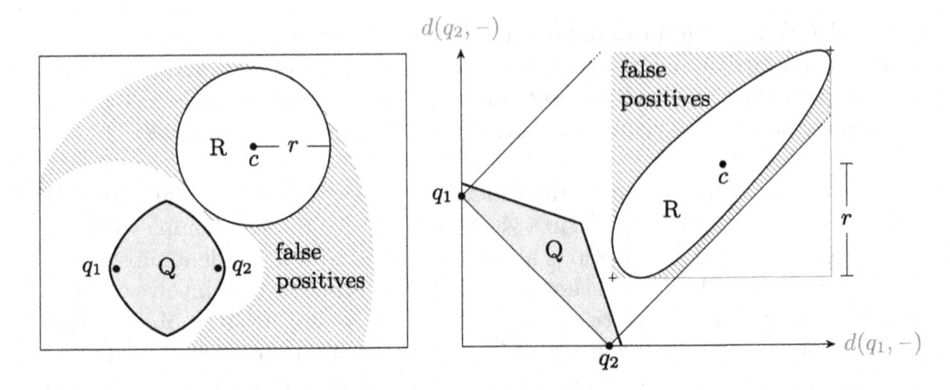

Fig. 1. A complex query Q with two query objects q_i and a ball region R. The right-hand subfigure shows the query and region in pivot space, where the two axes correspond to distances from the two query objects. The hatched areas could potentially fall within the ball R, depending on the nature of the metric; what remains outside R constitutes potential false positives. (Hetland discusses these concepts in depth [7].)

Two of the query types discussed explicitly by Ciaccia et al. are based on fuzzy logic, and one uses a weighted sum. These permit indicating degrees of relevance for the various query objects q_i, but may have *many equally good solutions*, with vastly different properties. What can be done if we wish to enforce some form of actual tradeoff? Consider a query predicate of the form $f(x) \leq s$. That is, we apply some monotone function f to the vector x of distance $d(q_i, o)$ and are only interested in objects o for which $f(x)$ falls below some search radius s. Different monotone functions f may yield very different query regions:

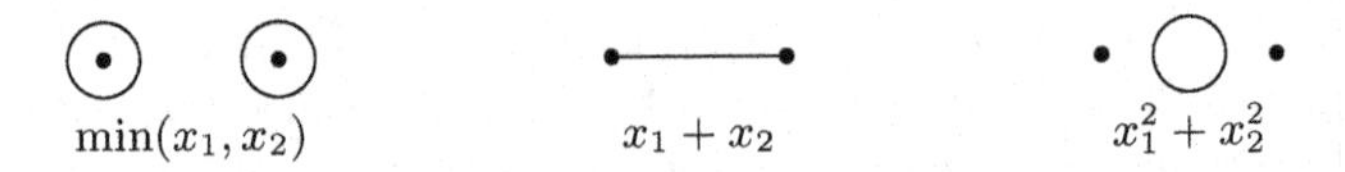

Minimum (corresponding to maximum, or standard fuzzy disjunction, in the similarity formalism of Ciaccia et al.) produces results that are close to one or the other of the two query points, but not both. A sum gives us points that can lie anywhere between the two (in general within an ellipsoid). A sum of positive powers, however, produces items that are *between* the query points—ideally in the middle (i.e., in their metric midset). This is the kind of query we want.

Using sums of powers to characterize tradeoffs is a common approach in cardinal welfarism, and it is one of a broader class of aggregation functions used in multicriteria decision making [6].[1] These are all generally monotonically increasing, with the optimum found for some fair tradeoff between their arguments. Applied to individual query distances $d(q_i, o)$, our measure will of course need

[1] Though Ciaccia et al. do not directly address fairness or tradeoffs, their *standard* and *algebraic fuzzy conjunctions*, correspond to the *maximin* and *Nash welfare* fairness measures, respectively, if applied, in isolation, to similarities [4].

to be *minimized*, and so must be an *unfairness* measure, rather than a *fairness* measure. In the following, we will focus on *ordered weighted average* (OWA), and its generalization, *weighted* OWA. The OWA of some vector x is based on a weighting of the elements of x, just like a weighted average, except that the weights are applied based on the rank of each element x_i. Given a weight vector $w \geq 0$, summing to 1, the OWA of x is $wx^{\uparrow}$, where, $x^{\uparrow}$ is a sorted version of x. As discussed in Sect. 3 (in a more general setting), by ensuring that w is also sorted, we get an unfairness measure. Our overlap check with an r-ball, using the complex query formalism, becomes:

$$f(x - r) \leq s \iff wx^{\uparrow} - r \leq s\,. \tag{1}$$

For some structures, such as VP-trees [9], LC [2] and HC [5], we also need to determine whether the query is *entirely inside* a ball region—or, equivalently, whether it intersects with the complement of the ball. Our lower bound on each distance between the query and the outside is $r - x_i$, and using monotonicity again, we get the criterion $r - wx^{\uparrow} < s$. If, however, we do not treat the query as a black-box monotone function, we can, as described in the following section, get the stronger criterion $r - wx^{\downarrow} < s$, where $x^{\downarrow}$ is x sorted in descending order. The difference between these two bounds can be *arbitrarily large*, even for just two query objects. The complemented ball is also just a particularly simple linear ambit with negative coefficients [7]; the situation is similar for other such regions.

3 Ordered Weighted Averages and Linear Ambits

It is possible to construct a weighted generalization of OWA, called *weighted* OWA (WOWA), where some individuals (i.e., query objects) get preferential treatment when determining a tradeoff [8]. The following definition is given by Gonzales and Perny [6].

Definition 1. *Let $p = [p_1, \ldots, p_m]$ and $w = [w_1, \ldots, w_m]$ be weighting vectors, where $p_i, w_i \in [0, 1]$ and $\sum_{i=1}^m p_i = \sum_{i=1}^m w_i = 1$. The* weighted ordered weighted average *(WOWA) of a vector $x \in \mathbb{R}^m$ with respect to p and w is defined by:*

$$\mathrm{WOWA}(x; p, w) = \sum_{i=1}^{m} \left[\varphi\left(\sum_{k=i}^{m} p_{\sigma(k)} \right) - \varphi\left(\sum_{k=i+1}^{m} p_{\sigma(k)} \right) \right] x_{\sigma(i)}\,, \tag{2}$$

where σ is a permutation of x in increasing order and $\varphi : [0, 1] \to [0, 1]$ is defined by linear interpolation between values $\varphi(i/m) = \sum_{k=1}^{i} w_{m-k+1}$ and $\varphi(0) = 0$.

With decreasing weights, WOWA is a fairness measure. This works well for similarities, but as discussed, for distances we need need *un*fairness. One way of achieving this is to use an *increasing* weight vector. This makes intuitive sense, and for similarities $s(u, v) = 1 - d(u, v)$, as used by Ciaccia et al. [4], we can show that the least unfair distance tradeoff is exactly the fairest similarity tradeoff.[2]

[2] Note that, following Ciaccia et al., we assume $s(u, v) \in [0, 1]$, which requires a bounded metric, with $d(u, v) \in [0, 1]$.

Proposition 1. *Let p, w and w' be WOWA weighting vectors, with $w'_i = w_{m-i+1}$ for all $i \in \{1, \ldots, m\}$. For any $x \in [0,1]^m$, we have that:*

$$\mathrm{WOWA}(x; p, w') = 1 - \mathrm{WOWA}(1 - x; p, w) \tag{3}$$

Proof. Let φ_w and $\varphi_{w'}$ be the function φ, as defined in Definition 1, for w and w', respectively. Also, let σ and σ' be permutations of, respectively, $1 - x$ and x in increasing order so that $\sigma'(i) = \sigma(m - i + 1)$. We have that:

$$\mathrm{WOWA}(1 - x; p, w) = 1 - \sum_{i=1}^{m} \left[\varphi_w \left(\sum_{k=i}^{m} p_{\sigma(k)} \right) - \varphi_w \left(\sum_{k=i+1}^{m} p_{\sigma(k)} \right) \right] x_{\sigma(i)} \tag{4}$$

One can easily verify that $\varphi_{w'}(b) - \varphi_{w'}(a) = \varphi_w(1-a) - \varphi_w(1-b)$ for $a, b \in [0,1]$ and that $\sum_{k=i}^{m} p_{\sigma(k)} = 1 - \sum_{k=1}^{i-1} p_{\sigma(k)}$. Thus:

$$\mathrm{WOWA}(1 - x; p, w) = 1 - \sum_{i=1}^{m} \left[\varphi_{w'} \left(\sum_{k=1}^{i} p_{\sigma(k)} \right) - \varphi_{w'} \left(\sum_{k=1}^{i-1} p_{\sigma(k)} \right) \right] x_{\sigma(i)} \tag{5}$$

$$= 1 - \sum_{i=1}^{m} \left[\varphi_{w'} \left(\sum_{k=i}^{m} p_{\sigma'(k)} \right) - \varphi_{w'} \left(\sum_{k=i+1}^{m} p_{\sigma'(k)} \right) \right] x_{\sigma'(i)} \tag{6}$$

$$= 1 - \mathrm{WOWA}(x; p, w') \tag{7}$$

Equation 3 can then easily be obtained from (7). □

For our overlap check, we wish to model a WOWA query as a *linear ambit* $\mathrm{B}[q, s; \mathrm{W}] = \{o : \mathrm{W}x_o \leq s\}$, where $x_o = [d(q_i, o)]_i$, as introduced by Hetland [7]. While WOWAs are not linear functions, we can emulate a query with m query objects as a linear ambit with $m!$ facets, one per possible permutation of x. Normally, the intersection check would require considering each facet in turn, which would quickly become computationally unfeasible with an increasing m, and could in theory lead to false positives.[3] However, when the weights for the WOWA representing our unfairness measure are in increasing order (corresponding to a fairness measure on similarities, per Proposition 1), membership and overlap checks need only consider *one* of the facets, eliminating both of these problems.

Proposition 2. *Let w and p be weighting vectors, where $w_1 \leq w_2 \leq \cdots \leq w_m$. Let W be a matrix with $m!$ rows, one for each possible permutation, σ, of an m-vector. For a permutation σ, the value in column i of the corresponding row is:*

$$\varphi \left(\sum_{k=j}^{m} p_{\sigma(k)} \right) - \varphi \left(\sum_{k=j+1}^{m} p_{\sigma(k)} \right), \tag{8}$$

where $\sigma(j) = i$ and φ is the function from Definition 1. For $x \in \mathbb{R}^m_{\geq 0}$ and $s \in \mathbb{R}$, let w_σ be the row in W *corresponding to σ. If σ puts x in increasing order,* $\mathrm{W}x \leq s$ *iff* $w_\sigma x \leq s$*. If σ puts x in decreasing order,* $\mathrm{W}x > s$ *iff* $w_\sigma x > s$.

[3] This is discussed by Hetland in Sect. 3.1 [7].

Proof. For any permutation σ, we can create a new permutation σ', with $\sigma'(i) = \sigma(i+1)$, $\sigma'(i+1) = \sigma(i)$ for some $i \in \{1, \ldots, m-1\}$ and $\sigma'(j) = \sigma(j)$ for all $j \notin \{i, i+1\}$. Since w is in increasing order, we know that the growth of φ is monotonically decreasing over $[0, 1]$. Combined with the fact that $\|w_\sigma\|_1 = \varphi(1) - \varphi(0) = \|w\|_1 = 1$ for all σ, we get that:

$$\begin{cases} w_\sigma x \geq w_{\sigma'} x & \text{if } x_{\sigma(i)} < x_{\sigma(i+1)} \\ w_\sigma x \leq w_{\sigma'} x & \text{if } x_{\sigma(i)} > x_{\sigma(i+1)} \\ w_\sigma x = w_{\sigma'} x & \text{otherwise} \end{cases} \tag{9}$$

If a permutation σ does not put x in increasing order, there is an i such that $x_{\sigma(i)} > x_{\sigma(i+1)}$. Thus, there is another permutation σ' with $w_{\sigma'} x \geq w_\sigma x$. Consequently, one of the permutations σ that maximizes $w_\sigma x$ must put x in increasing order. Note that by the third case in (9), if there are multiple permutations that put x in increasing order, the value of $w_\sigma x$ is the same for all of them. Similarly, any σ that puts x in decreasing order minimizes $w_\sigma x$. □

Using the construct in Proposition 2, we can for a WOWA-based unfairness measure, defined by weighting vectors w and p, create a linear ambit $\mathrm{B}[q, s; \mathrm{W}]$. As long as w is in increasing order, i.e., $w = w^{\uparrow}$, the membership check of this ambit, $\mathrm{W}x \leq s$, is equivalent to checking that $\mathrm{WOWA}(x; p, w^{\uparrow}) \leq s$. That is, this ambit is equivalent to a range query with the WOWA-based unfairness measure. And when checking whether this query ambit intersects the inverted r-ball round c, we can in principle perform $m!$ individual checks like $r - w_\sigma x < s$ (i.e., $w_\sigma x > r - s$), one per row σ.[4] Proposition 2 shows us that we need only consider the single row corresponding to a decreasing x. In other words, the overlap check is strengthened from $s > r - \mathrm{WOWA}(x; p, w^{\uparrow})$ to $s > r - \mathrm{WOWA}(x; p, w^{\downarrow})$.

4 Experiments

To demonstrate the practical feasibility of the method, even without any fine-tuning or high-effort optimization, we have tested it empirically on synthetic and real-world data, using the basic index structure *list of clusters* (LC), as described by Chávez and Navarro [2].[5] Briefly, the LC partitions the data set into a sequence of ball regions, each defined by a center, a covering radius, and a set of member items. A search progresses by detecting overlap with each ball in turn, potentially scanning its members for relevance. A defining feature of LC is that the points in later buckets fall entirely outside previous balls, so that if the query falls entirely *inside* one of the balls, the search process may be halted.

[4] This follows from the linear ambit overlap check described in Theorem 3.1.2 of Hetland [7], as well as from the monotonicity result of Ciaccia et al. [4], inserting the lower bound $r - x$ into the ambit membership predicate.

[5] The source code and raw experimental results are available as ancillary files for the preprint of this paper at https://arxiv.org/abs/2108.03621.

Table 1. Experimental results. For each of the double (two separate) and combined queries, the speedup (where higher is better) from the number of distance computations needed for linear scan is listed for each $k = 1, \ldots, 5$.

			Double					Combined				
Data set	Dim.	Scan	1	2	3	4	5	1	2	3	4	5
Colors	112	225 162	3.29	3.09	2.98	2.91	2.84	5.55	5.22	5.04	4.91	4.80
NASA	20	80 098	1.57	1.46	1.42	1.38	1.36	2.64	2.48	2.39	2.33	2.28
Uniform	4	200 000	2.17	2.14	2.12	2.10	2.09	7.13	6.94	6.82	6.72	6.65
	6	200 000	2.16	2.07	2.02	1.99	1.96	5.73	5.47	5.32	5.20	5.10
	8	200 000	2.01	1.89	1.82	1.78	1.74	4.20	3.95	3.79	3.68	3.59
	10	200 000	1.87	1.73	1.65	1.60	1.56	3.20	2.96	2.82	2.73	2.66
Clustered	4	200 000	2.28	2.25	2.22	2.21	2.19	7.62	7.47	7.37	7.30	7.23
	6	200 000	2.39	2.28	2.22	2.17	2.14	6.17	5.91	5.75	5.62	5.53
	8	200 000	2.07	1.93	1.87	1.82	1.79	4.41	4.12	3.95	3.84	3.75
	10	200 000	1.84	1.72	1.64	1.60	1.57	3.20	2.96	2.83	2.73	2.66
Listeria	—	41 118	1.25	1.17	1.16	1.13	1.06	1.28	1.28	1.27	1.27	1.27

More specifically, bucket centers were chosen to maximize distance to previous centers (heuristic $p5$ of Chávez and Navarro), with each ball constructed to contain the 20 closest points to the center, as well as any additional points that fall within the resulting radius.[6] The data sets used were:

- Synthetic: 100 000 uniformly random and clustered vectors from $[0, 1]^D$, for $D = 2, 4, \ldots, 10$. The clustered vectors were constructed by first generating 1000 cluster centers, uniformly at random, and then generating 100 vectors per cluster, by adding standard Gaussian noise.
- Real-world: The Colors, NASA and Listeria SISAP data sets.[7]

Euclidean and Levenshtein distance were used with vectors and strings, respectively. For the real-world data sets, the 101 first objects were taken as queries; for the synthetic ones, queries were generated in addition. These were used pairwise (1 and 2, 2 and 3, etc.) in an OWA query with weights 1 and 3 (like the Gini coefficient). Fairest neighbor queries (kFN) were run for $k = 1, \ldots, 5$. The number of distance computations was averaged over the 100 query object pairs.

Table 1 shows the results. As a baseline, the number of distance computations needed for a linear scan is listed, and the speedup for the combined kFN query is shown for each k. For comparison, we also performed a *double* query, where a separate k was found for each of the two query objects, to ensure that the true

[6] These choices were made based on the results of Chávez and Navarro [2], which indicate that $p5$ yields the best results overall, and a bucket size of 20 is a good tradeoff between filtering power and scanning time for a wide range of data sets.

[7] Available at https://sisap.org.

kFN would be returned,[8] and then two separate kNN queries were performed, with the fairest neighbors found in their intersection. The combined search used fewer distance computations than the alternatives for every data set and parameter setting. On average, the combined search used about half as many distance computations as the separate queries, and a quarter of a full linear scan.[9]

5 Conclusions and Future Work

Ordered weighted averages (OWA) and weighted OWAs (WOWA) may be used as a query modality with any current metric index, when tradeoffs between multiple query objects are needed, to find their k fairest neighbors, kFN. They provide a large degree of customizability, both in terms of their fairness profile and the relative weights of different query objects, and are easy to implement. Other monotone (un)fairness measures may also be used, though possibly with weakened overlap checks in some cases.

Future research might look into adapting index structures, e.g., by adjusting construction heuristics, to fair neighbor queries, and whether the requirements for efficiency in practice are different from, say, single-object ball queries. Generalizations of fairness might also be interesting, where one permits negative weights for certain objects, which is straightforward for weighted sum, but whose implementation is less obvious for WOWA. A more straightforward extension of this work would be to test on other data sets, perhaps with higher (intrinsic) dimensionality, using more advanced index structures.

References

1. Bustos, B., Skopal, T.: Dynamic similarity search in multi-metric spaces. In: Proceedings of the 8th ACM Intl. Workshop on Multimedia Information Retrieval, pp. 137–146 (2006)
2. Chávez, E., Navarro, G.: A compact space decomposition for effective metric indexing. Pattern Recognit. Lett. **26**(9), 1363–1376 (2005)
3. Chen, L., Lian, X.: Dynamic skyline queries in metric spaces. In: Proceedings of the 11th International Conference on Extending Database Technology, pp. 333–343 (2008)
4. Ciaccia, P., Patella, M., Zezula, P.: Processing complex similarity queries with distance-based access methods. In: Schek, H.-J., Alonso, G., Saltor, F., Ramos, I. (eds.) EDBT 1998. LNCS, vol. 1377, pp. 9–23. Springer, Heidelberg (1998). https://doi.org/10.1007/BFb0100974
5. Fredriksson, K.: Engineering efficient metric indexes. Pattern Recognit. Lett. **28**(1), 75–84 (2007)

[8] Of course, these individual k parameters would not be available when resolving a real query, but this gives an optimistic bound for the competition.

[9] More specifically, the average proportions, using the geometric mean, were 48.4% and 25.7%, respectively, corresponding to speedups of 2.1 and 3.9.

6. Gonzales, C., Perny, P.: Multicriteria decision making. In: Marquis, P., Papini, O., Prade, H. (eds.) A Guided Tour of Artificial Intelligence Research, pp. 519–548. Springer, Cham (2020). https://doi.org/10.1007/978-3-030-06164-7_16
7. Hetland, M.L.: Comparison-based indexing from first principles. arXiv preprint arXiv:1908.06318 (2019)
8. Torra, V.: The weighted OWA operator. Int. J. Intell. Syst. **12**(2), 153–166 (1997)
9. Yianilos, P.N.: Data structures and algorithms for nearest neighbor search in general metric spaces. In: Proceedings of the 4th Symposium on Discrete Algorithms, pp. 311–321 (1993)

Intrinsic Dimensionality

Local Intrinsic Dimensionality and Graphs: Towards LID-aware Graph Embedding Algorithms

Miloš Savić, Vladimir Kurbalija, and Miloš Radovanović(✉)

Department of Mathematics and Informatics, Faculty of Sciences,
University of Novi Sad, Trg Dositeja Obradovića 3, 21000 Novi Sad, Serbia
{svc,kurba,radacha}@dmi.uns.ac.rs

Abstract. Local intrinsic dimensionality (LID) has many important applications in the field of machine learning (ML) and data mining (DM). Existing LID models and estimators have mostly been applied to data points in Euclidean spaces, enabling LID-aware ML/DM algorithms for tabular data. To the best of our knowledge, prior research works have not considered LID for designing or evaluating graph-based ML/DM algorithms. In this paper, we discuss potential applications of LID to graph-structured data considering graph embeddings and graph distances. Then, we propose NC-LID – a LID-related measure for quantifying the discriminatory power of the shortest-path distance with respect to natural communities of nodes as their intrinsic neighborhoods. It is shown how NC-LID can be utilized to design LID-elastic graph embedding algorithms based on random walks by proposing two LID-elastic variants of Node2Vec. Our experimental evaluation on real-world graphs demonstrates that NC-LID can point to weak parts of Node2Vec embeddings that can be improved by the proposed LID-elastic extensions.

Keywords: Intrinsic dimensionality · Graph embeddings · Graph distances · Natural communities · LID-elastic Node2Vec

1 Introduction

The intrinsic dimensionality (ID) of a dataset is the minimal number of features that are needed to form a good lower-dimensional representation of the dataset without a large information loss. The estimation of ID is highly relevant for various machine learning and data mining tasks, especially when dealing with high-dimensional data. Namely, lower-dimensional data representations can be exploited to train machine learning models in order to improve their generalizability by alleviating negative effects of high dimensionality. Due to a smaller number of features, such models are more comprehensible and their training, tuning and validation is more time efficient.

N. Reyes et al. (Eds.): SISAP 2021, LNCS 13058, pp. 159–172, 2021.
https://doi.org/10.1007/978-3-030-89657-7_13

The notion of the local intrinsic dimensionality (LID) has been developed in recent years motivated by the fact that the ID may vary across a dataset. The main idea of LID is to focus the estimation of ID to a data space surrounding a data point. In a seminal paper, Houle [7] defined the LID considering the distribution of distances to a reference data point. Additionally, Houle showed that for continuous distance distributions with differentiable cumulative density functions the LID and the indiscriminability of the corresponding distance function are actually equivalent. Let x be a reference data point and let F denote the cumulative distribution function of distances to x. It can be said that the underlying distance function is discriminative at a given distance r if $F(r)$ has a small increase for a small increase in r. Thus, the indiscriminability of the distance function at r w.r.t x, denoted by $ID(r)$, can be quantified as the limit of the ratio of (a) the proportional rate of increase of $F(r)$, and (b) the proportional rate of increase in r. Then, the LID of x is given as $\lim_{r \to 0} ID(r)$. For practical applications, the LID of x can be estimated considering the distances of x to its k nearest data points [1,2]. Recent research works showed that the LID can be exploited for density-based clustering [9], outlier detection [9,10], training deep neural network classifiers on datasets with noisy labels [13], detection of adversarial data points when training deep neural networks [12], subspace clustering and estimating the local relevance of features [3] and similarity search [4,8].

The applications of machine learning and data mining algorithms designed for tabular datasets to graphs are enabled by various graph embedding algorithms [5]. Here we consider graph embedding algorithms translating graph nodes into n-dimensional real-valued vectors with the goal of preserving graph-based distances in the embedding space. Besides applications in node classification, node clustering and link prediction tasks, graph embeddings may be also utilized for similarity search applications. Namely, similarity search when performed directly on large-scale graphs may pose several difficulties due to the small-world phenomenon [16], i.e. for a given node (similarity search query) the number of nodes at a given shortest-path distance (potential similarity search hits) grows at a very fast rate with the shortest path distance.

In this paper we discuss potential applications of LID to graphs (Sect. 2). To the best of our knowledge, this is the first work considering LID for designing and evaluating ML/DM algorithms operating on graph-structured data. As the main contributions, we propose a LID-related measure called NC-LID to quantify the discriminability of the shortest-path distance locally per node with respect to their natural communities as intrinsic subgraph boundaries (Sect. 3) and two extensions of the Node2Vec graph embedding algorithm [6] that personalize and adjust Node2Vec parameters according to NC-LID values (Sect. 4). In the experimental evaluation presented in Sect. 5, it is demonstrated that NC-LID can indicate weak parts of Node2Vec embeddings prior to their construction and that our LID-elastic Node2Vec extensions provide better embeddings w.r.t. reconstruction errors. In the last section we discuss possible directions for future research.

2 LID and Graphs

Existing LID models and corresponding estimators have been designed for tabular datasets with real-valued features and smooth distance functions. There are two ways in which they can be applied to graphs: (a) by transforming graphs into tabular data representations using graph embedding algorithms, and (b) by using graph-based distances instead of distances of vectors in Euclidean spaces. To the best of our knowledge, we are not aware of any previous research study investigating the LID of graph embeddings or applying LID estimators to graph-based distances.

The first approach enables the LID-based evaluation of graph embeddings and their analysis in the context of distance-based machine learning and data mining algorithms. For example, Amsaleg et al. [1] proposed the maximum-likelihood LID estimator (MLE-LID). By computing MLE-LID for each node in a graph on embeddings produced by different graph embedding algorithms we can study which of the embeddings is the most effective for distance-based machine learning and data mining algorithms (under the assumption that the embeddings preserve the structure of the graph to a similar extent). Additionally, obtained MLE-LID values can indicate whether we can benefit from LID-aware data mining and machine learning algorithms for a concrete embedding.

LID estimates for graph nodes obtained by applying LID estimators on graph embeddings are relative to the selected graph embedding dimension that is explicitly required by graph embedding algorithms. Additionally, the usefulness of LID estimates depends on the ability of the selected graph embedding algorithm to preserve the structure of the input graph.

The MLE-LID estimator mentioned above (or any other LID estimator, e.g. the estimator also proposed by Amsaleg et al. [2] that estimates LID within tight localities) can be applied "directly" on graphs by taking shortest path distances instead of distances in Euclidean spaces (in the most general case since graph embedding algorithms try to preserve shortest path distances in embedded spaces). However, LID estimates based on shortest path distances will suffer from negative effects of the small-world property, i.e. for a randomly selected node n there will be an extremely large fraction of nodes at the same and relatively small shortest-path distance from n. The hubness property of large-scale real-world graphs (i.e., the existence of nodes with an extremely high degree that are called hubs) will also have a big impact on such LID estimates. For example, LID for hubs will be estimated as 0 by the MLE-LID estimator due to a large number of nearest neighbors at the shortest-path distance 1. Another problem with this approach is the shortest-path distance itself. The notion of LID is based on the assumption that the radius of a ball around a data point can be increased by a small value that tends to 0. However, the shortest-path distance does not have an increase that can go to 0 (the minimal increase is 1) in contrast to distances in Euclidean space.

3 NC-LID: LID-related Measure for Graph Nodes Based on Natural Communities

Following the discussion from the previous section, we consider a somewhat different conceptual approach to designing LID-related measures for nodes in a graph. The main idea is to substitute a ball around a data point with a subgraph around a node in order to estimate the discriminatory power of a graph-based distance of interest. Here we observe the most basic case which is a fixed subgraph that can be considered as the intrinsic locality of the node.

Let n denote a node in a graph $G = (V, E)$ and let S be a subgraph containing n. The graph-based distance of interest can be the shortest-path distance, but also any other node similarity function, including hybrid node similarity measures for attributed graphs. Assuming that S is a natural (intrinsic) locality of n, d can be considered as a perfectly discriminative distance measure if it clearly separates nodes in S from the rest of the nodes in G.

To measure the degree of discriminatory power of d considering S as the intrinsic locality of n we define a general limiting form of the local intrinsic discriminability of d as

$$\text{GB-LID}(n) = -\ln\left(\frac{|S|}{T(n, S)}\right), \tag{1}$$

where $|S|$ is the number of nodes in S. $T(n, S)$ is the number of nodes whose distance from n is smaller than or equal to r, where r is the maximal distance between n and any node from S:

$$T(n, S) = \left|\left\{y \in V \,:\, d(n, y) \leq \max_{z \in S} d(n, z)\right\}\right|. \tag{2}$$

Similarly to standard LID for tabular data, GB-LID assesses the local neighborhood size of n at two ranges: the number of nodes in a neighborhood of interest (S) and the total number of nodes that are within relevant distances from n considering distances from n to nodes in S. The more extreme the ratio between these two, the higher the intrinsic dimensionality (local complexity) of n. Unlike standard LID, GB-LID depends on the complexity of a fixed subgraph around the node rather than some measure reflecting the dynamics of expanding subgraphs (this will be part of our future work). Compared to other measures capturing the local complexity of a node (e.g., degree centrality and clustering coefficient), GB-LID is not restricted to ego-networks of nodes or regularly expanding subgraphs capturing all nodes within the given distance (e.g., LID-based intrinsic degree proposed by von Ritter et al. [15]).

GB-LID is a class of LID-related scores effectively parameterized by $\langle S_n, d\rangle$, where S_n is the subgraph denoting the intrinsic local neighborhood of node n and d is an underlying distance measure. From GB-LID we derive one concrete measure called NC-LID (NC is the abbreviation for "Natural Community"). In NC-LID we fix S_n to the natural (local) community of n determined by the fitness-based algorithm for recovering natural communities [11] and d is the shortest path distance.

A community in a graph is a highly cohesive subgraph. This means that the number of links within the community (so-called intra-community links) is significantly higher than the number of links connecting nodes from the community to nodes outside the community (so-called inter-community links). The natural or local community of node n is a community recovered from n. When computing NC-LID we use the fitness-based algorithm for identifying natural communities proposed by Lancichinetti et al. [11]. Starting from n, this algorithm recovers the natural community C of n by maximizing the community fitness function that is defined as:

$$f_C = \frac{k_{in}(C)}{(k_{in}(C) + k_{out}(C))^{\alpha}}, \tag{3}$$

where $k_{in}(C)$ is the total intra-degree of nodes in C, $k_{out}(C)$ is the total inter-degree of nodes in C, and α is a real-valued parameter controlling the size of C (larger α implies smaller C). The intra-degree and inter-degree of a node s are the number of intra-community and inter-community links incident to s, respectively. The most natural choice for α is $\alpha = 1$, which corresponds to the Raddichi notion of weak communities [14].

NC-LID(n) is equal to 0 if all nodes from the natural community of n are at shorter shortest-path distances to n than nodes outside its natural community. Higher values of NC-LID(n) imply that it is harder to distinguish the natural community of n from the rest of the graph using the shortest-path distance, i.e. the natural community of n tends to be more "concave" and elongated in depth with higher NC-LID(n) values. Nodes with such complex natural communities may also be brokers having large values of node centrality metrics that connect different parts of the graph by their long-range links (i.e., links whose removal significantly increase the average shortest path distance).

4 LID-elastic Node2Vec Variants

Having an appropriate LID-based score for graph nodes such as NC-LID, it is possible to design LID-aware or LID-elastic graph embedding algorithms. In this work we propose two LID-elastic variants of Node2Vec [6].

Node2Vec is a random-walk based algorithm for generating graph embeddings. The main idea of random-walk based graph embedding algorithms is to sample a certain number of random walks starting from each node in a graph. Sampled random walks are then treated as ordinary sentences over the alphabet encompassing node identifiers. This means that the problem of generating graph embeddings is reduced to the problem of generating text embeddings. Node2Vec relies on Word2Vec to produce node embedding vectors from random-walk sentences.

Node2Vec employs a second order random walk scheme with two parameters p and q which guide the walk. Let us assume that a random walk just transitioned from node t to node v. The parameter p (return parameter) controls the probability of intermediately returning back to t. The parameter q (in-out parameter) controls to what extent random walk resembles BFS or DFS graph exploration strategies. For $q > 1$, the random walk is more biased to nodes close to t (BFS-like graph exploration). If $q < 1$ then the random walk is more inclined to visit nodes that are further away from t (DFS-like graph exploration).

Our Node2Vec LID-elastic extensions are based on the premise that high NC-LID nodes have higher link reconstruction errors than low NC-LID nodes due to more complex natural communities. More specifically, the quality of graph embeddings can be assessed by comparing original graphs to graphs reconstructed from embeddings. Let G denote an arbitrary graph with L links and let E be an embedding constructed from G using some graph embedding algorithm. The graph reconstructed from E has the same number of links as G. The links in the reconstructed graph are formed by joining the L closest node vector pairs from E. Then, the following metrics quantifying the quality of E according to the principle that nodes close in G should be also close in E can be computed for each node n:

1. Link precision $P(n)$ is the number of correctly reconstructed links incident to n divided by the total number of links incident to n in the reconstructed graph.
2. Link recall $R(n)$ is the number of correctly reconstructed links incident to n divided by the total number of links incident to n in the original graph.
3. Link F_1 score $F_1(n)$ is a metric aggregating $P(n)$ and $R(n)$ into a single score that is defined as their harmonic mean: $F_1(n) = 2 \cdot P(n) \cdot R(n) / (P(n) + R(n))$.

Higher values of $P(n)$, $R(n)$ and $F_1(n)$ imply lower link reconstruction errors for n.

The sampling mechanism of Node2Vec is controlled by 4 parameters: the number of random walks sampled per node, the length of each random walk, p and q. The first two parameters are fixed for each node in a graph, while p and q are fixed for each pair of nodes. Our Node2Vec LID-elastic extensions are based on Node2Vec parameters personalized for nodes and pairs of nodes that are adjusted according to their NC-LID values.

The first LID-elastic variant of Node2Vec, denoted by `lid-n2v-rw`, personalizes the number of random walks sampled per node and the length of random walks according to the following rules:

1. The number of random walks sampled for n is equal to $\lfloor (1 + \text{NC-LID}(n)) \cdot B \rfloor$, where B is the base number of random walks (by default $B = 10$).
2. The length of random walks sampled for n is equal to $\lfloor W/(1 + \text{NC-LID}(n)) \rfloor$ (by default $W = 80$).

`lid-n2v-rw` samples a proportionally higher number of random walks for high NC-LID nodes while keeping the computational budget (the total number of random walk steps per node) approximately constant. The main idea is to increase the frequency of high NC-LID nodes in sampled random walks in order to better preserve their close neighborhood in formed embeddings. Additionally, the probability of the random walk leaving the natural community of the starting node is lowered for high NC-LID nodes due to shorter random walks.

The second LID-elastic variant of Node2Vec, denoted by `lid-n2v-rwpq`, extends `lid-n2v-rw` by personalizing p and q parameters controlling biases when sampling random walks. Let p_b and q_b denote the base values of p and q (by default $p_b = q_b = 1$). The `lid-n2v-rwpq` variant incorporates the following adjustments of p and q for a pair of nodes x and y, where x is the node on which the random walk currently resides and y is one of its neighbours:

1. If x is in the natural community of y then $p(x, y) = p_b$, otherwise $p(x, y) = p_b + \text{NC-LID}(y)$.
2. If y is in the natural community of x then $q(x, y) = q_b$, otherwise $q(x, y) = q_b + \text{NC-LID}(x)$

The first rule controls the probability of returning back from x to y if the random walk transitioned from y to x in the previous step. By increasing the base p value if x is not in the natural community of y `lid-n2v-rwpq` lowers the probability of making a transition between different natural communities. The increase is equal to NC-LID(y) which implies that the backtrack step is penalized more if y has a more complex natural community.

The second rule controls the probability of going to nodes that are more distant from the previously visited node in the random walk. The base q value is increased for nodes not belonging to the natural community of x meaning that again `lid-n2v-rwpq` penalizes transitioning between different natural communities. The increase in q_b is equal to NC-LID(x) implying that `lid-n2v-rwpq` biases the random walk to stay within more complex natural communities.

5 Experiments and Results

Our experimental evaluation of the NC-LID measure and LID-elastic Node2Vec extensions is performed on datasets (graphs) listed in Table 1. The experimental corpus encompasses three small social networks (Karate club, Les miserables and Florentine families), five paper citation networks (CORAML, CORA, CITESEER, PUBMED and DBLP) and two co-purchasing networks of Amazon products (AE photo and AE computers) that are commonly used to evaluate graph embedding methods. For each graph, Table 1 shows the number of nodes (N), the number of links (L), the number of connected components (C), the fraction of nodes in the largest connected component (F), the average degree ($\bar{d}$) and the skewness of the degree distribution (S). It can be observed that the experimental corpus encompasses both small and large sparse graphs ($\bar{d} \ll N - 1$). All graphs, except CITESEER, are either connected graphs ($C = 1$) or have a giant connected component ($F > 0.9$). The degree distribution of large graphs has a high positive skewness implying that those graphs contain so-called hubs (nodes whose degree is significantly higher than the average degree).

Table 1. Experimental datasets.

Graph	N	L	C	F	$\bar{d}$	S
Karate club	34	78	1	1.00	4.59	2.00
Les miserables	77	254	1	1.00	6.60	1.89
Florentine families	15	20	1	1.00	2.67	0.62
CORAML	2995	8158	61	0.94	5.45	12.28
CITESEER	4230	5337	515	0.40	2.52	8.44
AE photo	7650	119081	136	0.98	31.13	10.42
AE computers	13752	245861	314	0.97	35.76	17.34
PUBMED	19717	44324	1	1.00	4.50	5.21
CORA	19793	63421	364	0.95	6.41	7.87
DBLP	17716	52867	589	0.91	5.97	9.43

5.1 Natural Communities and NC-LID

Since natural communities are the base for the NC-LID measure, we first examine their characteristics. Figure 1 shows the complementary cumulative distribution (CCD) of the size of natural communities on a log-log plot. The size of a natural community is the number of nodes it contains. It can be seen that CCDs for large graphs have very long tails. This implies that a large majority of nodes have relatively small natural communities (10 or less nodes), but there are also nodes having exceptionally large natural communities (100 or more nodes). For example, 76.56% of CORA nodes have natural communities with 4 or less nodes, while the largest natural community in CORA contains 146 nodes.

The average NC-LID and the maximal NC-LID of nodes in examined graphs are presented in Fig. 2 sorted from the graph having the most compact natural communities to the graph with the most complex natural communities on average. The social network of Florentine families has the lowest average NC-LID equal to 0.48. This NC-LID level means that approximately 38% of nodes within the shortest-path radius of the natural community of a randomly selected node do not belong to its natural community. The largest average NC-LID for examined graphs is 5.12 (AE computers). This NC-LID value corresponds to situations in which approximately 0.6% of nodes within the shortest-path radius of a natural community belong to the natural community. It should be emphasized that NC-LID positively correlates with the size of the natural community (Spearman's correlations higher than 0.3) for 5 graphs, for 3 graphs negatively (correlations lower than −0.15), while for 2 graphs (PUBMED and AE Computers) significant correlations are absent.

5.2 Node2Vec Evaluation

Prior to evaluating LID-elastic Node2Vec modifications, we examine characteristics of Node2Vec embeddings. Graph reconstruction metrics (mean link precision,

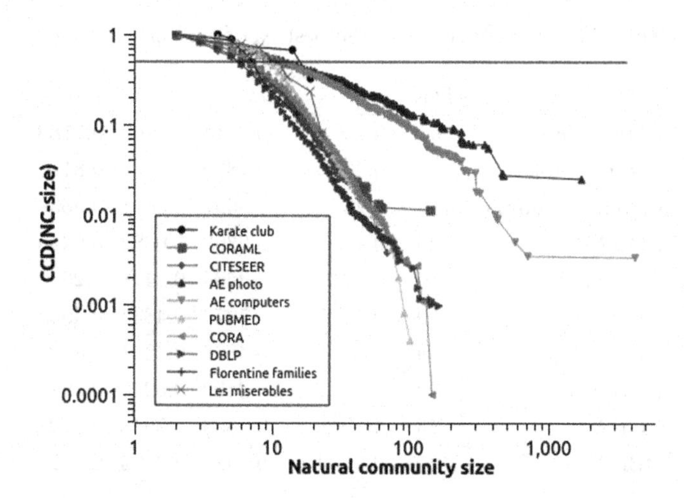

Fig. 1. The complementary cumulative distribution of sizes of natural communities. The solid line represents the 0.5 probability.

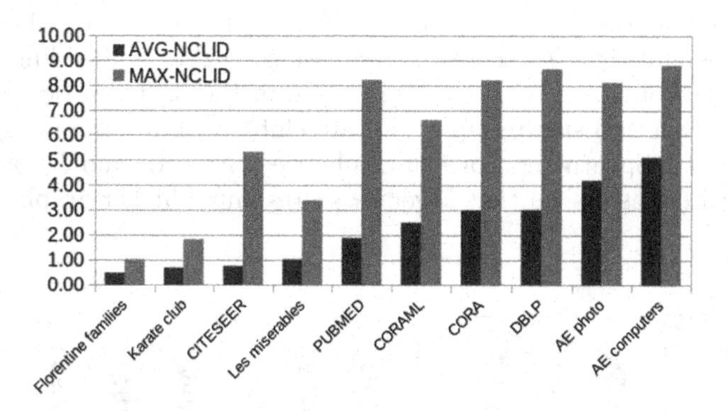

Fig. 2. The average and the maximal NC-LID for graphs from our experimental corpus.

recall and F_1 scores, see Sect. 4) were computed for 125 Node2Vec embeddings per graph in order to find the best embedding in the following parameter space: p and q were varied to take values in $\{0.25, 0.5, 1, 2, 4\}$, and the embedding dimension in $\{10, 25, 50, 100, 200\}$. The number of sampled random walks per node and the length of random walk was set to their default values (10 and 80, respectively) as suggested in [6]. The parameters for the best embeddings, selected according to the average F_1 score, are shown in Table 2 (P denotes the mean link precision and R the mean link recall). It can be seen that for all graphs except CITESEER, Node2Vec preserves the structure of examined graphs to a fairly good extent (F_1 in the range from 0.39 to 0.96).

The basic assumption of LID-elastic Node2Vec modifications is that high NC-LID nodes have higher graph reconstruction errors compared to low NC-LID

Table 2. Characteristics of the best Node2Vec embeddings.

Graph	Dim.	p	q	P	R	F_1
Karate club	100	0.25	4	0.7814	0.7762	0.7788
Les miserables	100	0.25	4	0.7889	0.8325	0.8101
Florentine families	100	0.25	4	0.9667	0.9611	0.9639
CORAML	25	0.5	0.25	0.6300	0.6682	0.6485
CITESEER	10	0.5	0.25	0.2284	0.2438	0.2359
AE photo	50	0.5	0.5	0.5076	0.4835	0.4953
AE computers	50	4	0.25	0.4856	0.4231	0.4522
PUBMED	50	4	0.25	0.3152	0.5245	0.3937
CORA	25	4	0.25	0.5803	0.5648	0.5724
DBLP	25	0.5	1	0.4431	0.3693	0.4029

nodes when applying the original Node2Vec to generate graph embeddings. To check this assumption we first examine Spearman's correlations between NC-LID of nodes and their F_1 scores in the best Node2Vec embeddings described in Table 2. The obtained results are presented in Fig. 3. It can be seen that for all graphs except two small graphs (Karate club and Les miserables) there are notable negative Spearman's correlations between NC-LID and F_1 ranging from -0.2 to -0.4 (please recall that lower F_1 scores imply higher graph reconstruction errors).

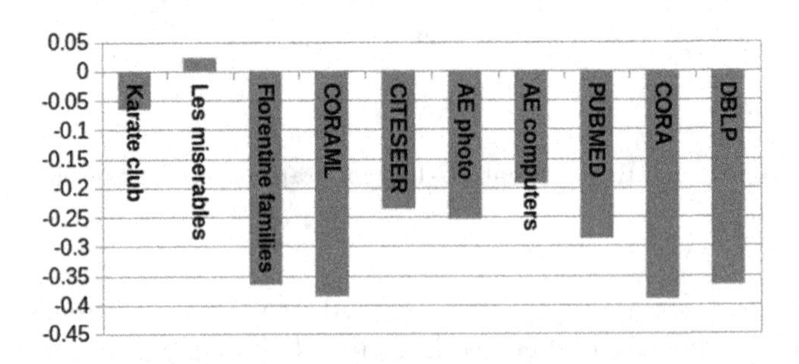

Fig. 3. The Spearman correlation between NC-LID of nodes and their F_1 scores in the best Node2Vec embeddings.

Second, we divide nodes into two groups: H – nodes that have high NC-LID values higher than the average NC-LID and L – nodes with low NC-LID values lower than the average NC-LID. Then, we apply the Mann-Whitney U (MWU) test to those two groups of nodes considering their F_1 scores. The MWU test checks the null hypothesis that scores in one group do not tend to be either higher or lower than scores in the other group. The results of conducted MWU tests are summarized in Table 3. The table shows the average F_1 score for H and

L ($F_1(H)$ and $F_1(L)$, respectively), the value of the MWU test statistic (U), the p-value of U (p) and values of two probabilities of superiority:

- PS(H) – the probability that the F_1 score of a randomly selected node from H (denoted by h) is strictly higher than the F_1 score of a randomly selected node from L (denoted by l), and
- PS(L) – the probability that the F_1 of l is strictly higher than the F_1 of h.

We accept the null hypothesis of MWU (no statistically significant differences between in F_1 scores of H and L) if $p > 0.01$ (column "acc." in Table 3). It can be observed that the null hypothesis of MWU is accepted only for the three smallest graphs from our experimental corpus. For large graphs we have that F_1 scores of high NC-LID nodes tend to be significantly lower than F_1 scores of low NC-LID nodes ($F_1(H) < F_1(L)$ and $\text{PS}(H) \ll \text{PS}(H)$).

Table 3. Comparison of F_1 scores of high NC-LID nodes (H) and low NC-LID nodes (L) using the Mann-Whitney U test.

Graph	$F_1(H)$	$F_1(L)$	U	p	acc.	PS(H)	PS(L)
Karate club	0.70	0.71	132	0.44	yes	0.44	0.48
Les miserables	0.76	0.76	734	0.50	yes	0.47	0.47
Florentine families	0.93	0.98	19	0.10	yes	0.07	0.39
CORAML	0.44	0.62	699380	$<10^{-2}$	no	0.29	0.67
CITESEER	0.10	0.25	1707420	$<10^{-2}$	no	0.19	0.31
AE photo	0.32	0.43	5239408	$<10^{-2}$	no	0.36	0.64
AE computers	0.29	0.38	17900546	$<10^{-2}$	no	0.38	0.61
PUBMED	0.19	0.32	31448278	$<10^{-2}$	no	0.28	0.59
CORA	0.36	0.54	29695497	$<10^{-2}$	no	0.28	0.68
DBLP	0.20	0.42	26684749	$<10^{-2}$	no	0.25	0.57

By taking into account both the observed Spearman's correlations and the results of the MWU tests it can be concluded that high NC-LID nodes tend to have significantly higher graph reconstruction errors than low NC-LID nodes. This implies that the NC-LID measure is able to point to "weak" parts of Node2Vec embeddings prior to their constructions. Consequently, Node2Vec embeddings could be possibly improved by adjusting Node2Vec parameters individually per node according to its NC-LID value.

5.3 LID-elastic Node2Vec Evaluation

Embeddings by LID-elastic Node2Vec variants proposed in Sect. 4 are generated according to the best configurations of original Node2Vec (Table 2). More specifically, for a given graph and embedding dimension we set base p and base q of

LID-elastic Node2Vec variants to p and q of the best corresponding Node2Vec embedding. As for Node2Vec embeddings examined in the previous section, the base number of random walks and the base length of random walks are set to their default values. The embedding dimension is varied in the same way as for Node2Vec. Then, we examine `lid-n2v-rw` and `lid-n2v-rwpq` embeddings by computing their average link F_1 scores, selecting the best embedding across considered embedding dimensions, and comparing the best LID-elastic Node2Vec embedding to the best embedding generated by Node2Vec (`n2v`). The obtained results are summarized in Table 4 showing the best F_1 score of `n2v` and the embedding dimension in which it is achieved and the best F_1 scores of LID-elastic Node2Vec variants and the corresponding embedding dimensions. The column "Best" indicates the best graph embedding algorithm according to F_1 and I is the percentage improvement in F_1 of a better LID-elastic Node2Vec variant over `n2v`.

Table 4. Comparison of Node2Vec and LID-elastic Node2Vec embeddings.

	`n2v`		`lid-n2v-rw`		`lid-n2v-rwpq`			
Graph	F_1	Dim.	F_1	Dim.	F_1	Dim.	Best	I [%]
Karate club	0.78	100	0.83	50	0.85	100	`lid-n2v-rwpq`	9.4
Les miserables	0.81	100	0.80	100	0.83	200	`lid-n2v-rwpq`	2.7
Florentine families	0.96	100	0.96	100	0.96	100	all	0.0
CORAML	0.65	25	0.66	50	0.63	25	`lid-n2v-rw`	1.3
CITESEER	0.24	10	0.25	10	0.28	10	`lid-n2v-rwpq`	18.7
AE photo	0.50	50	0.52	50	0.49	50	`lid-n2v-rw`	4.9
AE computers	0.45	50	0.47	100	0.42	50	`lid-n2v-rw`	4.7
PUBMED	0.39	50	0.43	50	0.42	50	`lid-n2v-rw`	9.4
CORA	0.57	25	0.60	50	0.59	50	`lid-n2v-rw`	3.9
DBLP	0.40	25	0.44	25	0.53	50	`lid-n2v-rwpq`	31.7

For Florentine families (the smallest graph in our experimental corpus) both LID-elastic Node2Vec variants achieve the same F_1 score as `n2v`. In all other cases at least one LID-elastic variant is better than `n2v`. For 5 graphs (out of 10) both LID-elastic variants have higher F_1 scores than `n2v`. The `lid-n2v-rw` variant achieves the highest F_1 score for 5 graphs, while `lid-n2v-rwpq` wins in 4 cases. The largest improvements in F_1 are achieved by `lid-n2v-rwpq` for DBLP and CITESEER. For those two graphs `lid-n2v-rwpq` significantly outperforms `n2v`: F_1 is improved by 31.7% and 18.7%, respectively. Significant improvements (approximately 5% or higher) are also present for 4 other graphs (Karate club, AE Photo, AE Computers and PUBMED).

6 Conclusions and Future Work

In this work we have discussed the notion of local intrinsic dimensionality in the context of graphs, which is the first step towards LID-aware ML/DL algorithms for graph-structured data. Since graphs are dimensionless objects, existing LID models could be applied to graphs by computing LID estimators either on graph embeddings or on graph-based distances.

Inspired by the fundamental connection between the local intrinsic dimensionality and the discriminability of distance functions in Euclidean spaces, we have proposed the NC-LID metric quantifying the discriminability of the shortest path distance considering natural communities of nodes in graphs. Then, we have suggested two LID-elastic modifications of the Node2Vec graph embedding algorithm in which Node2Vec parameters are personalized per node and adjusted according to their NC-LID values. Our experimental evaluation of the proposed LID-elastic Node2Vec modifications on 10 real-world graphs revealed that they are able to improve Node2Vec embeddings with respect to graph reconstruction errors.

The current work could be continued in two directions. One direction is to investigate possibilities for designing LID-related metrics reflecting the discriminability of graph-based distance functions considering expanding subgraph localities. In the same way as NC-LID, such metrics could be exploited to personalize and adjust parameters of graph embedding algorithms. Having in mind that nodes with complex intrinsic localities may have a significant brokerage role, it would also be interesting to examine correlations between LID-related scores and node centrality metrics.

The second research direction is related to natural communities. Namely, we will investigate alternative random walk strategies for graph embedding algorithms that explicitly take into account the inner structure of natural communities and characteristics of nodes within them.

Acknowledgments. This research is supported by the Science Fund of Republic of Serbia, #6518241, AI – GRASP. The authors would like to thank the anonymous reviewers for their insightful suggestions and comments that helped improve the quality of the paper.

References

1. Amsaleg, L., et al.: Estimating local intrinsic dimensionality. In: Proceedings of the 21th ACM SIGKDD International Conference on Knowledge Discovery and Data Mining, KDD 2015, pp. 29–38. Association for Computing Machinery, New York (2015). https://doi.org/10.1145/2783258.2783405
2. Amsaleg, L., Chelly, O., Houle, M.E., Kawarabayashi, K.I., Radovanović, M., Treeratanajaru, W.: Intrinsic dimensionality estimation within tight localities. In: Proceedings of the 2019 SIAM International Conference on Data Mining, pp. 181–189. Society for Industrial and Applied Mathematics, May 2019. https://doi.org/10.1137/1.9781611975673.21

3. Becker, R., Hafnaoui, I., Houle, M.E., Li, P., Zimek, A.: Subspace determination through local intrinsic dimensional decomposition: theory and experimentation. arXiv 1907.06771 (2019)
4. Casanova, G., et al.: Dimensional testing for reverse k-nearest neighbor search. Proc. VLDB Endow. **10**(7), 769–780 (2017). https://doi.org/10.14778/3067421.3067426
5. Goyal, P., Ferrara, E.: Graph embedding techniques, applications, and performance: a survey. Knowl. Based Syst. **151**, 78–94 (2018). https://doi.org/10.1016/j.knosys.2018.03.022
6. Grover, A., Leskovec, J.: Node2vec: scalable feature learning for networks. In: Proceedings of the 22nd ACM SIGKDD International Conference on Knowledge Discovery and Data Mining, KDD 2016, pp. 855–864. Association for Computing Machinery, New York (2016). https://doi.org/10.1145/2939672.2939754
7. Houle, M.E.: Dimensionality, discriminability, density and distance distributions. In: 2013 IEEE 13th International Conference on Data Mining Workshops, pp. 468–473 (2013). https://doi.org/10.1109/ICDMW.2013.139
8. Houle, M.E.: Local intrinsic dimensionality I: an extreme-value-theoretic foundation for similarity applications. In: Beecks, C., Borutta, F., Kröger, P., Seidl, T. (eds.) SISAP 2017. LNCS, vol. 10609, pp. 64–79. Springer, Cham (2017). https://doi.org/10.1007/978-3-319-68474-1_5
9. Houle, M.E.: Local intrinsic dimensionality III: density and similarity. In: Satoh, S., et al. (eds.) SISAP 2020. LNCS, vol. 12440, pp. 248–260. Springer, Cham (2020). https://doi.org/10.1007/978-3-030-60936-8_19
10. Houle, M.E., Schubert, E., Zimek, A.: On the correlation between local intrinsic dimensionality and outlierness. In: Marchand-Maillet, S., Silva, Y.N., Chávez, E. (eds.) SISAP 2018. LNCS, vol. 11223, pp. 177–191. Springer, Cham (2018). https://doi.org/10.1007/978-3-030-02224-2_14
11. Lancichinetti, A., Fortunato, S., Kertész, J.: Detecting the overlapping and hierarchical community structure in complex networks. New J. Phys. **11**(3), 033015 (2009). https://doi.org/10.1088/1367-2630/11/3/033015
12. Ma, X., et al.: Characterizing adversarial subspaces using local intrinsic dimensionality. In: International Conference on Learning Representations (2018). https://openreview.net/forum?id=B1gJ1L2aW
13. Ma, X., et al.: Dimensionality-driven learning with noisy labels. In: Dy, J.G., Krause, A. (eds.) Proceedings of the 35th International Conference on Machine Learning, ICML 2018, Stockholmsmässan, Stockholm, Sweden, 10–15 July 2018, vol. 80, pp. 3361–3370. PMLR (2018)
14. Radicchi, F., Castellano, C., Cecconi, F., Loreto, V., Parisi, D.: Defining and identifying communities in networks. Proc. Natl. Acad. Sci. **101**(9), 2658–2663 (2004). https://doi.org/10.1073/pnas.0400054101
15. von Ritter, L., Houle, M.E., Günnemann, S.: Intrinsic degree: an estimator of the local growth rate in graphs. In: Marchand-Maillet, S., Silva, Y.N., Chávez, E. (eds.) SISAP 2018. LNCS, vol. 11223, pp. 195–208. Springer, Cham (2018). https://doi.org/10.1007/978-3-030-02224-2_15
16. Watts, D.J., Strogatz, S.H.: Collective dynamics of 'small-world' networks. Nature **393**(6684), 440–442 (1998). https://doi.org/10.1038/30918

Structural Intrinsic Dimensionality

Stephane Marchand-Maillet[1(✉)], Oscar Pedreira[2], and Edgar Chávez[3]

[1] Department of Computer Science, University of Geneva, Geneva, Switzerland
stephane.marchand-maillet@unige.ch
[2] Universidade da Coruña, Coruña, Spain
oscar.pedreira@udc.es
[3] CICESE, Ensenada, Mexico
elchavez@cicese.mx

Abstract. The dimension of the space within which the data lives is a major driver for the performance of many processing operations. However, global dimensionality cannot be blindly trusted as the data may lie on structures of lower local dimensionality within the ambient space. Here, we address the problem of estimating the local dimensionality of the data space or to provide a consistent proxy for it.

The review of existing local dimensionality estimators shows the various types of geometric information they are based on. We propose the exploration of an alternative route using proximity constraints mapped into the structure of a spanner graph whose properties reflect the local geometry. We propose to adapt PageRank-like information propagation algorithms to infer the structural intrinsic dimensionality directly from the neighborhood structure of data points, taken as vertices. Further, the presence of the spanner over our dataset enables global operations to strengthen the coherence of our estimates and support similarity search.

Keywords: Local intrinsic dimension · Kissing number · Geometric graph spanner

1 Introduction

The dimension of the space containing the data generally refers to the geometric dimension corresponding to the number of linearly independent vectors the space can accommodate. Global data dimension is not a proper characteristic of the data. If the data is uniformly distributed within its ambient space, there is no structural pattern to exploit to construct index structures. The assumption is therefore that the global data dimension may simply represent the dimension of an ambient space within which the data lies over finer structures as a subspace of lower dimension. It is further assumed that the dimension of this subspace may vary locally, therefore defining the notion of local dimensionality. The intrinsic nature of this dimensionality attaches it to the data rather than to its representation and therefore makes it more of an invariant of that data.

N. Reyes et al. (Eds.): SISAP 2021, LNCS 13058, pp. 173–185, 2021.
https://doi.org/10.1007/978-3-030-89657-7_14

In this paper we investigate the nature of local dimensionality along with the proposed models for its estimation (Sect. 2). We then uncover the issues related to its estimation in a practical setup. In particular, we address the issue of the stability of the estimate in relation to the various parameters (Sects. 3 and 4).

The main contribution of this paper is to make proposals to break the paradox that local dimensionality is a local notion but the statistical nature of its estimation requires to extend its support beyond mere locality. We present experimental measures that support our proposals.

2 The Fundamental Information Behind Local Dimensionality

We define the *local dimensionality* at point $x \in \mathbb{R}^M$ as being a local indication of $\dim(x)$, the latent dimension at point x of a continuous information density distribution f immersed into the ambient space $\mathbb{R}^M$ ($f : \mathbb{R}^M \mapsto \mathbb{R}_+$; $\int_{\mathbb{R}} f = 1$). That is the lowest dimension of the subspace around x within which f could be embedded with no loss (isometrically).

f is thus a probability density function installed over the ambient space R^M from which we can sample discrete locations (the data). Call X a set of N points $X = \{x_i\}_{[\![N]\!]} \subset \mathbb{R}^M$ that is taken as a sample from the density distribution f. A metric (e.g. Euclidean distance) is used to define the neighborhood of every x_i. Then, the goal of *discrete local dimensionality estimation* is to infer the value of $\dim(x_i)$ at every x_i from the locations of points in the rest of X. In effect, the function dim(.) can take any positive scalar value ($\dim(x) \in \mathbb{R}^+$), i.e. "discrete" refers to the estimation being based on discrete point locations.

2.1 Motivation for an Estimation

Formally, the Nearest Neighbor Indexing (NNI) theorem [21] and subsequent works state that for a workload of vanishing variance in high dimensions, the performance of the class of convex indexes will approach that of sequential search (i.e. $O(N)$). This is clearly supported in practice when working with data of dimensions approaching 20.

Underlying the proof of the NNI theorem is the idea that the indexing covers the dataset X with potentially overlapping convex tiles. As the dimension increases, the vanishing variance of the distribution (D) of distances makes the width of the indexing tiles of the same order than the distance to the nearest neighbor (the best answer to the query). As a result, all tiles need to be fetched during any search process. In this situation, mostly all N points of X are explored as candidates for the result of any search.

The property of vanishing variance stating that

$$\exists \alpha \in \mathbb{R}^+_{\backslash\{0\}} \qquad \text{s.t.} \qquad \lim_{M \to \infty} \mathsf{Var}\left(\frac{D_M^\alpha}{\mathsf{E}[D_M^\alpha]}\right) = 0$$

is closely related to the so-called *concentration of distances* arising due to the Lipschitz structure of Minkowski distances (summing iid coordinates) and the Chebyshev inequality [3]. In other words, no convex indexing can provide exclusion, due to the concentration of distances. This makes this result rather universal in Data Analysis and motivates the quest for discrete local dimensionality estimation. In essence, by turning the argument upside down, we seek an estimate that correlates with the factor (called "dimensionality", $\dim(x)$) that influences the performance of any data nearest neighborhood indexing (and analysis).

Note further that most of the related literature focuses on estimating the dimensionality but a proper use of this characterization is yet to be proposed. This work is a step towards constructing a context where the analysis provides actionable tools to make effective similarity search in high dimensional spaces.

2.2 Expansion-Based Estimation

The class of expansion-based estimation techniques relies on the fact that the increase of volume of a M-dimensional hypersphere V_M is essentially related to the increase of its radius r by an exponential relationship.

$$V_M(r) = \frac{2\pi^{\frac{M}{2}}}{M\Gamma(\frac{M}{2})} r^M \quad \Rightarrow \quad \frac{\partial V_M}{\partial r} = \frac{2\pi^{\frac{M}{2}}}{\Gamma(\frac{M}{2})} r^{M-1}$$

This is exploited in the definition of the *Expansion Dimension* (ED) [18] and its generalization GED [12]. The strategy is to estimate a proxy for the volume of the hypersphere of radius r centered at x_i by counting the number n of data points in a r-range query from x_i. Hence, the dimension is estimated by a log of the relative increase of this number (from n_1 to n_2) versus the increase in radius (from r_1 to r_2): $\mathsf{GED}(x_i) = \frac{\log \frac{n_2}{n_1}}{\log \frac{r_2}{r_1}}$.

The above assumes (at least locally) a uniform distribution of data around x_i. It is further refined by considering (instead of the volume of the hypersphere) the cumulative function $F(r)$ of a distance distribution (whose 0 would be at -every- x_i). This allows to model a variable density within the space and to define the lID [13,14] that matches the GED for a uniform distribution.

2.3 Concentration of Correlates

Another route for exploring the local geometry of the space is to look at angles. Fixing one direction $\mathbf{u}_k$ from x_i (thru x_k say), one can study the distribution of the angles made between this direction and vectors $\mathbf{u}_j$ whose extremity x_j other than x_i is sampled over a hypersphere centered at x_i. Such an estimation amounts to compute the surface of spherical caps defined by the cones generated by $\mathbf{u}_k$ and angle $\theta \propto \angle(\mathbf{u}_k, \mathbf{u}_j)$. This distribution of correlates ($\cos(\theta) = \langle \mathbf{u}_k, \mathbf{u}_j \rangle$) is further known to concentrate with increasing dimensionality [5,6,9]:

$$\mathsf{P}(\theta) = \frac{\Gamma(\frac{M}{2})}{\Gamma(\frac{1}{2})\Gamma(\frac{M-1}{2})} \sin^{M-2}(\theta) \quad \text{and} \quad \mathsf{Var}(\cos(\theta)) = \frac{1}{M}$$

The latter second order information is then used to estimate the local dimensionality by local samples of angles [22].

2.4 Sphere Packing

Another possible approach is to also use the notion of sphere packing [6] but in relation to the *kissing number*. Here, the observed estimate is the number of non-intersecting hyperspheres of diameter r able to be tangent to (to "kiss") a hypersphere of radius r centered at x_i. This is known as the kissing number $\mathsf{Kiss}(M)$ whose exact values are known only for a select number of dimension values $M \in \{1, 2, 3, 4, 8, 24\}$. For other values, upper and lower bounds which illustrate the dynamic of the kissing number with respect to the dimension have been proposed [16]:

$$(1 + O(1))\sqrt{\frac{3\pi}{8}} \log\left(\frac{3}{2\sqrt{2}}\right) M^{\frac{3}{2}} \left(\frac{2}{\sqrt{3}}\right)^M \leq \mathsf{Kiss}(M) \leq (1 + O(1))\sqrt{\frac{\pi}{8}} M^{\frac{3}{2}} 2^{\frac{M}{2}}$$

The regular dependence of these bounds on dimension makes the kissing number another appropriate entry door to the estimation of the local dimensionality.

It can be noted that this information also relates to the above angle-based estimation in the sense that the kissing number counts the maximum number of non-intersecting spherical caps with total angle $\frac{\pi}{3}$ (generative angle $\frac{\pi}{6}$) one can segment the central hypersphere surface with. It is a particular instance of so-called "spherical codes" [4,16] with $\theta = \frac{\pi}{3}$. Equivalently, the kissing number counts the number of points a given point can be nearest neighbor of (so-called reverse nearest neighbor). For example, a point of a 2D plane can only be the nearest neighbor of $\mathsf{Kiss}(2)$=6 points (arranged as a hexagon).

2.5 Discussion

The above three approaches are different in their computation but rely on essentially the same information.

Expansion-based estimations explicitly use the shape of the density distribution along the distance axis. That is, from the central point x_i where the local dimensionality is to be estimated the hyperspherical shell of radius r around that point is integrated into the point of coordinate r on the distance axis. It is the growth rate of this value that is explicitly modeled by GED and lID. In the discrete version [1,2], the lID is a measure of the density of data within a thick spherical shell (from the 1-NN to the k-NN of x_i). The transition from the continuous model to the discrete estimation still imposes an assumption of local uniformity in the distribution of the k nearest neighbors.

The estimation of the ABID [22] is based on estimating the concentration of the cosine similarities between points on a hypersphere centered at x_i. Using fixed length vectors, the cosine similarities are known to correlate with squared distances (e.g. this is the basis for the MIPS problem [10]) [4]: $\langle \mathbf{u}_k, \mathbf{u}_j \rangle \propto d(x_k, x_j)^2$ In practice, the k nearest neighbors from x_i are used so that the ABID is also a reflection of the density of data within a thick spherical shell (from the 1-NN to the k-NN of x_i). The advantage of this estimation is that angles involve triplets of points and create a combinatoric volume of estimates, thus reducing the span of the neighborhood (value of k) required in practice to obtain a robust estimate.

Finally, using the kissing number to estimate the local dimensionality imposes complementary constraints: at fixed radius r from x_i (first constraint) the kissing number counts how many points can be organized so as to be at least r-distant from each other (second constraint). The first constraint may similarly be relaxed by exploring values of r along the distance axis. The second constraint may be imposed by selecting neighbors dispersed around x_i. This is handled via the generation of spanner graphs such as the reverse neighbor graph, the half-space proximal graph (HSP) [7] or the Yao and θ-graphs [20]. In earlier works [8,15], we explored the correlation between indicators of some of these graphs with local dimensionality to partition the dataset in view of improving its *indexability*.

3 Information Propagation on Neighbor Graphs

In their original presentations, the above measures essentially treat all points x_i individually and sequentially. They then operate some statistical analysis (e.g. mean or variance) on the distribution of the local dimensionality values throughout the dataset.

Considering the points independently creates the tension between the desire to compute a robust estimation over a large number of neighbors (large k in k-NN) and the intrinsic wish to stay local (small ε in ε-NN). The kissing number instructs us that for dimensions as limited as 20 the coverage of the hypersphere requires $O(10^4)$ neighbors already, which is beyond the density of any classical dataset[1]. One can therefore question the validity of the empirical estimates made using $k = O(10^2)$ neighbors. This is partly discussed in [22], for example.

In addition, the local dimensionality may vary from a point to another in X. Hence, computing global a posteriori statistics may not be so relevant for all datasets (e.g. a Saturn-shaped dataset). This can be related to the Yule-Simpson effect [11], which induces potentially contradictory interpretations, depending of the scale at which the data is studied[2].

In turn, a true local dimensionality estimate would enable operations like dimensionality-based clustering, and define *indexability* [15].

Here, we make steps in the direction pointed by the above remarks: using the global structural information provided by the full dataset X for estimating the local dimensionality at every $x_i \in X$. We relax the implicit above assumption of a constant local dimensionality by assuming that the local dimensionality bears

[1] We argue that estimating the $O(M)$ linearly independent vectors reflecting the geometric dimension (e.g. using rank-based methods such as local PCA) would not be reliable in that case due to quasi-orthogonality [17] and the issue of local neighborhood selection.

[2] This further relates the question of discrete local dimension estimation to local scale estimation, an important topic, addressed in [2], left for further investigation.

some smoothness of the form: $d(x_i, x_j) < r \Rightarrow |\dim(x_i) - \dim(x_j)| < \alpha$ for small values of r and α.

3.1 Structural Regression

Following the above discussion we propose to enforce that the dimensionality at x_i is the weighted average of the dimensionality of its neighbors x_j:

$$\dim(x_i) = \frac{1}{Z_i} \sum_{x_j \in \mathcal{N}(x_i)} w_{ij} \dim(x_j) \tag{1}$$

for some neighborhood $\mathcal{N}(x_i)$ and some influence weighting w_{ij} with proper normalization $Z_i = \sum_j w_{ij}$ (note that $x_i \notin \mathcal{N}(x_i)$ and $w_{ii} = 0$). This smoothness condition alone makes the dimensionality estimates prone to translation, as the above stays valid if $\dim'(x_i) = \dim(x_i) + K$ with any constant K. Hence, we apply this strategy starting from an estimate ϵ of the dimensionality (e.g. lID or ABID).

Given $\epsilon = [\epsilon_i]^\mathsf{T}$ as estimate for local dimensionality $\mathbf{d} = [d_i]^\mathsf{T}$, we resolve the classical regression:

$$\mathbf{d}^* = \min_{\mathbf{d} \in \mathbb{R}^N} \mathcal{L}(\mathbf{d}, \epsilon) \quad \text{where} \quad \mathcal{L}(\mathbf{d}, \epsilon) = \frac{1}{2}\|\mathbf{d} - \epsilon\|_2^2 + \frac{\lambda}{2}(d_i - \sum_{x_j \in \mathcal{N}(x_i)} d_j)^2$$

with $\lambda > 0$ controlling the smoothness when maintaining the volume $\sum_i d_i$ constant. The above is a classical convex minimization solved iteratively using:

$$d_i^{(t+1)} \leftarrow d_i^{(t)} - \eta \left[(d_i^{(t)} - \epsilon_i) + \lambda (d_i^{(t)} - \sum_{x_j \in \mathcal{N}(x_i)} w_{ij} d_j^{(t)}) \right] \tag{2}$$

for learning rate $0 < \eta < 1$. It is easy to see that this guarantees $\mathsf{Var}[\mathbf{d}^*] \leq \mathsf{Var}[\epsilon]$.

3.2 Experiments

We wish to validate empirically our analysis on the regularity of local dimensionality estimates. We use the lID [13] and its MLE estimate [2], and the HSP degree [15] over uniform datasets of known dimensionality to calibrate our study.

We generate datasets of various dimensionalities ($M \in \{3, 6, 10, 15, 20, 30, 50\}$) and various densities ($N \in \{10000, 20000, 30000, 40000, 50000, 100000, 150000\}$) to study the influence of dimension with respect to data density.

The datasets are composed of N samples of a distribution (Uniform or Gaussian) restricted to a M-dimensional hypersphere of radius 1. Spherical datasets are chosen to eliminate "corner" effects.

Figure 1(top) shows the estimates provided by the HSP built over an hypersphere filled with uniform (left) and Gaussian (right) sampling, using its degree as a function of the data density and for the dimensionalities listed above. One

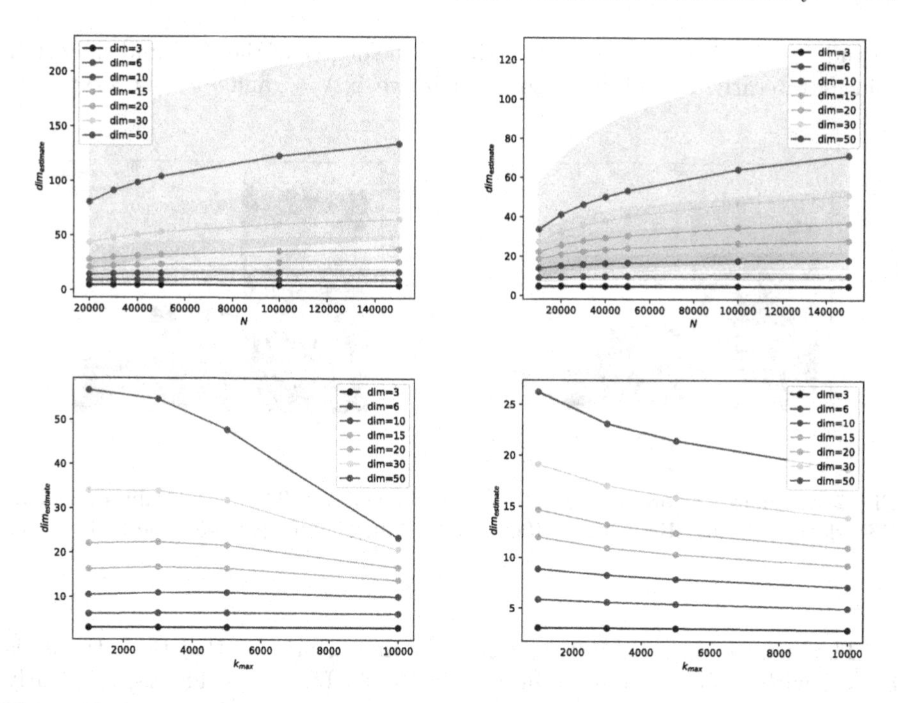

Fig. 1. Variation of dimensionality estimates versus parameters (N or $k_{\max}$) for [left: spherical data] [right: gaussian data] [up: HSP degree] [down: lID with $N = 50'000$]

clearly sees that although only correlated with the true dimensionality, the estimates stabilize with an increase of the density but that the variance augments with density.

Turning to the lID MLE estimate, we use the same datasets with fixing a high density with $N = 50'000$ and varying the size of the neighborhood over which the estimate is computed: $k_{\max} \in \{1000, 3000, 5000, 10000\}$. Clearly, the estimates get corrupted using a too large $k_{\max}$. Further, as dimensionality increases this phenomenon is more drastic (the variance here is too large to be properly displayed).

The above clearly illustrates the contradiction in extending the support of estimation of a local estimator. It shows that even though the estimates may be considered as overall reliable (e.g. when averaged), the sensitivity to their parameters and location of estimation is so that they cannot be blindly applied without some knowledge of an appropriate scale and the presence of a reasonable local data density.

The main issue lies in the variance of these estimates, as we seek a factor that correlates with what could be referred to as local dimensionality. We therefore look at the behavior of these measures over datasets of varying dimensionality to demonstrate the capability of our smoothing (Eq. 1) to reduce the variance of the estimates. We generate a dataset with 3 non-overlapping spherical uniform clusters containing 20'000, 15'000 and 15'000 points respectively and of dimensionality 20,

10 and 5 respectively. We initially estimate the degree of the HSP and smooth it using our iterative convolution (Eq. 2) where we fix $\lambda = 5$ and $\eta = 0.1$ everywhere.

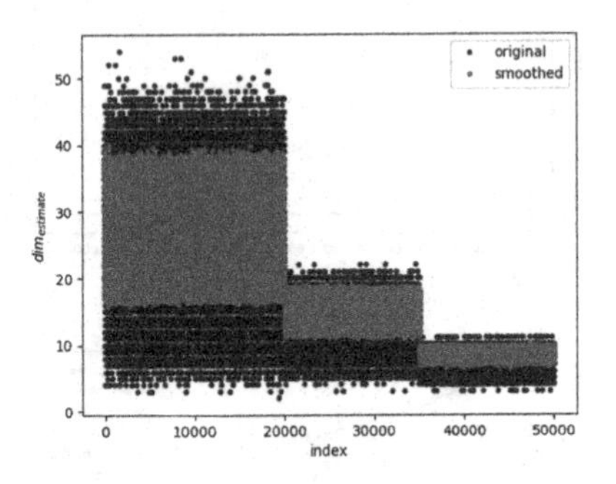

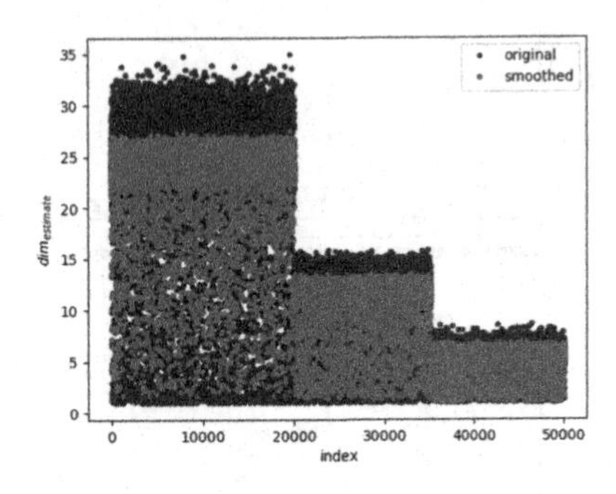

Fig. 2. Estimates before and after iterative convolution. The initial values are [left: HSP degree], [right: lID using k=1000]. Both diffusions happen over the HSP (Color figure online)

The result is reported in Fig. 2(left). We first clearly see that the estimate is correct with respect to our calibration in Fig. 1. The three clusters are clearly identified. However, the estimate vary significantly within clusters. As result of the regression, dimension estimates are corrected (from blue to red dots) and the variance diminish appropriately. Here, inspired by the gravitational physical model, we use $w_{ij} = \frac{1}{d^2(x_i, x_j)}$ as influence weight.

We use the same dataset to perform an estimate with the lID (fixing $k_{\max} = 1000$) (Fig. 2(right)). The regression clearly alleviates the problem of high variance in the estimates. However, due to the initial distribution, the estimates remain rather spread, indicating a need for exploring stronger constraints in the regression or using a stronger value for λ.

Table 1 gives the variation of mean and standard deviation (between brackets) of the estimates in both cases and per cluster. Note that the result of the convolution is not the mere mean and variance of the original, showing that the weighting structure does play a role.

4 Structural Intrinsic Dimensionality

We push the idea of information propagation over a neighborhood structure a step further. The geometry described in Sect. 2.4 suggests that dimensionality may be captured by the connectivity structure of the neighborhood graph itself.

We already demonstrated that the degree of the HSP whose construction relates to the kissing number correlates with dimensionality. It is known (notably from the development of the PageRank algorithm) that information propagation over directed graphs can provide essential information about the underlying connectivity structure. We therefore hypothesize that the local dimensionality may

Table 1. Mean and standard deviations for HSP degree and lID before and after iterative convolution ($\lambda = 5$ and $\eta = 0.1$)

	Cluster 1	Cluster 2	Cluster 3
HSP initial	28.02 (10.97)	13.89 (3.37)	7.54 (1.26)
HSP smoothed	26.08 (6.71)	15.29 (2.06)	8.71 (0.64)
lID initial	22.70 (6.51)	10.95 (2.83)	5.23 (1.06)
lID smoothed	22.74 (3.65)	10.93 (1.95)	5.21 (0.83)

be inferred via information diffusion, provided the graph encodes this information.

Given a directed graph $G = (X, E)$ with edge set E defined from geometric constraints, i.e. $(x_i, x_j) \in E$ iff $x_j \in \mathcal{N}(x_i)$, we define information propagation of value $d(x)$ as the convergence of the (directed) iterative process:

$$d_i^{(t+1)} \leftarrow \sum_{x_j \text{ s.t. } x_i \in \mathcal{N}(x_j)} w_{ji} d_j^{(t)} \tag{3}$$

Classically, the diffusion is done so as to preserve the value $\sum_i d(x_i)$ constant. The directed setup thus imposes $\sum_j w_{ij} = 1 \quad \forall i \in [\![N]\!]$, making matrix $W = [w_{ij}]_{ij \in [\![N]\!]}$ a row-stochastic matrix ($w_{ij} = 0$ if $(x_i, x_j) \notin E$). It is known that under proper conditions, this process converges to the principal eigenvector (with eigenvalue 1) of W, the weighted adjacency matrix of G. In PageRank-like diffusion algorithms, edge weights w_{ij} are tuned so as to distribute the value at node x_i to forward neighbors x_j based on the degree (e.g. $w_{ij} = \frac{1}{\deg^+(x_i)}$).

Adapting to our geometrical context we read $\deg^+(x_i) = \sum_{x_j \in \mathcal{N}(x_i)} 1$. That is, every outgoing edge from x_i counts 1, so that $w_{ij} = \frac{1}{\sum_{x_k \in \mathcal{N}(x_i)} 1}$. We transform this to influence by inserting an inverse dependence $\phi(.)$ to distance as edge weight, while preserving the row-stochasticity constraint:

$$w_{ij} = \frac{\phi(d(x_i, x_j))}{\sum_{x_k \in \mathcal{N}(x_i)} \phi(d(x_i, x_k))} \quad \text{where, for example} \quad \phi(x) = \frac{1}{x} \tag{4}$$

Using an inverse dependence as base edge weight (Eq. 4) therefore induces a softmax-like filter on edges, thus favoring the shortest edge emanating from every x_i. Combining this with our diffusion strategy (Eq. 3), every vertex x_i receives mostly influence from the other vertices x_j of which it is the closest neighbor. This corresponds exactly to the definition of the kissing number at x_i. We therefore expect diffusion over such graph structures to exhibit an information that correlates with local dimensionality and that we can refer to as *structural intrinsic dimensionality*.

The relationship with eigencentrality in graphs is also clear as it corresponds to the case where $\phi(x) = 1$. This nicely connects with and continues our earlier proposals [8,15], where we proposed graph centrality measures as indicators that correlate with local dimensionality. In this context, the degree of the HSP is seen as its degree centrality indicator, itself an approximation of the eigencentrality.

4.1 Experiments

We now propose results for our structural intrinsic dimensionality estimation using again the cluster dataset presented above. Our initial experiment confirms that a careful design of the edge weighting scheme is important. We found that using $\phi(x) = \frac{1}{\sqrt{x}}$ in Eq. 4 produces interesting results. As before, we smoothed these results via iterative convolution. The results are presented in Fig. 3.

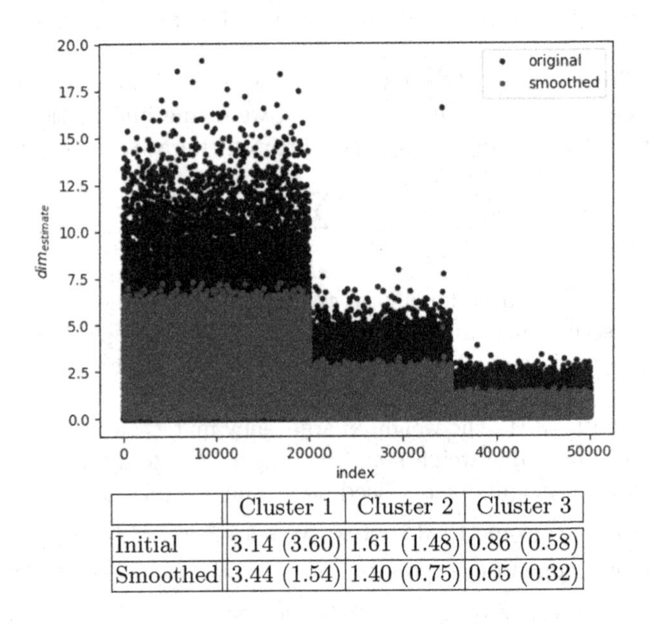

	Cluster 1	Cluster 2	Cluster 3
Initial	3.14 (3.60)	1.61 (1.48)	0.86 (0.58)
Smoothed	3.44 (1.54)	1.40 (0.75)	0.65 (0.32)

Fig. 3. Dimensionality indicator from information propagation over the HSP ($\phi(x) = \frac{1}{\sqrt{x}}$). Estimates before and after iterative convolution

Of course, the value of the estimate does not match the dimensionality as understood as space dimension. However, information propagation does produce a proper indicator of this dimensionality. More investigation is required to understand the most favorable structure of underlying spanner (HSP, k-NN, reverse-k-NN, ...) to use for propagation and the best weighting scheme. It can even be envisaged to join both, starting from the complete graph and decimating it with respect to the defined edge weight (i.e. removing edges with negligible transfer in order to reduce the computational cost).

Finally, it seems natural to target the integration of the iterative convolution process (Eq. 2) with the propagation (Eq. 3). We therefore believe that this graph-based strategy for the estimation of the structural intrinsic dimensionality opens many interesting questions.

5 Conclusion

Local dimensionality is a major driver for the performance of data processing techniques. Its effects are deeply rooted into statistics, as demonstrated by the concentration of distances that is one aspect of the curse of dimensionality. Obtaining indicators for local dimensionality in the discrete space is therefore of interest and most existing local dimensionality indicators are based on the estimation of the variation of local density.

Here, we consider any indicator that show a monotonic relationship with local dimensionality. We propose to exploit the definition of the kissing number to obtain such an indicator. Using a graph structure over the dataset, we show that information propagation can not only help strengthening classical indicators but also being used as an estimator itself. This work therefore gives a formal grounding for our earlier proposals [8,15].

The results open the question of the underlying graph structure that would be best suited for such an exploration. We suggest that this question is equivalent to defining a proper edge weight, capturing the geometry of the dataset in the graph structure. As this weighting naturally makes use of the underlying metric, this clearly relates to the construction of appropriate geometric t-spanners that will be one direction we wish to explore. One can note that graph-based computations also provide a computational solution to the problem of combining local and global structures. Computation can further be distributed using the tight equivalence between information propagation algorithms and random walk processes.

Finally, determining local dimensionality does not directly provide a solution to counter its adverse effects. We have proposed to use it to partition the dataset based on its *indexability* [15]. Following that route, we believe that the graphs arising from the estimation of the structural intrinsic dimensionality will be useful for constructing efficient indexing strategies in the line of recent graph-based indexing techniques using navigable structures [19].

Another option for using local dimensionality estimates in an operational setting may be their use to drive local embedding for adapting the indexing locally into a lower dimensional context.

Acknowledgments. O. Pedreira is partially funded by: MICIU/FEDER-UE BIZDEVOPSGLOBAL: RTI-2018-098309-B-C32 and Xunta de Galicia/FEDER-UE GRC: ED431C 2017/58.

References

1. Amsaleg, L., et al.: Extreme-value-theoretic estimation of local intrinsic dimensionality. Data Min. Knowl. Disc. **32**(6), 1768–1805 (2018)
2. Amsaleg, L., Chelly, O., Houle, M.E., Kawarabayashi, K., Radovanović, M., Treeratanajaru, W.: Intrinsic dimensionality estimation within tight localities. In: Proceedings of the SIAM International Conference on Data Mining (SDM). SIAM (2019)

3. Boucheron, S., Lugosi, G., Massart, P.: Concentration Inequalities: A Nonasymptotic Theory of Independence. OUP, Oxford (2013)
4. Boyvalenkov, P., Dodunekov, S., Musin, O.R.: A survey on the kissing numbers. CoRR arXiv:abs/1507.03631 (2015)
5. Cai, T., Fan, J., Jiang, T.: Distributions of angles in random packing on spheres. J. Mach. Learn. Res. **14**(21), 1837–1864 (2013). http://jmlr.org/papers/v14/cai13a.html
6. Cai, T., Jiang, T.: Phase transition in limiting distributions of coherence of high-dimensional random matrices. J. Multivariate Anal. **107**, 24–39 (2012). https://doi.org/10.1016/j.jmva.2011.11.008
7. Chavez, E., et al.: Half-space proximal: a new local test for extracting a bounded dilation spanner of a unit disk graph. In: Anderson, J.H., Prencipe, G., Wattenhofer, R. (eds.) OPODIS 2005. LNCS, vol. 3974, pp. 235–245. Springer, Heidelberg (2006). https://doi.org/10.1007/11795490_19
8. Chavez, E., Pedreira, O., Marchand-Maillet, S.: Reverse *k*-nearest neighbors centrality measures and local intrinsic dimension. In: Proceedings of the 13th International Conference on Similarity Search and Applications (SISAP 2020) (2020)
9. Connor, R., Dearle, A.: Sampled angles in high-dimensional spaces. In: Satoh, S., et al. (eds.) SISAP 2020. LNCS, vol. 12440, pp. 233–247. Springer, Cham (2020). https://doi.org/10.1007/978-3-030-60936-8_18
10. Ding, Q., Yu, H.F., Hsieh, C.J.: A fast sampling algorithm for maximum inner product search. In: Chaudhuri, K., Sugiyama, M. (eds.) Proceedings of the Twenty-Second International Conference on Artificial Intelligence and Statistics. Proceedings of Machine Learning Research, vol. 89, pp. 3004–3012. PMLR (2019)
11. Good, I.J., Mittal, Y.: The amalgamation and geometry of two-by-two contingency tables. Ann. Stat. **15**(2), 694–711 (1987)
12. Houle, M.E., Kashima, H., Nett, M.: Generalized expansion dimension. In: 2012 IEEE 12th International Conference on Data Mining Workshops, pp. 587–594 (2012)
13. Houle, M.E.: Local intrinsic dimensionality I: an extreme-value-theoretic foundation for similarity applications. In: Beecks, C., Borutta, F., Kröger, P., Seidl, T. (eds.) SISAP 2017. Lecture Notes in Computer Science, vol. 10609, pp. 64–79. Springer, Cham (2017). https://doi.org/10.1007/978-3-319-68474-1_5
14. Houle, M.E.: Local intrinsic dimensionality II: multivariate analysis and distributional support. In: Beecks, C., Borutta, F., Kröger, P., Seidl, T. (eds.) SISAP 2017. Lecture Notes in Computer Science, vol. 10609, pp. 80–95. Springer, Cham (2017). https://doi.org/10.1007/978-3-319-68474-1_6
15. Hoyos, A., Ruiz, U., Marchand-Maillet, S., Chávez, E.: Indexability-based dataset partitioning. In: Amato, G., Gennaro, C., Oria, V., Radovanović, M. (eds.) SISAP 2019. LNCS, vol. 11807, pp. 143–150. Springer, Cham (2019). https://doi.org/10.1007/978-3-030-32047-8_13
16. Jenssen, M., Joos, F., Perkins, W.: On kissing numbers and spherical codes in high dimensions. CoRR arXiv:abs/1803.02702 (2018)
17. Kainen, P., Kůrková, V.: Quasiorthogonal dimension of Euclidean spaces. Appl. Math. Lett. **6**(3), 7–10 (1993). https://doi.org/10.1016/0893-9659(93)90023-G
18. Karger, D.R., Ruhl, M.: Finding nearest neighbors in growth-restricted metrics. In: 34th ACM Symposium on Theory of Computing. ACM (2002)
19. Malkov, Y.A., Yashunin, D.A.: Efficient and robust approximate nearest neighbor search using hierarchical navigable small world graphs. IEEE Trans. Pattern Anal. Mach. Intell. **42**(04), 824–836 (2020). https://doi.org/10.1109/TPAMI.2018.2889473

20. Narasimhan, G., Smid, M.: Geometric Spanner Networks. Cambridge University Press, Cambridge (2007). https://doi.org/10.1017/CBO9780511546884
21. Shaft, U., Ramakrishnan, R.: Theory of nearest neighbors indexability. ACM Trans. Database Syst. **31**(3), 814–838 (2006)
22. Thordsen, E., Schubert, E.: ABID: angle based intrinsic dimensionality. In: Satoh, S., et al. (eds.) SISAP 2020. LNCS, vol. 12440, pp. 218–232. Springer, Cham (2020). https://doi.org/10.1007/978-3-030-60936-8_17

Relationships Between Local Intrinsic Dimensionality and Tail Entropy

James Bailey[1], Michael E. Houle[2(✉)], and Xingjun Ma[1,3]

[1] School of Computing and Information Systems, The University of Melbourne, 700 Swanston Street, Melbourne, VIC 3010, Australia
baileyj@unimelb.edu.au
[2] National Institute of Informatics, 2-1-2 Hitotsubashi, Chiyoda-ku, Tokyo 101-8430, Japan
meh@nii.ac.jp
[3] School of Information Technology, Deakin University, 221 Burwood Hwy., Burwood, VIC 3125, Australia
daniel.ma@deakin.edu.au

Abstract. The local intrinsic dimensionality (LID) model assesses the complexity of data within the vicinity of a query point, through the growth rate of the probability measure within an expanding neighborhood. In this paper, we show how LID is asymptotically related to the entropy of the lower tail of the distribution of distances from the query. We establish tight relationships for cumulative Shannon entropy, entropy power, and their generalized Tsallis entropy variants, all with the potential for serving as the basis for new estimators of LID, or as substitutes for LID-based characterization and feature representations in classification and other learning contexts.

1 Introduction

Assessing the complexity of high dimensional data is a fundamental task that underpins many activities in machine learning and data mining. One well-known measure of data complexity is the intrinsic dimensionality, a unitless quantity that can be interpreted as the minimum number of latent variables needed to describe the data.

The many extant formulations of intrinsic dimensionality can be divided into two broad groups, global and local. Global intrinsic dimensionality, which takes contributions from the full dataset to measure its complexity as a whole, has been more widely investigated. By contrast, local variants of intrinsic dimensionality assess the complexity of the data in the vicinity of a designated query location, most notably in terms of the growth rate in the probability measure captured by an expanding neighborhood. Local variants can therefore associate different intrinsic dimensional values to different locations in the data domain.

Our focus in this paper is on the local intrinsic dimension (LID) as formulated in [22,23], and in particular, establishing how it relates to entropy, perhaps the most fundamental and widely-used model of data complexity. In its essence,

N. Reyes et al. (Eds.): SISAP 2021, LNCS 13058, pp. 186–200, 2021.
https://doi.org/10.1007/978-3-030-89657-7_15

entropy can be regarded as a measure of the uncertainty of a distribution. Our study of entropy considers the distribution of distances to a query location, where the distances are induced by a global data distribution. In particular, we consider the entropy of the lower tail of the neighbor distance distribution (the *tail entropy*), and consider its asymptotic tendency as the neighborhood radius approaches zero.

Our analysis of the relationship between the tail entropy and local intrinsic dimensionality has further implications due to an established relationship between the latter and the statistical theory of extreme values (EVT) [2]. For any distribution of distances satisfying appropriate smoothness assumptions in the lower tail, as the neighborhood radius approaches zero, the tail distribution takes the form of a power law. Asymptotically, power law distributions can be said to arise naturally in the lower tail, with the exponent of the power law corresponding to the LID value.

We formulate asymptotic results that relate local intrinsic dimensionality with multiple variants of tail entropy. In particular, we relate LID to:

- **The cumulative tail entropy**. Cumulative entropy [17,35] is an information-theoretic measure popular in reliability theory, where it is used to model uncertainty over time intervals. It corresponds to the expected value of the mean inactivity time. Compared to ordinary Shannon differential entropy, cumulative entropy has certain attractive properties, such as non-negativity and ease of estimation.
- **The tail entropy power**. The entropy power is the exponential of the entropy, and is also known as *perplexity* in the natural language processing community. It corresponds to the volume of the smallest set that contains most of the probability measure [16], and can be interpreted as a measure of statistical dispersion [33]. It is also related to Fisher information via Stam's inequality [46].
- **Generalized tail entropies** (tail cumulative q-entropy and tail q-entropy power). Generalized Tsallis entropies [8,47] are a family of entropies characterized via an exponent parameter q applied to the probabilities, in which the traditional (Shannon) entropy variants are obtained as the special case $q \to 1$. The use of such a parameter can often facilitate more accurate fitting of data characteristics and robustness to outliers.

We believe our theoretical results are interesting in their own right, as they capture fundamental properties of local neighborhood geometry, and since they hold asymptotically for essentially all smooth data distributions. The relationships between LID and tail entropy formulations also have two interesting potential applications:

- **Estimation:** Our theory allows the development of new estimators for the LID of a query point, by applying existing estimators for cumulative entropy [17] and cumulative q-entropy [8] to samples of a sufficiently-small neighborhood of the query.

- **Feature representation**: LID estimates can be used as features or as characterizations within machine learning models, such as for the detection of adversarial examples [36] or overfitting during learning [37]. However, small errors in the estimation of LID can have a disproportionally large impact on learning models. In contrast, the tail entropy power has long been known to possess attractive properties for linear discrete systems [43], and thus has potential as a more robust substitute for LID when used as a feature in logistic regression models.

In summary, our key contributions are the development of new theory that asymptotically relates tail entropy and LID, with potential applications of this theory for estimation and feature representation. To the best of our knowledge, this is the first work relating intrinsic dimensionality and the asymptotic behavior of entropy within neighborhoods of a data domain.

2 Related Work

Our work relates to intrinsic dimensionality and its estimation, as well as tail entropy and its varieties such as generalized tail entropy and cumulative tail entropy. We briefly review these in turn.

Intrinsic dimensionality can be assessed either globally (for all data points) or locally (with respect to a chosen query point). Surveys of the field provide more detail [9,11,48]. In the global case, considerable work has focused on topological models, with accompanying estimation methods [7,38,41]. Examples here include PCA and its variants [29], graph based methods [15] and fractal models [9,20]. Other techniques such as IDEA [44,45] and DANCo [13] estimate the dimension based on concentration of norms and angles, or 2-nearest neighbors [18].

For local intrinsic dimensionality, a popular estimator is the maximum likelihood estimator, studied in the Euclidean setting by Levina and Bickel [34] and later formulated under the more general assumptions of EVT by Amsaleg et al. [2,23], who showed it to be equivalent to the classic Hill estimator [21]. Other local estimators include expected simplex skewness [28], the tight locality estimator [3], the MiND framework [44] and the manifold adaptive dimension [19].

Local intrinsic dimensionality has been used in a range of applications. These include modeling deformation in complex materials [49], dimension reduction via local PCA [30], similarity search [26], clustering [10], outlier detection [27], statistical manifold learning [12], adversarial example detection [36], and adversarial nearest neighbor characterization [1,4], and deep learning understanding [5,37]. In deep learning, it has been shown that adversarial examples are associated with high LID estimates, a characteristic that can be leveraged to build accurate adversarial example detectors [36]. It has also been found that the LID of deep representations [5] or input data [42] is an indicator of the generalization performance of deep neural networks (DNNs). A manifold 'dimensionality expansion' phenomenon has been observed when DNNs overfit to noisy class labels [37].

Cumulative entropy was formulated in [17] and is a variant of cumulative residual entropy [35]. Outside of reliability theory analysis, it has been used in such data mining tasks as dependency analysis [39] and subspace cluster analysis [6], where it has proved effective due to the existence of good estimators. Such investigation has been at a global level (over the entire data domain), rather than at the local level as in our study. Generalized variants based on Tsallis q-statistics have been developed for both entropy [47] and cumulative entropy [8].

The concept of tail entropy has been used in financial applications for assessing the expected shortfall [40] in the upper tail using quantization. This is different from our context, where we analyze lower tails and develop exact results for an asymptotic regime.

3 Local Intrinsic Dimensionality

In this section, we summarize the LID model using the formulation of [23].

LID can be regarded as a continuous extension of the expansion dimension due to Karger and Ruhl [25,32]. Like earlier expansion-based models of intrinsic dimension, it draws its motivation from the relationship between volume and radius in an expanding ball, where (as originally stated in [22]) the volume of the ball is taken to be the probability measure associated with the region it encloses. The probability as a function of radius—denoted by $F(r)$—has the form of a univariate cumulative distribution function (CDF). The model formulation (as stated in [23]) generalizes this notion to real-valued functions F for which $F(0) = 0$, under appropriate assumptions of smoothness.

Definition 1 ([23]). *Let F be a real-valued function that is non-zero over some open interval containing $r \in \mathbb{R}$, $r \neq 0$. The* intrinsic dimensionality of F at r *is defined as follows whenever the limit exists:*

$$\mathrm{IntrDim}_F(r) \triangleq \lim_{\epsilon \to 0} \frac{\ln\left(F((1+\epsilon)r)/F(r)\right)}{\ln(1+\epsilon)}.$$

When F satisfies certain smoothness conditions in the vicinity of r, its intrinsic dimensionality has a convenient known form:

Theorem 1 ([23]). *Let F be a real-valued function that is non-zero over some open interval containing $r \in \mathbb{R}$, $r \neq 0$. If F is continuously differentiable at r, then*

$$\mathrm{ID}_F(r) \triangleq \frac{r \cdot F'(r)}{F(r)} = \mathrm{IntrDim}_F(r).$$

Let $\mathbf{x}$ be a location of interest within a data domain $\mathcal{S}$ for which the distance measure d has been defined. To any generated sample $\mathbf{y} \in \mathcal{D}$ we can associate the distance $r = d(\mathbf{x}, \mathbf{y})$; in this way, the global distribution that produces samples $\mathbf{y}$ can be said to induce a local distance distribution with CDF F with respect to $\mathbf{x}$. In characterizing the local intrinsic dimensionality in the vicinity of location

x, we are interested in the limit of $\mathrm{ID}_F(r)$ as the distance r tends to 0, which we denote by

$$\mathrm{ID}^*_F \triangleq \lim_{r\to 0} \mathrm{ID}_F(r).$$

Henceforth, when we refer to the local intrinsic dimensionality (LID) of a function F, or of a point **x** whose induced distance distribution has F as its CDF, we will take 'LID' to mean the quantity ID^*_F. In general, ID^*_F is not necessarily an integer. In practice, estimation of the LID at **x** would give an indication of the dimension of the submanifold containing **x** that best fits the distribution.

The function ID_F can be seen to fully characterize its associated function F. This result is analogous to a foundational result from the statistical theory of extreme values (EVT), in that it corresponds under an inversion transformation to the Karamata representation theorem [31] for the upper tails of regularly varying functions. For more information on EVT and how the LID model relates to it, we refer the reader to [14,23,24].

Theorem 2 (LID Representation Theorem [23]**).** *Let* $F : \mathbb{R} \to \mathbb{R}$ *be a real-valued function, and assume that* ID^*_F *exists. Let* x *and* w *be values for which* x/w *and* $F(x)/F(w)$ *are both positive. If* F *is non-zero and continuously differentiable everywhere in the interval* $[\min\{x,w\}, \max\{x,w\}]$, *then*

$$\frac{F(x)}{F(w)} = \left(\frac{x}{w}\right)^{\mathrm{ID}^*_F} \cdot G_F(x,w), \quad \textit{where } G_F(x,w) \triangleq \exp\left(\int_x^w \frac{\mathrm{ID}^*_F - \mathrm{ID}_F(t)}{t}\,\mathrm{d}t\right),$$

whenever the integral exists.

In [23], conditions on x and w are provided for which the factor $G_F(x,w)$ can be seen to tend to 1 as $x, w \to 0$. The convergence characteristics of F to its asymptotic form are expressed by the factor $G_F(x,w)$, which is related to the slowly-varying component of functions as studied in EVT [14]. As we will shown in the next section, we make use of the LID Representation Theorem in our analysis of the limits of tail entropy variants under a form of normalization.

4 Tail Entropy and LID

In this section, we will establish relationships between local intrinsic dimensionality and several forms of entropy *conditioned* on the lower tails of smooth functions on domains bounded from below at zero. The results presented in this section all hold *asymptotically*, as the tail boundary tends toward zero, when *normalized* with respect to the length of the tail.

4.1 Definitions of Tail Entropy Variants

We begin with formal definitions of the tail entropies considered in this paper. In each case, we assume that we are given a non-negative real-valued function F whose restriction to $[0, w]$ satisfies the following smooth growth properties:

- $F(0) = 0$, and $F(t) > 0$ for $t \in (0, w]$;
- F is strictly monotonically increasing;
- F is continuously differentiable.

The function $\phi(t) \triangleq F(t)/F(w)$ thus satisfies the conditions of a cumulative distribution function over $t \in [0, w]$ (recall that $F(t|t \leq w) = F(t)/F(w)$ over $t \in [0, w]$), with the derivative $\phi'(t) = F'(t)/F(w)$ as its corresponding probability density function.

The following tail entropy formulations apply to any function F satisfying the conditions stated above. In their definitions, the only difference between the tail variants and the original versions is that the distribution is conditioned to the lower tail $[0, w]$. Consequently, in the tail variants, integration is performed over the lower tail and not the entire distributional range $[0, +\infty)$.

Definition 2 (Tail Entropy). *The entropy of F conditioned on $[0, w]$ is*

$$\mathrm{H}(F, w) \triangleq -\int_0^w \frac{F'(t)}{F(w)} \ln \frac{F'(t)}{F(w)} \, \mathrm{d}t \, .$$

The cumulative entropy is a variant of entropy proposed in [17,35] due to its attractive theoretical properties. Tail conditioning on the cumulative entropy has the same general form as that of the tail entropy.

Definition 3 (Cumulative Tail Entropy). *The cumulative entropy of F conditioned on $[0, w]$ is*

$$\mathrm{cH}(F, w) \triangleq -\int_0^w \frac{F(t)}{F(w)} \ln \frac{F(t)}{F(w)} \, \mathrm{d}t \, .$$

There are several standard definitions of entropy power in the research literature. For our purposes, we adopt the simplest—the exponential of Shannon entropy—for our definition conditioned to the tail.

Definition 4 (Tail Entropy Power). *The entropy power of F conditioned on $[0, w]$ is defined to be*

$$\mathrm{HP}(F, w) \triangleq \exp\left(\mathrm{H}(F, w)\right).$$

In the introduction, we briefly mentioned some motivation for the entropy power $\mathrm{HP}(F, w)$. We can add to this as follows:

- It can be interpreted as a diversity. Observe that when F is a (univariate) uniform distance distribution ranging over the interval $[0, w]$, we have $\mathrm{ID}^*_F = 1$ and $\mathrm{HP}(F, w) = w$. In other words, the entropy power is equal to the 'effective diversity' of the distribution (the number of neighbor distance possibilities).
- Given two different queries, each with its own neighborhood, one query with tail entropy power equal to 2 and the other with tail entropy power equal to 4, we can say that the distance distribution of the second query is twice as diverse as that of the first query.

For each of the tail entropy variants introduced above, we also propose analogous variants based on the q-entropy formulation due to Tallis [47]. In general, q-entropy formulations can be shown to be identical to their Shannon entropy analogues in the limit as q tends to 1.

Table 1. Asymptotic relationships between normalized tail entropy variants and local intrinsic dimensionality.

Entropy variant	Normalized tail entropy	Limit as $\mathbf{w} \to \mathbf{0}^+$
Cumulative entropy	$\mathrm{ncH}(F,w) \triangleq \frac{1}{w}\mathrm{cH}(F,w)$	$\frac{\mathrm{ID}_F^*}{(\mathrm{ID}_F^*+1)^2}$
Cumulative q-Entropy	$\mathrm{ncH}_q(F,w) \triangleq \frac{1}{w}\mathrm{cH}_q(F,w)$	$\frac{\mathrm{ID}_F^*}{(\mathrm{ID}_F^*+1)(q\,\mathrm{ID}_F^*+1)}$
Entropy power	$\mathrm{nHP}(F,w) \triangleq \frac{1}{w}\mathrm{HP}(F,w)$	$\frac{1}{\mathrm{ID}_F^*}\exp\left(1-\frac{1}{\mathrm{ID}_F^*}\right)$
q-Entropy power	$\mathrm{nHP}_q(F,w) \triangleq \frac{1}{w}\mathrm{HP}_q(F,w)$	$\left(\frac{(\mathrm{ID}_F^*)^q}{q\,\mathrm{ID}_F^*-q+1}\right)^{\frac{1}{1-q}}$

Definition 5 (Tail q-Entropy). *For any $q > 0$ ($q \neq 1$), the q-entropy of F conditioned on $[0, w]$ is defined to be*

$$\mathrm{H}_q(F,w) \triangleq \frac{1}{q-1}\left(1-\int_0^w\left(\frac{F'(t)}{F(w)}\right)^q \mathrm{d}t\right) = \frac{1}{q-1}\int_0^w \frac{F'(t)}{F(w)} - \left(\frac{F'(t)}{F(w)}\right)^q \mathrm{d}t\,.$$

Definition 6 (Cumulative Tail q-Entropy). *For any $q > 0$ ($q \neq 1$), the cumulative q-entropy of F conditioned on $[0, w]$ is defined to be*

$$\mathrm{cH}_q(F,w) \triangleq \frac{1}{q-1}\int_0^w \frac{F(t)}{F(w)} - \left(\frac{F(t)}{F(w)}\right)^q \mathrm{d}t\,.$$

We define the tail q-entropy power using the q-exponential function from Tsallis statistics [47], $\exp_q(x) \triangleq [1+(1-q)x]^{\frac{1}{1-q}}$. Note that L'Hôpital's rule can be used to show that $\exp_q(x) \to e^x$ as $q \to 1$.

Definition 7 (Tail q-Entropy Power). *For any $q > 0$ ($q \neq 1$), the q-entropy power of F conditioned on $[0, w]$ is defined to be*

$$\mathrm{HP}_q(F,w) \triangleq [1+(1-q)H_q(F,w)]^{\frac{1}{1-q}}\,.$$

For the cumulative tail entropy and tail entropy power variants, we will also consider a normalization given by the ratio of the entropy with w, the length of the tail. In the remainder of this section, we will show that as w tends to zero, the limits of these normalized entropies can be expressed in terms of the local intrinsic dimensionality of F. The notation for these normalized entropy variants, and our theorems for their limits in terms of LID, are summarized in Table 1.

4.2 Technical Preliminaries

Before presenting the main theoretical results of the paper, we begin with two technical lemmas. The first lemma concerns a slight generalization of the cumulative entropy formulation, that allows it to greatly facilitate the proofs for two tail entropy variants, the cumulative entropy and the entropy power.

Lemma 1. *Let $F : \mathbb{R}^{\geq 0} \to \mathbb{R}^{\geq 0}$ be a function such that $F(0) = 0$, and assume that ID_F^* exists and is positive. For some value of $r > 0$, let us further assume that within the interval $[0, r)$, F is continuously differentiable and strictly monotonically increasing, and ψ is positive. Then for any constant $u < \mathrm{ID}_F^*$,*

$$\begin{aligned}\lim_{w\to 0^+} w^{u-1} \int_0^w \frac{\psi(w)\, F(t)}{t^u F(w)} \ln \frac{\psi(w)\, F(t)}{t^u F(w)}\, \mathrm{d}t \\ = \lim_{w\to 0^+} \frac{\psi(w)}{\mathrm{ID}_F^* + 1 - u} \left[\ln \frac{\psi(w)}{w^u} - \frac{\mathrm{ID}_F^* - u}{\mathrm{ID}_F^* + 1 - u} \right]\end{aligned}$$

whenever the right-hand limit exists or diverges to $+\infty$ or $-\infty$.

Proof: Since the limit $\mathrm{ID}_F^* = \lim_{v\to 0^+} \mathrm{ID}_F(v)$ is assumed to exist, we have that for any real value $\epsilon > 0$ satisfying $\epsilon < \min\{r, \mathrm{ID}_F^* - u\}$, there must exist a value $0 < \delta < \epsilon$ such that $v < \delta$ implies that $|\,\mathrm{ID}_F(v) - \mathrm{ID}_F^*\,| < \epsilon$. Therefore, when $0 < t \leq w < \delta$,

$$|\ln G_F(t, w)| = \left| \int_t^w \frac{\mathrm{ID}_F^* - \mathrm{ID}_F(v)}{v}\, \mathrm{d}v \right| < \epsilon \cdot \left| \int_t^w \frac{1}{v}\, \mathrm{d}v \right| = \epsilon \cdot \ln \frac{w}{t}.$$

Exponentiating, we obtain the bounds

$$\left(\frac{w}{t}\right)^{-\epsilon} < G_F(t, w) < \left(\frac{w}{t}\right)^{\epsilon}. \tag{1}$$

For any real $x > 0$, we define $\mathrm{xlnx}(x) \triangleq x \ln x$. Applying Theorem 2 to $F(t)$, and making use of the upper bound on G_F, the integral becomes

$$\begin{aligned}&\int_0^w \mathrm{xlnx}\left(\frac{\psi(w)\, F(t)}{t^u F(w)}\right) \mathrm{d}t \qquad (2)\\ &= \int_0^w \mathrm{xlnx}\left(\frac{\psi(w)}{t^u}\left(\frac{t}{w}\right)^{\mathrm{ID}_F^*} G_F(t, w)\right) \mathrm{d}t < \int_0^w \mathrm{xlnx}\left(\frac{\psi(w)}{t^u}\left(\frac{t}{w}\right)^{\mathrm{ID}_F^*}\left(\frac{w}{t}\right)^{\epsilon}\right) \mathrm{d}t \\ &< \int_0^w \mathrm{xlnx}\left(\frac{\psi(w)}{t^u}\left(\frac{t}{w}\right)^{\mathrm{ID}_F^* - \epsilon}\right) \mathrm{d}t < \frac{\psi(w)}{w^{m+u}} \int_0^w t^m \cdot \left[m \ln t + \ln \frac{\psi(w)}{w^{m+u}} \right] \mathrm{d}t,\end{aligned}$$

where $m \triangleq \mathrm{ID}_F^* - u - \epsilon > 0$.

Noting that $m > 0$ implies that $\lim_{t\to 0} t^m \ln t = 0$, integration of Eq. 2 by parts yields an expression that depends on F only through its LID value.

$$
\begin{aligned}
w^{u-1} & \int_0^w \mathrm{xlnx}\left(\frac{\psi(w)\,F(t)}{t^u F(w)}\right) \mathrm{d}t \\
& < \frac{m w^{u-1}\psi(w)}{w^{m+u}} \left[\frac{t^{m+1}}{m+1} \ln t \Big|_0^w - \int_0^w \frac{t^{m+1}}{m+1} \cdot \frac{1}{t}\,\mathrm{d}t \right] \\
& \quad + \frac{w^{u-1}\psi(w)}{w^{m+u}} \ln \frac{\psi(w)}{w^{m+u}} \cdot \frac{w^{m+1}}{m+1}
\end{aligned}
$$

$$
\begin{aligned}
& < \frac{m\psi(w)}{w^{m+1}} \left[\frac{w^{m+1}}{m+1} \ln w - \frac{w^{m+1}}{(m+1)^2} \right] + \frac{\psi(w)}{m+1} \ln \frac{\psi(w)}{w^{m+u}} \\
& < \frac{\psi(w)}{m+1} \left[m \ln w - \frac{m}{m+1} + \ln \frac{\psi(w)}{w^{m+u}} \right] < \frac{\psi(w)}{m+1} \left[\ln \frac{\psi(w)}{w^u} - \frac{m}{m+1} \right] \\
& = \frac{\psi(w)}{\mathrm{ID}_F^* + 1 - u - \epsilon} \left[\ln \frac{\psi(w)}{w^u} - \frac{\mathrm{ID}_F^* - u - \epsilon}{\mathrm{ID}_F^* + 1 - u - \epsilon} \right].
\end{aligned}
$$

Similar arguments using the lower bound from Eq. 1 leads us to

$$
w^{u-1} \int_0^w \mathrm{xlnx}\left(\frac{\psi(w)\,F(t)}{t^u F(w)}\right) \mathrm{d}t > \frac{\psi(w)}{\mathrm{ID}_F^* + 1 - u + \epsilon} \left[\ln \frac{\psi(w)}{w^u} - \frac{\mathrm{ID}_F^* - u + \epsilon}{\mathrm{ID}_F^* + 1 - u + \epsilon} \right].
$$

Since ϵ can be chosen arbitrarily close to 0, and since $0 < w < \epsilon$ by construction, taking the limit as $w \to 0^+$ yields

$$
\lim_{w\to 0^+} w^{u-1} \int_0^w \mathrm{xlnx}\left(\frac{\psi(w)\,F(t)}{t^u F(w)}\right) \mathrm{d}t = \lim_{w\to 0^+} \frac{\psi(w)}{\mathrm{ID}_F^* + 1 - u} \left[\ln \frac{\psi(w)}{w^u} - \frac{\mathrm{ID}_F^* - u}{\mathrm{ID}_F^* + 1 - u} \right]
$$

whenever the right-hand limit exists, or diverges to $+\infty$ or $-\infty$. □

The second technical lemma follows as a corollary of Lemma 1, since it uses much of the same proof strategy, albeit more simply and directly. Analogous with Lemma 1, it concerns a slight generalization of the cumulative q-entropy formulation that facilitates the proof of the results for the q-entropy and q-entropy power variants.

Corollary 1. *Let $F : \mathbb{R}^{\geq 0} \to \mathbb{R}^{\geq 0}$ be a function such that $F(0) = 0$, and assume that ID_F^* exists and is positive. For some value of $r > 0$, let us further assume that within the interval $[0, r)$, F is continuously differentiable and strictly monotonically increasing, and ψ is positive. Then for any constants $u < \mathrm{ID}_F^*$ and $z > 0$,*

$$
\lim_{w\to 0^+} w^{zu-1} \int_0^w \left(\frac{\psi(w)\,F(t)}{t^u F(w)}\right)^z \mathrm{d}t = \frac{\lim_{w\to 0^+} \psi^z(w)}{z\,\mathrm{ID}_F^* - zu + 1}
$$

whenever the right-hand limit exists, or diverges to $+\infty$ or $-\infty$.

Proof: Following the same proof strategy of Lemma 1 that led to Eq. 2, we arrive at the following upper bound on the integral:

$$\int_0^w \left(\frac{\psi(w)\,F(t)}{t^u F(w)}\right)^z \mathrm{d}t < \frac{\psi^z(w)}{w^{z(m+u)}} \int_0^w t^{zm}\,\mathrm{d}t = \frac{\psi^z(w)}{(zm+1)w^{zu-1}},$$

where $m = \mathrm{ID}_F^* - u - \epsilon$ as before.

Continuing according to the proof strategy of Lemma 1, we use the lower bound from Eq. 1, let ϵ vanish, and then apply the limit $w \to 0^+$ with a factor of w^{zu-1}. This brings us to

$$\lim_{w\to 0^+} w^{zu-1} \int_0^w \left(\frac{\psi(w)\,F(t)}{t^u F(w)}\right)^z \mathrm{d}t$$
$$= \lim_{w\to 0^+} w^{zu-1} \frac{\psi^z(w)}{(z\,\mathrm{ID}_F^* - zu + 1)w^{zu-1}} = \frac{\lim_{w\to 0^+} \psi^z(w)}{z\,\mathrm{ID}_F^* - zu + 1},$$

as required. □

4.3 Cumulative Tail Entropy and LID

Using the technical lemmas established in Sect. 4.2, we present the main results for the cumulative tail entropy variants. The first result shows that as the tail length w tends to zero, the normalized cumulative entropy $\mathrm{ncH}(F,w) \triangleq \frac{1}{w}\mathrm{cH}(F,w)$ tends to a value entirely determined by the local intrinsic dimensionality associated with F.

Theorem 3. *Let $F : \mathbb{R}^{\geq 0} \to \mathbb{R}^{\geq 0}$ be a function such that $F(0) = 0$, and assume that ID_F^* exists and is positive. For some value of $r > 0$, let us further assume that within the interval $[0, r)$, F is continuously differentiable and strictly monotonically increasing. We have*

$$\lim_{w\to 0^+} \mathrm{ncH}(F,w) = \lim_{w\to 0^+} -\frac{1}{w}\int_0^w \frac{F(t)}{F(w)} \ln \frac{F(t)}{F(w)}\,\mathrm{d}t = \frac{\mathrm{ID}_F^*}{(\mathrm{ID}_F^* + 1)^2}.$$

Proof: Follows directly from Lemma 1, for the choices $u = 0$ and $\psi(w) = 1$. □

The second result uses Corollary 1 to show that as the tail length w tends to zero, the normalized cumulative q-entropy $\mathrm{ncH}_q(F,w) \triangleq \frac{1}{w}\mathrm{cH}_q(F,w)$ tends to a value determined by q together with the local intrinsic dimensionality associated with F.

Theorem 4. *Let $F : \mathbb{R}^{\geq 0} \to \mathbb{R}^{\geq 0}$ be a function such that $F(0) = 0$, and assume that ID_F^* exists and is positive. For some value of $r > 0$, let us further assume that within the interval $[0, r)$, F is continuously differentiable and strictly monotonically increasing. Then for $q > 0$ with $q \neq 1$,*

$$\lim_{w\to 0^+} \mathrm{ncH}_q(F,w)$$
$$= \lim_{w\to 0^+} \frac{1}{w(q-1)} \int_0^w \frac{F(t)}{F(w)} - \left(\frac{F(t)}{F(w)}\right)^q \mathrm{d}t = \frac{\mathrm{ID}_F^*}{(\mathrm{ID}_F^* + 1)(q\,\mathrm{ID}_F^* + 1)}.$$

Proof: Separating the integral and applying Corollary 1 twice,

$$\lim_{w\to 0^+} \frac{1}{w(q-1)} \int_0^w \frac{F(t)}{F(w)} - \left(\frac{F(t)}{F(w)}\right)^q \mathrm{d}t$$
$$= \frac{1}{q-1}\left(\frac{1}{\mathrm{ID}_F^* + 1} - \frac{1}{q\,\mathrm{ID}_F^* + 1}\right) = \frac{\mathrm{ID}_F^*}{(\mathrm{ID}_F^* + 1)(q\,\mathrm{ID}_F^* + 1)}$$

follows for the choices $u = 0$, $\psi(w) = 1$, and (respectively) $z = 1$ and $z = q$. □

Observe that as q tends to 1, the cumulative q-entropy variant $\mathrm{ncH}_q(F, w)$ does tend to the cumulative entropy $\mathrm{ncH}(F, w)$, as one would expect.

4.4 Tail Entropy Power and LID

We find that we encounter convergence issues when attempting to use the machinery of Lemma 1 to formulate a relationship between LID and either the tail entropy $\mathrm{H}(F, w)$ or the normalized tail entropy $\mathrm{nH}(F, w)$, the limits diverging as the tail size tends to zero.

Instead, we show that the entropy power, when normalized, does have a limit expressed as a function of the LID of F.

Theorem 5. *Let $F : \mathbb{R}^{\geq 0} \to \mathbb{R}^{\geq 0}$ be a function such that $F(0) = 0$, and assume that ID_F^* exists and is greater than 1. For some value of $r > 0$, let us further assume that within the interval $[0, r)$, F is continuously differentiable and strictly monotonically increasing. Then*

$$\lim_{w\to 0^+} \mathrm{nHP}(F, w)$$
$$= \lim_{w\to 0^+} \frac{1}{w} \exp\left(-\int_0^w \frac{F'(t)}{F(w)} \ln \frac{F'(t)}{F(w)}\,\mathrm{d}t\right) = \frac{1}{\mathrm{ID}_F^*} \exp\left(1 - \frac{1}{\mathrm{ID}_F^*}\right).$$

Proof: Due to space limitations, the details are omitted in this version. The proof is analogous to that of Theorem 3, and makes use of Theorem 1 and Lemma 1 with the choices $u = 1$ and $\psi(w) = \mathrm{ID}_F^*$. The choice of u is valid for Lemma 1 since by assumption $\mathrm{ID}_F^* > 1 = u$. □

For the case of the normalized tail q-entropy power $\mathrm{nHP}_q(F, w)$, we have the following result.

Theorem 6. *Let $F : \mathbb{R}^{\geq 0} \to \mathbb{R}^{\geq 0}$ be a function such that $F(0) = 0$, and assume that ID_F^* exists and is greater than 1. For some value of $r > 0$, let us further assume that within the interval $[0, r)$, F is continuously differentiable and strictly monotonically increasing. Then for $q > 0$ $(q \neq 1)$,*

$$\lim_{w\to 0^+} \mathrm{nHP}_q(F, w)$$
$$= \lim_{w\to 0^+} \frac{1}{w} \exp_q\left(\frac{1}{q-1}\left[1 - \int_0^w \left(\frac{F'(t)}{F(w)}\right)^q \mathrm{d}t\right]\right) = \left[\frac{(\mathrm{ID}_F^*)^q}{q\,\mathrm{ID}_F^* - q + 1}\right]^{\frac{1}{1-q}}.$$

Proof: Due to space limitations, the details are omitted in this version. The proof is analogous to that of Theorem 4, and makes use of Theorem 1 and Corollary 1 with the choices $u = 1$, $\psi(w) = \mathrm{ID}_F^*$, and $z = q$. The choice of u is valid for Corollary 1 since by assumption $\mathrm{ID}_F^* > 1 = u$. □

5 Conclusion

In this preliminary theoretical investigation, we have established an asymptotic relationship between tail entropy variants and the emerging theory of local intrinsic dimensionality. Our results provide new insights into the complexity of data within local neighborhoods, and how they may be assessed. These fundamental discoveries also open the door to cross-fertilization between intrinsic dimensionality research and entropy research, particularly as regards the potential for the use of robust estimators of tail entropy as substitutes for LID in learning contexts. Our results could also allow for applications and characterizations for DNNs based on LID to be extended to the field of information theory.

As future work, we plan to follow with in-depth experimental studies on the performance characteristics of cumulative entropy and entropy power as estimators or substitutes of LID for deep learning and data mining applications. We also plan to investigate the generalization and learning behaviors of DNNs based on both LID and tail entropy.

Acknowledgments. Michael E. Houle acknowledges the financial support of JSPS Kakenhi Kiban (B) Research Grant 18H03296.

References

1. Amsaleg, L., et al.: The vulnerability of learning to adversarial perturbation increases with intrinsic dimensionality. In: IEEE Workshop on Information Forensics and Security, pp. 1–6 (2017)
2. Amsaleg, L., et al.: Extreme-value-theoretic estimation of local intrinsic dimensionality. Data Min. Knowl. Disc. **32**(6), 1768–1805 (2018)
3. Amsaleg, L., Chelly, O., Houle, M.E., Kawarabayashi, K., Radovanović, R., Treeratanajaru, W.: Intrinsic dimensionality estimation within tight localities. In: Proceedings of 2019 SIAM International Conference on Data Mining, pp. 181–189 (2019)
4. Amsaleg, L., et al.: High intrinsic dimensionality facilitates adversarial attack: theoretical evidence. IEEE Trans. Inf. Forensics Secur. **16**, 854–865 (2021)
5. Ansuini, A., Laio, A., Macke, J.H., Zoccolan, D.: Intrinsic dimension of data representations in deep neural networks. In: Advances in Neural Information Processing Systems, pp. 6111–6122 (2019)
6. Böhm, K., Keller, F., Müller, E., Nguyen, H.V., Vreeken, J.: CMI: an information-theoretic contrast measure for enhancing subspace cluster and outlier detection. In: Proceedings of the 13th SIAM International Conference on Data Mining, pp. 198–206 (2013)

7. Bruske, J., Sommer, G.: Intrinsic dimensionality estimation with optimally topology preserving maps. IEEE Trans. Pattern Anal. Mach. Intell. **20**(5), 572–575 (1998)
8. Calì, C., Longobardi, M., Ahmadi, J.: Some properties of cumulative Tsallis entropy. Phys. A **486**, 1012–1021 (2017)
9. Camastra, F., Staiano, A.: Intrinsic dimension estimation: advances and open problems. Inf. Sci. **328**, 26–41 (2016)
10. Campadelli, P., Casiraghi, E., Ceruti, C., Lombardi, G., Rozza, A.: Local intrinsic dimensionality based features for clustering. In: International Conference on Image Analysis and Processing, pp. 41–50 (2013)
11. Campadelli, P., Casiraghi, E., Ceruti, C., Rozza, A.: Intrinsic dimension estimation: relevant techniques and a benchmark framework. Math. Prob. Eng. **2015**, 759567 (2015). https://doi.org/10.1155/2015/759567
12. Carter, K.M., Raich, R., Finn, W.G., Hero, A.O., III.: FINE: fisher information non-parametric embedding. IEEE Trans. Pattern Anal. Mach. Intell. **31**(11), 2093–2098 (2009)
13. Ceruti, C., Bassis, S., Rozza, A., Lombardi, G., Casiraghi, E., Campadelli, P.: DANCo: an intrinsic dimensionality estimator exploiting angle and norm concentration. Pattern Recogn. **47**, 2569–2581 (2014)
14. Coles, S., Bawa, J., Trenner, L., Dorazio, P.: An Introduction to Statistical Modeling of Extreme Values. Springer Series in Statistics, vol. 208, p. 209. Springer, London (2001). https://doi.org/10.1007/978-1-4471-3675-0
15. Costa, J.A., Hero, A.O., III.: Entropic graphs for manifold learning. In: The 37th Asilomar Conference on Signals, Systems & Computers, vol. 1, pp. 316–320 (2003)
16. Cover, T.M., Thomas, J.A.: Elements of Information Theory. Wiley Series in Telecommunications and Signal Processing, Wiley, USA (2006)
17. Di Crescenzo, A., Longobardi, M.: On cumulative entropies. J. Stat. Plan. Inference **139**(12), 4072–4087 (2009)
18. Facco, E., d'Errico, M., Rodriguez, A., Laio, A.: Estimating the intrinsic dimension of datasets by a minimal neighborhood information. Sci. Rep. **7**, 12140 (2017)
19. Farahmand, A.M., Szepesvári, C., Audibert, J.Y.: Manifold-adaptive dimension estimation. In: Proceedings of the 24th International Conference on Machine Learning, pp. 265–272 (2007)
20. Hein, M., Audibert, J.Y.: Intrinsic dimensionality estimation of submanifolds in R^d. In: Proceedings of the 22nd International Conference on Machine Learning, pp. 289–296 (2005)
21. Hill, B.M.: A simple general approach to inference about the tail of a distribution. Ann. Stat. **3**(5), 1163–1174 (1975)
22. Houle, M.E.: Dimensionality, discriminability, density and distance distributions. In: IEEE 13th International Conference on Data Mining Workshops, pp. 468–473 (2013)
23. Houle, M.E.: Local intrinsic dimensionality I: an extreme-value-theoretic foundation for similarity applications. In: International Conference on Similarity Search and Applications, pp. 64–79 (2017)
24. Houle, M.E.: Local intrinsic dimensionality II: multivariate analysis and distributional support. In: International Conference on Similarity Search and Applications, pp. 80–95, (2017)
25. Houle, M.E., Kashima, H., Nett, M.: Generalized expansion dimension. In: IEEE 12th International Conference on Data Mining Workshops, pp. 587–594 (2012)

26. Houle, M.E., Ma, X., Nett, M., Oria, V.: Dimensional testing for multi-step similarity search. In: IEEE 12th International Conference on Data Mining, pp. 299–308 (2012)
27. Houle, M.E., Schubert, E., Zimek, A.: On the correlation between local intrinsic dimensionality and outlierness. In: International Conference on Similarity Search and Applications, pp. 177–191 (2018)
28. Johnsson, K., Soneson, C., Fontes, M.: Low bias local intrinsic dimension estimation from expected simplex skewness. IEEE TPAMI **37**(1), 196–202 (2015)
29. Jolliffe, I.T.: Principal Component Analysis. Springer Series in Statistics, Springer, New York (2002). https://doi.org/10.1007/b98835
30. Kambhatla, N., Leen, T.K.: Dimension reduction by local principal component analysis. Neural Comput. **9**(7), 1493–1516 (1997)
31. Karamata, J.: Sur un mode de croissance régulière. Théorèmes fondamentaux. Bull. Soc. Math. Fr. **61**, 55–62 (1933)
32. Karger, D.R., Ruhl, M.: Finding nearest neighbors in growth-restricted metrics. In: Proceedings of the 34th ACM Symposium on Theory of Computing, pp. 741–750 (2002)
33. Kostal, L., Lansky, P., Pokora, O.: Measures of statistical dispersion based on Shannon and Fisher information concepts. Inf. Sci. **235**, 214–223 (2013). https://doi.org/10.1016/j.ins.2013.02.023
34. Levina, E., Bickel, P.J.: Maximum likelihood estimation of intrinsic dimension. In: Advances in Neural Information Processing Systems, pp. 777–784 (2004)
35. Rao, M., Chen, Y., Vemuri, B.C., Wang, F.: Cumulative residual entropy: a new measure of information. IEEE Trans. Inf. Theor. **50**(6), 1220–1228 (2004)
36. Ma, X., et al.: Characterizing adversarial subspaces using local intrinsic dimensionality. In: International Conference on Learning Representations, pp. 1–15 (2018)
37. Ma, X., et al.: Dimensionality-driven learning with noisy labels. In: International Conference on Machine Learning, pp. 3361–3370 (2018)
38. Navarro, G., Paredes, R., Reyes, N., Bustos, C.: An empirical evaluation of intrinsic dimension estimators. Inf. Syst. **64**, 206–218 (2017)
39. Nguyen, H.V., Mandros, P., Vreeken, J.: Universal dependency analysis. In: Proceedings of the 2016 SIAM International Conference on Data Mining, pp. 792–800 (2016)
40. Pele, D.T., Lazar, E., Mazurencu-Marinescu-Pele, M.: Modeling expected shortfall using tail entropy. Entropy **21**(12), 1204 (2019)
41. Pettis, K.W., Bailey, T.A., Jain, A.K., Dubes, R.C.: An intrinsic dimensionality estimator from near-neighbor information. IEEE Trans. Pattern Anal. Mach. Intell. **1**, 25–37 (1979)
42. Pope, P., Zhu, C., Abdelkader, A., Goldblum, M., Goldstein, T.: The intrinsic dimension of images and its impact on learning. In: International Conference on Learning Representations (2021)
43. Ratz, H.C.: Entropy power factors for linear discrete systems. Can. Electr. Eng. J. **8**(2), 73–78 (1983)
44. Rozza, A., Lombardi, G., Ceruti, C., Casiraghi, E., Campadelli, P.: Novel high intrinsic dimensionality estimators. Mach. Learn. **89**(1–2), 37–65 (2012)
45. Rozza, A., Lombardi, G., Rosa, M., Casiraghi, E., Campadelli, P.: IDEA: Intrinsic dimension estimation algorithm. In: International Conference on Image Analysis and Processing, pp. 433–442 (2011)
46. Stam, A.J.: Some inequalities satisfied by the quantities of information of Fisher and Shannon. Inf. Control **2**, 101–112 (1959)

47. Tsallis, C.: Possible generalization of Boltzmann-Gibbs statistics. J. Stat. Phys. **52**, 479–487 (1988)
48. Verveer, P.J., Duin, R.P.W.: An evaluation of intrinsic dimensionality estimators. IEEE Trans. Pattern Anal. Mach. Intell. **17**(1), 81–86 (1995)
49. Zhou, S., Tordesillas, A., Pouragha, M., Bailey, J., Bondell, H.: On local intrinsic dimensionality of deformation in complex materials. Nat. Sci. Rep. **11**, 10216 (2021)

The Effect of Random Projection on Local Intrinsic Dimensionality

Michael E. Houle(✉) and Ken-ichi Kawarabayashi

National Institute of Informatics, 2-1-2 Hitotsubashi,
Chiyoda-ku, Tokyo 101-8430, Japan
{meh,k_keniti}@nii.ac.jp

Abstract. Much attention has been given in the research literature to the study of distance-preserving random projections of discrete data sets, the limitations of which are established by the classical Johnson-Lindenstrauss existence lemma. In this theoretical paper, we analyze the effect of random projection on a natural measure of the local intrinsic dimensionality (LID) of smooth distance distributions in the Euclidean setting. The main contribution of the paper consists of upper and lower bounds on the LID in the vicinity of a reference point after random projection. The bounds depend only on the LID in the original data domain and the target dimension of the projection; as the difference between the target and intrinsic dimensionalities grows, these bounds converge to the LID of the original domain. The paper concludes with a brief discussion of the implications for applications in databases, machine learning and data mining.

1 Introduction

In an attempt to alleviate the effects of high dimensionality, and thereby improve the discriminability of data, simpler representations of the data are often sought by means of a number of supervised or unsupervised learning techniques. One of the earliest and most well-established simplification strategies is dimensional reduction, which seeks a projection to a lower-dimensional subspace that minimizes the distortion of the data. Dimensional reduction has applications throughout machine learning and data mining: these include feature extraction, such as in PCA and its variants [6,39]; multidimensional scaling [38,41]; manifold learning [38,40,42]; and regression-based similarity learning [43].

Among the various approaches to dimensional reduction, much attention has been given to the study of projections that approximately preserve all pairwise distances within discrete point sets. The limitations of such projections have been established by the classical Johnson-Lindenstrauss (JL) existence lemma [30], which can be stated as follows: given some distortion threshold $0 < \varepsilon < 1$, a set of n points in $\mathbb{R}^m$, and a target dimension $t > (8 \ln n)/\varepsilon^2$, there exists a linear projection $f : \mathbb{R}^m \to \mathbb{R}^t$ such that

$$(1-\varepsilon) \cdot \|u-v\|^2 \leq \|f(u)-f(v)\|^2 \leq (1+\varepsilon) \cdot \|u-v\|^2$$

N. Reyes et al. (Eds.): SISAP 2021, LNCS 13058, pp. 201–214, 2021.
https://doi.org/10.1007/978-3-030-89657-7_16

for all pairs of points u and v in the set. This bound on the target dimension has been shown to be asymptotically worst-case optimal for linear projection [33].

Subsequent research has focused on the determination of data transforms that satisfy the JL bounds. Early approaches (such as in [11,16,29]) were based on projection to spherically random subspaces; however, the associated transform matrices were dense and expensive to compute. Achlioptas [1] showed that the entries of a projection matrix could be randomly selected from among $\{-1, 0, 1\}$ so as to satisfy the bounds with high probability (after the introduction of a scaling factor). More recent work has been devoted to improving the speed and sparsity of JL transforms [2,10,31]. Variants of the JL lemma have also been applied to subspace- and manifold-structured continuous point sets [4,5].

In general, dimensional reduction requires that an appropriate dimension for the reduced space (or approximating manifold) must be either supplied or learned, ideally so as to minimize the error or loss of information incurred. The dimension of the surface that best approximates the data can be regarded as an indication of the intrinsic dimensionality (ID) of the data set, or of the minimum number of latent variables needed to represent the data. ID thus serves as an important natural measure of the complexity of data.

Over the past decades, many characterizations of the ID of sets have been proposed: classical measures (primarily of theoretical interest), including the Hausdorff dimension, Minkowski-Bouligand or 'box counting' dimension, and packing dimension (for a general reference, see [14,36]); the correlation dimension [18]; 'fractal' measures of the space-filling capacity or self-similarity of the data [7,15,19]; topological estimation of the basis dimension of the tangent space of a data manifold from local samples [6,39]. Projection-based learning methods such as PCA can produce as a byproduct an estimate of ID.

The aforementioned ID measures can be described as 'global', in that they consider the dimensionality of the set in its entirety. However, when the data set resides on a collection of manifolds, or is distributed according to a mixture of underlying models, global measures may not be indicative of the intrinsic dimensionality in all regions of the set. In order to assess the intrinsic dimensionality in the vicinity of a specified reference point, 'local' ID measures have been proposed that are defined solely in terms of the distances to a set of near neighbors of the reference point. Expansion models, in particular, assess ID in terms of the rate at which the number of encountered objects grows as the considered range of distances expands from the reference location. Such models include the expansion dimension (ED) [32], the generalized expansion dimension (GED) [23], Levina and Bickel's estimator [34], the minimum neighbor distance (MiND) [39], and the local intrinsic dimenension (LID) [3,20]. The correlation dimension can also be regarded as an expansion model, albeit one that takes into account the growth rates from all points [22,37]. Local expansion models of ID have also been used in the analysis of a projection-based heuristic for outlier detection [12], and of the complexity of search queries in indexing [8,24–27,32].

In this paper, we will be concerned with the LID model of intrinsic dimensionality, which can be regarded as an extension of the (generalized) expansion

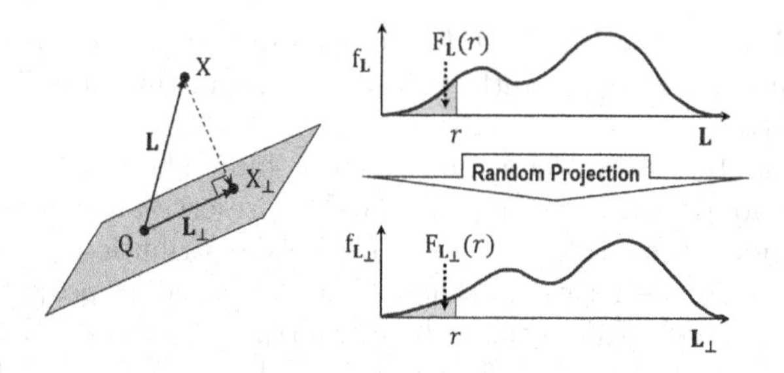

Fig. 1. Random projection of the distribution associated with the smooth random distance variable $\mathbf{L}$, defined as the Euclidean distance from reference point Q to a sample point X drawn from some domain in $\mathbb{R}^m$. The projection induces a new random distance variable $\mathbf{L}_\perp$, defined as the distance from Q to the projection of X in a randomly-oriented t-dimensional subspace.

dimension to the statistical setting of smooth distributions over the non-negative reals [3,20–22]. Instead of regarding intrinsic dimensionality as a characteristic of a collection of data points (as evidenced by their distances from a supplied reference location), the LID is a direct characterization of the complexity of the underlying distribution itself. With this latter perspective, an original data set drawn from a metric space defines a sample of distances from this underlying distribution, from which one can seek the intrinsic dimensionality of the distribution of distances to some fixed reference location. Note that the model does not require that the sample data be constrained to lie on a manifold.

The LID formulation can be shown to be equivalent to a formulation of the indiscriminability of the underlying smooth distance distribution as evidenced by its cumulative distribution function F. The indiscriminability is modeled as a function $\mathrm{ID}_F(r)$ of the distance $r \in [0, \infty)$, which tends to the local intrinsic dimension value $\mathrm{ID}^*_F \triangleq \lim_{r\to 0^+} \mathrm{ID}_F(r)$ as the radius r vanishes. ID^*_F has been shown to be equivalent to the notion of the 'degree' or 'index' in the statistical theory of extreme values (EVT); indeed, the EVT index has been interpreted as a form of dimension within statistical contexts [9]. Practical methods that have been developed within EVT for the estimation of the index, including the well-known Hill estimator and its variants obtained through maximum likelihood estimation [3,28,34], can all be applied to LID (for a survey, see [17]).

In this theoretical paper, we will be concerned with the effect on LID when the distance distribution is subjected to a random projection (as illustrated in Fig. 1). As the main contribution of our paper, we prove that under reasonable assumptions, for the LID formulation for Euclidean distance distributions derived from a reference location within a global data distribution, a randomly-oriented linear projection produces a distance distribution (relative to the projective subspace) whose LID value at the reference location, $\mathrm{ID}^*_{F_\perp}$, satisfies

$$\frac{t \cdot \mathrm{ID}^*_F}{t + \mathrm{ID}^*_F} \leq \mathrm{ID}^*_{F_\perp} \leq \mathrm{ID}^*_F \ .$$

The result indicates that LID is stable under random projection whenever the projection dimension t significantly exceeds ID_F^*, and that stability is lost as t approaches ID_F^*.

Whereas the JL lemma determines a lower limit above which a projection can always be found so as to (approximately) preserve all pairwise distances, our result considers the effect of projection on the distribution of distances from an individual reference location. Bounds on the ID after projection are also known for the Hausdorff dimension: here, Mattila [35] proved that for almost all projections of an analytic set E from $\mathbb{R}^m$ to the set $E_\perp$ in $\mathbb{R}^t$, the Hausdorff dimension $\mathrm{HD}(E_\perp)$ of the projection equals $\min\{\mathrm{HD}(E), t\}$. However, the Hausdorff dimension is a global measure of ID defined on analytic sets—it does not give any insight into the problem considered in this paper: the effect of projection on the local ID of distance distributions, and the discriminability of distance measures.

The remainder of the paper is organized as follows. In the next section, we give an overview of LID and its properties. In Sect. 3, we give a proof of our main result. For the initial part, we borrow the projection framework and Chernoff bound employed by [11] in their proof of the JL lemma. In Sect. 4, we conclude with a discussion of the implications of the LID projection bounds.

2 Local Intrinsic Dimensionality

In this section, we present an overview of the measure of local ID for distance distributions as formulated in [21].

2.1 Intrinsic Dimensionality and Indiscriminability

The LID model as first proposed in [20] takes a distributional view of data—instead of inferring dimensional characteristics from a sample of points, dimensionality is modeled in terms of a distribution of non-negative scalar values, as one would expect to see from the distances calculated from a reference location to points generated according to some hidden process.

As a motivating example from m-dimensional Euclidean space, consider the situation which the volumes V_1 and V_2 are known for two balls of differing radii r_1 and r_2, respectively, centered at a common reference point. The dimension m can be deduced from the ratios of the volumes and the distances to the reference point, as follows:

$$\frac{V_2}{V_1} = \left(\frac{r_2}{r_1}\right)^m \implies m = \frac{\ln(V_2/V_1)}{\ln(r_2/r_1)} .$$

For finite data sets, GED formulations are obtained by estimating the volume of balls by the numbers of points they enclose [23]. In contrast, for continuous real-valued random distance variables, the notion of volume is naturally analogous to that of probability measure. ID can then be modeled as a function of distance $\mathbf{X} = r$, by letting the radii of the two balls be $r_1 = r$ and $r_2 = (1 + \epsilon)r$, and letting $\epsilon \to 0$. The following definition generalizes this notion even further, to any real-valued function (not necessarily a cumulative distribution function) that non-zero in the vicinity of $r \neq 0$.

Definition 1. *Let F be a real-valued function that is non-zero over some open interval containing $r \in \mathbb{R}$, $r \neq 0$. The* intrinsic dimensionality of F at r *is defined as*

$$\mathrm{IntrDim}_F(r) \triangleq \lim_{\epsilon \to 0} \frac{\ln\left(F((1+\epsilon)r)/F(r)\right)}{\ln(1+\epsilon)},$$

whenever the limit exists.

The intrinsic dimensionality of the cumulative distribution function F of a distance distribution has also been shown in [20,21] to be equivalent to a measure of its indiscriminability. The discriminability of a random distance variable $\mathbf{X}$ is assessed in terms of the relative rate at which probability measure increases as the distance increases.

Definition 2. *Let F be a real-valued function that is non-zero over some open interval containing $r \in \mathbb{R}$, $r \neq 0$. The* indiscriminability of F at r *is defined as*

$$\mathrm{InDiscr}_F(r) \triangleq \lim_{\epsilon \to 0} \frac{F((1+\epsilon)r) - F(r)}{\epsilon \cdot F(r)},$$

whenever the limit exists.

The following fundamental theorem adapted from [21] shows that for distance distributions with continuously differentiable cumulative distribution functions, the notions of indiscriminability and intrinsic dimensionality are in fact one and the same. The proof follows by applying l'Hôpital's rule to the limits in Definitions 1 and 2.

Theorem 1 ([21])**.** *Let F be a real-valued function that is non-zero over some open interval containing $r \in \mathbb{R}$, $r \neq 0$. If F is continuously differentiable at r, then*

$$\mathrm{ID}_F(r) \triangleq \frac{r \cdot F'(r)}{F(r)} = \mathrm{IntrDim}_F(r) = \mathrm{InDiscr}_F(r).$$

When considering the local intrinsic dimensionality of a distance distribution, the question arises as to how the choice of r should be made. Asymptotically, as the number of data samples rise, for any fixed positive integer k the k-nearest neighbor radius can be seen to tend to zero. For this reason, we are especially interested in the case where $r \to 0$. Accordingly, we define the local intrinsic dimensionality (LID) to be the limit of the indiscriminability as $r \to 0$, whenever the limit exists:

$$\mathrm{ID}^*_F \triangleq \lim_{r \to 0^+} \mathrm{ID}_F(r).$$

2.2 Two Properties of Local ID

We now state (without proof) two technical results from [20] needed for the proof of the main theorem of this paper.

In the context of distance distributions with smooth cumulative distribution functions, the indiscriminability of a cumulative distribution function after transformation can be decomposed into two factors: the indiscriminability of the cumulative distribution function before transformation, and the indiscriminability of the transform itself.

Theorem 2 ([20]). *Let g be a real-valued function that is non-zero and continuously differentiable over some open interval containing $r \in \mathbb{R}$, except perhaps at r itself. Let f be a real-valued function that is non-zero and continuously differentiable over some open interval containing $g(r) \in \mathbb{R}$, except perhaps at $g(r)$ itself. Then*

$$\mathrm{ID}_{f \circ g}(r) = \mathrm{ID}_g(r) \cdot \mathrm{ID}_f(g(r))$$

whenever $\mathrm{ID}_g(r)$ and $\mathrm{ID}_f(g(r))$ are defined. If $r = f(r) = g(r) = 0$, then

$$\mathrm{ID}^*_{f \circ g} = \mathrm{ID}^*_f \cdot \mathrm{ID}^*_g$$

whenever ID^*_f and ID^*_g are defined.

The second technical result needed establishes upper and lower bounds on the expansion of probability measure over a fixed range of distances, in terms of upper and lower bounds on LID values over the range.

Theorem 3 ([20]). *Let F be a real-valued function that is non-zero and continuously differentiable over some open interval containing $[a, b] \subset \mathbb{R}$, where $0 < a \leq b$. Let $\overline{\mathrm{ID}}_F(a, b)$ and $\underline{\mathrm{ID}}_F(a, b)$ be the supremum and infimum of $\mathrm{ID}_F(r)$ taken over the range $r \in [a, b]$. Then*

$$\left(\frac{b}{a}\right)^{\underline{\mathrm{ID}}_F(a,b)} \leq \frac{F(b)}{F(a)} \leq \left(\frac{b}{a}\right)^{\overline{\mathrm{ID}}_F(a,b)} .$$

3 Intrinsic Dimensionality After Projection

In this section, as the main contribution of this paper, we examine the effect of random projection on the distribution of distances to a reference point induced by a data distribution in Euclidean space. In particular, we prove the following upper and lower bounds on the LID of the reference point after projection of the data distribution to a t-dimensional subspace (which we will refer to as $\mathrm{ID}^*_{F_\perp}$), in terms of both t and the original LID value (which we will refer to as ID^*_F).

Theorem 4. *Let $\mathbf{L}$ be a random variable representing the Euclidean distance from some fixed reference point Q to a randomly-generated point $X \in \mathbb{R}^m$. Also, let $\mathbf{L}_\perp$ be the random variable representing the Euclidean distance between the images of these points under a uniform random projection $\psi : \mathbb{R}^m \to \mathbb{R}^t$ to an* *arbitrarily-oriented subspace of target dimension* $t < m$. *Let* F *and* $F_\perp$ *be the respective cumulative distribution functions of* $\mathbf{L}$ *and* $\mathbf{L}_\perp$. *If there exists some* $\epsilon > 0$ *such that both* F *and* $F_\perp$ *are continuously differentiable over the interval* $(0, \epsilon)$, *and if the limits* ID_F^* *and* $\mathrm{ID}_{F_\perp}^*$ *both exist, then*

$$\frac{t \cdot \mathrm{ID}_F^*}{t + \mathrm{ID}_F^*} \leq \mathrm{ID}_{F_\perp}^* \leq \mathrm{ID}_F^* .$$

It should be emphasized that the theorem is a statement concerning the distribution of $\mathbf{L}_\perp$, and not the random projection of a particular fixed data set. The random variable $\mathbf{L}_\perp$ follows the distribution of distances to Q obtained when generating a data point, and then subjecting the point to a random projection before measuring its distance to Q (see Fig. 1). We do not reason in terms of collections of data samples, but rather on the effect of projection on the distance to Q of a single data sample.

3.1 Random Projection

For the initial part of our proof, we borrow the projection framework and Chernoff error bound formula employed by Dasgupta and Gupta [11] in their proof of the Johnson-Lindenstrauss Theorem (an excellent treatment of which can also be found in [13]). However, instead of using their framework to analyze the probability of obtaining a low-distortion embedding of a fixed data set, we will use it to bound the growth rates within neighborhoods of the reference point before and after projection.

Without loss of generality, we may assume that our reference point Q coincides with the origin of our original Euclidean space $\mathbb{R}^m$. Under this assumption, all the distances of interest coincide with the length of randomly-generated vectors from either the original data distribution, or the data distribution obtained after random projection.

The proof framework of [11] considers the effect of random projection on the length of a fixed vector, by first considering the effect on the associated normalized (unit length) vector. The authors note that the distribution of lengths of projection of this fixed unit vector onto a randomly-selected space is the same as the distribution of the lengths of the projection of a randomly-selected unit vector onto a fixed space (see Fig. 2). Within this setting, the expected length of projection of a random m-dimensional unit vector to a t-dimensional subspace, as well as Chernoff-style bounds on the probability of the length varying from this expected length, are established by the following two lemmas.

Lemma 1 ([11]). *Let* $Y = (Y_1, \ldots, Y_m)$ *be a vector selected uniformly at random from the unit sphere in* $\mathbb{R}^m$. *Let* $Y_\perp \in \mathbb{R}^t$ *be the projection of* Y *onto any* t *of its coordinates, where* $0 < t < m$. *Then*

$$\mathbf{E}[\|Y_\perp\|^2] = \frac{t}{m}.$$

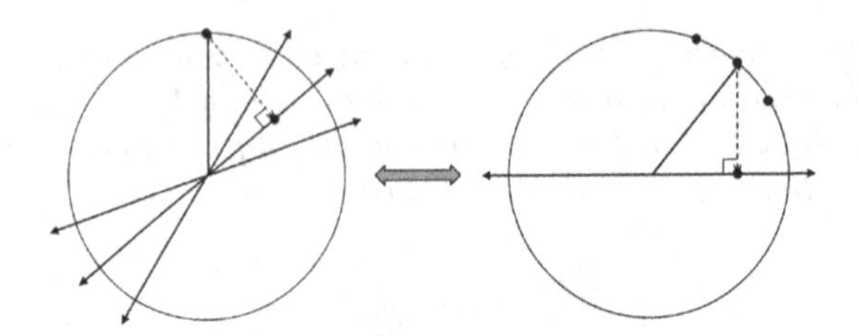

Fig. 2. The distribution of lengths of projection of a fixed unit vector onto a randomly-selected space containing the origin (as shown on the left) is the same as the distribution of the lengths of the projection of a randomly-selected unit vector onto a fixed space (as shown on the right).

Proof: Since the Y_i are identically distributed, we may without loss of generality assume that the projection spans the first t coordinates of $\mathbb{R}^m$. Thus,

$$\mathbf{E}[\|Y_\perp\|^2] = \mathbf{E}\left[\sum_{i=1}^{t} Y_i^2\right] = \sum_{i=1}^{t} \mathbf{E}[Y_i^2] = t \cdot \mathbf{E}[Y_j^2]$$

for any choice of $j \in \{1, \ldots, m\}$. Since Y is a unit vector, we also have

$$1 = \mathbf{E}[\|Y\|^2] = \mathbf{E}\left[\sum_{i=1}^{m} Y_i^2\right] = m \cdot \mathbf{E}[Y_j^2].$$

Combining these two expressions, the result follows. □

Lemma 2 ([11]). *Let $Y = (Y_1, \ldots, Y_m)$ be a vector selected uniformly at random from the unit sphere in $\mathbb{R}^m$. Let $Y_\perp \in \mathbb{R}^t$ be the projection of Y onto any t of its coordinates, where $0 < t < m$. If $\beta < 1$, then*

$$\Pr\left[\|Y_\perp\|^2 \le \beta\frac{t}{m}\right] \le \beta^{\frac{t}{2}}\left(1 + \frac{t(1-\beta)}{m-t}\right)^{\frac{m-t}{2}},$$

and if $\beta > 1$, then

$$\Pr\left[\|Y_\perp\|^2 \ge \beta\frac{t}{m}\right] \le \beta^{\frac{t}{2}}\left(1 + \frac{t(1-\beta)}{m-t}\right)^{\frac{m-t}{2}}.$$

For details of this latter proof, we refer the reader to [11]. Here, we note only that the proof of the Chernoff-style bound of Lemma 2 relies heavily on the expected squared vector length stated in Lemma 1.

3.2 Proof of Theorem 4

Although the proof of our main result makes use of Lemma 2 from [11], the proof strategy thereafter is quite different, and considerably more complex. Whereas their proof of the Johnson-Lindenstrauss lemma made direct use of the Chernoff-style bound of Lemma 2, our result instead uses Theorem 3 to establish a Chernoff-style bound in terms of the local IDs of the cumulative distribution functions associated with $\mathbf{L}$ and $\mathbf{L}_\perp$. It then relies on a careful choice of the parameter β so as to guide the convergence of a double-limit process towards the desired bounds relating $\mathrm{ID}^*_{F_\perp}$ to ID^*_F.

For any vector X, let $Y = X/\|X\|$ be the unit vector obtained by the normalization of X, and let $Y_\perp$ and $X_\perp$ be the projections of X and Y under ψ.

To prove the lower bound, we first consider the cumulative probability of the squared distance distribution after projection by ψ. Since ψ is not known *a priori*, for any squared distance threshold $r > 0$, this probability can be expressed as

$$F_\perp(\sqrt{r_\perp}) \;=\; \mathbf{Pr}[\mathbf{L}_\perp^2 \le r_\perp] \;=\; \mathbf{Pr}[\|X_\perp\|^2 \le r_\perp] \;=\; \mathbf{Pr}\left[\|Y_\perp\|^2 \le \frac{r_\perp}{\|X\|^2}\right].$$

For any choice of $0 < \beta < 1$, this probability can be bounded by

$$\begin{aligned}\mathbf{Pr}[\mathbf{L}_\perp^2 \le r_\perp] &= \mathbf{Pr}\left[\|Y_\perp\|^2 \le \frac{r_\perp}{\|X\|^2} \bigwedge \|Y_\perp\|^2 \ge \beta\frac{t}{m}\right] \\ &\qquad + \mathbf{Pr}\left[\|Y_\perp\|^2 \le \frac{r_\perp}{\|X\|^2} \bigwedge \|Y_\perp\|^2 < \beta\frac{t}{m}\right] \\ &\le \mathbf{Pr}\left[\beta\frac{t}{m} \le \frac{r_\perp}{\|X\|^2}\right] + \mathbf{Pr}\left[\|Y_\perp\|^2 \le \beta\frac{t}{m}\right] \\ &\le \mathbf{Pr}\left[\mathbf{L}^2 \le \frac{m}{\beta t} r_\perp\right] + \mathbf{Pr}\left[\|Y_\perp\|^2 \le \beta\frac{t}{m}\right] \\ &\le \mathbf{Pr}\left[\mathbf{L}^2 \le \frac{1}{\beta} r\right] + \mathbf{Pr}\left[\|Y_\perp\|^2 \le \beta\frac{t}{m}\right],\end{aligned}$$

where $r_\perp = \frac{t}{m}r$, the expected squared length of the projection under ψ of a vector of squared length r.

Recall that the length of a fixed unit vector after uniform random projection to a t-dimensional space follows the same distribution as the length of a uniform random unit vector after a fixed projection to $\mathbb{R}^t$. Lemma 2 can therefore be applied to yield

$$\begin{aligned}\mathbf{Pr}[\mathbf{L}_\perp^2 \le r_\perp] &\le \mathbf{Pr}\left[\mathbf{L}^2 \le \frac{1}{\beta} r\right] + \beta^{\frac{t}{2}} \left(1 + \frac{t(1-\beta)}{m-t}\right)^{\frac{m-t}{2}} \\ &\le \mathbf{Pr}[\mathbf{L}^2 \le r/\beta] + \beta^{\frac{t}{2}} \left(e^{\frac{(1-\beta)t}{m-t}}\right)^{\frac{m-t}{2}} \\ &\le \mathbf{Pr}[\mathbf{L}^2 \le r/\beta] + \beta^{\frac{t}{2}} e^{\frac{t}{2}(1-\beta)} \;\le\; \mathbf{Pr}[\mathbf{L}^2 \le r/\beta] + \beta^{\frac{t}{2}} e^{\frac{t}{2}}, \qquad (1)\end{aligned}$$

since $0 < \beta < 1$.

Since F and $F_\perp$ are assumed to be continuously differentiable over the range of distances $(0, \epsilon)$, the cumulative distribution functions of $\mathbf{L}^2$ and $\mathbf{L}_\perp^2$ must also be continuously differentiable over $(0, \epsilon^2)$. Let $\mathrm{ID}\Box$ and $\mathrm{ID}\Box_\perp$ denote the LID values of the cumulative distribution functions of $\mathbf{L}^2$ and $\mathbf{L}_\perp^2$, respectively. We can therefore apply Theorem 3 to obtain

$$\mathbf{Pr}[\mathbf{L}^2 \leq r/\beta] \leq \mathbf{Pr}[\mathbf{L}^2 \leq \delta] \cdot \left(\frac{1}{\beta} \cdot \frac{r}{\delta}\right)^{\underline{\mathrm{ID}\Box}_\delta}, \quad \text{and}$$

$$\mathbf{Pr}[\mathbf{L}_\perp^2 \leq r_\perp] \geq \mathbf{Pr}[\mathbf{L}_\perp^2 \leq \delta_\perp] \cdot \left(\frac{r_\perp}{\delta_\perp}\right)^{\overline{\mathrm{ID}\Box}_{\delta_\perp}},$$

where $\underline{\mathrm{ID}\Box}_\delta$ denotes the infimum $\underline{\mathrm{ID}\Box}(0,\delta)$ of $\mathrm{ID}\Box$ over the range $[0,\delta]$, and where $\overline{\mathrm{ID}\Box}_{\delta_\perp}$ denotes the supremum $\overline{\mathrm{ID}\Box}_\perp(0,\delta_\perp)$ of $\mathrm{ID}\Box_\perp$ over the range $[0,\delta_\perp]$. Here, we assume that the variables have been chosen such $\delta_\perp = \frac{t}{m}\delta$, and furthermore that r/β, $r_\perp$, δ and $\delta_\perp$ are all strictly less than ϵ^2 (this latter condition will be enforced later, as more constraints on these variables are introduced).

Since by construction $\frac{r_\perp}{\delta_\perp} = \frac{r}{\delta}$, substituting the above inequalities into Inequality 1 yields

$$\mathbf{Pr}[\mathbf{L}_\perp^2 \leq \delta_\perp] \cdot \left(\frac{r}{\delta}\right)^{\overline{\mathrm{ID}\Box}_{\delta_\perp}} \leq \mathbf{Pr}[\mathbf{L}^2 \leq \delta] \cdot \left(\frac{1}{\beta} \cdot \frac{r}{\delta}\right)^{\underline{\mathrm{ID}\Box}_\delta} + \beta^{\frac{t}{2}} e^{\frac{t}{2}}. \tag{2}$$

Consider now a new interpolation parameter c, whose role will be explained below. In terms of r, δ, and c, we fix the parameter β as follows:

$$\beta = \left(\frac{r}{\delta}\right)^c.$$

Note that under these conditions, for any $0 < c < 1$ and $\delta > 0$, choosing r such that $0 < r < \delta$ ensures that $0 < \beta < 1$.

Substitution into Inequality 2 gives

$$\mathbf{Pr}[\mathbf{L}_\perp^2 \leq \delta_\perp] \cdot \left(\frac{r}{\delta}\right)^{\overline{\mathrm{ID}\Box}_{\delta_\perp}} \leq \mathbf{Pr}[\mathbf{L}^2 \leq \delta] \cdot \left(\frac{r}{\delta}\right)^{(1-c)\cdot\underline{\mathrm{ID}\Box}_\delta} + e^{\frac{t}{2}} \cdot \left(\frac{r}{\delta}\right)^{\frac{ct}{2}}.$$

Next, we balance the contributions of the terms on the right-hand side of the inequality, by choosing c such that $(1-c)\cdot\underline{\mathrm{ID}\Box}_\delta = ct/2$. This produces

$$\mathbf{Pr}[\mathbf{L}_\perp^2 \leq \delta_\perp] \cdot \left(\frac{r}{\delta}\right)^{\overline{\mathrm{ID}\Box}_{\delta_\perp}} \leq \left(\mathbf{Pr}[\mathbf{L}^2 \leq \delta] + e^{\frac{t}{2}}\right) \cdot \left(\frac{r}{\delta}\right)^{\frac{\underline{\mathrm{ID}\Box}_\delta\cdot(t/2)}{\underline{\mathrm{ID}\Box}_\delta + t/2}}. \tag{3}$$

Note that from Theorem 1, the existence of the limits ID_F^* and $\mathrm{ID}_{F_\perp}^*$ implies that $F(\sqrt{\delta}) = \mathbf{Pr}[\mathbf{L} \leq \sqrt{\delta}] > 0$ and $F_\perp(\sqrt{\delta_\perp}) = \mathbf{Pr}[\mathbf{L}_\perp \leq \sqrt{\delta_\perp}] > 0$ whenever δ and $\delta_\perp$ are chosen to be sufficiently small. Taking the logarithms of both sides of Inequality 3, and dividing by $\ln(r/\delta)$, leads us to the following:

$$\overline{\mathrm{ID}\Box}_{\delta_\perp} \geq \frac{\underline{\mathrm{ID}\Box}_\delta\cdot(t/2)}{\underline{\mathrm{ID}\Box}_\delta + t/2} - \frac{\ln(\mathbf{Pr}[\mathbf{L}^2 \leq \delta] + e^{\frac{t}{2}})}{\ln(\delta/r)} + \frac{\ln \mathbf{Pr}[\mathbf{L}_\perp^2 \leq \delta_\perp]}{\ln(\delta/r)}.$$

Fixing δ and $\delta_\perp$, and letting $r \to 0$, the inequality has the limit

$$\overline{\mathrm{ID}\Box}_{\delta_\perp} \geq \frac{\underline{\mathrm{ID}\Box}_\delta\cdot(t/2)}{\underline{\mathrm{ID}\Box}_\delta + t/2}.$$

Next, letting $\delta \to 0$, we observe that $\delta_\perp \to 0$, $\underline{\mathrm{ID}\square}_\delta \to \mathrm{ID}\square^*$ and $\overline{\mathrm{ID}\square}_{\delta_\perp} \to \mathrm{ID}\square^*_\perp$, and thus

$$\mathrm{ID}\square^*_\perp \geq \frac{\mathrm{ID}\square^* \cdot (t/2)}{\mathrm{ID}\square^* + t/2}\,. \tag{4}$$

Up until now we have assumed that the quantities r/β, $r_\perp$, δ and $\delta_\perp$ were all within the interval $(0, \epsilon^2)$. Here, as $r \to 0$ and $\delta \to 0$, it can be verified that the aforementioned quantities all tend to 0 as well, and that this assumption is therefore eventually justified.

Finally, we transform the bound of Inequality 4 for the ID of the squared distance distributions (before and after projection) to one involving the original distributions. It follows from Theorem 2 that $\mathrm{ID}_F = 2{\cdot}\mathrm{ID}\square$ and $\mathrm{ID}_{F_\perp} = 2{\cdot}\mathrm{ID}\square_\perp$, from which we see that the lower bound

$$\mathrm{ID}^*_{F_\perp} \geq \frac{\mathrm{ID}^*_F \cdot t}{\mathrm{ID}^*_F + t}$$

holds as required.

We now turn our attention to the proof of the upper bound $\mathrm{ID}^*_{F_\perp} \leq \mathrm{ID}^*_F$. The proof is similar to (but much simpler than) that of the lower bound. Since $\|X\|^2 \geq \|X_\perp\|^2$, for any $r > 0$,

$$\mathbf{Pr}[\mathbf{L}^2 \leq r] \;=\; \mathbf{Pr}[\|X\|^2 \leq r] \;\leq\; \mathbf{Pr}[\|X_\perp\|^2 \leq r] \;=\; \mathbf{Pr}[\mathbf{L}^2_\perp \leq r]\,.$$

Applying Theorem 3, and choosing $\delta_\perp = \delta > r$, we obtain

$$\mathbf{Pr}[\mathbf{L}^2 \leq \delta] \cdot \left(\frac{r}{\delta}\right)^{\underline{\mathrm{ID}\square}_\delta} \leq \mathbf{Pr}[\mathbf{L}^2_\perp \leq \delta] \cdot \left(\frac{r}{\delta}\right)^{\overline{\mathrm{ID}\square}_{\delta_\perp}}\,,$$

Taking the logarithms of both sides, and dividing by $\ln(r/\delta)$, leads us to:

$$\underline{\mathrm{ID}\square}_\delta \;\geq\; \overline{\mathrm{ID}\square}_{\delta_\perp} + \frac{\ln \mathbf{Pr}[\mathbf{L}^2 \leq \delta]}{\ln(\delta/r)} - \frac{\ln \mathbf{Pr}[\mathbf{L}^2_\perp \leq \delta_\perp]}{\ln(\delta/r)}\,.$$

Fixing δ and $\delta_\perp$, and letting $r \to 0$, the inequality has the limit $\overline{\mathrm{ID}\square}_{\delta_\perp} \leq \underline{\mathrm{ID}\square}_\delta$. Next, letting $\delta \to 0$, we observe that $\delta_\perp \to 0$, $\underline{\mathrm{ID}\square}_\delta \to \mathrm{ID}\square^*$ and $\overline{\mathrm{ID}\square}_{\delta_\perp} \to \mathrm{ID}\square^*_\perp$, and thus $\mathrm{ID}\square^*_\perp \leq \mathrm{ID}\square^*$, which in turn implies that $\mathrm{ID}^*_{F_\perp} \leq \mathrm{ID}^*_F$ as required.

4 Conclusion

Theorem 4 has important implications for the theory and practice of databases, machine learning, data mining, and other areas in which similarity information plays a role. Under a reasonable assumption of the continuity of the local data distribution, random projection in Euclidean vector spaces cannot be relied upon to significantly improve the discriminability of a distance measure as the number of data samples tends to infinity, nor can it be counted upon to greatly alleviate the asymptotic effects of the curse of dimensionality.

To see this, let us assume that we have a reference point Q within the domain of a global data distribution, whose distance distribution has a local intrinsic dimensionality of ID^*_F. For a random projection to a subspace of dimension $t \gg \mathrm{ID}^*_F$, the bounds of Theorem 4 become almost tight, showing that the local ID of the distribution is essentially unchanged after projection. As increasingly larger data samples are drawn from the distribution, the k-nearest neighbor distance r_k tends to 0, and thus the discriminability of the distance measure at r_k tends to ID^*_F. Thus, under this scenario, as the data set size scales, the discriminability of the distance measure over fixed-cardinality neighborhoods is less and less affected by random projection.

On the other hand, when the projection dimension t is of the same order as ID^*_F (or of lower order), Theorem 4 implies that the local ID of the projected distribution is no smaller than t, and no greater than ID^*_F. However, the information loss associated with projection to dimensionalities below that of the intrinsic dimension would make any improvements in discriminability a moot point, as the distance distribution would no longer be well-preserved in the vicinity of Q.

These implications together are of particular importance when randomly projecting data drawn from a mixture of distributions, where the local intrinsic dimensionality can vary greatly from location to location. Theorem 4 indicates that the target dimension for projection should be chosen to be substantially larger than the LID estimate at locations of particular interest or importance. Using existing estimators of LID [3,17], appropriate target dimensions for dimensional reduction can be determined locally, without the need to construct an explicit embedding of the data, or an explicit representation of the projective subspace.

These conclusions should not be taken to mean that random projection is never capable of reducing the number of latent variables of a user-supplied data set, or improving the discriminability of distances within the set. Theorem 4 addresses only the *asymptotic* effect of random projection on the LID of continuous Euclidean distance distributions, as the number of data instance rises. In applications where the number of data instances can scale into the billions or more, it is possible that these asymptotic effects could become more and more evident. However, any attempt to empirically verify the predictions of Theorem 4 would face significant obstacles, due to the limits in precision and stability exhibited by all existing estimators of ID (not just LID) [3,17], and due to the difficulty in determining an appropriate locality size—if too large, locality is violated; if too small, there are too few samples for the estimators to converge. For this reason, the development of more effective ID estimation methods for small local data samples is an important topic for future research.

Acknowledgments. Michael E. Houle acknowledges the financial support of JSPS Kakenhi Kiban (B) Research Grant 18H03296. Ken-ichi Kawarabayashi is supported by JSPS Kakenhi Research Grant JP18H05291.

References

1. Achlioptas, D.: Database-friendly random projections: Johnson-Lindenstrauss with binary coins. J. Comput. Sys. Sci. **66**(4), 671–687 (2003)
2. Ailon, N., Chazelle, B.: The fast Johnson-Lindenstrauss transform and approximate nearest neighbors. SIAM J. Comput. **39**(1), 302–322 (2009)
3. Amsaleg, L.: Extreme-value-theoretic estimation of local intrinsic dimensionality. Data Min. Knowl. Disc. **32**(6), 1768–1805 (2018). https://doi.org/10.1007/s10618-018-0578-6
4. Baraniuk, R.G., Wakin, M.B.: Random projections of smooth manifolds. Found. Comput. Math. **9**(1), 51–77 (2009)
5. Bartal, Y., Recht, B., Schulman, L.J.: Dimensionality reduction: beyond the Johnson-Lindenstrauss bound. In: Proceedings of the Twenty-second Annual ACM-SIAM Symposium on Discrete Algorithms, SODA 2011, pp. 868–887 (2011)
6. Bruske, J., Sommer, G.: Intrinsic dimensionality estimation with optimally topology preserving maps. IEEE Trans. Pattern Anal. Mach. Intell. **20**(5), 572–575 (1998)
7. Camastra, F., Vinciarelli, A.: Estimating the intrinsic dimension of data with a fractal-based method. IEEE Trans. Pattern Anal. Mach. Intell. **24**(10), 1404–1407 (2002)
8. Casanova, G., Englmeier, E., Houle, M.E., Kröger, P., Nett, M., Zimek, A.: Dimensional testing for reverse *k*-nearest neighbor search. PVLDB **10**(7), 769–780 (2017)
9. Coles, S.: An Introduction to Statistical Modeling of Extreme Values. Springer Series in Statistics, Springer, London (2001). https://doi.org/10.1007/978-1-4471-3675-0
10. Dasgupta, A., Kumar, R., Sarlos, T.: A sparse Johnson-Lindenstrauss transform. In: STOC, pp. 341–350 (2010)
11. Dasgupta, S., Gupta, A.: An elementary proof of a theorem of Johnson and Lindenstrauss. Random Struct. Algorithms **22**(1), 60–65 (2003)
12. de Vries, T., Chawla, S., Houle, M.E.: Density-preserving projections for large-scale local anomaly detection. Knowl. Inf. Syst. **32**(1), 25–52 (2012)
13. Dubhashi, D.P., Panconesi, A.: Concentration of Measure for the Analysis of Randomized Algorithms. Cambridge University Press, Cambridge (2009)
14. Falconer, K.: Fractal Geometry: Mathematical Foundations and Applications. Wiley (2003)
15. Faloutsos, C., Kamel, I.: Beyond uniformity and independence: analysis of R-trees using the concept of fractal dimension. In: PODS 1994, pp. 4–13 (1994)
16. Frankl, P., Maehara, H.: The Johnson-Lindenstrauss lemma and the sphericity of some graphs. J. Comb. Theor. Ser. B **44**(3), 355–362 (1988)
17. Gomes, M.I., Canto e Castro, L., Fraga Alves, M.I., Pestana, D., Laurens de Haan leading contributions: Statistics of extremes for IID data and breakthroughs in the estimation of the extreme value index. Extremes **11**, 3–34 (2008)
18. Grassberger, P., Procaccia, I.: Measuring the strangeness of strange attractors. Phys. D **9**(1–2), 189–208 (1983)
19. Gupta, A., Krauthgamer, R., Lee, J.R.: Bounded geometries, fractals, and low-distortion embeddings. In: FOCS 2003, pp. 534–543. IEEE Computer Society (2003)
20. Houle, M.E.: Dimensionality, discriminability, density & distance distributions. In: Proceedings of the ICDMW 2013, pp. 468–473 (2013)

21. Houle, M.E.: Local intrinsic dimensionality I: an extreme-value-theoretic foundation for similarity applications. In: International Conference on Similarity Search and Applications, pp. 64–79 (2017)
22. Houle, M.E.: Local intrinsic dimensionality II: multivariate analysis and distributional support. In: International Conference on Similarity Search and Applications, pp. 80–95 (2017)
23. Houle, M.E., Kashima, H., Nett, M.: Generalized expansion dimension. In: Proceedings of the ICDMW 2012, pp. 587–594 (2012)
24. Houle, M.E., Ma, X., Nett, M., Oria, V.: Dimensional testing for multi-step similarity search. In: ICDM 2012, pp. 299–308 (2012)
25. Houle, M.E., Ma, X., Oria, V.: Effective and efficient algorithms for flexible aggregate similarity search in high dimensional spaces. IEEE TKDE **27**(12), 3258–3273 (2015)
26. Houle, M.E., Ma, X., Oria, V., Sun, J.: Efficient algorithms for similarity search in user-specified projective subspaces. Inf. Syst. **59**, 2–14 (2016)
27. Houle, M.E., Nett, M.: Rank-based similarity search: Reducing the dimensional dependence. IEEE TPAMI **37**(1), 136–150 (2015)
28. Huisman, R., Koedijk, K.G., Kool, C.J.M., Palm, F.: Tail-index estimates in small samples. J. Bus. Econ. Stat. **19**(2), 208–216 (2001)
29. Indyk, P., Motwani, R.: Approximate nearest neighbors: towards removing the curse of dimensionality. In: STOC, pp. 604–613 (1998)
30. Johnson, W.B., Lindenstrauss, J.: Extensions of Lipschitz mappings into a Hilbert space. In: AMS Conference in Modern Analysis and Probability, pp. 189–206 (1982)
31. Kane, D.M., Nelson, J.: Sparser Johnson-Lindenstrauss transforms. J. ACM **61**(1), 4:1-4:23 (2014)
32. Karger, D.R., Ruhl, M.: Finding nearest neighbors in growth-restricted metrics. In: STOC 2002, pp. 741–750 (2002)
33. Larsen, K.G., Nelson, J.: The Johnson-Lindenstrauss lemma is optimal for linear dimensionality reduction. arXiv.org, cs.IT (2014)
34. Levina, E., Bickel, P.J.: Maximum likelihood estimation of intrinsic dimension. In: Advances in Neural Information Processing Systems 17 (NIPS 2004) (2004)
35. Mattila, P.: Hausdorff dimension, orthogonal projections and intersections with planes. Ann. Acad. Sci. Fenn. A Math. **1**, 227–244 (1975)
36. Navarro, G., Paredes, R., Reyes, N., Bustos, C.: An empirical evaluation of intrinsic dimension estimators. Inf. Syst. **64**, 206–218 (2017)
37. Romano, S., Chelly,O., Nguyen, V., Bailey, J., Houle, M.E.: Measuring dependency via intrinsic dimensionality. In: ICPR, pp. 1207–1212 (2016)
38. Roweis, S.T., Saul, L.K.: Nonlinear dimensionality reduction by locally linear embedding. Science **290**(5500), 2323–2326 (2000)
39. Rozza, A., Lombardi, G., Ceruti, C., Casiraghi, E., Campadelli, P.: Novel high intrinsic dimensionality estimators. Mach. Learn. J. **89**(1–2), 37–65 (2012)
40. Schölkopf, B., Smola, A.J., Müller, K.-R.: Nonlinear component analysis as a Kernel eigenvalue problem. Neural Comput. **10**(5), 1299–1319 (1998)
41. Tenenbaum, J., Silva, V.D., Langford, J.: A global geometric framework for non linear dimensionality reduction. Science **290**(5500), 2319–2323 (2000)
42. Venna, J., Kaski, S.: Local multidimensional scaling. Neural Netw. **19**(6–7), 889–899 (2006)
43. Xing, E.P., Ng, A.Y., Jordan, M.I., Russell, S.J.: Distance metric learning with application to clustering with side-information. In: NIPS 2002, pp. 505–512 (2002)

Clustering and Classification

Accelerating Spherical k-Means

Erich Schubert(✉), Andreas Lang, and Gloria Feher

TU Dortmund University, Dortmund, Germany
{erich.schubert,andreas.lang,gloria.feher}@tu-dortmund.de

Abstract. Spherical k-means is a widely used clustering algorithm for sparse and high-dimensional data such as document vectors. While several improvements and accelerations have been introduced for the original k-means algorithm, not all easily translate to the spherical variant: Many acceleration techniques, such as the algorithms of Elkan and Hamerly, rely on the triangle inequality of Euclidean distances. However, spherical k-means uses cosine similarities instead of distances for computational efficiency. In this paper, we incorporate the Elkan and Hamerly accelerations to the spherical k-means algorithm working directly with the cosines instead of Euclidean distances to obtain a substantial speedup and evaluate these spherical accelerations on real data.

1 Introduction

Clustering textual data is an important task in data science with applications in areas like information retrieval, topic modeling, and knowledge organization. Spherical k-means [8] is a widely used adaptation of the k-means clustering algorithm to high-dimensional sparse data, such as document vectors where cosine similarity is a popular choice. While it is generally used for clustering documents, it has also been applied to medical images [2,20], multivariate species occurrence data [14], and plant leaf images [1]. Because of its importance, several improvements and extensions have been suggested. Many optimizations improve the initialization of k-means cluster centers, such as k-means++ [3] and k-means|| [5,6], some of which have also been adapted to spherical k-means [11,19,22].

A key area of optimizations is focussed on the iterative optimization phase of k-means. The standard algorithm computes the distance of every point to every cluster in each iteration. Many of these computations are not necessary if cluster centers have not moved much, and hence a lot of research has been on how to avoid computing distances. The central work in this domain is the algorithm of Elkan [10], which is the base for many other variants such as Hamerly's algorithm [12], but also recently the Exponion algorithm [21], the Shallot algorithm [7], and the variants of Yu et al. [26], all of which rely on the Euclidean triangle inequality to avoid distance computations.

Part of the work on this paper has been supported by Deutsche Forschungsgemeinschaft (DFG), project number 124020371, within the Collaborative Research Center SFB 876 "Providing Information by Resource-Constrained Analysis", project A2.

N. Reyes et al. (Eds.): SISAP 2021, LNCS 13058, pp. 217–231, 2021.
https://doi.org/10.1007/978-3-030-89657-7_17

This paper studies how to adapt such acceleration techniques to spherical k-means, thus providing a more efficient approach for clustering text documents.

2 Foundations

Cosine similarity (which we will simply denote using sim in the following) is commonly defined as the cosine of the angle θ between two vectors $\mathbf{x}$ and $\mathbf{y}$:

$$\mathrm{sim}(\mathbf{x},\mathbf{y}) := \mathrm{sim}_{\mathrm{cosine}}(\mathbf{x},\mathbf{y}) := \frac{\langle \mathbf{x},\mathbf{y}\rangle}{\|\mathbf{x}\|_2 \cdot \|\mathbf{y}\|_2} = \frac{\sum_i x_i y_i}{\sqrt{\sum_i x_i^2} \cdot \sqrt{\sum_i y_i^2}} = \cos\theta$$

In the following, we will only consider vectors normalized to unit length, i.e., with Euclidean norm $\|\mathbf{x}\|_2 = 1$. It is trivial to see that on such vectors, the cosine similarity is simply the dot product. Consider the Euclidean distance of two *normalized* vectors $\mathbf{x}$ and $\mathbf{y}$, and expand using the binomial equations, we obtain:

$$d_{\mathrm{Euclidean}}(\mathbf{x},\mathbf{y}) := \sqrt{\sum\nolimits_i (x_i - y_i)^2} = \sqrt{\sum\nolimits_i (x_i^2 + y_i^2 - 2x_i y_i)} \tag{1}$$

$$= \sqrt{\|\mathbf{x}\|^2 + \|\mathbf{y}\|^2 - 2\langle \mathbf{x},\mathbf{y}\rangle} = \sqrt{2 - 2\cdot \mathrm{sim}(\mathbf{x},\mathbf{y})} \tag{2}$$

where the last step relies on the vectors being normalized. Hence we have an extremely close relationship between cosine similarity and squared Euclidean distance on normalized vectors: $\mathrm{sim}(\mathbf{x},\mathbf{y}) = 1 - \frac{1}{2} d^2_{\mathrm{Euclidean}}(\mathbf{x},\mathbf{y})$.

k-means minimizes the *squared* Euclidean distances of points to their cluster centers and hence can be used to maximize cosine similarities. Because the total variance of a data set is constant, by minimizing the within-cluster squared deviations, k-means also maximizes the between-cluster squared deviations. By adapting this to cosine, we obtain clusters where objects in the same cluster have to be more similar, while objects in different clusters are less similar.

Dhillon and Modha [8] popularized this idea as "spherical k-means" for clustering text documents and exploited exactly the above relationship between the squared Euclidean distance and cosine similarity. Only a tiny modification of the standard k-means algorithm is necessary to obtain the desired results: the arithmetic mean of a cluster usually does not have unit Euclidean length. Hence, after recomputing the cluster mean, we normalize it accordingly. This constrains the clustering to split the data at great circles (i.e., hyperplanes through the origin), rather than arbitrary Voronoi cells as with regular k-means.

On text data, computing the cosine similarity is more efficient than computing Euclidean distance because of sparsity: rather than storing the vectors as a long array of values, most of which are zeros, only the non-zero values can be encoded as pairs (i, v) of an index i and a value v, and stored and kept in sorted order. The dot product of two such vectors can then be efficiently computed by a *merge* operation, where only those indexes i need to be considered that are contained in both vectors, because in $\langle \mathbf{x},\mathbf{y}\rangle = \sum_i x_i y_i$ only those terms matter where both x_i and y_i are not zero. A merge is most efficient if both vectors are sparse, but even the dot product of a sparse and a dense vector is often much faster than that of two dense vectors. While we can also compute Euclidean distance this way (using Eq. (1)), this computation is prone to the numerical problem called "catastrophic cancellation" for small distances that can be problematic in clustering (see, e.g., [16,24]). Hence, working with cosines directly is preferable.

Instead of recomputing the distances to all cluster centers, the idea of algorithms such as Elkan's is to keep an upper bound on the distance to the nearest cluster, and one or more lower bounds on the distances to the other centers. Let d_n be the distance to the nearest center, $d_n \leq u$ an upper bound, d_s the distance to the second nearest, and $l \leq d_s$ a lower bound. If we have $u \leq l$, then the nearest cluster must still be the same since $d_n \leq u \leq l \leq d_s$. Updating the distance bounds uses the triangle inequality: if the nearest center μ_n has moved to μ'_n, then $d(x, \mu'_n) \leq d(x, \mu_n) + d(\mu_n, \mu'_n)$; and we hence can obtain an upper bound u by adding every movement of a cluster center to the previous distance. Lower bounds are obtained similarly: starting with the initial distance as the lower bound, we subtract the distance the other center has moved to obtain a provable new lower bound. While Elkan stored a lower bound for each cluster (which needs $O(N \cdot k)$ memory), Hamerly [12] reduced the memory usage by using just one lower bound to the second nearest cluster, updated by the largest distance moved. Additional pruning rules involve the pairwise distances of centers [10], annuli around centers [21], and the relative movement of centers [26].

In the following, we describe how such accelerations can be applied to spherical k-means, i.e., for cosine similarity and high-dimensional data.

3 Pruning with Cosine Similarity

Many acceleration techniques rely on the triangle inequality of the (non-squared) Euclidean distance. Hence, we can adapt these methods by computing Euclidean distances from our cosine similarities using $d_{\text{Euclidean}}(\mathbf{x}, \mathbf{y}) = \sqrt{2 - 2 \cdot \text{sim}(\mathbf{x}, \mathbf{y})}$, but we wanted to avoid this because of (i) the square root takes 10–50 CPU cycles (depending on the exact CPU, precision, and input value) and (ii) the risk of numerical instability because of catastrophic cancellation. Hence we develop techniques that directly use similarities instead of distances, yet allow a similar pruning to these (very successful) acceleration techniques of regular k-means.

The arc length (i.e., the angle θ itself, rather than the cosine of the angle) satisfies the triangle inequality and hence we could use

$$\text{sim}(\mathbf{x}, \mathbf{y}) \geq \cos(\arccos(\text{sim}(\mathbf{x}, \mathbf{z})) + \arccos(\text{sim}(\mathbf{z}, \mathbf{y}))) \ , \tag{3}$$

but unfortunately the trigonometric functions in here are even more expensive (60–100 CPU cycles each). Schubert [23] recently proposed reformulations avoiding the expensive trigonometric functions (but still using the square root):

$$\text{sim}(\mathbf{x}, \mathbf{y}) \geq \text{sim}(\mathbf{x}, \mathbf{z}) \cdot \text{sim}(\mathbf{z}, \mathbf{y}) - \sqrt{(1 - \text{sim}(\mathbf{x}, \mathbf{z})^2) \cdot (1 - \text{sim}(\mathbf{z}, \mathbf{y})^2)} \tag{4}$$

$$\text{sim}(\mathbf{x}, \mathbf{y}) \leq \text{sim}(\mathbf{x}, \mathbf{z}) \cdot \text{sim}(\mathbf{z}, \mathbf{y}) + \sqrt{(1 - \text{sim}(\mathbf{x}, \mathbf{z})^2) \cdot (1 - \text{sim}(\mathbf{z}, \mathbf{y})^2)} \tag{5}$$

In this paper, we explain how to integrate these triangle inequalities into spherical k-means, and discuss an easily overlooked pitfall therein.

4 Upper and Lower Bounds

In the following, we orient ourselves on the very concise presentation and notation of Hamerly [12] as well as Newling and Fleuret [21], except that we swap the names of u and l, because switching from distance to similarity requires us to swap the roles of upper and lower bounds. We will assume that all points are normalized to unit length, and hence $\text{sim}(\mathbf{x}, \mathbf{y}) = \langle \mathbf{x}, \mathbf{y} \rangle = \mathbf{x}^T \cdot \mathbf{y}$.

The algorithms we discuss will employ upper and lower bounds for the similarities of each sample $x(i)$ to the cluster centers $c(j)$. $l(i)$ is a lower bound for the similarity to the current cluster $a(i)$, $u(i,j)$ are upper bounds on the similarity of each point to each cluster center, respectively $u(i)$ is an upper bound on the similarity to all other cluster centers ($u(i,j)$ and $u(i)$ are used in different variants, not at the same time). These bounds are maintained to satisfy:

$$l(i) \leq \langle x(i), c(a(i)) \rangle \qquad u(i,j) \geq \langle x(i), c(j) \rangle \qquad u(i) \geq \max_{j \neq a(i)} \langle x(i), c(j) \rangle$$

The central idea of all the discussed variants is that if we have $l(i) \geq u(i,j)$, then $\langle x(i), c(a(i)) \rangle \geq l(i) \geq u(i,j) \geq \langle x(i), c(j) \rangle$ implies that the current cluster assignment of object $x(i)$ is optimal, and we do not need to recompute the similarities.

The bounds $l(i)$ and $u(i,j)$, can be maintained using above triangle inequality if we know how much the cluster centers $c(j)$ moved from their previous location $c'(j)$. Let $p(j) := \langle c(j), c'(j) \rangle$ denote this similarity. Based on the triangle inequalities Eqs. (4) and (5), we obtain the following bound update equations:

$$l(i) \leftarrow l(i) \cdot p(a(i)) - \sqrt{(1 - l(i)^2) \cdot (1 - p(a(i))^2)} \tag{6}$$

$$u(i,j) \leftarrow u(i,j) \cdot p(j) + \sqrt{(1 - u(i,j)^2) \cdot (1 - p(j)^2)} \tag{7}$$

5 Accelerated Spherical k-Means

The algorithms discussed here all follow the outline of the standard k-means algorithm of alternating optimization. During initialization, all data samples $x(i)$ are normalized to have length $\|x(i)\| = 1$. In the first step, all objects are reassigned to the nearest cluster, in the second step, the cluster center is optimized. However, we switch the notation from distance to similarity. Let the variable $a(i)$ denote the current cluster assignment of sample $x(i)$, and denote the current cluster centers using $c(j)$, the two steps can be written as:

$$a(i) \leftarrow \arg\max_j \langle x(i), c(j) \rangle \qquad i \in 1..N$$

$$c(j) \leftarrow \frac{\sum_{i|a(i)=j} x(i)}{\left\|\sum_{i|a(i)=j} x(i)\right\|} \qquad j \in 1..k$$

When computing $a(i)$ we maximize the cosine similarity instead of the squared Euclidean distance in regular k-means. For $c(j)$, note that the denominator is different here, as we want to have $\|c(j)\| = 1$ for all j. We hence do not need to compute the arithmetic mean, but we can scale the sum directly to length 1.

There are several optimizations we can do for the baseline algorithm that make a difference: (i) By normalizing the vectors, we do not have to take the vector lengths of $x(i)$ into account when updating $c(j)$, and by also normalizing the $c(j)$ we can use the dot product when computing $a(i)$. (ii) Both the dot product as well as the sum operation when computing $c(j)$ can be optimized for sparse data. (iii) Instead of recomputing $c(j)$ each time, it is better to store the sums before normalization and update them when a cluster assignment changes.

5.1 Spherical Simplified Elkan's Algorithm

As the name suggests, this algorithm is a simplified version of Elkan's approach, introduced by Newling and Fleuret [21]. As it uses a subset of the pruning rules, we introduce it before Elkan's full algorithm. Both are presented directly in the adaptation for spherical k-means.

Simplified Elkan uses the test $u(i,j) \leq l(i)$ to skip computing the similarity between $x(i)$ and $c(j)$ when it is not necessary. If this test fails, $l(i) \leftarrow \langle x(i), c(a(i)) \rangle$ is updated first (as the current assignment is clearly the best guess), and only if the condition still is violated, $u(i,j) \leftarrow \langle x(i), c(j) \rangle$ is computed next, and the point is reassigned if necessary (updating $l(i)$ and $a(i)$ then).

5.2 Spherical Elkan's Algorithm

Elkan's algorithm [10] uses additional tests based on the pairwise distance of centers, respectively pairwise cluster similarities here. The idea is that cluster centers are supposedly well separated, whereas points are close to their nearest cluster, and we can use half the distance between two centers as a threshold. We simplify the computation of half of the angle per $\cos(\frac{1}{2}\arccos(x)) = \sqrt{(x+1)/2}$. Let $cc(i,j) := \sqrt{(\langle c(i), c(j) \rangle + 1)/2}$ be this lower bound (cc for center-center bounds, as in [21]). Let $s(i) := \max_{j \neq i} cc(i,j)$ denote the maximum such bound for each i.

Suppose that $cc(a(i),j) \leq l(i)$ and $l(i) \geq 0$, then $\langle c(i), c(j) \rangle \leq 2l(i)^2 - 1$. We can then use Eq. (5) to bound the distance to another cluster $c(j) \neq c(a(i))$ per

$$
\begin{aligned}
\langle x(i), c(j) \rangle &\leq \langle x(i), c(a(i)) \rangle \cdot \langle c(a(i)), c(j) \rangle \\
&\quad + \sqrt{(1 - \langle x(i), c(a(i)) \rangle^2) \cdot (1 - \langle c(a(i)), c(j) \rangle^2)} \\
&\leq l(i)(2l(i)^2 - 1) + \sqrt{(1 - l(i)^2) \cdot (1 - (2l(i)^2 - 1)^2)} \\
&= 2l(i)^3 - l(i) + \sqrt{(1 - l(i)^2) \cdot 4l(i)^2 (1 - l(i)^2)} \\
&= 2l(i)^3 - l(i) + 2l(i)(1 - l(i)^2) = l(i) \quad ,
\end{aligned}
$$

and hence do not have to consider other cluster centers $c(j)$ *if* $cc(a(i),j) \leq l(i)$. Because $s(i)$ is the maximum of these values, we can skip iterating over the means if $s(i) \leq l(i)$ altogether. While these additional tests are fairly cheap to compute, they were found to not always be effective by Newling and Fleuret [21] (who, hence, suggested the simplified variant discussed in the previous section).

For spherical k-means clustering, these bounds may not be very effective because of the high dimensionality. Using these bounds adds $k \cdot (k-1)/2 = O(k^2)$

similarity computations to each iteration. Furthermore, the necessary computations can become more expensive because the centers are best stored using dense vectors because (i) we aggregate many vectors into each center, and only attributes zero in all of the assigned vectors will be zero in the resulting center, i.e., the sparsity decreases often to the point where a dense representation is more compact, and (ii) the efficient sparse data structures we use for the $x(i)$ are not well suited for adding and removing attributes. We could aggregate into a dense vector and convert it to a sparse representation when normalizing the center, but the resulting vectors will still often be too dense to be efficient.

5.3 Spherical Hamerly's Algorithm

Where Elkan's algorithm used one upper bound for each cluster, Hamerly [12] only uses a single bound for all clusters. This does not only saves memory (for large k, memory consumption of Elkan's algorithm can be an issue) but updating the $N{\cdot}k$ bounds each iteration even if the clusters change only very little takes a considerable amount of time. Hamerly's idea is to make a worst-case assumption, where we use the distance to the second nearest center as the initial bound, and update it based on the largest cluster movement (of all clusters, except the one currently assigned to). Because of this, the bound will become loose much faster, and hence we need to recompute more often (and then we need to recompute the distances to all clusters). Because of this, it is hard to predict which algorithm works better, we are trading reduced memory and fewer bound updates against additional distance computations. Nevertheless, many later works have confirmed that it is often favorable to only keep one bound.

At first, adapting Hamerly to cosine similarity appears to be straightforward. To obtain the lowest upper bound per object $u(i) \leq \min_{j \neq a(i)} u(i,j)$, we would compute the smallest similarity of a cluster center to its previous location (as well as the second smallest, in case the point is currently assigned to that center), then use Eq. (7) with $p'(i) := \min_{j \neq i} p(j)$ (which is either the smallest or the second smallest $p(j)$). Most of the time this is fine, but there is a subtly hidden catch here because of the underlying non-monotone trigonometric functions.

Recall the update equation (7), rewritten to $u(i)$ instead of $u(i,j)$ already:

$$u(i) \leftarrow u(i) \cdot p(j) + \sqrt{(1 - u(i)^2) \cdot (1 - p(j)^2)}$$

This equation is not necessarily minimized by the smallest $p(j)$, because of the square root term. For large $u(i)$ (e.g., 1), the result will be determined by the first term, and a smaller $p(j)$ is what is needed. But for small $u(i)$ (e.g., 0), the second term becomes influential, and a larger $p(j)$ causes a smaller bound. This is because we are working with the cosines $\cos\theta$, not the angles θ themselves. Unfortunately, this depends on the previous value of $u(i)$, and we probably cannot use just one $p(j)$ for all points.

One option would be to use both the minimum $p'(i) := \min_{j \neq i} p(j)$ and the maximum $p''(i) := \max_{j \neq i} p(j)$ to update the bound with:

$$(7) \leq u(i) \cdot p''(a(i)) + \sqrt{(1 - u(i)^2) \cdot (1 - p'(a(i))^2)} \ . \tag{8}$$

Because $p''(j) \to 1$ as the algorithm converges, we may omit this term entirely:

$$(8) \leq u(i) + \sqrt{(1 - u(i)^2) \cdot (1 - p'(a(i))^2)} \tag{9}$$

This has almost identical pruning power once $p''(j)$ becomes large enough in later iterations. As we can precompute $(1 - p'(j))$ for all j, this is quite efficient. We cannot rule out that a tighter and computationally efficient bound exists.

If the condition $l(i) \geq u(i)$ is violated, first $l(i)$ is made tight again, and if it still is violated, all remaining similarities are computed to update $u(i)$, or to potentially obtain a new cluster assignment (updating $a(i)$, $l(i)$, and $u(i)$).

5.4 Spherical Simplified Hamerly's Algorithm

Hamerly's algorithm contains a bounds test similar to Elkan's algorithm, but using only the distance of each center to its nearest neighbor center instead of keeping all pairwise center distances, i.e., only the threshold $s(i) := \max_{j \neq i} cc(i,j)$ to prune objects with $l(i) \geq s(a(i))$. We also consider a "simplified" variant of Hamerly's algorithm in our experiments with this bound check removed for the same reasons as discussed with Elkan's algorithm.

5.5 Further k-Means Variants

An obvious candidate to extend this work is Yin-Yang k-means [9], which groups the cluster centers and uses one bound for each group. This is a compromise between Elkan's and Hamerly's approaches, encompassing both as extreme cases (k groups respectively one group). The results of this paper will trivially transfer to this method. The Annulus algorithm [13] additionally uses the distance from the origin for pruning. As all our data is normalized to unit length, this approach clearly will not help for spherical k-means. The Exponion [21] and Shallot [7] algorithms transfer this idea to using pairwise distances of cluster centers, where our considerations may be applicable again.

5.6 Spherical k-means++

We experiment with the canonical adaptation of k-means++, using the analogy with squared Euclidean distance. The first sample is chosen uniformly at random, the remaining instances are sampled proportional to $1 - \max_c \langle x(i), c \rangle$ which is proportional to the squared Euclidean distance used by k-means++. This can be done in $O(nk)$ by caching the previous maximum, and the scalar product is efficient for two sparse vectors. Endo and Miyamoto [11] prove theoretical guarantees for a slight modification of spherical k-means using the dissimilarity of $\alpha - \langle \mathbf{x}, \mathbf{y} \rangle$ with $\alpha \geq \frac{3}{2}$ to make it metric, and hence sample proportionally to $1 - \max_c \langle x(i), c \rangle$. Pratap et al. [22] use the same trick to apply the AFK-MC^2 algorithm [4] to spherical k-means-clustering.

Table 1. Data sets used in the experiments.

Data set	Rows	Columns	Non-zero
DBLP Author-Conference	1842986	5236	0.056%
DBLP Conference-Author	5236	1842986	0.056%
DBLP Author-Venue	2722762	7192	0.099%
Simpsons Wiki	10126	12941	0.463%
20 Newsgroups	11314	101631	0.096%
Reuters RCV-1	804414	47236	0.160%

6 Experiments

We implemented our algorithms in the Java framework ELKI [25], which already contained a large collection of k-means variants. By keeping the implementation differences to a minimum, we try to make the benchmark experiments more reliable (c.f., [15]), but the caveats of Java just-in-time compilation remain.

As our method is designed for sparse and high-dimensional data sets, we focus on textual and graph data as input. Table 1 summarizes the data sets used in the experiments. From the Digital Bibliography & Library Project (DBLP, [18]) we extracted graphs that connect authors and conferences. As this includes many authors with just a single paper, the data set is very sparse. We can either use the authors as samples and the conferences as columns or transposed. But because we use TF-IDF weighting afterward the semantics will be different. Spherical k-means clustering has been used successfully for community detection on such data sets (although we have to choose the number of communities as a parameter). If we also include journals, the data set becomes both larger and denser. A second data set was obtained from the Simpsons Fandom Wiki,[1] from which we extracted the text of around 10000 articles. The text was tokenized and lemmatized, stop words were removed as well as infrequent tokens (reducing the dimensionality from 42124 to 12941, and increasing the density of non-zero values from 0.153% to 0.463%). This data set is more typical of a smaller domain-specific text corpus. The 20 Newsgroups data set is a classic, popularized by the textbook of Tom Mitchell. We use a version available via scikit-learn, with headers, footers, and quotes removed and vectorized using the default settings (i.e., TF-IDF weighting). This is much more sparse than the Simpsons wiki because of the poor input data quality (including Base64-encoded attachments). After removing stop words and rare words as above, the density would have been 0.317%, but we opted for the default scikit-learn version instead. Reuters RCV-1 [17] is another classic text categorization benchmark, with a density between the Simpsons and the 20news data.

We first discuss the algorithms on a single data set, with a single random seed, averaged over 10 re-runs, to observe some characteristic behavior. The reason that we do not average over different random initializations is that we

[1] https://simpsons.fandom.com/wiki/Simpsons_Wiki.

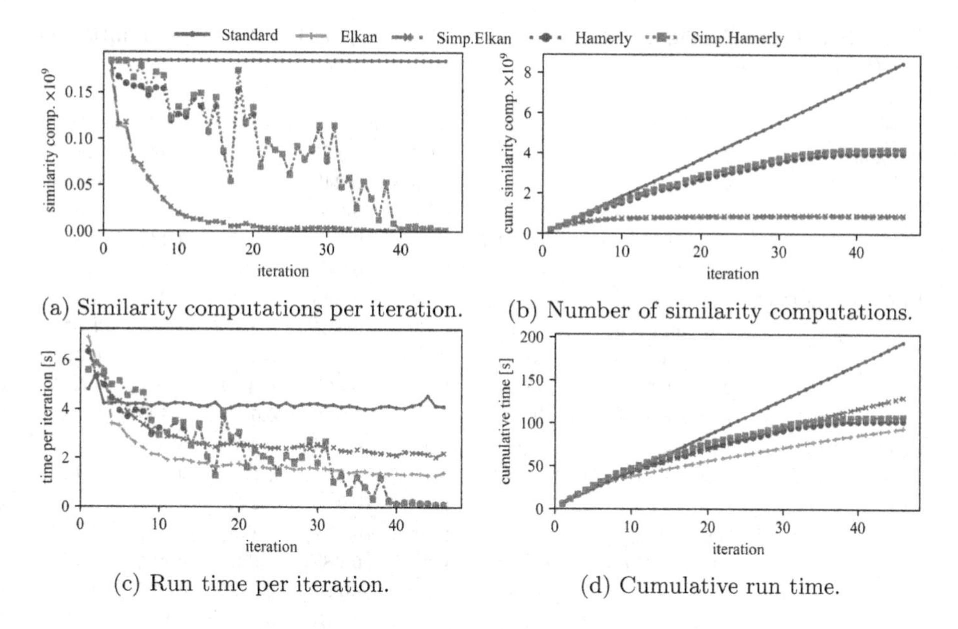

(a) Similarity computations per iteration. (b) Number of similarity computations.

(c) Run time per iteration. (d) Cumulative run time.

Fig. 1. Distance computations and resulting run time for one initialization.

want to observe individual iterations of the algorithms, which depend on the initialization. Figure 1 shows the results on the DBLP author-conference data set with a large $k = 100$. Considering only similarity computations (Fig. 1a and 1b), both Elkan and Simplified Elkan shine (as expected) and use the fewest computations as they have the tightest bounds. There is next to no difference among the two concerning the number of computations, but considering the run time (Fig. 1c and 1d) the simplified variant is much worse. Perhaps unexpectedly, this can be explained by the high k. The additional pruning rule of the full algorithm allows skipping the loop over all clusters k, which would otherwise each have to be checked against their bounds. The behavior of the Hamerly variants is much more chaotic because it only depends on the cluster center that changed most. Because of this, Hamerly computes many more similarities than Elkan until the last few iterations. Nevertheless, its total run time is initially similar to that of Simplified Elkan, and after around 30 iterations its run time per iteration (c.f., Fig. 1c) becomes even lower than the full Elkan algorithm's. These savings arise once clusters do not change much anymore because only 2 bounds need to be updated instead of $k+1$ bounds per iteration. For $k = 10$ (not shown in the figures), both Hamerly variants outperform Elkan, while for $k = 1000$ even Simplified Elkan clearly outperforms both Hamerly variants. Note that we used random sampling as initialization. If we had known the optimal initial cluster centers, all methods would have converged instantly.

Next, we compare the quality and run time of the initialization methods. Table 2 shows the difference in the sum of variances, averaged over 10 random seeds, compared to the uniform random initialization. It shows that the quality difference of the converged solutions between all initialization methods is small

Table 2. Relative change in the objective function compared to the random initialization (lower is better).

Data set	Initialization	k = 2	k = 10	k = 20	k = 50	k = 100	k = 200
Simpsons Wiki	Uniform	0.00%	0.00%	0.00%	0.00%	0.00%	0.00%
	k-means++ $\alpha = 1$	−0.27%	**−0.16%**	**−0.24%**	−0.07%	−0.18%	−0.07%
	k-means++ $\alpha = 1.5$	−0.16%	−0.13%	−0.17%	−0.01%	−0.18%	**−0.09%**
	AFK-MC^2 $\alpha = 1$	**−0.44%**	0.12%	−0.15%	**−0.15%**	**−0.24%**	−0.08%
	AFK-MC^2 $\alpha = 1.5$	−0.31%	0.21%	0.09%	0.09%	−0.05%	−0.02%
DBLP Author-Conf.	Uniform	0.00%	0.00%	0.00%	0.00%	0.00%	0.00%
	k-means++ $\alpha = 1$	**−0.11%**	0.12%	−0.07%	0.27%	0.14%	**−1.67%**
	k-means++ $\alpha = 1.5$	−0.03%	0.11%	0.33%	0.68%	0.53%	−0.74%
	AFK-MC^2 $\alpha = 1$	−0.01%	**−0.06%**	**−0.87%**	**−0.47%**	−0.48%	−1.03%
	AFK-MC^2 $\alpha = 1.5$	−0.03%	0.34%	−0.32%	0.09%	**−0.56%**	−1.10%
DBLP Author-Venue	Uniform	0.00%	0.00%	0.00%	0.00%	0.00%	0.00%
	k-means++ $\alpha = 1$	−0.13%	0.09%	−0.12%	0.13%	**−0.74%**	**−1.70%**
	k-means++ $\alpha = 1.5$	−0.01%	0.18%	0.00%	0.23%	0.39%	−0.17%
	AFK-MC^2 $\alpha = 1$	−0.17%	0.10%	−0.05%	0.47%	−0.20%	−0.68%
	AFK-MC^2 $\alpha = 1.5$	**−0.19%**	**−0.33%**	**−0.68%**	**−0.04%**	−0.50%	−1.41%
DBLP Conf.-Author	Uniform	**0.00%**	**0.00%**	**0.00%**	0.00%	0.00%	0.00%
	k-means++ $\alpha = 1$	0.01%	0.04%	0.05%	−0.02%	−0.13%	−0.09%
	k-means++ $\alpha = 1.5$	**0.00%**	0.11%	0.08%	**−0.15%**	−0.18%	**−0.13%**
	AFK-MC^2 $\alpha = 1$	0.04%	**0.00%**	0.05%	−0.10%	**−0.19%**	−0.02%
	AFK-MC^2 $\alpha = 1.5$	0.04%	**0.00%**	0.06%	−0.12%	−0.15%	−0.06%
20 Newsgroups	Uniform	**0.00%**	**0.00%**	**0.00%**	**0.00%**	**0.00%**	**0.00%**
	k-means++ $\alpha = 1$	0.38%	0.52%	0.78%	1.83%	4.09%	7.34%
	k-means++ $\alpha = 1.5$	0.72%	0.93%	0.89%	2.39%	4.65%	7.87%
	AFK-MC^2 $\alpha = 1$	0.24%	0.31%	0.31%	0.41%	0.11%	0.23%
	AFK-MC^2 $\alpha = 1.5$	0.37%	0.17%	0.26%	0.30%	0.08%	0.23%
RCV-1	Uniform	0.00%	0.00%	0.00%	0.00%	0.00%	**0.00%**
	k-means++ $\alpha = 1$	0.13%	**−0.11%**	0.08%	**−0.25%**	**−0.17%**	0.06%
	k-means++ $\alpha = 1.5$	−0.03%	0.21%	0.53%	0.44%	0.04%	0.16%
	AFK-MC^2 $\alpha = 1$	**−0.24%**	−0.01%	**−0.03%**	0.39%	0.05%	0.24%
	AFK-MC^2 $\alpha = 1.5$	0.13%	0.07%	**−0.03%**	0.22%	−0.09%	0.15%

except for the 20-news data set where k-means++ performs up to 8% worse. Supposedly, because this data set contains anomalies. AFK-MC^2 [4] with $\alpha = 1$ finds the best initialization most of the time. While k-means++ with $\alpha = 1.5$ does not quite reach same the quality, it performs generally better than uniform random. With $\alpha = 1.5$, both initialization methods are worse and more often than not are below the quality of the random uniform initialization. The run time behavior is similar on all data sets. The uniform initialization is nearly instantaneous, while the $k$means++ and AFK-MC^2 initialization generally stay below the time needed for one iteration. They only have a small impact on the overall run time. Usually, $\alpha = 1$ seems to work better than $\alpha = 1.5$, where the first is the standard cosine similarity, while the latter was used in the proofs to obtain a metric.

Table 3. Run times of all k-means variants in milliseconds.

Data set	Algorithm	k = 2	k = 10	k = 20	k = 50	k = 100	k = 200
Simpsons Wiki	Standard	166	457	845	1,646	3,015	10,047
	Elkan	161	352	532	1,198	2,657	8,247
	Simp.Elkan	**145**	**312**	**436**	**800**	**1,230**	**3,100**
	Hamerly	171	434	732	1,860	3,976	14,386
	Simp.Hamerly	166	421	657	1,450	2,471	9,858
DBLP Author-Conf.	Standard	32,228	29,865	24,687	42,229	50,851	80,553
	Elkan	5,675	9,650	12,366	39,652	54,901	82,732
	Simp.Elkan	5,732	10,841	15,514	44,991	66,731	105,905
	Hamerly	**4,220**	7,072	9,834	19,988	**30,846**	55,687
	Simp.Hamerly	4,285	**7,002**	**9,810**	**19,690**	31,589	**55,250**
DBLP Author-Venue	Standard	33,359	46,328	50,596	70,772	80,218	199,230
	Elkan	5,730	14,593	22,733	59,725	84,011	165,756
	Simp.Elkan	5,986	16,822	27,200	68,577	103,678	209,835
	Hamerly	4,321	11,410	18,056	33,881	**51,242**	125,066
	Simp.Hamerly	**4,188**	**11,096**	**17,799**	**33,017**	52,593	**123,931**
DBLP Conf.-Author	Standard	1,149	6,017	9,672	20,908	33,973	61,680
	Elkan	943	5,549	11,907	41,078	108,028	32,103
	Simp.Elkan	**894**	**4,018**	**6,184**	**10,998**	**16,435**	**29,093**
	Hamerly	944	6,840	14,760	50,282	125,513	347,668
	Simp.Hamerly	944	5,347	9,158	20,115	32,640	55,421
20 Newsgroups	Standard	**101**	**234**	1,223	6,755	16,394	38,131
	Elkan	118	269	498	6,683	19,917	83,407
	Simp.Elkan	118	251	**342**	**1,876**	**3,915**	**7,891**
	Hamerly	111	272	536	9,542	28,005	109,204
	Simp.Hamerly	121	266	443	5,298	12,653	29,915
RCV-1	Standard	24,569	153,170	224,939	917,894	2,669,733	6,064,203
	Elkan	7,639	38,199	47,963	**115,275**	**260,924**	547,110
	Simp.Elkan	8,825	**41,162**	**50,161**	123,428	263,728	**474,800**
	Hamerly	**5,424**	49,041	80,793	325,433	1,132,352	3,181,667
	Simp.Hamerly	5,498	47,977	81,593	320,677	1,144,947	3,266,234

At last, we discuss the achieved improvements in run time for the accelerated spherical k-means algorithms. As with the other experiments, each one was repeated 10 times with various random seeds. Table 3 shows that for most data sets the simplified Elkan algorithm is the fastest, but there are several interesting observations to be made. On the Author-Conference data set, which has the most rows of all data sets but also the lowest number of columns, the normal Elkan and both Hamerly variants are faster. Interestingly, this changes when we increase the number of columns in relation to the number of rows by transposing the data (before applying TF-IDF), shown in Fig. 2. Here, the normal Elkan and Hamerly variants increase drastically in their run time when k increases. This effect originates in the increasing cost of calculating the distances between

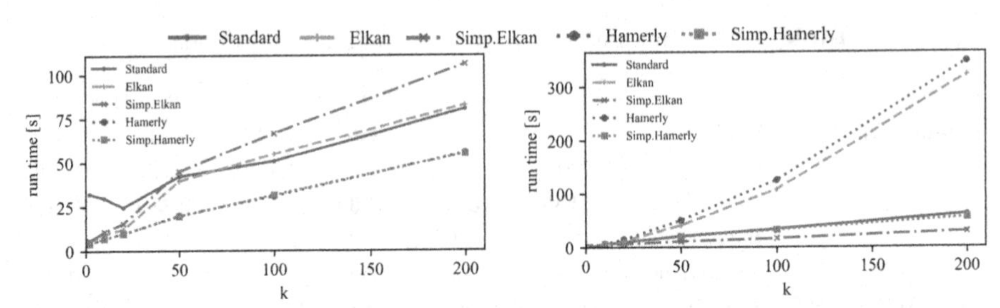

(a) Run time of the different algorithms on Authors-Conf. (higher N, lower d).

(b) Run time of the different algorithms on Conf.-Authors (lower N, higher d).

Fig. 2. Run times of the different algorithms on the DBLP author-conference data set, and its transpose, with very different characteristics.

cluster centers for the additional pruning step. By transposing the data (to cluster conferences, not authors), we increased the dimensionality by 350×, while at the same time reducing the number of instances by the same factor. Computing the pairwise cluster distances now became a substantial effort. This shows that there is no "one size fits all", but the best k-means variant needs to be chosen depending on data characteristics such as dimensionality and the number of instances. While Simplified Hamerly is among the best methods in both situations, it barely outperforms the standard algorithm on the latter data set. Supposedly because of the very high dimensionality, its pruning power is rather limited. While the spherical Hamerly and Elkan implementations can be faster than the standard spherical k-means algorithms, this depends on the data, and with an unfavorable data set they can be much worse. The simplified version of spherical Hamerly seems to be a reasonable default choice, but for small k, it may often be outperformed by the Elkan variants. On the well-known RCV-1 data set, speedups of over 10× are achievable for $k \geq 100$. It may be a bit disappointing that there is no "winner" solution, but data sets simply may have very different characteristics. Possibly some simple heuristics can be identified to automatically choose an appropriate alternative based on empirical thresholds (which need to be determined for a particular implementation and hence are outside the scope of a scientific paper) on the data dimensionality and data set size. In many cases, the limiting factor may be the memory usage and bandwidth for the Elkan variants. Consider the DBLP authors-conference data set with $k = 100$, the bounds used by Elkan with double precision require 2 GB of RAM for the bounds alone, and have to be read and written each iteration. The Hamerly variants only add an overhead of 44 MB. The Yin-Yang variant which we did not yet implement allows choosing the number of bounds to use, and hence make better use of the available RAM.

7 Conclusions

In this article, we use the triangle inequality for cosine similarity of Schubert [23], to accelerate spherical k-means clustering by avoiding unnecessary similarity computations. We were able to adapt the well-known algorithms of Elkan and Hamerly (along with some simplified variants) to work with similarities rather than distances throughout the algorithm. This is desirable because the similarities are more efficient to compute, and the trigonometric bounds are tighter than the Euclidean bounds [23] (with the first corresponding to the arc length, the latter to the chord length). Both require the computation of a square root and hence require similar effort.

We integrated the new triangle inequality into Elkan's and Hamerly's algorithm as two prominent and popular choices, but acknowledge there exist further improved algorithms such as the Yin-Yang, Exponion, and Shallot algorithms that deserve attention in future work. The purpose of this paper is to demonstrate that we can perform pruning directly on the cosine similarities now and that it can speed up the algorithm run times considerably (we observed speedups of over 10× for the well-known RCV-1 data set).

For further speedups, the new technique can also be combined with improved initialization methods from literature. There exists a synergy between some initialization methods that we are not yet exploiting in our implementation, where, e.g., the k-means++ initialization can pre-initialize the bounds used here, and will then allow pruning computations already in the first iteration of the main algorithm.

We hope that this article spurs new research on further accelerating spherical k-means clustering using the triangle inequality, similar to Euclidean k-means.

Acknowledgments. A simpler approach of adapting Hamerly's and Elkan's algorithms for spherical k-means clustering still using Euclidean distances and not the cosine triangle inequalities was explored by our student, Alexander Voß, in his bachelor thesis.

References

1. Alamoudi, S., Hong, X., Wei, H.: Plant leaf recognition using texture features and semi-supervised spherical k-means clustering. In: IJCNN, pp. 1–8 (2020). https://doi.org/10.1109/IJCNN48605.2020.9207386
2. Arfiani, R.Z., Pandelaki, J., Siahaan, A.: Kernel spherical k-means and support vector machine for acute sinusitis classification. IOP Conf. Ser. Mater. Sci. Eng. **546**, 052011 (2019). https://doi.org/10.1088/1757-899x/546/5/052011
3. Arthur, D., Vassilvitskii, S.: K-means++: the advantages of careful seeding. In: ACM-SIAM Symposium on Discrete Algorithms, SODA (2007)
4. Bachem, O., Lucic, M., Hassani, S.H., Krause, A.: Fast and provably good seedings for k-means. In: Neural Information Processing Systems (2016)
5. Bachem, O., Lucic, M., Krause, A.: Distributed and provably good seedings for k-means in constant rounds. In: International Conference on Machine Learning (2017)

6. Bahmani, B., Moseley, B., Vattani, A., Kumar, R., Vassilvitskii, S.: Scalable k-means++. Proc. VLDB Endow. **5**(7), 622–633 (2012). https://doi.org/10.14778/2180912.2180915
7. Borgelt, C.: Even faster exact k-means clustering. In: Int. Symp. Intelligent Data Analysis, IDA, pp. 93–105 (2020). https://doi.org/10.1007/978-3-030-44584-3_8
8. Dhillon, I.S., Modha, D.S.: Concept decompositions for large sparse text data using clustering. Mach. Learn. **42**(1/2), 143–175 (2001). https://doi.org/10.1023/A:1007612920971
9. Ding, Y., Zhao, Y., Shen, X., Musuvathi, M., Mytkowicz, T.: Yinyang k-means: A drop-in replacement of the classic k-means with consistent speedup. In: International Conference on Machine Learning (2015)
10. Elkan, C.: Using the triangle inequality to accelerate k-means. In: International Conference on Machine Learning (2003)
11. Endo, Y., Miyamoto, S.: Spherical k-Means++ clustering. In: Torra, V., Narukawa, Y. (eds.) MDAI 2015. LNCS (LNAI), vol. 9321, pp. 103–114. Springer, Cham (2015). https://doi.org/10.1007/978-3-319-23240-9_9
12. Hamerly, G.: Making k-means even faster. In: SIAM Int. Conf. Data Mining, pp. 130–140 (2010). https://doi.org/10.1137/1.9781611972801.12
13. Hamerly, G., Drake, J.: Accelerating Lloyd's algorithm for k-means clustering. In: Celebi, M.E. (ed.) Partitional Clustering Algorithms, pp. 41–78. Springer, Cham (2015). https://doi.org/10.1007/978-3-319-09259-1_2
14. Hill, M.O., Harrower, C.A., Preston, C.D.: Spherical k-means clustering is good for interpreting multivariate species occurrence data. Methods Ecol. Evol. **4**(6), 542–551 (2013). https://doi.org/10.1111/2041-210X.12038
15. Kriegel, H.-P., Schubert, E., Zimek, A.: The (black) art of runtime evaluation: are we comparing algorithms or implementations? Knowl. Inf. Syst. **52**(2), 341–378 (2016). https://doi.org/10.1007/s10115-016-1004-2
16. Lang, A., Schubert, E.: BETULA: numerically stable CF-trees for BIRCH clustering. In: International Conference on Similarity Search and Applications, SISAP, pp. 281–296 (2020). https://doi.org/10.1007/978-3-030-60936-8_22
17. Lewis, D.D., Yang, Y., Rose, T.G., Li, F.: RCV1: a new benchmark collection for text categorization research. J. Mach. Learn. Res. **5**, 361–397 (2004)
18. Ley, M.: The DBLP computer science bibliography: evolution, research issues, perspectives. In: Laender, A.H.F., Oliveira, A.L. (eds.) SPIRE 2002. LNCS, vol. 2476, pp. 1–10. Springer, Heidelberg (2002). https://doi.org/10.1007/3-540-45735-6_1
19. Li, M., Xu, D., Zhang, D., Zou, J.: The seeding algorithms for spherical k-means clustering. J. Glob. Optim. **76**(4), 695–708 (2020). https://doi.org/10.1007/s10898-019-00779-w
20. Moriya, T., et al.: Unsupervised pathology image segmentation using representation learning with spherical k-means. In: Medical Imaging 2018: Digital Pathology. vol. 10581, p. 1058111 (2018). https://doi.org/10.1117/12.2292172
21. Newling, J., Fleuret, F.: Fast k-means with accurate bounds. In: International Conference on Machine Learning (2016)
22. Pratap, R., Deshmukh, A.A., Nair, P., Dutt, T.: A faster sampling algorithm for spherical k-means. In: Asian Conference on Machine Learning (2018)
23. Schubert, E.: A triangle inequality for cosine similarity. In: International Conference on Similarity Search and Applications, SISAP (2021). https://doi.org/10.1007/978-3-030-89657-7_3
24. Schubert, E., Gertz, M.: Numerically stable parallel computation of (co-)variance. In: International Conference on Scientific and Statistical Database Management, pp. 10:1–10:12 (2018). https://doi.org/10.1145/3221269.3223036

25. Schubert, E., Zimek, A.: ELKI: a large open-source library for data analysis - ELKI release 0.7.5 "Heidelberg". CoRR abs/1902.03616 (2019). http://arxiv.org/abs/1902.03616
26. Yu, Q., Chen, K., Chen, J.: Using a set of triangle inequalities to accelerate k-means clustering. In: International Conference on Similarity Search and Applications, SISAP, pp. 297–311 (2020). https://doi.org/10.1007/978-3-030-60936-8_23

MESS: Manifold Embedding Motivated Super Sampling

Erik Thordsen(✉) and Erich Schubert

TU Dortmund University, Dortmund, Germany
{erik.thordsen,erich.schubert}@tu-dortmund.de

Abstract. Many approaches in the field of machine learning and data analysis rely on the assumption that the observed data lies on lower-dimensional manifolds. This assumption has been verified empirically for many real data sets. To make use of this manifold assumption one generally requires the manifold to be locally sampled to a certain density such that features of the manifold can be observed. However, for increasing intrinsic dimensionality of a data set the required data density introduces the need for very large data sets, resulting in one of the many faces of the curse of dimensionality. To combat the increased requirement for local data density we propose a framework to generate virtual data points faithful to an approximate embedding function underlying the manifold observable in the data.

1 Introduction

It is generally assumed that data is not entirely random but rather obeys some internal laws that can be considered a generative mechanism. Whenever each sample is a numerical vector, one key observation is almost self-evident: Correlating dimensions hint at possible causalities and are thereby useful for understanding the underlying generative mechanism. In the most basic approach, one can apply principal component analysis to find the strongest correlations throughout the entire dataset and use, e.g., the five largest components as a simplifying model. Semantically this corresponds to a generative mechanism that takes a five-dimensional vector and embeds it alongside the principal components in whatever many features are observed. Therein already lie the limitations of this idea: We can only represent linear embedding functions in this way.

Considering the case of a sine wave on a plot in two dimensions, the observed data is using only one parameter. It would hence be reasonable to use a model that only uses a single parameter as well. As a simple relaxation, we can assume the data to lie on any manifold. This generalizes the concept of our generative mechanism to any *locally* linear function. However, this introduces a novel problem: If we wish to use the methods developed for linear functions on locally linear functions, we introduce an error that grows with the observed locality. In practice, that means if we analyze the manifold surrounding a single point, we can not take as many of the surrounding points as possible but must restrict ourselves to all points within some maximum radius. In the field of intrinsic dimensionality (ID) estimation that focuses on estimating the number of parameters

N. Reyes et al. (Eds.): SISAP 2021, LNCS 13058, pp. 232–246, 2021.
https://doi.org/10.1007/978-3-030-89657-7_18

of one such embedding function, it is therefore common to use the k-nearest-neighbors for estimating the ID [1,2,5,12,17]. Yet, ID estimators require a sufficient amount of points in a neighborhood to be stable, which contradicts the requirement to observe as small a neighborhood as possible. Ideally, we would like to use infinitesimal small neighborhoods containing infinitely many neighbors as in this case ID estimation can be investigated purely analytically [9–11]. Yet, as this is practically impossible, ID estimation is inherently approximate.

In this paper, we introduce a novel framework to generate additional data points from the existing ones that are meant to lie on, or very close to, the manifold of the generating mechanism. We intend to increase the local point density on the manifold, thus decreasing the radius required for stable ID estimation. Further applications, for example in machine learning, which could benefit from additional training data faithful to the generative mechanism, are outside the focus of this paper, yet certainly of interest for further research.

The remainder of this paper is structured as follows: Sect. 2 surveys related work and the ID estimation approaches used in this paper. The novel supersampling framework with its underlying theory is described in Sect. 3. The impact of the supersampling on ID estimators under varying parameterization is highlighted in Sect. 4. The closing Sect. 5 gives an overview of the current state of the method and provides an outlook on future work.

2 Related Work

In the past, the concept of intrinsic dimensionality has been analyzed within two major categories. In the first category lie the geometrically motivated ID estimators like the PCA estimator, its local variant the lPCA estimator, the FCI estimator [6], or the recent ABID estimator [17], which uses the pairwise angles of neighbor points. These estimators use measures like the local covariance of (normalized) points to estimate the spectrum of the neighborhoods, from which they deduce an estimate of the ID that mostly ignores weaker components as observational noise. The second category of analytically motivated ID estimators contains, e.g., the Hill estimator [8], the Generalized Expansion Dimension estimator [12], or more recent variants like the ALID [5] and TLE estimator [2]. These are based on the idea that if the data lies uniformly in the parameter space, i.e. the preimage of the data under the embedding function, then distances in the data should increase exponentially in the ID at any point on the manifold. These estimators can be considered the discrete continuation of analytical approaches like the Hausdorff dimension. However, they are vulnerable to varying sampling densities. As this work focuses on the improvement of estimates by supersampling the data and not on the comparative performance between estimators, we will use only a selection of estimators as prototypes for their categories. The supersampling framework itself rests somewhere between these two categories: It is generally motivated geometrically but the supersampled data can be understood as uncertainty added to the data set whose influence decreases over distance. The framework can thus introduce information orthogonal to the ID estimation approaches and potentially enrich estimators from both categories.

The concept of enriching data sets by introducing new points is by itself not a new idea. Domain-specific supersampling techniques are very common for example in image-based machine learning, where images are created by adding noise, scratches, color or brightness shifts, flipping or rotating the image, and many more. The first few can be thought of as adding noise to (some of) the features while the latter correspond to geometric operations like rotation or translation. These operations, although possibly faithful to the manifold of the generative mechanism (e.g., amount of noise in the imaging hardware), are not necessarily faithful to the manifold presented in the analyzed data set.

There are, however, approaches that in a very general manner attempt to mimic the data with additional points. One of these is the highly-cited SMOTE approach introduced by Chawla et al. [4] which introduces new points by adding a series of linear interpolations of points in the feature space. On locally non-convex manifolds this approach creates points that do not lie on the manifold. For machine learning tasks, this deviation is likely tolerable as, e.g., barely overlapping classes are still separated. In terms of ID estimation, however, even small systematic deviations from the manifold can introduce new orthogonal components resulting in overestimation. Later variants of the SMOTE approach have tried to work around this issue but focus on the machine learning objective of not mixing classes.

The objectives of supersampling (or oversampling in the words of SMOTE) for machine learning and ID estimation are quite similar yet inherently different. For machine learning, a more redundant sampling of the manifold is useful for parameter optimization yet possibly diminishes the model robustness compared to data scattered around the manifold. In ID estimation, we need an exact representation of the manifold without additional orthogonal components. One can therefore consider the constraint set of supersampling for ID estimation as stronger than that for machine learning. Solutions for ID estimation are applicable for machine learning but not vice versa.

The framework by Bellinger et al. [3] is similar to the framework introduced here, as it also attempts to take additional samples from the manifold itself. They, however, base their method on manifold learning approaches, which to give decent results, already require quite dense data on the manifold and additionally a parameterization of the intrinsic dimensionality. Starting with a method that requires ID estimates to improve ID estimates is a circular argument that makes the framework of Bellinger et al. inapplicable for ID estimation. This problem also affects manifold-based variants of SMOTE, e.g., the one by Wang et al. [18] which involves locally linear embeddings.

At last, multiple techniques in computer graphics have been studied to improve, e.g., the resolution of point cloud scans of objects. These approaches, however, mostly focus on improving 2D surfaces in 3D space and do not generalize to arbitrary dimensions of both parameter and feature space, partly as information like surface normals used in these methods do not generalize either.

3 Manifold Faithful Supersampling

As mentioned in Sect. 2, the framework for supersampling introduced in this paper is based on geometric observations and aims to create samples from the manifold that is created by the generative mechanism via an embedding function from some parameter space. It hence shares the manifold assumption implied by many downstream applications like autoencoders and ID estimators like ABID [17] or ALID [5]. Whilst Chawla et al. [4] and Bellinger et al. [3] named their methods over-sampling, it will here be called supersampling in line with the computer graphics semantics: We increase the amount of data to reduce undesired artifacts and improve the "bigger picture" of the original data set.

The remainder of this section will consist of the geometric observations underlying the framework, an explanation of the generation pipeline, followed by different options for the modules that make up the pipeline.

3.1 Approaching the Embedding Function

We assume that the data lies on some manifold. To be more precise, we assume that our data X obeys some locally linear[1] embedding function E such that for every $x \in X$ we have some preimage $\tilde{x}$ such that $E(\tilde{x}) = x$. Yet data is often noisy, which is why this will likely not hold for all $x \in X$, and for the rest, we probably only have $E(\tilde{x}) \approx x$. While we can tolerate that the embedding function is only approximate, we will have to accept that we can not know which $x \in X$ are noise points and are unrelated to the embedding function. It is, therefore, reasonable to assume that this assumption holds for all x, as we cannot correct it for the other points either way, except for pre-filtering X with some outlier detection method. In the remainder of this paper, the tilde over a data point or data set will represent the preimage as per the assumed embedding function. We denote the ground truth ID as δ and the observed feature dimension as d.

Assuming $E : \mathbb{R}^\delta \to \mathbb{R}^d$ to be locally linear, for sufficiently small ε we obtain

$$\mathrm{Cov}[E(B_\varepsilon(\tilde{x}))] \approx \frac{\varepsilon^2}{\delta + 2} \nabla E(\tilde{x}) \nabla E(\tilde{x})^T \tag{1}$$

where $\mathrm{Cov}[E(B_\varepsilon(\tilde{x}))] \in \mathbb{R}^{d \times d}$ is the covariance matrix of all points from an ε-ball around $\tilde{x}$ embedded with the embedding function E, $\nabla E(\tilde{x}) \in \mathbb{R}^{d \times \delta}$ is the Jacobian of E evaluated at $\tilde{x}$. This formula can be easily derived from the propagation of uncertainty [16] taking the Jacobian as the linear approximation of E. The covariance matrix of the δ-ball with radius ε can therein be substituted by $\frac{\varepsilon^2}{\delta+2}$ as it is a diagonal matrix with $\frac{\varepsilon^2}{\delta+2}$ on the diagonal [14].

Naturally, we do not have such a high sampling density on real data sets that we can consider arbitrarily small neighborhoods. For larger neighborhoods, the

[1] The aspect of local linearity is disputable, as it does not allow for singularities and is weak on functions with high curvature. However, the typically low sampling density of real data sets makes geometric methods ineffective on non-locally-linear functions.

manifold curvature and sampling density of the data set can affect the covariance matrix and act on its eigenvalues. When projecting points onto one of the components, strong curvature contracts the neighborhoods, reducing the scale along this component. Dependent on the curvature, the effect is independent for each component giving the simpler approximation

$$\mathrm{Cov}[N_k(x)] \approx \nabla E(\tilde{x}) C(\tilde{x}) \nabla E(\tilde{x})^T \tag{2}$$

where $\mathrm{Cov}[N_k(x)]$ is the covariance matrix of the k-nearest-neighbors of x in X, and $C(\tilde{x})$ is a diagonal matrix with scaling factors for each component gradient. As we do not know the curvature, we can not give an estimate for C. Solving for $\nabla E(\tilde{x})$ is, therefore, futile, as it is underdetermined from this approximation. Yet, the rough approximation of aspects of E via covariance matrices gives a glimpse at the manifold structure underlying the data set. In the remainder, we will omit $C^{\frac{1}{2}}$ in benefit of readability whenever using ∇E as they are inseparable when analyzing how the embedding function acts on points in the neighborhood.

To stretch data according to some covariance matrix Σ, it is well known that the lower diagonal matrix L of the Cholesky decomposition ($LL^T{=}\Sigma$) can be used. Starting with a set of points x_i sampled from $\mathcal{N}(0,1)$ we can, hence, obtain points $x'_i = Lx_i$ which are distributed according to $\mathcal{N}(0,\Sigma)$. That is, we can use the Cholesky decomposition to mimic the local linear approximation of the embedding function ∇E. Using that approximation of ∇E we can then compute locally linear embeddings of arbitrary distributions in parameter space.

To compute the covariance matrices of the k-nearest-neighbors $N_k(x)$ of some point x, we use the biased formula $\frac{1}{k}N^TN$ where $N{\in}\mathbb{R}^{k\times d}$ contains the k-nearest-neighbors subtracted by x as row vectors. This virtually puts x in the center of the distribution and is less susceptible to introducing additional orthogonal components due to manifold curvature. Using the query point x as the center of the distribution is common practice in ID estimation [2,5,17]. In addition, we add a small constant to the diagonal of all covariance matrices whenever computing any Cholesky or eigendecomposition to avoid numerical errors.

3.2 Supersampling Pipeline

As explained in the previous section, we can use the covariance matrices of neighborhoods to describe local linear approximations of the embedding function. From these local approximations, we can then sample some points that are already close to or even on the manifold. However, contrary to Chawla et al. [4] and Bellinger et al. [3] we will afterward modify each of the samples to better constrain them to the manifold. If we were to use the generated points without further processing, they would frequently lie outside the manifold whenever it has non-zero curvature, or when our observations are noisy. This can be seen in Fig. 1 as the surface of the corrected supersampling is much smoother than the raw supersampling with less spread orthogonal to the swiss roll. To move the supersampled points onto the manifold, we use weighted means of candidate points that are more aligned with the covariances of the nearby original data

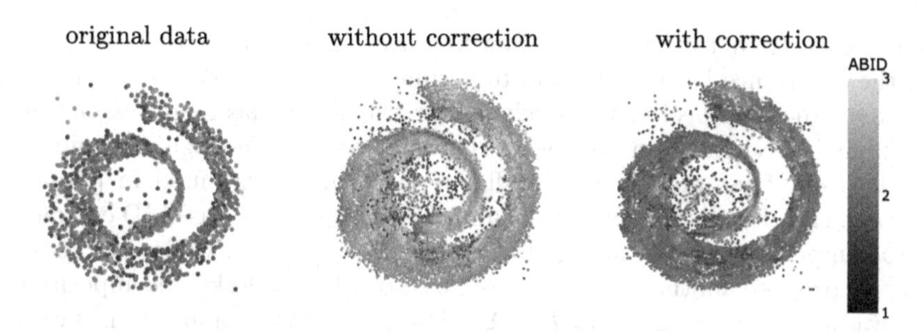

Fig. 1. Supersampling a swiss roll with normal noise in feature space ($\sigma = 0.15$) and added uniform noise points. The left image shows the original data, the middle image is supersampled 50× using covariances and the right is the supersampled data "corrected" onto the manifold. All plots are colored by ABID estimates with 10 neighbors for the original data and 500 for the others.

points. For ID estimation we then chose the k-nearest-neighbors from the supersampled points rather than the original data set. The k used for the ID estimates can then be larger by a factor up to the number of samples per original point without increasing the neighborhood radius. We, hence, start with some data set X, supersample it by a factor of `ext` to X_{ext}, move the supersamples onto the manifold obtaining X_{corr} and then compute ID estimates for each $x \in X$ using the $(\texttt{ext} \cdot k)$-nn of x in X_{corr}.

3.3 Supersampling Modules

The approach in this general form allows for many variants of each step. We can, for example, generate initial samples from a multivariate normal distribution around the original samples using the locally estimated covariance matrix. The resulting points would certainly lie close to, or on, the local linear approximation of the manifold. To structure this section, we will first describe the proposed generation modules, followed by the correction modules.

Sample Generation: The generated samples can have two kinds of "closeness" to the manifold: Either close to the original points in feature space, or close in the hypothesized parameter space.

The first type of generation rules generates samples close to the original points in feature space. Samples can be drawn from normal distributions or d-balls around our original points. For the radius of the d-ball or the standard deviation of the normal distribution, we can, e.g., use distances to the k-th nearest neighbor. These approaches, however, generally introduce additional orthogonal components as they do not obey the shape of the manifold and in initial experiments performed quite poorly. We, therefore, did not further analyze these generation rules.

The second type of generation rules mimics the local linear approximation of the manifold, thereby generating points close in the hypothesized parameter space rather than feature space. The first rule of this type that we propose uses the covariance-based multivariate normal distribution. Points generated by this rule can be expected to closely follow the local linear approximation of the manifold, yet have an increasing density towards the original points. The resulting set of samples, hence, is all but uniformly distributed in parameter space and can thereby potentially encumber expansion rate-based ID estimators.

To compensate for the non-uniform density of multivariate normal distributions, we propose another rule which is based on the Cholesky decomposition of the covariance matrix (L with $LL^T{=}\Sigma$). The Cholesky decomposition is easy to compute and gives a linear map that maps a unit d-ball with a Mahalanobis distance of ≤ 1 with respect to Σ. Using this map, we can then transform uniform samples from a unit d-ball into samples within one standard deviation according to Σ. Assuming the Cholesky decomposition to be a linear approximation of the embedding function, this produces points uniformly at random in parameter space whenever $\delta \approx d$. Yet ID estimation is generally performed on data sets where the ID is much lower than the number of features ($\delta \ll d$), where the points would again be very concentrated around the original points.

The problem therein is simple: To generate points at the exact expansion rate of the parameter space, one needs to know the dimension of the parameter space. We can, in general, not know the true ID, yet an initial estimate could be close enough for generated data to be approximately uniform in parameter space. But we cannot simply scale the lengths of samples uniformly from the unit d-ball by exponentiating with d/δ as the lengths are not equally distributed along the $\delta \ll d$ components of ∇E. To scale along the components of ∇E according to their lengths, we can use the eigenvalues of the Cholesky decomposition. However, we can also skip the Cholesky decomposition and use the eigendecomposition of the covariance matrix ($V\Lambda V^T{=}\Sigma$), where $V\Lambda^{\frac{1}{2}}$ is assumed to be approximately ∇E (extended to $d \times d$ with zeros). If Eq. (2) holds, $V\Lambda^{\frac{1}{2}}$ and ∇E must be approximately equal up to rotations and reflections, as they decompose the positive semi-definite Σ in the same fashion. These two matrices, therefore, act equally on any radially symmetric distribution like the uniformly sampled d-ball considered here. In doing so, the neighborhood radius-, sampling density- and curvature-based scaling factors from Eq. (2) as well as the component gradient norms are included in the eigenvalues. Starting with samples uniform at random in a d-ball, we can multiply them with $\Lambda^{\frac{1}{2}}$ and project them onto the sphere. The resulting angular distribution is then compliant to that of a multivariate normal distribution using $\Lambda^{\frac{1}{2}}$ as the covariance matrix. By scaling each of these points with $r^{\delta'}$, where r is uniform at random from 0 to 1 and δ' is the initial ID estimate we obtain a distribution that is compliant to the expansion of a δ'-ball. Finally multiplying with $\Lambda^{\frac{1}{2}}V^T$ (or L, though we already have the Eigen- but not the Cholesky decomposition) embeds the points according to ∇E. Scaling with $\Lambda^{\frac{1}{2}}$ before normalization does not eliminate $d - \delta'$ components nor give equal weight to the remaining components, thereby not creating a uniform distribution in δ' components.However, it is less reliant on knowing the exact δ and works for fractal δ' and is hence preferable in this early stage. The resulting samples are approximately uniformly distributed in parameter space whenever $\delta \approx \delta'$.

As the regions in which samples are generated are overlapping, it is not obvious whether the increased effort from multivariate normal distributions to Eigendecomposition-based samples yields a clear improvement of sampling density in parameter space. Besides that, not all ID estimators are inherently sensitive to misleading expansion rates. In contrast, the more we enforce a proper uniform δ-ball sampling in parameter space, the more we increase the average distance from our original points. Improving on the sampling density might thus even result in more points that are further away from the manifold. We, therefore, propose all three generation rules (covariance, Cholesky, δ-ball) and provide a comparison in the evaluation section.

Due to the overlapping sampling regions, using an ID estimate for the sample generation is not the same circular argument mentioned in Sect. 2. Even if our ID estimate is quite different from the true ID, the generated data will still be more or less dense around the original points. When using $(\texttt{ext} \cdot k)$-nearest neighbors for estimation, the average expansion rate will remain similar to that of the original data. We neither drop nor induce additional orthogonal components, which a badly parameterized manifold learning might do. The ID estimates used for sample generation, therefore, have much less influence on the manifold shape and the final ID estimates than in the approach of Bellinger et al. [3].

Sample Correction: Once we have created some samples for each of our original points, we could immediately compute ID estimates. However, as displayed in Fig. 1, the samples might be too noisy to describe the manifold. Aside from the linear approximation diverging too strongly from the manifold in areas of high curvature, the covariance matrices in areas of high curvature contain non-zero orthogonal components to the manifold. Thereby, the approximation of ∇E via the covariance matrix adds additional noise. Original points that are not exactly on the manifold but have additional noise in the observed feature space introduce additional errors in our approximation. It is, therefore, safe to assume, that on curved manifolds, or in the presence of high dimensional observational noise, supersamples frequently lie outside the manifold. To constrain the generated samples back onto the manifold, we propose the following general approach: For each supersample $p \in X_{ext}$ we search for its k-nn $x_1, \ldots, x_k \in X$ in the original data. We then generate candidates c_i and weights w_i dependent on x_i and p for each $1 \le i \le k$ and use the weighted mean over the c_i as a corrected supersample. By combining possible realizations of the maps $x, p \mapsto c$, and $x, p \mapsto w$, we obtain a set of correction rules. As candidate maps, we propose:

$$C_1 : x, p \mapsto \Sigma_x (p - x) \frac{\|p-x\|}{\|\Sigma_x (p-x)\|} + x$$

$$C_2 : x, p \mapsto L_x (p - x) \frac{\|p-x\|}{\|L_x (p-x)\|} + x$$

where Σ_x is the covariance matrix for x and L_x is its Cholesky decomposition. The purpose of these maps is to rotate p around x towards a direction at which the probability density of the estimated multivariate normal is higher. Each candidate is hence a more likely observation at a fixed distance for the k-nn. The map C_1 multiplies each point with the covariance matrix, which equates to scaling the points along the principal components with a factor of the variance in these directions. Components orthogonal to the manifold should have a comparably very small variance, which nearly eliminates all components weaker than those tangential to the manifold.

The second candidate map C_2 gives a similar result. The Mahalanobis distance, which gives the distance in units of "standard deviations in that direction", of the candidates prior to rescaling can be written as

$$\begin{aligned}\sqrt{(L_x(p-x))^T \Sigma_x^{-1} (L_x(p-x))} &= \sqrt{(p-x)^T L_x^T (\Sigma_x^T)^{-1} L_x (p-x)} \\ &= \sqrt{(p-x)^T L_x^T (L_x^T L_x)^{-1} L_x (p-x)} \\ &= \sqrt{(p-x)^T (p-x)} = \|p-x\|.\end{aligned}$$

This candidate map, therefore, maps the Euclidean unit sphere onto the Mahalanobis unit sphere, effectively equalizing the influence of different components for p to their relative strengths in Σ_x.

Where using the Cholesky decomposition neutralizes components orthogonal to the manifold approximated by Σ_x, the covariance matrix actively reinforces the components tangential to the manifold. Both maps move samples onto the manifold. C_1 increases the density along the larger components of Σ_x, whereas C_2 leaves the density about equal at the cost of being less strict.

For the weights, we propose an inverse distance weighted (IDW) scheme:

$$\begin{aligned} W_1 &: x, p \mapsto \|p-x\|^{-1} \\ W_2 &: x, p \mapsto \sqrt{(p-x)^T \Sigma_x^{-1} (p-x)}^{-1} \\ W_3 &: x, p \mapsto \sqrt{(x-p)^T \Sigma_p^{-1} (x-p)}^{-1} \end{aligned}$$

Using IDW means enforces the corrected supersample set to be interpolating for the original data set, that is, if we sampled all possible points, the sample would pass through all of our original points. This property is necessary to use the original points as centers for ID estimation as otherwise, they could be outside of the manifold spun by the supersampling. In our experiments, we observed that W_2 and W_3 gave largely similar results, as the samples were already close to the manifold resulting in locally similar covariance matrices. With W_1 the original points have a very strong pull on the samples, occasionally introducing dents in the manifold when the original data is noisy. With W_2 and W_3, the corrected points give a smooth manifold approximation that is robust against noise on the original data set. If the covariance matrices have a high variance among the x_i and p, W_3 moves the samples into a slightly more compact shape than W_2 as single outlying x_i have less impact on the correction. These effects are displayed in Fig. 2. However, it vanishes for increasing neighborhood sizes.

In addition to using IDW means, we used powers of the weights to simulate the increase of the Mahalanobis sphere (points with equal Mahalanobis distance) surface area. The exponents, as with the δ-balls described for sample generation,

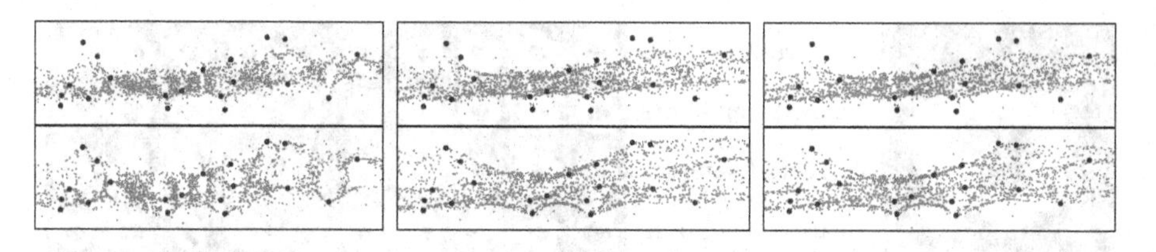

Fig. 2. Corrected supersampling of the dark original points for different weighting rules. From left to right are the weighting rules W_1, W_2, and W_3. The top row is without powers and the bottom row with a constant power of 2.

should ideally be equal to the ID δ. Figure 2 displays the impact of both the different weighting rules and powers. With powered weights, each original point has a much stronger pull on the nearby samples, reducing the tendency to interpolate larger trends. The estimate quality seemed to rather decrease in initial experiments with powered weights even with ground truth ID values, which is why we did not further pursue this.

4 Evaluation

In the experiments, we applied the supersampling framework to the Hill estimator [8] and the ABID estimator [17] as prototypes for expansion- and geometry-based estimators.[2] We chose these two estimators specifically because they can be evaluated very quickly even on very large neighborhoods. This is very important, as upscaling the data set size by a factor of, e.g., 200 also increases the neighborhoods used for ID estimation by up to the same scale. If we consider 20 neighbors for ID estimation and supersampling covariances, the closest 4000 points in the supersampling should describe the same geometry as the initial 20 neighbors. Estimators with super-linear complexity in the size of the neighborhood like the ALID [5] or TLE estimators [2] are slower when using supersampling.

As initial ID estimates, we used the ABID estimator [17] as it is geometrically motivated like the MESS framework and should give a close enough ID estimate even for smaller neighborhoods on which the Hill estimator can give too large estimates especially for outlying points [1,5,13,17]. The number of neighbors to compute the covariance matrices of the original points is called $\texttt{k}_1$, the number of neighbors for the covariances of samples as well as the correction rules is called $\texttt{k}_2$ and the number of samples to use for ID estimation of the original points is called $\texttt{k}_3$. The number of samples generated per original point is `ext`. Unless otherwise stated we used $\texttt{k}_2 := \texttt{k}_1$ and $\texttt{k}_3 := \texttt{k}_1 \cdot \texttt{ext}$.

We made experiments on multiple synthetic data sets like the `m1` through `m13` sets introduced by Rozza et al. [15] in part using generators by Hein et al. [7], which have been used repeatedly to evaluate ID estimate quality and the toy examples from evaluations on the ABID estimator [17]. In addition to that, we

[2] A demo implementation of the MESS framework with these estimators is available at https://github.com/eth42/mess-demo.

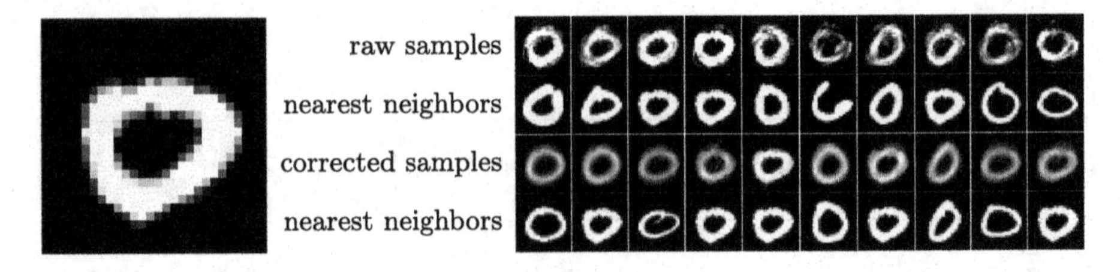

Fig. 3. Supersamples around one of 1000 random MNIST images with $k_1 = 20$. The original image is displayed on the left. The top two rows are raw covariance-based supersamples with their nearest neighbor in the original data set. The bottom two rows are the corrected ($C_1 + W_3$) supersamples with their nearest neighbor in the original data set. Values were cropped to $[0, 1]$ for rendering.

experimented on a few real data sets like point clouds from 3D scans or MNIST which consists of 28×28 grayscale images which we interpret as 784-dimensional vectors. In our experiments, we could observe qualitative improvement due to the proposed correction. On MNIST, the corrected samples are less distorted than the raw samples as can be seen in Fig. 3. The corrected samples tend to be "between" the original image and their nearest neighbor in the original set. Their shape is similar to the original images yet they are not mere linear interpolations which suggests that they lie on the manifold. Qualitative observations about the correction step like those in Fig. 1 and Fig. 3 were made throughout all humanly visualizable data sets. The visualizing approach for quality comparison entails a subjective component, which is arguably undesired. Yet, only arguing ID estimate quality in terms of histograms and summary statistics can be misleading. Figure 4 displays ID estimates without and with supersampling on the `m11` data set. The median of the ABID estimates moves away from the ground truth ($\delta = 2$) while the interquartile range barely changes, which would hint at a lower estimate quality. The 3D plots however show that the ABID estimates are better fitted to the geometry with less local variance. The better summary statistics without supersampling hide the fact that these estimates are less helpful in understanding the data set. The Hill estimates are also improved by using supersampling.

As for the generation variants (covariance, Cholesky, Eigendecomposition), we could observe that on the 1000 point subset of MNIST only the samples generated by the covariance-based approach were sufficiently far away from their original samples to be interesting for small k_1. The Cholesky decomposition-based approach generates samples very close to the original points as expected for $\delta \ll 784$ (δ of MNIST is suspected to be below 30 [1,2,5,17]). As for the Eigendecomposition-based approach, either the initial ID estimate ($\approx$4 as per ABID) was below δ or the distance bound of one standard deviation is too low. Both would lead to an increased supersampling density close to the original points. Using varying initial ID estimates and radius scales, more interesting supersamples can be created for, e.g., an ID of 4 with 5σ radius, or an ID of 12 with 3σ. "Good" radii appear to be both dependent on k_1 and the initial

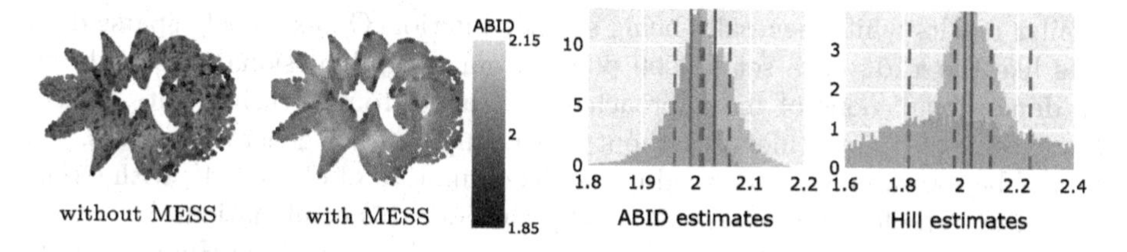

Fig. 4. 3D plots of the `m11` data set colored by ABID estimates without (left) and with (right) supersampling and ID histograms without (orange) and with (blue) supersampling. The solid lines are at the median and the dashed lines at the first and third quartile. Parameters for these plots are $\mathtt{k}_1 = \mathtt{k}_2 = 50, \mathtt{k}_3 = 5000, \mathtt{ext} = 100$ with covariance generation and $C_2 + W_3$ correction. (Color figure online)

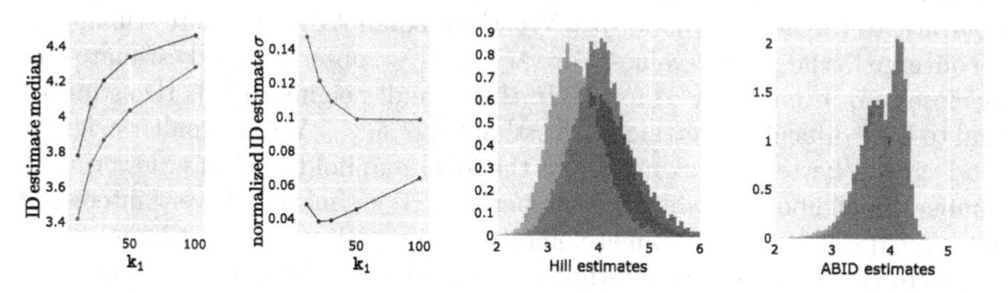

Fig. 5. The plots on the left show the impact of varying $\mathtt{k}_1$ for $\mathtt{ext} = 75$ on ABID (blue) and Hill (green) estimates on the `m4` data set ($d = 8, \delta = 4$). The histograms on the right display the raw estimates (blue) and those obtained with MESS (green) and SMOTE (pink), respectively for the parameters with the lowest local ID standard deviation: $\mathtt{k}_1 = 40$ for Hill and $\mathtt{k}_1 = 20$ for ABID. (Color figure online)

ID estimates, which makes a good choice difficult. The covariance-based sample generation, therefore, appears to be the overall most promising approach and has been used in all following experiments.

The correction schemes, consisting of a correction and a weighting rule, give different qualitative results for different estimators. For the ABID estimator, we observed, that the covariance-based correction C_1 can be too aggressive in constraining points onto a sort of "mean surface", whereby the original points in noisy or strongly curved data sets tend to lie outside the supersampled set. In these cases, the estimates become too low, since the ABID estimator gives lower estimates to points on the margin of manifolds. Using the Cholesky decomposition-based correction C_2 does not introduce this effect as strongly. For the Hill estimator, the correction rules C_1 and C_2 both give comparable results, since the estimator solely analyzes the expansion rate which is unaffected by points lying on the surface of a manifold. As for weighting rules, the Mahalanobis-based rules W_2 and W_3 appear to give the best results whenever the original data set has a high enough sampling density to obtain a good enough approximation of the local gradient via the covariance matrix. The euclidean W_1 rule, which depends less on the original data sampling density, mostly gives similar results whilst generally being slightly inferior. On extremely sparse data sets like the `m10c` data set (10000 points from a 24-dimensional hypercube in 25 dimensions), none of the approaches gave near-ground-truth results, which, however, is also not achieved without supersampling. For general purposes, the overall best results were achieved using a combination of C_2 and W_3 with either estimator. The following paragraphs only consider this combination.

The `ext` factor does not appear to have much of an impact on Hill and ABID estimates for $\texttt{ext} \geq 25$. Larger `ext` values of course yield better results at the cost of additional computation time, as $\texttt{k}_3$ should be increased with `ext`. This observation, however, does not readily generalize to other estimators. For the ALID estimator, the choice of `ext` can have a larger impact even for $\texttt{ext} > 100$, which largely increases computational cost, as ALID has a computational cost quadratic to the neighborhood size. Yet, the median ALID estimate still appears to converge for larger `ext` values. For varying $\texttt{k}_1$ we observed good estimate quality close to or even below δ^2 for $\texttt{k}_1$. In these small neighborhoods ID estimators tend to have difficulties getting good estimates [2,5,17]. Yet, as smaller neighborhoods give a better approximation of the local manifold structure, lowering the number of neighbors on which we can perform ID estimation, is very interesting. In most experiments, the estimate quality for increasing $\texttt{k}_1$ at first improves and afterward deteriorates. ABID estimates for too low a $\texttt{k}_1$ tend to be too low. This can be explained by the neighborhoods being too small to encompass neighbors along each of the orthogonal components of the manifold. For larger $\texttt{k}_1$ the ABID estimates can decrease as the original points tend to lie "on the surface" of the generated manifold in at least one dimension as weaker features of the manifold structure are suppressed. This is even observable without any candidate correction, albeit at much larger $\texttt{k}_1$ values. While that is not necessarily a problem for other applications like classification, it largely affects geometry-based ID estimation. They can also steadily increase as the entire curved manifold might geometrically appear up to full dimensional. The Hill estimator appears to give ever-growing median ID estimates for increased $\texttt{k}_1$ values approaching the ground truth ID with a growing upper tail. For very large $\texttt{k}_1$, the Hill estimator can overestimate the ground truth ID with MESS. The lower end of the Hill estimates, however, appears not to exceed the ground truth ID. The on-average growth of the estimates can likely be attributed to fluctuations in the sampling density. The Hill estimate distributions, hence, are highly skewed and require a visual interpretation. By additionally considering the mean standard deviation of ID estimates of the neighbors divided by the ID estimate of the central point, we can examine the local smoothness of ID estimates. In our experiments, the $\texttt{k}_1$ with the smallest mean ID deviation indicated the best ID estimates for both ABID and Hill estimators, except for `m10c`, where the estimates did not reach 24. The median Hill estimate of about 20 with MESS was at least closer than the median of about 17 without MESS. Figure 5 showcases the impact of $\texttt{k}_1$ choices and also gives a qualitative comparison to supersampling with SMOTE.

In summary, `ext` values of 100 or above are beneficial for ID estimation, although smaller values can suffice for ABID or Hill estimates, and the choice of k_1 largely depends on the data set, where there might be a "sweet spot" that is large enough to encompass the full complexity of the manifold whilst not exceeding local structures. Finding that "sweet spot" can be achieved by analyzing the local deviation of estimates.

5 Conclusions

In this paper, we introduced a supersampling technique motivated by the geometric form of the embedding function defining the generative mechanism of observed manifolds. It contrasts preceding methods like those similar to SMOTE by Chawla et al. [4] by using the manifold structure. Yet, it also opposes the method by Bellinger et al. [3] as it does not require an explicit model of the manifold but uses covariances to mimic the Jacobian of the embedding function. Additional modular correction rules allow compensating for high dimensional noise on the data. In our experiments, we have shown that the novel approach is capable of generating samples that mimic the manifold structure so that it can even improve ID estimates if sufficiently many supersamples are drawn. From that, we conclude, that the MESS framework can sample data from the manifold populated by the given data. It allows to analyze data sets on smaller scales that satisfy the locality assumption of ID estimators. The improved ID estimates in return support the claim, that the manifold has been properly supersampled, for other applications, like classification. Additionally, a scan of different k_1 values can help to find reasonable neighborhood sizes using the mean ID deviation. The major drawback of this technique is the increased time and space requirements due to the increase in the data set and neighborhood sizes. We did not perform experiments in that regard, but we expect the MESS framework to benefit other machine learning tasks, such as classification, which would be a nearby application beyond ID estimation.

References

1. Amsaleg, L., et al.: Estimating local intrinsic dimensionality. In: Knowledge Discovery in Databases, KDD, pp. 29–38 (2015). https://doi.org/10.1145/2783258.2783405
2. Amsaleg, L., Chelly, O., Houle, M.E., Kawarabayashi, K., Radovanovic, M., Treeratanajaru, W.: Intrinsic dimensionality estimation within tight localities. In: SIAM Data Mining, SDM, pp. 181–189 (2019). https://doi.org/10.1137/1.9781611975673.21
3. Bellinger, C., Drummond, C., Japkowicz, N.: Manifold-based synthetic oversampling with manifold conformance estimation. Mach. Learn. **107**(3), 605–637 (2017). https://doi.org/10.1007/s10994-017-5670-4
4. Chawla, N., Bowyer, K., Hall, L., Kegelmeyer, W.P.: SMOTE: synthetic minority over-sampling technique. J. Artif. Intell. Res. **16**, 321–357 (2002). https://doi.org/10.1613/jair.953
5. Chelly, O., Houle, M.E., Kawarabayashi, K.: Enhanced estimation of local intrinsic dimensionality using auxiliary distances. Technical Report NII-2016-007E. National Institute of Informatics (2016)

6. Erba, V., Gherardi, M., Rotondo, P.: Intrinsic dimension estimation for locally undersampled data. Sci. Rep. **9**, 1–9 (2019). https://doi.org/10.1038/s41598-019-53549-9
7. Hein, M., Audibert, J.: Intrinsic dimensionality estimation of submanifolds in $\mathbb{R}^d$. In: International Conference on Machine Learning, ICML, pp. 289–296 (2005). https://doi.org/10.1145/1102351.1102388
8. Hill, B.M.: A simple general approach to inference about the tail of a distribution. Ann. Stat. **3**(5), 1163–1174 (1975). https://doi.org/10.1214/aos/1176343247
9. Houle, M.E.: Local intrinsic dimensionality I: an extreme-value-theoretic foundation for similarity applications. In: Beecks, C., Borutta, F., Kröger, P., Seidl, T. (eds.) Similarity Search and Applications, pp. 64–79. Springer, Cham (2017). https://doi.org/10.1007/978-3-319-68474-1_5
10. Houle, M.E.: Local intrinsic dimensionality II: multivariate analysis and distributional support. In: Beecks, C., Borutta, F., Kröger, P., Seidl, T. (eds.) Similarity Search and Applications, pp. 80–95. Springer, Cham (2017). https://doi.org/10.1007/978-3-319-68474-1_6
11. Houle, M.E.: Local intrinsic dimensionality III: density and similarity. In: Satoh, S., et al. (eds.) SISAP 2020. LNCS, vol. 12440, pp. 248–260. Springer, Cham (2020). https://doi.org/10.1007/978-3-030-60936-8_19
12. Houle, M.E., Kashima, H., Nett, M.: Generalized expansion dimension. In: ICDM Workshops, pp. 587–594 (2012). https://doi.org/10.1109/ICDMW.2012.94
13. Houle, M.E., Schubert, E., Zimek, A.: On the correlation between local intrinsic dimensionality and outlierness. In: Marchand-Maillet, S., Silva, Y.N., Chávez, E. (eds.) SISAP 2018. LNCS, vol. 11223, pp. 177–191. Springer, Cham (2018). https://doi.org/10.1007/978-3-030-02224-2_14
14. Joarder, A., Al-Sabah, W.S., Omar, M.H., Fahd, K.: On the distributions of norms of spherical distributions. J. Probab. Stat. Sci. **6**(1), 115–123 (2008)
15. Rozza, A., Lombardi, G., Ceruti, C., Casiraghi, E., Campadelli, P.: Novel high intrinsic dimensionality estimators. Mach. Learn. **89**(1–2), 37–65 (2012). https://doi.org/10.1007/s10994-012-5294-7
16. Sengupta, D., Jammalamadaka, S.R.: Linear Models: An Integrated Approach. World Scientific, Singapore (2003)
17. Thordsen, E., Schubert, E.: ABID: angle based intrinsic dimensionality. In: Satoh, S., et al. (eds.) SISAP 2020. LNCS, vol. 12440, pp. 218–232. Springer, Cham (2020). https://doi.org/10.1007/978-3-030-60936-8_17
18. Wang, J., Xu, M., Wang, H., Zhang, J.: Classification of imbalanced data by using the SMOTE algorithm and locally linear embedding. In: International Conference on Signal Processing, vol. 3 (2006). https://doi.org/10.1109/ICOSP.2006.345752

Handling Class Imbalance in k-Nearest Neighbor Classification by Balancing Prior Probabilities

Jonatan Møller Nuutinen Gøttcke(✉) and Arthur Zimek(✉)

Institute of Mathematics and Computer Science,
University of Southern Denmark, Odense, Denmark
{goettcke,zimek}@imada.sdu.dk

Abstract. It is well known that recall rather than precision is the performance measure to optimize in imbalanced classification problems, yet most existing methods that adjust for class imbalance do not particularly address the optimization of recall. Here we propose an elegant and straightforward variation of the k-nearest neighbor classifier to balance imbalanced classification problems internally in a probabilistic interpretation and show how this relates to the optimization of the recall. We evaluate this novel method against popular k-nearest neighbor-based class imbalance handling algorithms and compare them to general oversampling and undersampling techniques. We demonstrate that the performance of the proposed method is on par with SMOTE yet our method is much simpler and outperforms several competitors over a large selection of real-world and synthetic datasets and parameter choices while having the same complexity as the regular k-nearest neighbor classifier.

Keywords: Class imbalance · Bayesian learning · k-Nearest neighbor classification

1 Introduction

In classification problems, skewed class distributions often result in poor accuracy when predicting instances of the minority classes. The problem is common and found in substantially different areas such as fraud detection, propaganda detection, and medical diagnosis. The typical class imbalance problem is often presented as a dichotomous classification problem, where it is crucial that the minority class is predicted correctly. A common example would be data about a rare disease, where the available training data contains many instances without the disease, and only few with the disease. The majority class dominates the training and potentially also the evaluation, if done naively. The imbalance ratio, IR, captures the severity of a problem by the ratio of the size of the (largest) majority class (c_{maj}) over the size of the (smallest) minority class (c_{min}):

$$IR = \frac{|c_{maj}|}{|c_{min}|} \tag{1}$$

N. Reyes et al. (Eds.): SISAP 2021, LNCS 13058, pp. 247–261, 2021.
https://doi.org/10.1007/978-3-030-89657-7_19

Of course, the problem can vary in complexity by having multiple minority classes with different degrees of importance for each of the minority classes. The problem also becomes more difficult when the imbalance ratio (IR) is increased. In the literature, a value of $IR > 3.5$ is seen as signalling a high degree of imbalance in a dataset [26].

Besides the absolute value of IR, we can also distinguish *absolute* imbalance and *relative* imbalance. Absolute imbalance describes the case where there is a small absolute amount of minority instances. For example, if there are 5 instances in the minority class and 95 in the majority class, we get an IR of 19. Relative imbalance simply means that the IR is large, e.g., if there are 5000 instances in the minority class and 95000 in the majority. This would also result in an IR of 19. However, a re-sampling strategy should most likely be different for these two cases. Absolute and relative imbalance has been discussed in more detail by Bellinger et al. [3].

When studying the performance of algorithms on imbalanced classification problems, it is a fallacy to just report the accuracy, error rate, precision, or f1 measure. The ROC curve, although inherently accounting for imbalance, comes with its own problems as well [8,13].

The *precision* measure reports, for one class against all other classes, the number of true positives (TP) divided by the false positives (FP) plus the true positives, i.e., $\frac{TP}{FP+TP}$. In class imbalanced problems, precision is not a viable measure since the majority classes will have relatively few false positives no matter how many of the minority points they predict as majority points. Examples for the minority classes on the other hand are rarely mistaken for majority points. The true positives will thus typically be divided by almost the same number, because there are no or few false positives, which leads to a high precision. The harmonic mean between precision and recall (f1) is also reported in some studies [10], but since precision is not a meaningful measure in class imbalanced problems its presence in the f1 measure only hides the algorithms' performance in terms of recall.

Some argue [2,16] that accuracy and error rate are strongly biased to favor the majority class. The problem with accuracy and error rate is obvious when the class imbalances are extreme. If in a binary classification problem 99.9% belongs to the minority class, and only 0.01% belongs to the minority, then if we completely fail to predict the minority class the classifier still has an accuracy of 99.9%. Thus the G-mean score is a popular measure [4,16,18], that is the geometric mean over recall: $(\prod_{i=1}^{n} r_i)^{\frac{1}{n}} = \sqrt[n]{r_1 \cdot r_2 \cdot \ldots \cdot r_n}$, where r_i is the recall for class c_i.

This results in a quality measure that heavily penalizes a low recall for any of the classes, reflecting the algorithm's inability to hypothesize that point x belongs to the minority class. Failing completely on one class results in a zero value for the overall G-mean score. We therefore argue for the sensibility in reporting the macro-averaged recall in addition to the G-mean score. For further discussion on performance assessment of imbalanced classification problems see the work by Japkowicz [16].

Considering the appropriateness of recall when working with imbalanced classification problems, in this paper we introduce a variant of the k-nearest neighbor classifier that balances the prior class probabilities. We show how this effectively resembles using a local recall as a classification rule, while the standard k-nearest neighbor classifier effectively uses the local precision as a decision criterion.

The remainder of the paper is organized as follows. In Sect. 2, we discuss related work. In Sect. 3 we introduce our method and discuss some properties. In Sect. 4 we perform an experimental evaluation of our method against state-of-the-art competitors. In Sect. 5 we summarize and give perspectives for future work.

2 Related Work

There exist many approaches to handling imbalanced classification problems. The approaches can be divided into three main categories, namely external, internal, and cost-sensitive approaches.

2.1 External Approaches

External approaches alter the dataset a classifier is trained on. The alterations are typically different re-sampling techniques such as majority undersampling, minority oversampling, or a combination of both. In oversampling, the minority class domain is extended by adding real points or synthetic points to the existing training data. The simplest such approaches are Random Undersampling (RUS) [15,25] which randomly removes points from the majority class until a uniform class distribution is reached. A similar oversampling method exists, namely random oversampling (ROS) [15], which picks random real points from the minority classes and oversamples these until a uniform class distribution is reached.

SMOTE [6] is likely the most popular oversampling technique. It adds additional data points to the dataset by inserting new synthetic samples within the convex hull of the minority class. The synthetic sample is positioned on the straight line between two minority points, i.e., $a + (b - a) \cdot \alpha$, where a is the feature vector of the point under consideration, b is a feature vector of a random instance of the nearest neighbors and $\alpha \in [0, 1]$ is a random value determining where on the line the point should be positioned. The algorithm iterates through all data points of the minority class and oversamples each point by finding the k nearest neighbors and picking n randomly, where n is the oversample rate. A data point is inserted at a random point on each line between these n sets of two. Chawla and Nitesh demonstrated that SMOTE improved performance over random oversampling and that SMOTE results in reduced decision tree sizes, when used in combination with C4.5 [24]. A plethora of slightly modified SMOTE variations have been developed since the original. Noteworthy mentions are Borderline SMOTE [12] and ADASYN [14].

2.2 Internal Approaches and Modifications of the kNN Classifier

Internal approaches modify an existing classifier to account for the class imbalance. Over the past two decades, several attempts have been proposed to modify specifically the kNN classifier to account for an imbalanced class distribution. Song et al. [27] introduced IkNN, where they employ information about the distance from a query point to its nearest neighbors to determine the most informative nearest neighbors. Kriminger et al. [17] created a class imbalance handling kNN variation which also works on imbalanced data streams. Liu and Chawla [20] introduce a class confidence weighted kNN rule by employing Gaussian mixture models, and Bayesian networks to estimate class weights.

Dubey and Pudi [10] claim to improve performance over these previous internal modifications. They modify the existing kNN algorithm to be sensitive to class imbalance by observing the class distributions within the kNN hyperballs of a subset of the neighbors' neighbors. This information is used to determine if the query point is a local minority given this new sense of locality and its new prior probability distribution.

2.3 Cost-Sensitive Learning and k-Nearest Neighbors

A way of making an arbitrary classifier sensitive to the class imbalance problem without modifying the dataset as done in the external approaches, is by employing *cost-sensitive learning*. In cost-sensitive learning each possible prediction is associated with some misclassification cost. The goal is then to minimize the total misclassification cost over the test dataset. Elkan [11] generalized cost-sensitive learning to the goal of minimizing the conditional risk as:

$$R(x, c_i) = \sum_j \Pr(c_j|x)C(i, j) \tag{2}$$

where x is an example, c_i and c_j class labels, and the $C(i, j)$ entry in the cost matrix is the cost of predicting class c_j when the true class is c_i. Picking the prediction that minimizes the conditional risk leads to decisions that are not necessarily the most probable outcome. Improving the sensitivity towards the minority class in cost sensitive learning can be achieved by increasing the cost of misclassifying minority instances. Domingos [9] proposed a method which can make any classifier cost sensitive, by employing ensemble learning. Qin et al. [23] and Zhang [28] proposed a cost-sensitive kNN classifier based directly on Elkan's formulation of conditional risk. The schema for their *Direct-CS-kNN* classifier is given by

$$\mathcal{L}(x, c_i) = \sum_{c_j \in C} \Pr(c_j|x)C(i, j) \tag{3}$$

The conditional risk is described as a loss function $\mathcal{L}(x, c_i)$, describing the loss of predicting class c_i, given query point x:

$$h(x) = \arg\min_{c_i \in C} \mathcal{L}(x, c_i) \tag{4}$$

If we alter Eq. 3 to sum over the probabilities that it is not class c_j multiplied by a cost of predicting class c_j instead of predicting class c_i, the weights can become more understandable and we obtain some nice properties with respect to the minimization:

$$h(x) = \arg\min_{c_i \in C} \sum_{c_j \in C} 1 - \frac{\Pr(c_j|x)}{\sum_{c \in C} 1 - \Pr(c_c|x)} \cdot C(i,j) \tag{5}$$

The modification ensures several nice properties such as, if the cost matrix is equal to the identity matrix $C = I$, then we get the conventional majority vote kNN decision rule. If we use a diagonal matrix $\mathcal{D}$ with positive weights greater than or equal to 1, then if we pick $w_i = \frac{1}{D(i,i)}$ we obtain the basic weighted kNN decision rule.

2.4 Summary

In summary, although various methods exist to adjust classifiers in general or the kNN classifier in particular to imbalanced classification problems, none of these methods tackles the particular problem of considering the recall as an important objective of imbalanced classification. In the following, we introduce an elegant and straightforward way to do so.

3 Class-Balanced k-Nearest Neighbors Classification

The kNN classifier is an instance-based learning method that classifies an instance x from the input space by applying a decision rule to the set of k-nearest neighbors of x in the training data space. The decision rule is conventionally the majority vote, but could also be a weighted majority vote to handle a difference in the importance of attributes or to give higher weight to closer neighbors. As we have seen above, weights can also be associated with different class labels as an approach to handle the class imbalance problem [10].

3.1 Basic Weighted kNN

The most intuitive approach to handling the class imbalance problem is perhaps to add importance to instances belonging to the minority classes. This can be done with a modification of the decision rule by multiplying the observed number of minority instances with a positive weight greater than 1, or the majority instances with a weight between 0 and 1, or a combination of both. To ensure the *relative importance* of each class is uniform despite accounting for different proportions of the dataset, the weight could be defined as:

$$w_i = \frac{|\{x|x \in c_{maj}\}|}{|\{x|x \in c_i\}|} \tag{6}$$

where c_{maj} is the majority class. This weighting scheme was generally proposed by Japkowicz [15].

3.2 Balancing a Probabilistic k-Nearest Neighbor Classifier

Choosing the class of the majority among the k nearest neighbors is from the point of view of probabilistic learning equivalent to choosing the maximum a posteriori (MAP) hypothesis when estimating the class probabilities for the different classes $c_i \in C$, given the query instance:

$$h_{\text{MAP}}(x) = \arg\max_{c_i \in C} \Pr(c_i|x) \tag{7}$$

where we can find $\Pr(c_i|x)$ by Bayes' rule as:

$$\Pr(c_i|x) = \frac{\Pr(x|c_i) \cdot \Pr(c_i)}{\sum_{j=1}^{m} \Pr(x|c_j) \cdot \Pr(c_j)} \tag{8}$$

From this it is obvious that the prior class probabilities have some influence on the decision rule, and the core idea for our method is to not estimate these prior probabilities from the training sample but to define them as required by fairness. Intuitively, balancing an imbalanced classification problem means in this perspective to require uniform prior class probabilities, i.e., the decision of the classifier (Eq. 7) for m classes should use

$$\Pr(c_i|x) = \frac{\Pr(x|c_i) \cdot \frac{1}{m}}{\sum_{j=1}^{m} \Pr(x|c_j) \cdot \frac{1}{m}} \tag{9}$$

in order to treat all classes fair in a balanced way.

Interestingly, this addresses, locally, the need for optimizing the recall instead of the precision, as the decision rule turns out to be choosing the class that is captured to the largest proportion among the k nearest neighbors:

Theorem 1. *Given some query object x in a classification problem with a set C of m classes, let k_i be the number of instances among the k nearest neighbors of x that belong to class c_i, let n_i be the number of instances that belong to class c_i overall (i.e., $n_i = |c_i|$). For the k nearest neighbor classifier, adjusting the prior class probabilities such that all classes are equally likely, i.e., $\forall_i \Pr(c_i) = \frac{1}{m}$, is equivalent to choosing $\arg\max_{c_i \in C} \left(\frac{k_i}{n_i}\right)$, which is the local recall for x.*

Proof. The proxy for the probability $\Pr(x|c_i)$ is the density estimation given by the k nearest neighbors, conditional on class c_i, that we can describe as

$$\Pr(x|c_i) \propto \frac{k_i}{n_i V(x)} \tag{10}$$

where $V(x)$ is the volume, centered at x, required to capture k nearest neighbors of x. We can therefore rewrite Eq. 8 as follows:

$$\Pr(c_i|x) \propto \frac{\frac{k_i}{n_i V(x)} \cdot \Pr(c_i)}{\sum_{j=1}^{m} \frac{k_j}{n_j V(x)} \cdot \Pr(c_j)} \tag{11}$$

Choosing equal prior class probabilities results in:

$$\Pr(c_i|x) \propto \frac{\frac{k_i}{n_i V(x)} \cdot \frac{1}{m}}{\sum_{j=1}^{m} \frac{k_j}{n_j V(x)} \cdot \frac{1}{m}} \tag{12}$$

which simplifies to

$$\Pr(c_i|x) \propto \frac{\frac{k_i}{n_i}}{\sum_{j=1}^{m} \frac{k_j}{n_j}} \tag{13}$$

where the denominator is obviously identical for all classes. We therefore have

$$\arg\max_{c_i \in C} \Pr(c_i|x) = \arg\max_{c_i \in C} \left(\frac{k_i}{n_i}\right) \tag{14}$$

□

The novel decision rule described in Eq. 9 has therefore a straightforward practical interpretation and is easy to compute. The decision rule determines how large a fraction of the points with a specific class label in the domain is present in a given neighborhood query. Effectively, this decision rule adds a variable, local neighborhood-dependent class weight.

3.3 On the Difference Between Weighted *k*NN and Adjustment of Prior Class Probabilities

In the following we show that the same effect of modifying the prior probabilities in the probabilistic interpretation cannot, in general, be achieved by using any weight in the basic weighted *k*NN approach discussed in Sect. 3.1.

Theorem 2. *A probabilistic kNN classifier with uniform prior probabilities is not equivalent to a class-based weighted kNN classifier.*

Proof. A weighted version of the probabilistic interpretation of *k*NN, taking Eq. 11 as a starting point, can be formulated as:

$$w_i \cdot \Pr(c_i|x) = w_i \cdot \frac{\frac{k_i}{n_i} \cdot \Pr(c_i)}{\sum_{j=1}^{m} \frac{k_j}{n_j} \Pr(c_j)} \tag{15}$$

However, unless the prior probabilities are equal for all classes, there is no possible choice of w_i that also balances all $\Pr(c_j)$ in the denominator. □

Intuitively, our proposed method is effectively employing locally adaptive weights, which is not possible to model with a weight- or cost-based approach.

In a polytomous and heavily imbalanced classification problem where the largest majority class contains 1000 points, one of two minority classes, say c_2, contains 10 points, and the other minority class contains 50 points, a query with

Table 1. Dataset information for the real datasets. Abbreviations: dimensionality (Dim.), imbalance ratio (IR), number of classes (CL) and number of points (n)

Dataset	n	Dim	IR	Cl
appendicitis	106	7	4.05	2
balance	625	4	5.88	3
cleveland	297	13	12.31	5
coil 2000	9822	85	15.76	2
dermatology	358	34	5.55	6
ecoli	336	7	71.5	7
glass	214	9	8.44	6
haberman	306	3	2.78	2
hayes roth	160	4	2.10	3
hepatitis	80	19	5.15	2
marketing	6876	13	2.49	9
new thyroid	215	5	5.00	3

Dataset	n	Dim	IR	Cl
page-blocks	5472	10	175.46	5
phoneme	5404	5	2.41	2
satimage	6435	36	2.45	6
spectfheart	267	44	3.85	2
shuttle	58000	9	4558.60	7
thyroid	7200	21	40.16	3
titanic	2201	3	2.10	2
wine-red	1599	11	68.10	6
wine-white	4898	11	439.60	7
yeast	1484	8	92.60	10
usps	1500	50	4.00	2

2 minority points will be weighted differently dependent on which of the classes are present in the query:

$$\Pr(c_2|x_1) = \frac{\frac{2}{10}}{\frac{2}{10} + \frac{1}{1000} + \frac{0}{50}} = 0.995, \ \Pr(c_2|x_2) = \frac{\frac{2}{10}}{\frac{2}{10} + \frac{0}{1000} + \frac{1}{50}} = 0.909$$

This exemplifies how the weight for some class is dependent on the neighborhood of the query point.

4 Experimental Evaluation

4.1 Datasets

The datasets have been picked from the Keel dataset repository [1], and the USPS dataset is from Chapelle & Schölkopf [5]. The datasets were picked from these repositories, based on their imbalance ratio (IR) which is larger than 2. All datasets have numerical attributes. An overview on the used datasets is given in Table 1.

4.2 Compared Methods

We compare our method "k-Nearest Neighbors with Balanced Prior Probabilities" (*k*NN-BPP) against representatives for the different categories of approaches, as discussed in Sect. 2. An overview is given in Table 2. As competitors, we include the conventional *k*NN classifier as well as more complex *k*NN-based algorithms that have been designed with the class imbalance problem in mind. The algorithms are also evaluated against the most common resampling strategies with a regular *k*NN-classifier. As can be seen in Table 2, several of the algorithm implementations are currently not publicly available.

Table 2. Compared methods

Method	Short name	Impl. Source
k-Nearest Neighbors classifier [7]	kNN	Scikit-Learn [22]
Random undersampling [15,25]	RUS	Imblearn [19]
Synthetic Minority Oversampling Technique [6]	SMOTE	Imblearn [19]
Class Based Weighted k-Nearest Neighbor over Imbalance Dataset [10]	CW-kNN	https://github.com/Goettcke/kNN_BPP
Direct Cost Sensitive kNN Classifier [23,28]	Direct-CS-kNN	
k-Nearest Neighbors with Balanced Prior Probabilities	kNN-BPP	

The original authors were contacted but did not provide the implementations. All implemented algorithms will be made available with this paper. The algorithms were written following the requirements for *Scikit-Learn* implementations and written in Python.

4.3 Parameter Selection

We evaluate the methods using stratified 10-fold cross-validation. The k-value in these experiments varied for each dataset between 3 and 35.

For the *Direct-CS-kNN* classifier, a cost-matrix has to be defined. Since no method to generate such a cost matrix was proposed in the original papers we use the following cost-matrix: We construct an asymmetrical matrix that ensures the cost of predicting class c_j instead of c_i is proportional to the imbalance ratio between these two classes $C(i,j) = \frac{|c_i|}{|c_j|}$. Notice that, if class c_i is larger than c_j then the weight is greater than 1 which can be interpreted as it being costly to make this mistake. However, if c_i is a minority class, then $C(i,j) \in [0,1]$. This can be interpreted as a discount to making this type of mistake.

For *SMOTE* the number of nearest neighbors to include in the oversampling set was set to 6 as per the default in the Imblearn package. For random undersampling (RUS) and SMOTE, re-sampling was done to achieve uniform class distributions.

4.4 Evaluation Measures

We evaluate the methods in terms of recall, taking the geometric mean over the classes (G-mean) and the macro average. We also tested precision, where all the methods show only minor differences, thus these results are not included.

4.5 Results

Ranking Distribution over Datasets and Parameter Values. As a first overview we compare the methods performance as the average ranking for each choice of k over all datasets. The rankings over all tested values of k are shown in

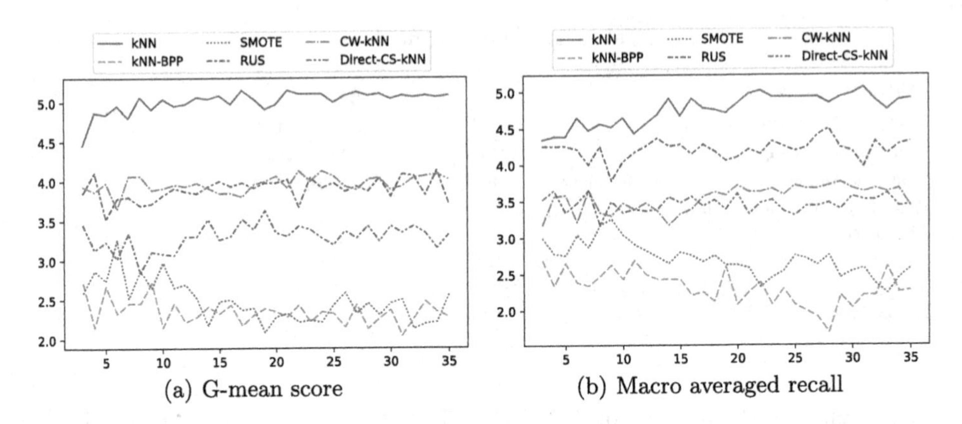

(a) G-mean score

(b) Macro averaged recall

Fig. 1. Performance for $k \in [3, 35]$ in terms of the mean rank over all datasets

Fig. 1. Notice that a lower rank indicates better performance. The plots show how the evaluated algorithms seem to form 3 groups. The first group with the highest rank is the unmodified kNN algorithm. The second group contains the modified kNN algorithms from [10] and [28] as well as random undersampling [15]. In the third and best performing group we have SMOTE [6] and the proposed kNN-BPP method.

Analysis of Statistical Performance Differences. We performed a statistical analysis of the ranking differences for $k = 10, 20, 30$. The results are shown in critical difference plots in Fig. 2. The plots show the average ranking of methods over all datasets together with a bar connecting methods that are not performing differently with statistical significance. We see our method on top or, in one case, second to SMOTE, although their performance is only different with statistical significance from some other methods in most cases. The classic, unchanged kNN classifier is typically worst, and several of the improved versions are not better with statistical significance. kNN-BPP is significantly better than the unchanged kNN classifier in all tests and better than RUS and CW-kNN in several, also in cases where SMOTE is not significantly better. In summary, the critical difference (CD) plots indicate that kNN-BPP performs as well or slightly better than a non-trivial oversampling technique, but without adding runtime to the original k-nearest neighbor classifier.

Distribution of Raw Performance. To complement the picture, we also show the distribution of performance in terms of raw recall (G-mean and macro average) values over the datasets for $k = 10, 20, 30$ in Fig. 3 (here higher values are better). We see that the distributions overlap strongly, but we can identify a tendency of SMOTE and our method to perform better than others.

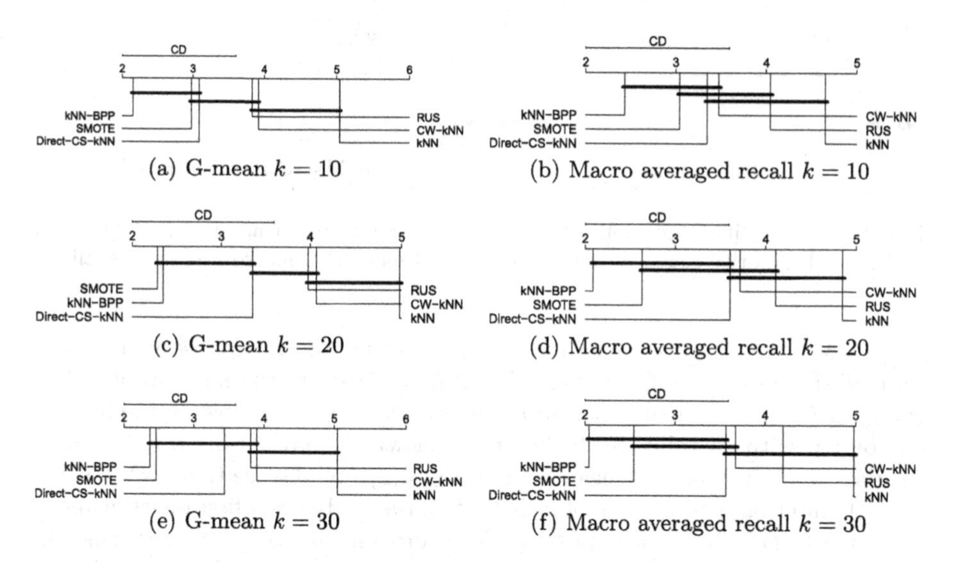

(a) G-mean $k = 10$ (b) Macro averaged recall $k = 10$

(c) G-mean $k = 20$ (d) Macro averaged recall $k = 20$

(e) G-mean $k = 30$ (f) Macro averaged recall $k = 30$

Fig. 2. Statistical assessment of performance differences: critical difference plots

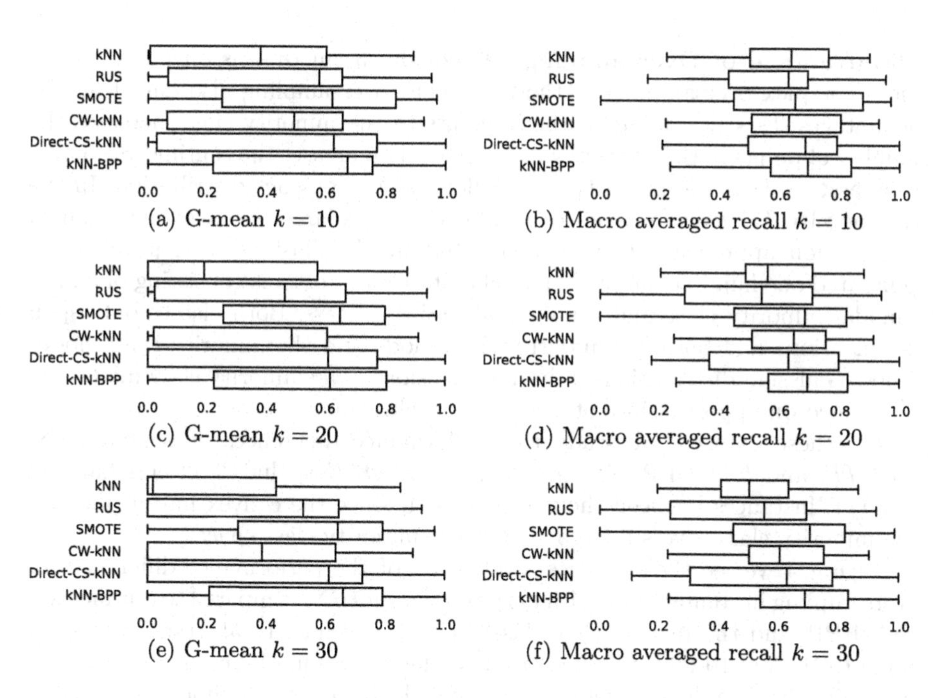

(a) G-mean $k = 10$ (b) Macro averaged recall $k = 10$

(c) G-mean $k = 20$ (d) Macro averaged recall $k = 20$

(e) G-mean $k = 30$ (f) Macro averaged recall $k = 30$

Fig. 3. Distribution of recall (G-mean, macro average) over the datasets

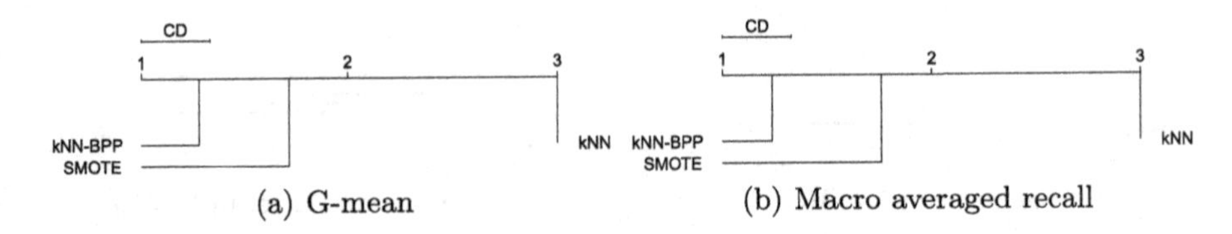

(a) G-mean (b) Macro averaged recall

Fig. 4. Critical difference plots showing the significant disadvantage of SMOTE on multimodal minority class datasets in terms of G-mean and macro averaged recall

In the box plots over macro-averaged recall in Fig. 3, we observe that the result distribution of kNN, CW-kNN and kNN-BPP are the most stable when changing k, but the median performance of kNN-BPP is also overall the highest and on par with SMOTE. In the box plots showing the result distributions over the datasets in terms of G-mean score in Fig. 3(e), we observe that kNN-BPP is the only internal modification of kNN that consistently has a first quartile above 0. Conventional kNN is the most stable algorithm but the worst performer in terms of the distribution over G-mean score. We also observe that SMOTE and kNN-BPP are the best performers and almost as stable as kNN.

Disadvantage of Oversampling. Although our method is on par with the more complex oversampling method SMOTE, oversampling also has clear disadvantages. Firstly, adding synthetic points to the minority class increases the number of points in the dataset, which obviously increases the runtime. Secondly, they assume that the minority class follows some compact distribution. In the case of *SMOTE* it depends on the k value chosen, a larger k makes the assumed distribution approach a Gaussian distribution. To illustrate this problem we generated one hundred simple binary classification problems consisting of a multimodal minority class and a unimodal majority class. Both classes only span 2-dimensions and both the majority class mode and the minority class modes follow Gaussian distributions in both dimensions. The minority class modes are positioned on opposite sides of the majority class mode.

On these datasets we study the performance differences of regular kNN, *SMOTE*, and kNN-BPP. The problem for *SMOTE* is that it inserts harmful SMOTE instances between the two modes [4] since the convex hull defined by the minority class covers a large area of the majority class space.

In Fig. 4 we see the statistical evaluation of the performance differences of oversampling in combination with majority voting kNN, compared to our method kNN-BPP, and the regular kNN classifier as a baseline. In all tests, k was set equal to 10, and 10-fold stratified cross-validation was used. For the re-sampling strategies the default parameters were used which ensures a uniform class distribution. In this simple test, kNN-BPP is significantly better than SMOTE. The reason behind the clear win is that the number of instances in the two modes is approximately the same, which means that oversampling within one mode by inserting a point on the vector between two of the minority points is approximately as likely as inserting a harmful SMOTE instance between the two modes.

This is of course dependent on which points are picked at random so if SMOTE is extremely lucky it can perform better than kNN-BPP. However repeating this experiment a hundred times shows that this is typically not the case, hence the significant difference.

5 Conclusion

In this paper we addressed the importance of considering the recall when tackling imbalanced classification problems. We developed an elegant and straightforward kNN classifier, kNN-BPP, that balances prior class probabilities and thus treats imbalanced classes in a fair manner. The proposed kNN-BPP algorithm shows performance on par with a popular oversampler applied to the datasets in combination with the conventional kNN-algorithm for all measured k-values, while having the same computational complexity as regular kNN. The algorithm's difference from a weighted kNN-algorithm has been shown. kNN-BPP's advantage over other recent internal modifications of kNN over a wide set of k-values has been established.

For future work it could be interesting to investigate this idea for other classifiers that are amenable to a Bayesian probabilistic interpretation, and to perform case studies in application scenarios requiring special attention to fairness and bias [21].

References

1. Alcalá-Fdez, J., Fernández, A., Luengo, J., Derrac, J., García, S.: KEEL data-mining software tool: data set repository, integration of algorithms and experimental analysis framework. J. Multiple Valued Log. Soft Comput. **17**(2–3), 255–287 (2011)
2. Batista, G.E.A.P.A., Prati, R.C., Monard, M.C.: A study of the behavior of several methods for balancing machine learning training data. SIGKDD Explor. **6**(1), 20–29 (2004)
3. Bellinger, C., Drummond, C., Japkowicz, N.: Beyond the boundaries of SMOTE. In: Frasconi, P., Landwehr, N., Manco, G., Vreeken, J. (eds.) ECML PKDD 2016. LNCS (LNAI), vol. 9851, pp. 248–263. Springer, Cham (2016). https://doi.org/10.1007/978-3-319-46128-1_16
4. Bellinger, C., Sharma, S., Japkowicz, N., Zaïane, O.R.: Framework for extreme imbalance classification: SWIM - sampling with the majority class. Knowl. Inf. Syst. **62**(3), 841–866 (2020)
5. Chapelle, O., Schölkopf, B., Zien, A.: Introduction to semi-supervised learning. In: Semi-Supervised Learning, pp. 1–12. The MIT Press (2006)
6. Chawla, N.V., Bowyer, K.W., Hall, L.O., Kegelmeyer, W.P.: SMOTE: synthetic minority over-sampling technique. J. Artif. Intell. Res. **16**, 321–357 (2002)
7. Cover, T.M., Hart, P.E.: Nearest neighbor pattern classification. IEEE Trans. Inf. Theory **13**(1), 21–27 (1967)
8. Davis, J., Goadrich, M.: The relationship between precision-recall and ROC curves. In: Cohen, W.W., Moore, A.W. (eds.) Proceedings of ICML, pp. 233–240 (2006). https://doi.org/10.1145/1143844.1143874

9. Domingos, P.M.: MetaCost: a general method for making classifiers cost-sensitive. In: KDD, pp. 155–164. ACM (1999)
10. Dubey, H., Pudi, V.: Class based weighted K-nearest neighbor over imbalance dataset. In: Pei, J., Tseng, V.S., Cao, L., Motoda, H., Xu, G. (eds.) PAKDD 2013. LNCS (LNAI), vol. 7819, pp. 305–316. Springer, Heidelberg (2013). https://doi.org/10.1007/978-3-642-37456-2_26
11. Elkan, C.: The foundations of cost-sensitive learning. In: IJCAI, pp. 973–978. Morgan Kaufmann (2001)
12. Han, H., Wang, W.-Y., Mao, B.-H.: Borderline-SMOTE: a new over-sampling method in imbalanced data sets learning. In: Huang, D.-S., Zhang, X.-P., Huang, G.-B. (eds.) ICIC 2005. LNCS, vol. 3644, pp. 878–887. Springer, Heidelberg (2005). https://doi.org/10.1007/11538059_91
13. Hand, D.J.: Measuring classifier performance: a coherent alternative to the area under the ROC curve. Mach. Learn. **77**(1), 103–123 (2009). https://doi.org/10.1007/s10994-009-5119-5
14. He, H., Bai, Y., Garcia, E.A., Li, S.: ADASYN: adaptive synthetic sampling approach for imbalanced learning. In: IJCNN, pp. 1322–1328. IEEE (2008)
15. Japkowicz, N.: The class imbalance problem: significance and strategies. In: Proceedings of the International Conference on Artificial Intelligence, vol. 56 (2000)
16. Japkowicz, N.: Assessment metrics for imbalanced learning. In: He, H., Ma, Y. (eds.) Imbalanced Learning: Foundations, algorithms, and applications, chap. 8, pp. 187–206. Wiley, Hoboken (2013)
17. Kriminger, E., Príncipe, J.C., Lakshminarayan, C.: Nearest neighbor distributions for imbalanced classification. In: IJCNN, pp. 1–5. IEEE (2012)
18. Kubat, M., Matwin, S.: Addressing the curse of imbalanced training sets: One-sided selection. In: ICML, pp. 179–186. Morgan Kaufmann (1997)
19. Lemaitre, G., Nogueira, F., Aridas, C.K.: Imbalanced-learn: a python toolbox to tackle the curse of imbalanced datasets in machine learning. J. Mach. Learn. Res. **18**, 17:1–17:5 (2017)
20. Liu, W., Chawla, S.: Class confidence weighted *k*NN algorithms for imbalanced data sets. In: Huang, J.Z., Cao, L., Srivastava, J. (eds.) PAKDD 2011. LNCS (LNAI), vol. 6635, pp. 345–356. Springer, Heidelberg (2011). https://doi.org/10.1007/978-3-642-20847-8_29
21. Ntoutsi, E., et al.: Bias in data-driven artificial intelligence systems - an introductory survey. Wiley Interdiscip. Rev. Data Min. Knowl. Discov. **10**(3), e1356 (2020)
22. Pedregosa, F.: Scikit-learn: machine learning in python. J. Mach. Learn. Res. **12**, 2825–2830 (2011)
23. Qin, Z., Wang, A.T., Zhang, C., Zhang, S.: Cost-sensitive classification with k-nearest neighbors. In: Wang, M. (ed.) KSEM 2013. LNCS (LNAI), vol. 8041, pp. 112–131. Springer, Heidelberg (2013). https://doi.org/10.1007/978-3-642-39787-5_10
24. Quinlan, J.R.: C4.5: Programs for Machine Learning. Morgan Kaufmann, Burlington (1993)
25. Rodieck, R.W.: The density recovery profile: a method for the analysis of points in the plane applicable to retinal studies. Vis. Neurosci. **6**(2), 95–111 (1991)
26. Siddappa, N.G., Kampalappa, T.: Imbalance data classification using local Mahalanobis distance learning based on nearest neighbor. SN Comput. Sci. **1**(2), 76 (2020)

27. Song, Y., Huang, J., Zhou, D., Zha, H., Giles, C.L.: IKNN: informative K-nearest neighbor pattern classification. In: Kok, J.N., Koronacki, J., Lopez de Mantaras, R., Matwin, S., Mladenič, D., Skowron, A. (eds.) PKDD 2007. LNCS (LNAI), vol. 4702, pp. 248–264. Springer, Heidelberg (2007). https://doi.org/10.1007/978-3-540-74976-9_25
28. Zhang, S.: Cost-sensitive KNN classification. Neurocomputing **391**, 234–242 (2020)

Applications of Similarity Search

Similarity Search for an Extreme Application: Experience and Implementation

Vladimir Mic[1(✉)], Tomáš Raček[2], Aleš Křenek[2], and Pavel Zezula[1]

[1] Faculty of Informatics, Masaryk University, Brno, Czech Republic
xmic@fi.muni.cz

[2] Institute of Computer Science, Masaryk University, Brno, Czech Republic

Abstract. Contemporary challenges for efficient similarity search include complex similarity functions, the curse of dimensionality, and large sizes of descriptive features of data objects. This article reports our experience with a database of *protein chains* which form (almost) metric space and demonstrate the following extreme properties. Evaluation of the pairwise similarity of protein chains can take even tens of minutes, and has a variance of six orders of magnitude. The minimisation of a number of similarity comparisons is thus crucial, so we propose a generic three stage search engine to solve it. We improve the median searching time 73 times in comparison with the search engine currently employed for the protein database in practice.

Keywords: Similarity search in metric space · Efficiency · Distance distribution · Dimensionality curse · Extreme distance function

1 Introduction

The similarity search is quite well developed for traditional domains like texts, images, videos, and many of their sub-domains like photos of human faces and irises. Contemporary challenges can be seen in complex and quickly developing data domains studied within interdisciplinary research. This article describes our experience with the similarity search in *protein chains*. However, apart of the similarity (*distance*) function which is domain specific, we approach the problem in a generic way, as a similarity search with difficult distance distribution and expensive distance computation.

We address the search in the worldwide *Protein Data Bank* (PDB, [4]), specifically in its European version (PDBe) [2] which contains about 500,000 protein chains, and tens of thousands are added every year. The protein chain is a long

V. Mic and P. Zezula—This research was supported by ERDF "CyberSecurity, CyberCrime and Critical Information Infrastructures Center of Excellence" (No. CZ.02.1.01/0.0/0.0/16_019/0000822). Computational resources were supplied by the project "e-Infrastruktura CZ" (e-INFRA LM2018140) provided within the program Projects of Large Research, Development and Innovations Infrastructures.

N. Reyes et al. (Eds.): SISAP 2021, LNCS 13058, pp. 265–279, 2021.
https://doi.org/10.1007/978-3-030-89657-7_20

Fig. 1. 3D shape of a protein (PDB ID: 1L2Y) with a single chain. The green ribbon built upon the CA atoms presents a simplification of a complex shape of the protein chain. (Balls = atoms, sticks between them = bonds) (Color figure online).

sequence of *amino acids* connected with chemical bonds, entangled into a complex 3D shape (see Fig. 1 with an example). The 3D shape of a protein chain is sufficiently described by coordinates of carbon atoms of the chain backbone – the *alpha carbons* (CA). Consequently, the similarity of two protein chains can be assessed by finding matching pairs of CA atoms, aligning them in 3D Euclidean space in the best possible way, and computing their distances.

Searching for protein chains with similar 3D structure is of utmost importance, since similar proteins are likely to have a similar biological function. The current similarity search [18] employed at the PDBe database [2] is slow as it does not use any index. Instead, it scans the whole dataset, and the distance computation is skipped just if the sizes of compared chains are as different as they prevent the chains from being similar.

An efficient protein chains search is very challenging for several reasons:

- sizes of descriptors of protein chains vary by two orders of magnitude,
- computation times of chains similarity vary by six orders of magnitude,
- the distribution of protein chains distances is extremely skewed, making the similarity search difficult,

These features make an efficient generic similarity search even impossible for some query objects.

This article presents the novel search engine that runs a three-step gradual search and is available at https://similar-pdb.cerit-sc.cz. While all search techniques have been published in the past, we elaborate on their novel combination within a search engine to maximize user contentment. We define and justify our design choices to maximize the search speed and achieve top search quality.

The article is organised as follows. The similarity of protein chains and challenges of the search are described in Sect. 2. Section 3 describes our approach to build the search engine that maximises the user satisfaction, Sect. 4 summarizes our experiments and Sect. 5 provides conclusions of the article.

2 Similarity of Protein Chains

To formalise a pairwise similarity of protein chains, we use the metric space (D, d) defined by the data domain D and the *distance function* $d : D \times D \mapsto \mathbf{R}^+$.

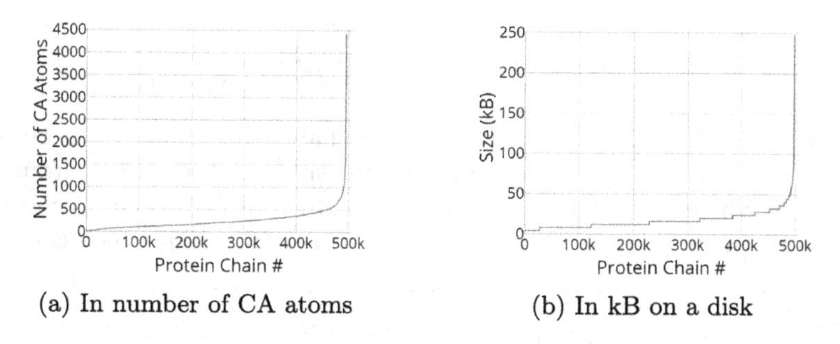

(a) In number of CA atoms (b) In kB on a disk

Fig. 2. Sizes of protein chains – all 0.5 million chains are sorted according to their size, and x-axis depicts their order

The interpretation is that the bigger the distance, the less similar chains, and distances meet the properties of non-negativity, symmetry, identity, and triangle inequality [21]. This article describes the search for the most similar protein chains to an arbitrary given $q \in D$ in the dataset $X \subset D$ which consists of 495,085 protein chains from the PDBe database [2]. A snapshot was taken in September 2020.

2.1 Properties of Descriptors

The pairwise similarity of complex objects, e.g., multimedia, is not usually evaluated directly using the raw data. Instead, the characteristic features (*descriptors*) are extracted to describe objects from a specific perspective. Most of the contemporary descriptors have a fixed size which brings advantages for their processing. Especially, distances of descriptors are evaluated in almost the same time.

Descriptors which are sufficiently rich to express the 3D shape of protein chains tend to be of size that follows the size of chains. Big variance of descriptors sizes causes extreme differences in distance computation times [8]. Figure 2a depicts the number of CA atoms in all 495,085 chains $o \in X$ after dropping extremely small chains, i.e., with less than 10 CA atoms. This is a common practice as such chains are biologically irrelevant. The median protein chain size is 207 CA atoms, but 0.97 % of chains are bigger than 1,000 CA atoms, and 0.03 % of them are bigger than 4,000 CA atoms. Figure 2b depicts the size of protein chains on SSD, which is varying from 4 kB to 248 kB with a median 16 kB. The total size of the dataset is 8.2 GB. The variance of protein chain size has important consequences for assessing their similarity, as we discuss in the following section.

2.2 Similarity Score

There is no general agreement on a universal measure of the protein chain similarity [8,19,20]. We follow the method [10] implemented in the current search service of PDBe, which is based on the *Qscore*:

$$\text{Qscore}(o_1, o_2) = \frac{N_{align}^2}{(1 + (\text{RMSD}/R_0)^2) \cdot N_1 \cdot N_2} \tag{1}$$

where N_1, N_2 are numbers of CA atoms in chains o_1, o_2, R_0 is an empirical constant ($3\,\text{Å} = 3 \times 10^{-10}\,\text{m}$), and N_{align} is the size of subset of CA atoms from both chains which are aligned on each other (by shifting and rotating in 3D space) to minimize their root mean square distance (RMSD), which is defined:

$$RMSD = \sqrt{\sum_{i=1}^{N_{align}} \delta_i / N_{align}}$$

where δ_i is the actual distance (in 3D Euclidean space) between corresponding CA atoms.

The alignment and RMSD computation is fairly easy [9] – the difficult part is the choice of CA subsets to minimize Eq. 1. Systematic search of all possible subsets is practically impossible due to its $O(2^N)$ complexity, therefore heuristics must be use. We use the heuristic of [11] as the current PDBe search. Review of protein alignment algorithms in [8] concludes that virtually all modern methods follow the pattern of minimizing some metric (the score) over all possible subsets of residues (i.e. 1:1 with CA), hence they have to overcome the same complexity problem, and they are comparable in speed.

2.3 Transformation of Qscores to Distances

The range of the Qscore is $[0, 1]$, so we can easily transform it to the distance:

$$d(o_1, o_2) = 1 - \text{Qscore}(o_1, o_2) \tag{2}$$

This function is not a metric distance function due to the imperfection of the heuristic that estimates the Qscore. We examined 250 million pairs $d(o_1, o_2)$, $d(o_2, o_1)$ to reveal that they are equal just in 86,8 % of cases. Differences $d(o_1, o_2) - d(o_2, o_1)$ are rather small: 96.3 % of pairs differ by at most 0.01, and 99.7 % differ up to 0.05. These differences cause violations of the triangle inequality rule, and thus the similarity search based on the filtering by triangle inequalities introduces false negatives errors. While these imperfections could be almost fixed by a double distance computation: $d_m = \min(d(o_1, o_2), d(o_2, o_1))$, it does not pay off due to the complexity of distance evaluations.

Small violations of metric postulates motivate us to use techniques based on *pivot permutations* since they are robust enough to deal small violations of metric postulates. Pivot permutation based techniques use just the order of several closest reference chains (*pivots*) to each chain to approximate its position in a space [1,6,16]. Usage of this type of techniques has an important connotation with the Qscore-to-distance transformation given by Eq. 2. Many transformations of a score to distance exist, and they usually swap the order:

$$(\text{Qscore}(o_1, o_2) < \text{Qscore}(o_2, o_1)) \implies (d(o_1, o_2) > d(o_2, o_1)) \tag{3}$$

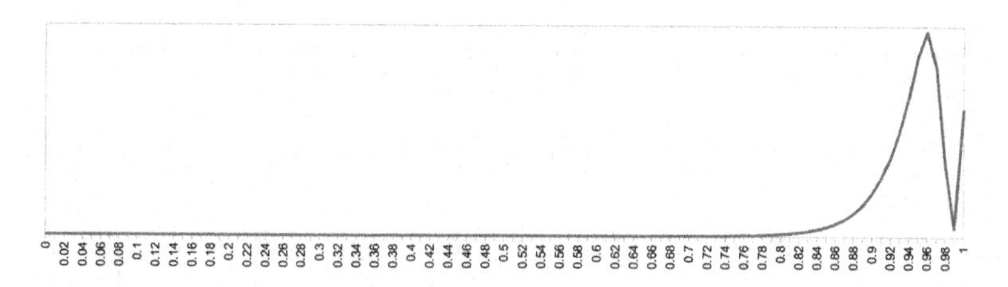

Fig. 3. Distance density

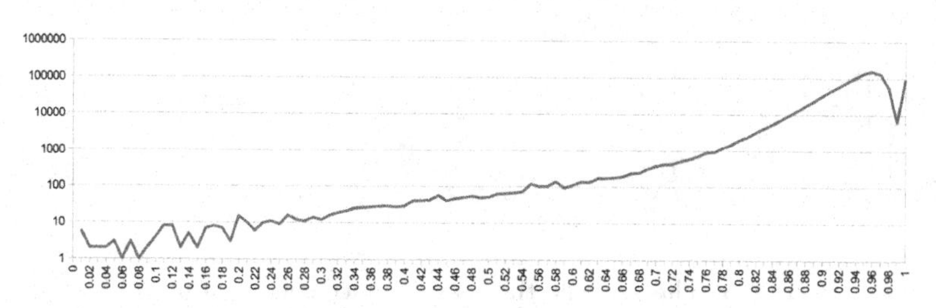

Fig. 4. Histogram of sampled distances with the logarithmic scale of y-axis

Different score-to-distance transformations just change the distribution of distances – see e.g. *the convex transforms* of distances [5,7,17]. Ordering of the closest pivots to an arbitrary given chain $o \in D$ is, however, the same for all score-to-distance transformations that meet Eq. 3. It is thus meaningless to elaborate on a more sophisticated Qscore-to-distance transformation, if we always use the pivot-permutation-based techniques to search the protein chains.

Beside the solutions described in this article, we also tried the filtering based on triangle inequalities. It is ineffective due to the dimensionality curse described in the following section. We also tried to apply convex transforms to the distance function given by Eq. 2. We observed a small ratio of triangle violations which, however, leads to an inadequate false reject rate despite a slow searching.

2.4 Curse of Dimensionality

The efficiency of the similarity search in complex data suffers from the "*curse of dimensionality*": The volume of the searched space quickly increases with increasing distances of nearest neighbours to query chain q. The efficiency of the similarity search thus decreases. Besides, the efficient pruning of the searched dataset is getting harder with decreasing variance of distances $d(q, o), o \in X$. Figure 3 illustrates the density of the distance function defined by Eq. 2 – the curve is made of a sample of million distances $d(o_1, o_2), o_1, o_2 \in X$. The distribution of distances is as skewed, as 98.86 % of distances are bigger than 0.8, and 89.8 % of them are bigger than 0.9. The variance of distances is just 0.002. The *k-nearest neighbours* (kNN) queries searching for k objects $o \in X$ with minimum distances $d(q, o)$ thus cannot be evaluated efficiently in practice for query chains $q \in D$ that have kth nearest neighbour in a large distance.

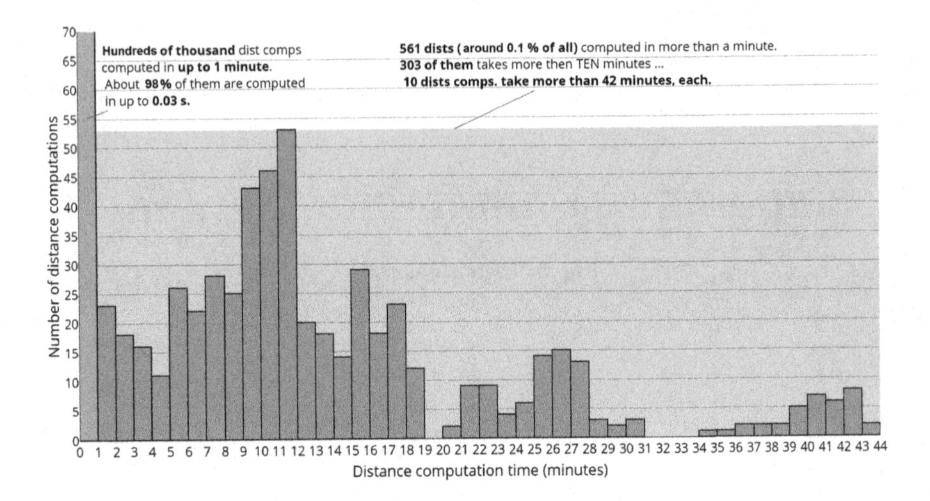

Fig. 5. The extreme times of distance computations

The protein chain descriptors are, however, actual 3D models of protein chains, and the distance function directly expresses their best alignment. The distance of chains thus well corresponds to the perceived similarity of protein chains – which often is not a case of contemporary metric space similarity models with descriptors from neural networks. The searching radius 0.5 thus figures an acceptable limitation for practitioners searching protein chains, since more distant chains are always too dissimilar. We thus focus on the searching for chains $o \in X$ that are within distance $d(q, o) \leq 0.5$, and we consider at most $k = 30$ of them. Similarity queries in this article are meant mainly as the 30NN query limited by range 0.5. In the web application, we allow redefining the k value arbitrarily, but the maximum searching radius is 0.7 to basically limit the query execution times.

Figure 4 illustrates the distribution of the distances with the logarithmic scale of the y-axis. This plot depicts the same curve as Fig. 3, but bins of the width 0.01 are used to create the histogram of distances for which the range of the y-axis can be meaningfully depicted. The figure reveals that there are protein chains within small distances, and thus the similarity search with a limited range, e.g., 0.5, can be meaningful. This is experimentally confirmed in Sect. 4.

2.5 Distance Function Complexity

We evaluate the distance of protein chains by a heuristic [11] that estimates the Qscore. Its evaluation is more efficient than the precise Qscore evaluation, but still, it has a complexity $\mathcal{O}(N_1 \cdot N_2)$ where N_1, N_2 are the chain sizes. Therefore the distance computation time may explode if two extremely big protein chains are compared – see Fig. 2a with the protein chain sizes. Figure 5 illustrates the extreme times of distance computations. These data are gathered from our online running search engine which temporarily stores all distances evaluated in more than 30 ms, and persistently stores those evaluated in more than a second. Figure 5 depicts the stored distance computations that relate to 460 different query chains – please notice that the vast majority of query executions do not store any distance computation. Axis x and y depicts the times in minutes per one distance computation and the number of observed distances, respectively. The first column visualising the distances evaluated below a minute should have a height of around half a million samples[1], and approximately 98 % of these distance evaluations take less than 30 ms. We observed 5,600 distance computations that take more than a second and less than a minute. The biggest problem is the extreme tail of around 0.1 % of distance computations which take minutes or even tens of minutes, each. Until now, we observed 10 distance computations which took more than 42 min, each.

We tried to skip all these extreme distance evaluations, but this results in an inability to find even very similar protein chains to some of the biggest query chains. We decided not to employ such skipping since the newly identified protein chains in the PDBe database are rather bigger ones, so the quality of the search engine could be perceived as decreasing in the future. We also observed a moderate Pearson correlation +0.46 between the distance computation times and the returned distances, which could be a motivation to skip long-lasting distances. Nevertheless, this correlation seems insufficient and influenced by an inability to effectively sample the extreme values: for instance, 6 out of 10 observed distance evaluations that take more than 42 min are returning distances smaller than 0.21. These are thus distances between very similar protein chains – see Fig. 4.

The best way to search the protein chains that we found is to minimize the number of the Qscore evaluations and to cache expensive distance computations.

3 Gradual Similarity Search

We provide users with three gradual query answers of increasing quality to maximise the search engine usefulness. We denote these three consecutive parts of the query execution as *phases*, and each of them returns a query answer.

- The first phase is always finished in a few seconds since it evaluates just 61 distances $d(q, p)$ of q to pivots. It is usually evaluated below 0.5 s.
- The second phase uses just 489 distances $d(q, p)$ to pivots, including those 61 from the first phase. It takes usually less than 4 s including the first phase.
- The third phase requires a variable number of distance computations with the median value 702, including those 489 from previous phases. The whole search usually takes less than 8 s, but with an extreme and necessary tail.

The results of the first and second phases thus add a negligible overhead to the third phase, since the IDs of chains likely to be similar to q figure more or less the intermediate result of the query execution.

[1] This is an estimation made as an extrapolation from other query executions. Our search engine evaluates approximately 1,000 distances per average query. The vast majority of distances is not stored, so we do not know the precise number of distances evaluated in less than a second.

3.1 Data Preprocessing and Sketches

The whole search engine is based on 512 pivots $p \in D$ that approximate the position of each protein chain in a space. We select the pivots *uniformly randomly* with respect to the number of CA atoms in protein chains. Specifically, we sort the chains $o \in X$ according to the number of CA atoms, split this list into 512 parts of the same size, and randomly pick one protein chain from each part as a pivot. All distances $d(o, p), o \in X$ between chains o and pivots p are precomputed and stored in the DB during the data preprocessing.

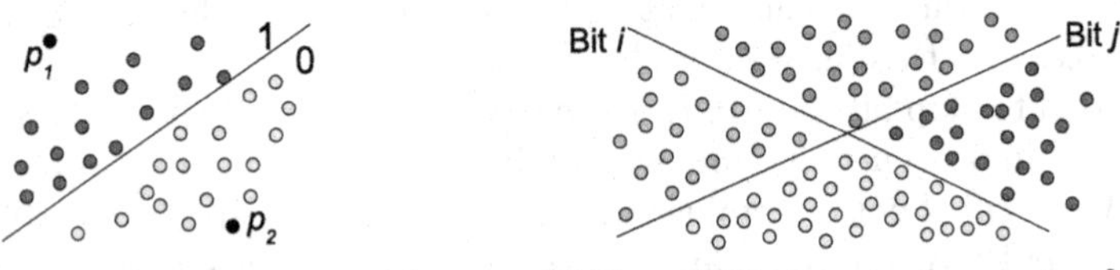

(a) GHP to set values in one bit (b) Two GHPs to set values in two bits

Fig. 6. The generalised hyperplane partitioning to define bits of sketches of chains

We also create and store the *sketches* of protein chains. Sketch $sk(o), o \in X$ of protein chain o is a bit-string in the Hamming space $(\{0,1\}^\lambda, h)$, and the Hamming distance of sketches approximates the distance of protein chains. We use the sketching technique denoted here as *GHP_50*, which is defined in [12]. The GHP_50 uses a single instance of the *generalised hyperplane partitioning* [21] to define each bit of all sketches $sk(o), o \in X$: a given bit of sketches $sk(o), o \in X$ expresses which of the two pivots is closer to o, so λ hyperplanes define sketches $sk(o), o \in X$ of length λ bits – see schema in Fig. 6. Pivot pairs are selected by a heuristic from a set of pivots to define approximately *balanced* and *low-correlated* bits of sketches $sk(o), o \in X$, i.e., each bit of sketches splits dataset X approximately into halves, and bits of sketches $sk(o), o \in X$ have pairwise Pearson correlations as close to 0 as possible [12].

We use two types of sketches for each protein chain $o \in X$. The *small sketches* have length 320 bits, and they are defined using just 61 pivots out of those 512 preselected. We offered 64 pivots to the heuristic to define the hyperplanes, and it did not use 3 of them. The probable cause is a vast majority of distances 1 between these 3 pivots and protein chains from the dataset X which prevent all hyperplanes defined by these pivots from defining the balanced bits. The 64 pivots that we offered were selected again *uniformly randomly* according to the size from 512 preselected pivots. In practice, sketches maximizes the memory usefulness iff λ is a multiple of 64, since we use java class *BitSet* to store sketches as an array of *longs*. For the first search phase, sketches of 320 bits created from 64 pivots provide a suitable trade-off between the time needed to create sketches, and their quality.

Table 1. The mapping of distances that we use for sketches of length 1,024 bits and $\pi = 0.75$ (the majority of lines is omitted)

Ham. dist. b	orig. dist. t	Ham. dist. b	orig. dist. t	Ham. dist. b	orig. dist. t
0 - 144	0	211	0.03	238	0.1
145	0.0003	...	...	...	...
146	0.0006	222	0.04	270	0.2
147	0.001	223	0.05	...	...
...	...	...	...	290	0.3
149	0.002	227	0.06	...	...
...	...	...	...	307	0.4
172	0.01	231	0.07	...	...
...	...	...	...	**340**	**0.5**
199	0.02	236	0.08	...	...
...	...	237	0.09	**562 - 1024**	1

We also use the *large sketches* of $o \in X$ that have length 1,024 bits and are defined by 489 out of 512 preselected pivots. Similarly, we offered all 512 pivots to the heuristic to define large sketches, and it used just some of them. Our database thus contains the following metadata for each chain $o \in X$: (1) 512 chain-to-pivots distances, (2) small sketch of o, and (3) large sketch of o.

3.2 First Phase of the Query Execution

Following sections describe the query execution, so we consider an arbitrary given query chain $q \in D$. The first phase of the search evaluates just 61 distances $d(q, p)$ to create the small sketch of q, and executes the kNN query on the small sketches. We use just a sequential evaluation of all 495,085 Hamming distances $h(sk(q), sk(o)), o \in X$; such evaluation takes about 0.15 s (per query), so the execution time of the first phase is practically given by the evaluation of 61 distances $d(q, p)$ to create small sketch $sk(q)$. Since none of these pivots is extremely big, the first phase is evaluated in up to a few seconds for an arbitrary $q \in D$.

The first phase answer consists of IDs of k chains $o \in X$ with the smallest Hamming distances $h(sk(q), sk(o))$. These IDs are immediately shown in the GUI, and we start the parallel and asynchronous evaluation of k distances $d(q, o) = 1 - \text{Qscore}(q, o)$, as well as the second phase of the search. When Qscore(q, o) is evaluated, the alignment of the protein chains q and o is displayed since the Qscore computation involves the best alignment of protein chains in 3D Euclidean space. Asynchronous evaluation allows to provide some results quickly, and we stop remaining evaluations when the results of the second phase are delivered. The results are sorted dynamically according to the distances $d(q, o)$. If the distance is bigger than the specified radius of the query, the ID of chain o is hidden from the results.

3.3 Second Phase of the Query Execution

The second phase of the query execution is similar to the first one, but it uses large sketches $sk(o), o \in X$. Specifically, 489 distances $d(q,p)$ between q and pivots p are evaluated to create a large sketch of q, and all 495,085 Hamming distances $h(sk(q), sk(o)), o \in X$ are evaluated to return the result of the second phase. We, however, utilise also the minimum required Qscore to define the searching radius in the Hamming space of large sketches, and we evaluate the kNN queries with the limited searching radius in the Hamming space. We use the probabilistic model from the article [13] that approximates the mapping of distances $t = d(q,o)$ to the Hamming distances $b = h(sk(q), sk(o))$ of sketches created by the *GHP_50* sketching technique, such that:

$$\mathbf{P}\left(d(q,o) \geq t \mid h(sk(q), sk(o)) = b\right) \approx \pi \tag{4}$$

where π is the probability empirically set to 0.75 [13]. Intuitively, having the Hamming distance of sketches, the mapping estimates the minimum probable distance of the protein chains. Table 1 gives examples of the mapping that is used in our web application. While the mapping of distances is used just to set a disposable Hamming radius to search the large sketches in the second phase, it plays a crucial role in the third phase.

3.4 Third Phase of the Query Execution

We use a high-quality pivot permutation based index called *the PPP-codes* [16] and the *secondary filtering* by large sketches [13] in the third phase.

The *PPP-codes* index [15,16] uses 4 independent Voronoi partitionings [21] of the metric space (D, d). Each Voronoi partitioning uses 128 pivots that are disjunctive subsets of all 512 preselected pivots. Each protein chain is indexed using all these partitionings in the main memory. Given a query chain q, the chains $CandSet(q)$ that are likely to be similar to q are determined (*the candidate set*) by a selective combination of individual Voronoi partitionings [16]. The candidate set size is given as a parameter in advance.

Usage of PPP-codes clarifies the number of 512 pivots that we use. Each Voronoi partitioning uses $512/4 = 128$ pivots to approximate position of each chain, and we need to have a few distances between 128 pivots and each protein chain smaller than 1. We found 512 pivots as the optimum number, since 768 and 124 pivots provide practically the same results as presented with 512 pivots.

Particular position of the query chain q significantly infers the performance of similarity indexes. A fixed candidate set size used for all query chains $q \in D$ does not take into account different density of chains $o \in X$ around given $q \in D$, and thus decreases the quality, or unnecessarily increases the number of distance computation in case of many query chains [13]. The *secondary filtering* of the $CandSet(q)$ by sketches can effectively reduce the $CandSet(q)$ dynamically, using the current searching radius given either by the query assignment or by the distance to the kth nearest neighbour, found so far.

We evaluate the third phase of the query execution as follows. When the second phase of the query execution is finished, we evaluate all the remaining distances of q to 512 pivots p, and search for the candidate set $CandSet(q)$ of size 5,000 ($\approx$1% of the dataset) by the PPP-codes. When a protein chain $o \in X$ is confirmed to be in the $CandSet(q)$, we evaluate the Hamming distance $h(sk(q), sk(o))$ of large sketches. This Hamming distance is used together with the mapping illustrated by Table 1 to estimate the minimum probable distance of protein chains. If this estimated distance is bigger than the current searching radius, o is discarded. Otherwise, we evaluate the distance $d(q, o)$, and we gradually build the answer of the third phase. Building the answer is finished when the whole $CandSet(q)$, i.e. 5,000 chains, is processed.

The need to minimise the number of distance computations also clarifies the need to limit the searching radius from the beginning, i.e., to evaluate kNN + range queries instead of kNN queries. If a kNN query is evaluated, the secondary filtering is not utilised until k distances are evaluated to fill the query answer. Then the searching radius decreases gradually, usually from a high value. Immediate range limitation thus enables an effective secondary filtering from the beginning of the query execution which effectively decreases the number of distance computations. Similarly, the k value improves the effectiveness of the secondary filtering since if the query answer is full, the searching radius is given by the distance to the kNN instead of the original query range, so the secondary filtering power increases dynamically.

Table 1 illustrates an extreme power of the secondary filtering with the *GHP_50* sketches. The implicit searching radius in our application is 0.5, which is a very small distance considering the distance density depicted in Fig. 3. Distance 0.5 is mapped to the Hamming distance 340, i.e., if sketches $sk(q)$, $sk(o)$ of length 1024 bits differs in at least 340 bits, we skip the evaluation of distance $d(q, o)$. Lemma 1 and Theorem 2 in article [12] defines the mean and the variance of the Hamming distances on *GHP_50* sketches: the mean is $\lambda/2$, i.e. 512, and the variance decreases towards $\lambda/4$, i.e. 256, with the decreasing pairwise bit correlations – and they are minimised by the *GHP_50* sketching technique. Therefore, the Hamming distance of large sketches $sk(q)$ and $sk(o)$ as small as 340 is very improbable, and the secondary filtering usually discards a vast majority of the $CandSet(q)$ identified by the PPP-codes.

4 Experiments

Since we focus on kNN queries with limited radius 0.5, we report the number of nearest neighbours within this radius. We use 1,000 query chains in our experiments, that are selected in the same way as pivots, i.e., uniformly randomly with respect to the protein chain size. None of the query chains is used also as a pivot. Figure 7 illustrates the number of nearest neighbours in the *ground truth* (i.e., the precise answer) for all 218 query chains q that have less than 30 nearest neighbours within distance 0.5. All other 782 query chains have at least 30 nearest neighbours within the distance 0.5, and we use 30 closest of them

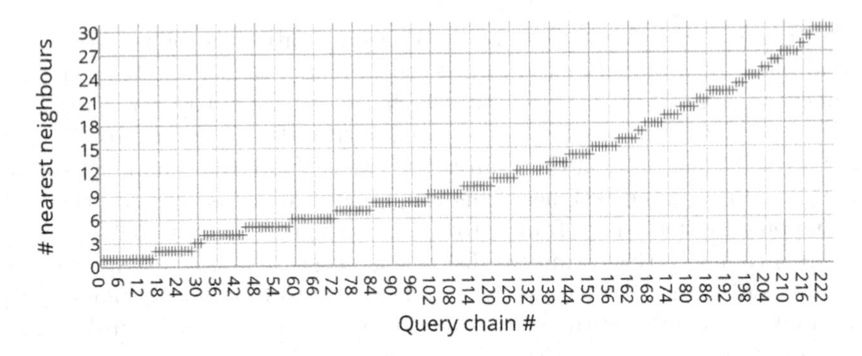

Fig. 7. Number of nearest neighbours within distance 0.5 to each query chain q. 218 out of 1,000 tested q do not have 30 nearest neighbours up to distance 0.5.

as their ground truth. The searching quality is described by the *accuracy*, i.e., the ratio of nearest neighbours from the ground-truth that is found. Figure 7 clarifies that the similarity queries with the search radius 0.5 should provide non-empty answers for around 98 % of query chains despite an extreme distance density illustrated by Fig. 3. We point out that an empty answer provides useful information due to a strong relation of the distance function to the protein chain structures, as discussed in Sect. 2.4.

Box-plots in Fig. 8a depict the searching accuracy over particular query chains. The first and second phases search with a median accuracy 0.467 and 0.667, respectively. Both have a huge variance – the differences between the first and the third quartiles are 0.44 and 0.43, respectively. The third phase evaluates 700 out of 1,000 queries precisely, so it has the median accuracy 1. The average accuracy is 0.937 due to the worst query chains – averages are depicted by the dashed line for each box-plot in Fig. 8.

We evaluate 30NN queries with the radius 0.5 also by the existing PDBe search engine, but we use the setting which guarantees to find all nearest neighbour up to distance 0.3. This setting provides much faster search than the precise one, and it is still of a slightly better accuracy than our inherently approximate search with the radius 0.5 (see Fig. 8a). The speed comparisons of the search engines with these settings are thus quite fair.

Figure 8b relates to the third phase of the query execution which uses the PPP-codes to select 5,000 candidate chains $CandSet(q)$, and filters [13] them with the large sketches. Figure 8b illustrates that just 190 out of 5,000 candidate chains remain per median q after this filtering. The first and third quartiles are 87 and 465, respectively, the minimum number of remained chains is 2 and the maximum is 4,688. These numbers thus correspond to the only distance $d(q, o)$ evaluations conducted apart from 512 query-to-pivots distance evaluations to return the query answers with median accuracy 1 and average accuracy 0.937. Please see that this highest number of distance computations, 4,688, clarifies our choice of the candidate set size, i.e., 5,000 selected by the PPP-codes – we have observed query chains that utilises the vast majority of such candidate set.

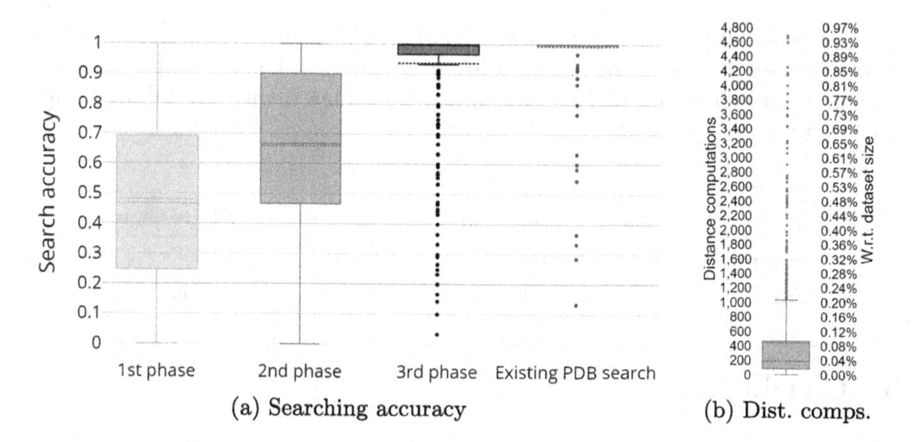

(a) Searching accuracy (b) Dist. comps.

Fig. 8. Searching accuracy and number of distance computations after the secondary filtering with large sketches in the 3rd phase

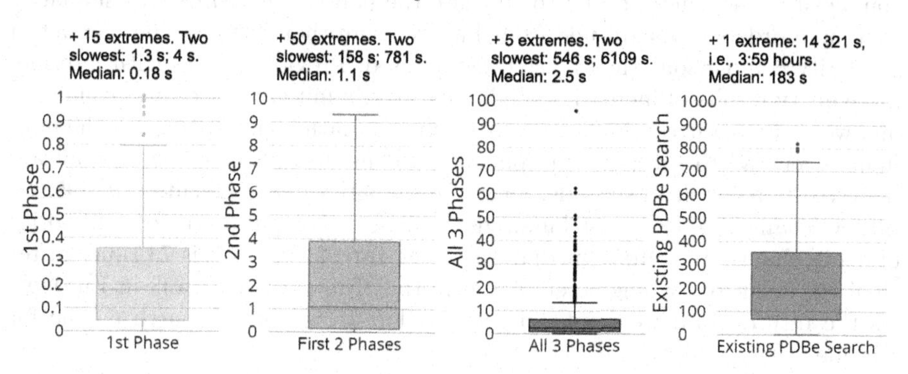

Fig. 9. Query execution times in seconds

Figure 9 depicts the searching times in seconds[2]. Box-plots again describe the distribution of values over particular query chains q. The scale of the times is multiplied by 10 for each box-plot, and there are a few outliers that do not fit the scale and are described above each box-plot. All 3 phases provide their answers within 5 s for almost 70 % of query chains (see the third box-plot), but there is an extreme tail clearly explained by the previous analysis. 1 % of the slowest query executions (i.e., 10 queries) requires more than 49 s, each. The median searching time of all 3 phases is 2.5 s. The fourth box-plot depicts the searching times of the current PDBe search engine. The median is 183 s, i.e., 73 times slower.

[2] No caching is used here except a re-using the distances evaluated in the previous phases of the same query execution.

The suitability of our approach is confirmed by an ability of the first and second phases to deliver results of a lower accuracy but much faster. The most difficult query chain out of 1,000 examined is evaluated with the following accuracy and times:

	1st phase	1st and 2nd phase	All 3 phases	PDBe engine
Accuracy	0.46	0.66	0.93	1
Time	4 s	13 min	1:40 h	3:59 h

5 Conclusions

We described our experience with the similarity search in extreme metric space which strongly suffers from the variance of objects size and the similarity function complexity. The times needed to evaluate the pairwise *protein chain* similarity vary by 6 orders of magnitude from 1 ms to more than 43 min. The number of similarity comparisons thus must be minimised. Providing users with intermediate query results of increasing quality effectively mitigates user inconvenience, and we evaluate queries in 3 consecutive phases. Each phase introduces the minimum overhead to the following phases, and the first query results are always delivered in a couple of seconds. Since the similarity query execution in 495,085 protein chains evaluates just hundreds of distance computations for a majority of query chains, we achieve median searching time 2.5 s which is 73 times faster than the result of the engine employed in the "Protein Data Bank in Europe". As the future work, we would like to develop a distributed search engine for real-life usage [3,14].

References

1. Amato, G., Savino, P.: Approximate similarity search in metric spaces using inverted files. In: 3rd International ICST Conference on Scalable Information Systems, INFOSCALE 2008, Vico Equense, Italy, 2008. p. 28. ICST / ACM (2008)
2. Armstrong, D.R., et al.: PDBe: improved findability of macromolecular structure data in the PDB. Nucleic Acids Res. **48**(D1), D335–D343 (2019)
3. Batko, M., Novak, D., Falchi, F., Zezula, P.: Scalability comparison of peer-to-peer similarity search structures. Future Gener. Comput. Syst. **24**(8), 834–848 (2008)
4. Berman, H.M., et al.: The protein data bank. Nucleic Acids Res. **28**(1), 235–242 (2000)
5. Bernhauer, D., Skopal, T.: Analysing indexability of intrinsically high-dimensional data using TriGen. In: Satoh S., et al. (eds.) Similarity Search and Applications. SISAP 2020. Lecture Notes in Computer Science, **12440**, 261-269. Springer, Cham (2020). https://doi.org/10.1007/978-3-030-60936-8_20
6. Chávez, E., Figueroa, K., Navarro, G.: Effective proximity retrieval by ordering permutations. IEEE Trans. Pattern Anal. Mach. Intell. **30**(9), 1647–1658 (2008)
7. Connor, R.C.H., Dearle, A., Mic, V., Zezula, P.: On the application of convex transforms to metric search. Pattern Recognit. Lett. **138**, 563–570 (2020)

8. Deng, L., et al.: MADOKA: an ultra-fast approach for large-scale protein structure similarity searching. BMC Bioinform. **20**, 662 (2019)
9. Kearsley, S.K.: On the orthogonal transformation used for structural comparisons. Acta Crystallogr. **A45**, 208–210 (1989)
10. Krissinel, E., Henrick, K.: Secondary-structure matching (SSM), a new tool for fast protein structure alignment in three dimensions. Acta Crystallogr. Sect. D Biol. Crystallogr. **60**(12), 2256–2268 (2004)
11. Krissinel, E.: Enhanced fold recognition using efficient short fragment clustering. J. Mol. Biochem. **1**(2), 76 (2012)
12. Mic, V., Novak, D., Zezula, P.: Designing sketches for similarity filtering. In: 2016 IEEE 16th International Conference on Data Mining Workshops (ICDMW), pp. 655–662 (2016)
13. Mic, V., Novak, D., Zezula, P.: Binary sketches for secondary filtering. ACM Trans. Inf. Syst. **37**(1), 1:1–1:28 (2018)
14. Novak, D., Batko, M., Zezula, P.: Large-scale similarity data management with distributed metric index. Inf. Process. Manag. **48**(5), 855–872 (2012)
15. Novak, D., Zezula, P.: Performance study of independent anchor spaces for similarity searching. Comput. J. **57**(11), 1741 (2014)
16. Novak, D., Zezula, P.: PPP-codes for large-scale similarity searching. Trans. Large-Scale Data Knowl. Cent. Syst. **24**, 61–87 (2016)
17. Skopal, T.: Unified framework for fast exact and approximate search in dissimilarity spaces. ACM Trans. Database Syst. **32**(4), 29 (2007)
18. Velankar, S., et al.: PDBe: protein data bank in Europe. Nucleic Acids Res. **38**(suppl_1), D308–D317 (2009)
19. Winn, M.D., et al.: Overview of the CCP4 suite and current developments. Acta Crystallogr. **D67**, 235–242 (2011)
20. Yang, A., Honig, B.: An integrated approach to the analysis and modeling of protein sequences and structures. i. protein structural alignment and a quantitative measure for protein structural distance. J. Mol. Biol. **301**, 665–678 (2000)
21. Zezula, P., Amato, G., Dohnal, V., Batko, M.: Similarity Search - The Metric Space Approach, Advances in Database Systems, **32**, Springer, Heidelberg (2006)

What Makes a Good Movie Recommendation? Feature Selection for Content-Based Filtering

Maciej Gawinecki(✉), Wojciech Szmyd(✉), Urszula Żuchowicz(✉), and Marcin Walas(✉)

Samsung R&D Institute Poland (SRPOL), Bora Komorowskiego Street 25C, 31-416 Cracow, Poland
{m.gawinecki,w.szmyd,u.zuchowicz,m.walas}@samsung.com

Abstract. Nowadays, recommendation systems are becoming ubiquitous, especially in the entertainment industry, such as movie streaming services. In *More-Like-This* recommendation approach, movies are suggested based on attributes of a currently inspected movie. However, it is not obvious which features are the best predictors for similarity, as perceived by users. To address this problem, we developed and evaluated a recommendation system consisting of nine features and a variety of their representations. We crowdsourced relevance judgments for more than 5 thousand movie recommendations to evaluate the configurations of several dozen of movie features. From five embedding techniques for textual attributes, we selected Universal Sentence Encoder model as the best representation method for producing recommendations. Evaluation of movie features relevance showed that summary and categories extracted from Wikipedia led to the highest similarity on user perceptions in comparison to other analyzed features. We applied the feature weighting methods, commonly used in classification tasks, to determine optimal weights for a given feature set. Our results showed that we can reduce features to only *genres*, *summary*, *plot*, *categories*, and *release year* without losing the quality of recommendations.

Keywords: Recommender system · Content-based filtering · Feature selection · Feature weighting

1 Introduction

Selecting interesting movies in TV streaming services can be time-consuming for users due to the increasing amount of available content. Therefore, recommender systems are utilized broadly to assist users in handling information overload by suggesting new similar content. Conventional recommendation methods are classified into Collaborative Filtering (CF), Content-Based Filtering (CBF), and Hybrid systems, combining both methods. Recommending in CF is based on the similarity between users' preferences. In CBF, the retrieval process is driven by the characteristics of the products, providing items similar to items that the

N. Reyes et al. (Eds.): SISAP 2021, LNCS 13058, pp. 280–294, 2021.
https://doi.org/10.1007/978-3-030-89657-7_21

user selects or liked in the past. In this paper, we focused on the so-called *More-Like-This* (MLT) version of CBF recommenders that suggests more content similar to particular items and does not consider user profile. A prerequisite for CBF is the availability of information about relevant content attributes of the items. Attributes in the movie domain mostly comprise of structured information, e.g., genres, release year, and unstructured information, such as plot. The relevance of each feature is not obvious in the context of the recommendation system. Typically, the features are chosen based on their relative usefulness at hand [5,17], not using some external heuristic, e.g. optimization techniques [19]. In a long run, using wrong features can lead to inaccurate recommendations and unnecessary engineering costs, like acquiring a source of data for a feature that does not pay off.

In this paper, we asked: (1) Which feature representation is most effective? (2) Which single feature provides the best recommendations? (3) Do we need all features to get good recommendations? (4) How to combine them to provide the most relevant recommendations? To answer those questions, we crowdsourced a large volume of recommendation relevance judgments. We released part of the collected dataset to the public as a benchmark for evaluation in the movie recommendation domain. For textual features, we compared the quality of recommendations from various novel embedding methods and found the most promising representations. Then, for all the features, we applied well-known feature selection algorithms to find which ones are relevant and which can be omitted without losing recommendation performance. We conclude with a summary of our findings.

2 Related Works

There is a substantial body of work on *feature selection* algorithms for machine learning and statistics (see [26] for a survey). However, for most algorithms, it is unclear how to extend them to the case of a recommendation system. For instance, *filter methods* use similarity measures such as Spearman Correlation, to score features based on their information content concerning the prediction task. Yet, filter methods cannot be naturally extended to recommender systems, in which the prediction target varies because it depends both on the input item (selected movie in our case) and on the item under consideration (recommended movie). Ronen et al. [22] addressed this limitation by scoring not single items but the similarity between pairs of items.

The more frequent approach for feature selection in recommender systems is to use domain knowledge and non-systematic trial-and-error method that, as some authors like Colucci et al. [6] admit, is a naïve eyeball technique. Another approach is to run *online evaluation*, where users are asked to assess recommendations from multiple recommenders but are not told where each recommendation comes from [6,13]. While online evaluation can provide the most credible results by simulating real conditions and getting real user's decisions, it does not scale well for several dozen of possible recommender versions to evaluate. To address these limitations, recommenders are evaluated in offline setup against previously collected *relevance judgments*.

There is a number of public datasets with movie ratings, like Netflix [1] or MovieLens [10]. However, those datasets describe whether a movie is a good recommendation for a user rather than for a selected movie. One could argue that two movies are similar, if they were liked by same users. However, users tend to like a variety of films, e.g., both comedies and thrillers, and that does not automatically make them similar. To address this lack, authors of [6,13] collected two datasets[1] from online evaluation of multiple MLT recommenders. There is a risk that subsequent recommender systems generate recommendations that are not present in the dataset. The authors address this problem by ignoring movie pairs without relevance judgments when measuring performance. Research in information retrieval has shown that such an approach may lead to unfair performance results: it may happen that top search results should be treated as relevant but are considered as non-relevant if they were left unjudged, as reported by Webber et al. in [28]. However, it is impractical to obtain relevance judgments for all items. Tonon et al. [27] handle this problem by introducing *iterative pooling*: for each new retrieval system missing relevance judgments are obtained and added to the existing dataset.

In the context of movie recommenders, there has been little research on which features are best at predicting, and the results are often contradictory. For instance, in offline evaluation Soares et al. [24] found that *director* feature alone can provide better recommendations than other features like *actors* and *genres*, while the order starting from the most important (*title*, *genre*, *cast*, *screenwriter*, *director*, and *plot*) is suggested by Colucci et al. in [6]. There have been significantly more systematic research on which *feature representation* provides best recommendations, especially on representing textual content such as a *movie/book plot*. LSI and LDA were evaluated in [2], LSI and Random Indexing in [16], TF/IDF, Word2Vec, GloVe and Doc2Vec in [25], Word2Vec alone in [18] and Doc2Vec in [23]. Their results show that the quality of recommendations depends not only on the topic model used but also on the type and size of data used for training the embedding model.

3 Recommender System Used in Experiments

To perform experiments, we developed a prototypical recommendation system built using the dataset that we collected. Further details are described in this section.

3.1 Recommending Approach

The system used in our experiments was designed to show five unordered recommendations next to a *selected movie* on a user's TV screen. Each movie is represented as a set of encoded features. The system calculates *similarity* between vectors for each feature separately, using suitable distance metrics, and takes

[1] http://moviesim.org/.

a weighted sum of them. Weights enable us to control how much each feature contributes to the final output distance.

3.2 Movies Dataset

For experiments, we built the Movies Dataset of over 20K movies that were used as the input for the recommender. The movies came from intersecting internal Samsung dataset[2] and Wikipedia dump[3]. We also added additional movie ratings from MovieTweetings[4] [8]. For each title we extracted the following attributes: *release year*, *genre*, *language*, *screenwriter*, *director*, *summary* (merged Wikipedia introduction page and distributor description), *plot* (Wikipedia "Plot" section), *category* (from Wikipedia categories, e.g., "1990s black comedy films" or "Films about psychopaths"), and *popularity* (ratings from IMDb and metacritic.com). Three attributes in the dataset had incomplete values: language (1.5%), director (34.0%), and popularity (70.0%). One entry of Movies Dataset is included as an example in our public repository[5].

3.3 Features Representations

For each movie attribute in the Movies Dataset, we developed the following *features representations* and *distance metrics*:

- **Release year**. The intuition is that when a user is looking for an old movie, we should recommend him/her a similarly old movie. However, when a user is looking for an old movie, it doesn't matter if it is from the 1930s or 1950s, but when looking for more contemporary movies, the subjective difference between the 2000s and 2020 movies may seem to be much bigger. To express this intuition, we represent it in a logarithmic form and use the Euclidean distance metric.
- **Language**. The expectation is that when a user is looking for a movie originally spoken in French, they might be interested in other French movies as well. Since a movie can have more than one language assigned, we used Jaccard distance to measure language overlap between two movies.
- **Genre**. For genres, we proposed two representations: a simple sparse label vector with a Jaccard distance metric and Word2Vec embeddings [11], trained over genres co-occurring in our dataset. The latter can capture the perceived similarity between genres, e.g., thriller can be considered to be more similar to action than to cartoon. We applied cosine distance between embedding vectors of movie genres, and we took a mean vector over embeddings for movies with multiple genres.

---

[2] Accessed on May 25, 2020.
[3] https://dumps.wikimedia.org, accessed on August 25, 2020.
[4] https://github.com/sidooms/MovieTweetings, accessed on August 27, 2020.
[5] https://github.com/la-samsung-poland/more-like-this-dataset/blob/main/sample_movie.json.

- **Category**. We extracted about 60K Wikipedia categories. To express similarity between categories we embedded them using Word2Vec with negative sampling [11]. The network used for machine learning was fed with movie titles as target words and corresponding categories as context words. We used the cosine metric to calculate the distance between vectors.
- **Director and screenwriter**. We suspected that directors and screenwriters often produce movies of a similar style and topic. To express similarity between movies we applied sparse vectors with the Jaccard distance.
- **Popularity**. A user may search for a blockbuster, a movie that is both highly rated and popular, or a niche movie, appreciated by critics but not so popular. To support such use cases, we constructed a vector of popularity indices and averaged ratings among users and critics, where distance is measured with the cosine metric.
- **Summary and plot**. For these textual features we experimented with a number of topic modelling approaches: Doc2Vec [11], LDA [3], LSI with TF-IDF [7], USE [4]. USE is a transformer-based model pre-trained on a variety of NLP tasks with large multi-domain datasets. We did not fine-tune the model on our dataset. Doc2Vec, LSI, LDA models were trained on the collected plot and summary.

4 Evaluation and Feature Selection Methods

To find meaningful answers to questions posed in this paper we ran experiments with the tools and methods described below.

4.1 Comparing Recommenders Performance

We carried out an *offline evaluation*, where recommendations for a given input movie are compared against ground truth ratings (relevance judgments). For input movies, we have manually selected 153 movies from Movies Dataset (described in Sect. 3.2). We strove to achieve high diversity in content – from Marvel blockbusters through computer animations, European and Asian art-house cinema to silent movies. Diversification of evaluation set allows to measure how recommender system performs for different inputs but also describes partially general diversity of recommender. List of all the movies from Evaluation Dataset is available publicly[6]

For each of the selected input movies, a recommender produced five recommendations. Input movies, together with their recommendations, were sent to annotation for collecting ratings. Given the relevance judgments, we were able to calculate metrics to assess and compare recommenders with various configurations of features and features weights.

The order of recommendations in our system was irrelevant because they were displayed in unordered series of tiles. Therefore, using any rank-based metric

[6] https://github.com/la-samsung-poland/more-like-this-dataset/blob/main/evaluation_set.tsv.

was pointless. Thus, we decided to measure the performance of the system using *Precision@5* defined as:

$$Precision@5 = \frac{\#true\_positives}{\#true\_positives + \#false\_positives} \quad (1)$$

Because of the limited annotation budget and lack of agreement between annotators in some cases, we were not able to collect ratings for all possible pairs. Only explicitly labeled examples were considered during calculations. Hence, the denominator can sometimes be lower than 5. However, comparing systems using just this metric could lead to incorrect conclusions due to differences in the number of rated pairs. For this reason, we introduced another metric, *coverage*:

$$Coverage = \frac{\#(rated\_pairs \cap generated\_pairs)}{\#generated\_pairs} \quad (2)$$

For instance, coverage of 0.3 means that we know whether recommendation is good or bad for only 30% of recommendations returned by the recommender. It would assure us that the two systems were evaluated using a sufficient number of test examples. Low coverage might lead to erroneous interpretation of results. Features relevances can be inferred from comparing recommender systems with different sets of features and features weights.

4.2 Collecting Relevance Judgments

Evaluating recommender performance requires a binary label that denotes whether a recommendation is good or bad for the selected movie.

To collect relevance judgments, for each movie pair generated by the recommender, we submitted a rating task to the *crowdsourcing system*. To simulate real recommendation context, we started each task with the short introduction: *"Imagine you have been searching for a movie with Smart TV and the TV has recommended you another one in the 'More-Like-This' section. Would you be interested in watching the recommended movie, given the movie you were searching for?"*. For both selected and recommended movies, we showed only the information that would be present in the recommendation application: movie title, release year, summary, and poster.

We asked users to rate recommendations using 5-point scale of answers: *definitely interested*, *rather interested*, *rather uninterested*, *definitely uninterested* or *don't know*. 3-point scale, applied in similar studies [6,13], may result in losing some information due to a *rounding error* [12,20] and thus leave some less known movie pairs unrated.

Given the fixed budget, annotating all possible movie pairs upfront is impractical. To control the cost we employed *iterative pooling* [27]. Each time a new version of a recommender system was tested, we submitted recommendations, generated and not rated yet, to the crowdsourcing system. We also limited the number of annotators per movie pair to two. Those two users had to agree about the rating: whether they rated it positively or negatively. Answers *rather interested* and *definitely interested* were considered a positive rating, while *rather uninterested* and *definitely uninterested* were considered a negative one. Only in case of disagreement, the movie pair was submitted for the third annotation to decide. If that did not help, i.e., when an annotator answered with *don't know*, the movie pair was not included in the relevance judgments. This strategy offered us a good trade-off between annotation cost and confidence in ratings.

4.3 Selecting Optimal Weights

To assess the relevance of features and to tune the recommender, we proposed three methods for determining an optimal set of features' weights. Since the order of recommendations is irrelevant, we can describe searching for nearest neighbors as a binary classification problem where the objective is to classify a pair of movies as good or bad recommendations. We experimentally showed that coefficients obtained by feature selection and classification algorithms can be used as weights in CBF. Input to algorithms were movie pairs represented by vectors of normalized distances between consecutive features of both movies.

Since there is no universal feature weighting method that works for all features configurations, we tried three different methods:

- Coefficients of a linear classifier such as Support Vector Machine (SVM) with linear kernel, perceptron, or logistic regression. In [15] similar procedure was developed for a classification task. This set of models works analogously to the presented recommender system. A linear combination of features is calculated. In the classification case, the calculated sum is compared against a fixed threshold to determine an output label. In our system, a linear combination is interpreted as a distance between movies in a pair, and only five movies closest to the input are selected. We used SVM as a representative of linear classifiers based on results from our initial experiments.
- ReliefF [21], a filter method for feature selection and feature weighting [29].
- Mean Decrease Impurity (MDI) [14] of variables in a random forest classifier.

4.4 Finding the Best Set of Features

Our goal was to find a subset of features that are most relevant in predicting the target variable, i.e., whether a recommendation is good or bad. Evaluating recommendations for all possible feature subsets (2^N) is intractable. To address the problem, we used Recursive Feature Elimination (RFE) with SVM, originally proposed by [9] for feature selection in classifiers. RFE is a greedy algorithm that helps find a subset of a *given size* by recursively removing the least important features (i.e., the least informative during SVM classification). To find an *optimal size* of a subset, we combined RFE with 10-fold cross-validation. Since we optimized our recommender towards high precision rather than high recall, we used precision as a scoring function. For the same reason, we penalized the SVM training cost function for returning bad recommendations (false positives errors) more (4 times) than for missing good ones (false negative errors). Once we found an optimal features subset, we trained SVM over it and treated its coefficients as *optimal weights* for our recommender. Finally, we validated those weights with a recommender against relevance judgments.

5 Experimental Results

As a result of crowdsourcing, we collected 15334 annotations from 33 annotators for 6901 movie pairs. Out of those we got 5500 rated movie pairs (for the remaining ones users could not agree about the rating). The split between good and bad recommendations is 2988/2512 (54%/46%). We published 30% of collected ratings as MoreLikeThis dataset[7].

5.1 How to Represent Features Effectively?

We reviewed the set of representation methods for the following features: *genre*, *plot*, and *summary*. We asked how the chosen representation affects recommendation performance, as different techniques have their own characteristics. In the first experiment, we evaluated how those selected features work together with other features (*features collectively*). The results of experiments are depicted in Table 1.

For *plot* and *summary* we tested the following document representation models: Doc2Vec [11], LDA [3], LSI with TF-IDF [7], and USE [4]. We collected ratings for output recommendations achieving coverage of about 0.6 for each model, which was acceptable for comparison of precision. The results showed that USE and Doc2Vec models provide higher precision (0.62 and 0.60 respectively).

In the second part of the experiment we analysed only textual features – plot and summary – evaluating representations *in isolation* as the single-feature recommender. We narrowed methods in this analysis to two the most promising semantic models – USE and Doc2Vec. Sufficient coverage was obtained only for analysis of summary feature (0.64 for Doc2Vec and 0.72 for USE). For remaining configurations (plot, and summary+plot), the gap in coverage between the two represenations was too be big to make any conclusions about them. Still, higher precision for USE in case of summary demonstrates that it outperforms Doc2Vec even though USE was trained on external data and was not even fine-tuned to our Movies Dataset.

All of the above confirms that transformer-based models pre-trained on large datasets are more powerful and generalizable.

Precision for Word2Vec embedding method applied to genre data was relatively lower than sparse vector (Table 1). Using not only exact genre matches, but also similar ones, probably increased the number of false positives.

Overall, our experiments showed that representation strategies can play a crucial role in the final recommendations.

[7] https://github.com/la-samsung-poland/more-like-this-dataset.

Table 1. Performance of recommenders with various representations for a selected feature: (1) features collectively, (2) features in isolation. Weights were split evenly across features. In the first step, for uninvestigated features we used encoding methods depicted in 3.3, specifically USE representation for *plot* and *summary*, and Word2Vec representation for *genre*.

Feature(s)	Method	Precision	Coverage
Features collectively			
Summary +Plot	Doc2Vec	0.78	0.60
	USE	0.81	0.62
	LDA	0.80	0.57
	LSI	0.79	0.56
Genres	Simple Vector	0.86	0.62
	Word2Vec	0.81	0.62
Features in isolation			
Summary	Doc2Vec	0.45	0.64
	USE	0.86	0.72
Plot	Doc2Vec	0.52	0.19
	USE	0.71	0.71
Summary +Plot	Doc2Vec	0.56	0.24
	USE	0.90	0.61

5.2 Which Features Are Relevant?

In the next experiments, we used only selected representations and distance metrics: the logarithm version of *release year* with the Euclidean distance, averaged Word2Vec vector with the cosine distance for *genre*, a vector of metrics with the Euclidean distance for *popularity*, USE model for textual features *plot* and *summary*, and sparse vectors with the Jaccard distance for *director*, *screenwriter*, and *language* features.

The main goal of this section is to gain an initial insight into the relevance of particular features. We ranked those features from the most relevant using the weight optimization procedure described in Sect. 4.3. Results are summed up in Fig. 1.

All three algorithms are almost consistent – *summary*, *category*, *plot*, and *genre* were important according to all of them. Two of these features, *summary* and *category*, contain a wide range of information, e.g., brief storyline, cast, awards, or influence of the film on popular culture. We suspected them to be correlated with each other and with remaining features. Then, we looked for an optimal subset of features analyzing whether they are redundant (described in Sect. 5.4).

We evaluated recommender with three different sets of weights obtained by SVM, ReliefF and MDI, and one set of equal weights. Results are available in

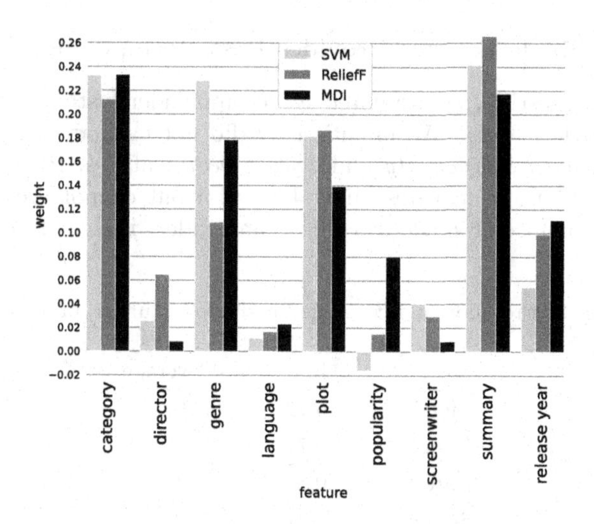

Fig. 1. Normalized (sum up to 1) weights for each feature obtained by SVM, ReliefF, and MDI.

Table 2. Coverage for each recommender was above 0.60 which allowed us to make further conclusions.

Precisions of recommenders with weights optimized by all methods are similar to each other and exceptionally higher than in the equal-weights scenario. That confirms the effectiveness of SVM, ReliefF, and MDI as weights selection algorithms.

Table 2. Evaluation of the recommender with weight obtained by SVM, ReliefF and MDI compared with setting all weights to the same value.

	Precision	Coverage
SVM	0.92	0.65
ReliefF	0.91	0.63
MDI	0.90	0.66
Equal weights	0.81	0.62

Some features turned out to be more informative, i.e., they are better at *discriminating* good recommendations from bad ones. However, we do not know which provide better recommendations.

5.3 Which Single Feature Provides Best Recommendations?

To answer this question, we evaluated six recommender systems, each consisting of a single movie feature. We intentionally did not evaluate recommendations created by *language*, *popularity* and *release year* features. By common sense, we assumed that these features might be useful but cannot create standalone systems. Performance of single-feature recommenders is shown in Table 3.

Table 3. Precision and coverage for single-feature recommenders.

	Precision	Coverage
Plot	0.72	0.70
Genre	0.77	0.39
Category	0.87	0.53
Summary	0.87	0.72
Director	0.64	0.37
Screenwriter	0.65	0.35

The results confirmed our findings from Sect. 5.2: *category* and *summary* are great standalone features while *plot* and *genre* give slightly worse but still decent performance.

5.4 Do We Need All Features to Get Good Recommendations?

We used RFE with cross-validation (Sect. 4.4) to find the smallest subset of features. The results are shown in Fig. 2. It can be observed that RFE classification precision for five and nine features is almost the same (0.828 and 0.820 respectively). It demonstrates that we can remove four features and still get the best recommendations for a given setup. The best subset contains the following features (together with their optimal weights): *genre* (0.37), *plot* (0.20), *summary* (0.20), *categories* (0.20), and *release year* (0.03). The validation of this configuration with a recommender got 0.93 precision for 0.55 coverage. The validation of configuration with all features got the precision of 0.92 as well with coverage of 0.65.

If we compare the recommendation performance of this combination with the performance of single features (see Sect. 5.3), we can see that no feature alone can achieve such a high precision. The best single features, *summary* or *category*, have precision of 0.87. The results also indicate that it pays off to engineering features with more complex representations (*summary*, *plot*, and *categories*).

The remaining features: *director*, *languages*, *popularity* and *screenwriter* were found redundant.

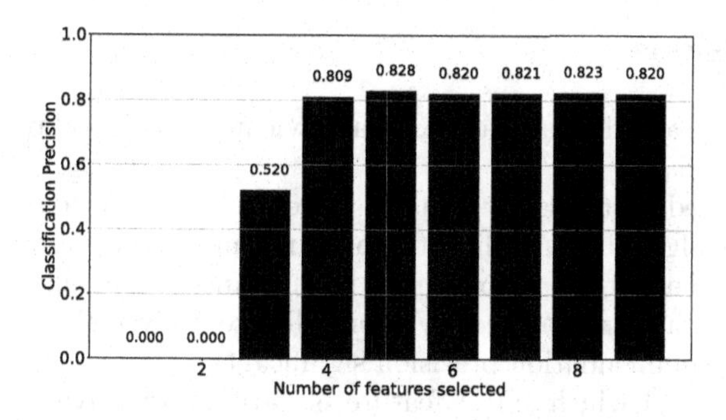

Fig. 2. The results for RFE with SVM and 10-fold cross-validation over relevance judgments.

5.5 Are Relevance Judgments Credible?

We asked whether collected relevance judgments were *credible* and *consistent.*

We pushed random movie pairs and regular recommendations to the same group of annotators. The distributions of ratings for both recommendation types are presented in Fig. 3. It showed that random recommendations were mostly rated as bad ones, whereas regular recommendations as good ones. That is in line with our intuition and demonstrates that our annotators provide credible ratings. That gives us trust in the results of the experiments.

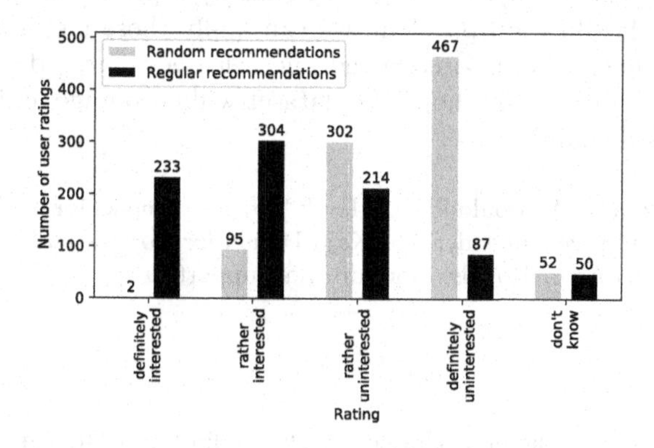

Fig. 3. Distribution of user ratings for regular and random recommendations in the first round of ratings collection.

Additionally, we checked whether users rated the same movie pair consistently and found that 70% of movie pairs had a complete agreement, i.e., all annotators agreed that a recommendation is good or bad. Remaining movie pairs had $\frac{2}{3}$ agreement. This demonstrates that ratings were relatively consistent across users.

6 Conclusions

In this paper, we asked what features of a movie make it a good recommendation for another movie.

We researched methods for feature representation. We found that various representation algorithms applied to the same movie attribute may result in different recommendation performance. For instance, changing movie *summary* or *plot* representations from Doc2Vec or LSI into Universal Sentence Encoder can increase recommendation precision significantly.

We also studied which single feature is most informative and can provide the best recommendations. We found that users rate recommendations generated solely from a movie *summary* or *category* higher than when using other features. We also found that combining these features with others improves recommendations quality, but not significantly. This suggests there is potential for single-feature recommenders if the feature is represented properly.

We also looked for the smallest set of features with high recommendation performance. We found that using only five features (*genres*, *plot*, *summary*, *categories*, and *release year*) we can provide as good recommendations as using all nine proposed features. Surprisingly, other features such as movie *director*, *screenwriter*, or *language*, often mentioned in the literature [6,13], were found redundant. Removing those features can shorten recommender response time, save storage space, and cut costs of acquiring data for those features.

Machine learning has a long-standing list of methods for feature ranking, weighting, and selection. We have shown that by representing a recommendation task as a classification problem, we can apply those methods for content-based recommenders. We also collected and released a large dataset of movie-recommendation ratings. We hope the dataset will encourage and enable future research in this domain.

Acknowledgments. We would like to thank Wojciech Smołka for his substantial help in implementing the recommender and Kaja Pękala for work in building datasets. We are also grateful to Artur Rogulski for editorial support.

References

1. Bennett, J., et al.: The netflix prize. In: Proceedings of KDD Cup and Workshop, vol. 2007, p. 35. New York (2007)
2. Bergamaschi, S., Po, L.: Comparing LDA and LSA topic models for content-based movie recommendation systems. In: Monfort, V., Krempels, K.H. (eds.) Web Information Systems and Technologies. WEBIST 2014. Lecture Notes in Business Information Processing, textbf226, 247–263. Springer, Cham(2014). https://doi.org/10.1007/978-3-319-27030-2_16
3. Blei, D.M., Ng, A.Y., Jordan, M.I.: Latent Dirichlet allocation. J. Mach. Learn. Res. **3**(Jan), 993–1022 (2003)
4. Cer, D., et al.: Universal Sentence Encoder (2018). arXiv preprint: arXiv: 1803.11175

5. Chen, H.W., Wu, Y.L., Hor, M.K., Tang, C.Y.: Fully content-based movie recommender system with feature extraction using neural network. In: 2017 International Conference on Machine Learning and Cybernetics (ICMLC), vol. 2, pp. 504–509. IEEE (2017)
6. Colucci, L., et al.: Evaluating item-item similarity algorithms for movies. In: Proceedings of the 2016 CHI Conference Extended Abstracts on Human Factors in Computing Systems, pp. 2141–2147 (2016)
7. Deerwester, S., Dumais, S., Landauer, T., Furnas, G., Beck, L.: Improving information-retrieval with latent semantic indexing. In: Proceedings of the ASIS Annual Meeting, vol. 25, pp. 36–40 (1988)
8. Dooms, S., De Pessemier, T., Martens, L.: Movietweetings: a movie rating dataset collected from Twitter. In: Workshop on Crowdsourcing and Human Computation for Recommender systems, CrowdRec at ACM RecSys, vol. 2013, p. 43 (2013)
9. Guyon, I., Weston, J., Barnhill, S., Vapnik, V.: Gene selection for cancer classification using support vector machines. Mach. Learn. **46**(1–3), 389–422 (2002)
10. Harper, F.M., Konstan, J.A.: The MovieLens datasets: history and context. ACM Trans. Interact. Intell. Syst. (TiiS) **5**(4), 1–19 (2015)
11. Le, Q., Mikolov, T.: Distributed representations of sentences and documents. In: International Conference on Machine Learning, pp. 1188–1196 (2014)
12. Lehmann, D.R., Hulbert, J.: Are three-point scales always good enough? J. Mark. Res. **9**(4), 444–446 (1972)
13. Leng, H., et al.: Finding Similar Movies: Dataset, Tools, and Methods (2018)
14. Louppe, G., Wehenkel, L., Sutera, A., Geurts, P.: Understanding variable importances in forests of randomized trees. In: Advances in Neural Information Processing Systems,. pp. 431–439 (2013)
15. Mladenić, D., Brank, J., Grobelnik, M., Milic-Frayling, N.: Feature selection using linear classifier weights: interaction with classification models. In: Proceedings of the 27th Annual International ACM SIGIR Conference on Research and Development in Information Retrieval, pp. 234–241 (2004)
16. Musto, C., Semeraro, G., de Gemmis, M., Lops, P.: Learning word embeddings from Wikipedia for content-based recommender systems. In: Ferro, N., et al. (eds.) Advances in Information Retrieval. ECIR 2016. Lecture Notes in Computer Science, **9626**, 729–734. Springer, Cham (2016). https://doi.org/10.1007/978-3-319-30671-1_60
17. Nasery, M., Elahi, M., Cremonesi, P.: Polimovie: a feature-based dataset for recommender systems. In: ACM (2015)
18. Nguyen, L.V., Nguyen, T.H., Jung, J.J.: Content-based collaborative filtering using word embedding: a case study on movie recommendation. In: Proceedings of the International Conference on Research in Adaptive and Convergent Systems (ACM RACS), pp. 96–100. ACM (2020)
19. Odić, A., Tkalčič, M., Tasič, J.F., Košir, A.: Predicting and detecting the relevant contextual information in a movie-recommender system. Interact. Comput. **25**(1), 74–90 (2013)
20. Preston, C.C., Colman, A.M.: Optimal number of response categories in rating scales: reliability, validity, discriminating power, and respondent preferences. Acta Psychol. **104**(1), 1–15 (2000)
21. Robnik-Šikonja, M., Kononenko, I.: An adaptation of relief for attribute estimation in regression. In: Machine Learning: Proceedings of the Fourteenth International Conference (ICML1997), vol. 5, pp. 296–304 (1997)

22. Ronen, R., Koenigstein, N., Ziklik, E., Nice, N.: Selecting content-based features for collaborative filtering recommenders. In: Proceedings of the 7th ACM Conference on Recommender Systems, pp. 407–410 (2013)
23. Singla, R., Gupta, S., Gupta, A., Vishwakarma, D.K.: FLEX: a content based movie recommender. In: International Conference for Emerging Technology (INCET), pp. 1–4. IEEE (2020)
24. Soares, M., Viana, P.: Tuning metadata for better movie content-based recommendation systems. Multimed. Tools Appl. **74**(17), 7015–7036 (2015)
25. Suglia, A., Greco, C., Musto, C., De Gemmis, M., Lops, P., Semeraro, G.: A deep architecture for content-based recommendations exploiting recurrent neural networks. In: Proceedings of the 25th Conference on User Modeling, Adaptation and Personalization, pp. 202–211 (2017)
26. Tang, J., Alelyani, S., Liu, H.: Feature Selection for Classification: A Review. Data classification: Algorithms and Applications p. 37 (2014)
27. Tonon, A., Demartini, G., Cudré-Mauroux, P.: Pooling-based continuous evaluation of information retrieval systems. Inf. Retr. J. **18**(5), 445–472 (2015)
28. Webber, W., Park, L.A.: Score adjustment for correction of pooling bias. In: Proceedings of the 32nd International ACM SIGIR Conference on Research and Development in Information Retrieval, pp. 444–451 (2009)
29. Wettschereck, D., Aha, D.W., Mohri, T.: A review and empirical evaluation of feature weighting methods for a class of lazy learning algorithms. Artif. Intell. Rev. **11**(1–5), 273–314 (1997)

Indexed Polygon Matching Under Similarities

Fernando Luque-Suarez[1], J. L. López-López[2], and Edgar Chavez[1(✉)]

[1] CICESE Ensenada, Ensenada, Mexico
elchavez@cicese.mx
[2] UMSNH, Morelia, Mexico
jlopez@umich.mx

Abstract. Polygons appear as constructors in many applications and deciding if two polygons match under similarity transformations and noise is a fundamental problem. Solutions in the literature consider only matching pairs of polygons, implying a sequential comparison when we have a collection. In this paper, we present the first algorithm allowing indexed retrieval of polygons under similarities. We reduce the problem to searching points in the plane, exact searching in the absence of noise, and approximate searching for similar noisy polygons. The above gives a $O(n + \log(m))$ time algorithm to find the matching polygons under noise and $O(1)$ time for exact similar polygons. We tested our heuristic for indexed polygons in an extensive collection of convex, star-shaped, simple, and self-intersecting polygons. For small amounts of noise, we achieve perfect recall for all polygons. For large amounts of noise, the lowest recall is for convex polygons, while attaining the highest recall is for general (self-intersecting) polygons. The above is not a significant limitation. To recover convex polygons efficiently before indexing, we define a random permutation of the vertices, converting all input polygons to a general polygon and achieving the same successful recovery rates, which is a perfect recall for high noise levels.

Keywords: Polygon matching · Shape matching · Shape indexing

1 Introduction

Shapes appear in many applications fields like computer-aided design, computer-aided manufacturing, computer vision [18], medical imaging [12] and even archaeology [16]. Shape analysis deals with the concept of matching shapes. The definition of matching changes with the application field. It ranges from congruence transformations, where the shapes could be rotated, translated, or reflected without being scaled, to similarity transformations, which includes scaling to the previous set of transformations, affine transformations, projective transformations, to Riemannian isometries (for curved surfaces), and conformal mappings, or more general transformations. Matching could also include partial matching, where only a portion of the shape has a match. As a rule of thumb, the more general the transformation, it is more difficult to find a fast algorithm to find the best match. For an arbitrary transformation, the problem is NP-complete.

N. Reyes et al. (Eds.): SISAP 2021, LNCS 13058, pp. 295–306, 2021.
https://doi.org/10.1007/978-3-030-89657-7_22

We fix our attention on the fundamental problem of complete polygon matching, as opposed to partial matching. The problem of partial matching, or dealing with insertions and deletions, can be handled by fragmenting the polygons being compared. Moreover, we are especially interested in the indexed version of the problem instead of just comparing two polygons for matching.

1.1 The Problem: Indexed Polygon Matching

We define the problem of *indexed matching.* A collection of polygons is preprocessed and stored, and a query is presented to the system. The outcome will be all the matching shapes in the collection.

We shall identify points (x, y) in the plane with corresponding complex numbers $z = x + y\sqrt{-1}$. A polygon in the plane will be an ordered set of points, or complex numbers, where the *order specifies consecutive vertices.* Self-intersection are allowed since the order is arbitrary. Therefore, the cyclic shifts $(z_2, z_3, \ldots, z_n, z_1)$, $(z_3, z_4, \ldots, z_1, z_2), \ldots, (z_n, z_1, \ldots, z_{n-2}, z_{n-1})$ and the reversed labeling $(z_n, z_{n-1}, \ldots, z_2, z_1)$ determine different labels for the same polygon $(z_1, z_2, \ldots, z_{n-1}, z_n)$. But a general permutation $\mathfrak{p} : \{1, 2, \ldots, n\} \to \{1, 2, \ldots, n\}$ could determine a different polygon $(z_{\mathfrak{p}(1)}, \ldots, z_{\mathfrak{p}(n)})$ because the consecutive vertices vary and therefore the edges are different.

An affine transformation $f : \mathbb{R}^2 \to \mathbb{R}^2$ can be (uniquely) written in terms of sums and products of complex numbers as

$$f(z) = \alpha z + \beta \bar{z} + \gamma,$$

where $\alpha, \beta, \gamma \in \mathbb{C}$ and $|\alpha|^2 - |\beta|^2 = \det f \neq 0$. Here $\bar{z}$ stands for the complex conjugated of z. When $\beta = 0$ the affine transformation is a similarity transformation.

Given polygons $Z = (z_1, z_2, \ldots, z_n) \in \mathbb{C}^n$ and $W = (w_1, w_2, \ldots, w_n) \in \mathbb{C}^n$, our problem consists of determining if there exists an affine transformation f such that $Z = f(W)$. Since affine transformations have three complex parameters, finding two corresponding triples of consecutive points in both polygons is enough. A naïve procedure will be to fix a triplet in Z and try all the cyclic shifts in W to find the correspondence, which takes $O(n)$ operations for one triplet. Since there are $O(n)$ consecutive triplets, the entire process takes $O(n^2)$ operations.

Now assume we have a given collection of polygons $Z_1, Z_2, \ldots, Z_m$ and a query polygon W, and we want to know which of the Z_ℓ are images of W under a similarity. Using a sequential approach and the naïve procedure above, the solution can be found in $O(mn^2)$ operations. In general, without an index, the complexity will depend linearly on the number of polygons in the collection, multiplied by the complexity of an individual match. We will show how to improve this complexity using the defined invariants and a two-dimensional index for querying.

1.2 Summary of Results

For polygons $Z \in \mathbb{C}^n$, we construct complex scalar functions $\varphi_j : \mathbb{C}^n \to \mathbb{C}, j = 1, \ldots, \lfloor (n-1)/2 \rfloor$ with the following properties

1. $\varphi_j^n(Z) = \varphi_j^n(f(Z))$ with $f : \mathbb{C}^n \to \mathbb{C}^n$ an arbitrary similarity function, including mirroring and cyclic shifts.
2. φ_j is analytic, that is for $\Delta Z = (\Delta z_1, \ldots, \Delta z_n)$, an unknown *bounded* additive noise, we have $|\varphi_j(Z)^n - \varphi_j(Z + \Delta(Z))^n| \leq r$.
3. If $Z, W \in \mathbb{C}^n$ and Z and W are not similar, then $\varphi_j(Z) \neq \varphi_j(W)$ almost surely.
4. For a collection of polygons $Z_1, \ldots, Z_m$ and a query polygon W, we show how to use the previous properties to preprocess $Z_1, \ldots, Z_m$ to quickly find all Z_i such that $Z_i = f(W)$. This is done by using a two-dimensional spatial index to store $\varphi_j(Z_1)^n, \ldots, \varphi_j(Z_m)^n$ at preprocessing time, and finding the nearest neighbor of $\varphi_j(W)^n$, as $NN(\varphi_j(W)^n$ at query time.
5. The above procedure has high recall only for general, auto-intersecting polygons when the amount of noise Δ is above a certain threshold. We show that if we permute the polygons Z_i before indexing, using an arbitrary but fixed permutation Π, we can obtain the same high recall results even for convex polygons, which had the lowest recall rates without permutations.

This paper is an experimental report of a previous theoretical paper [7]. We reproduce here the mains results to make this contribution self-contained. The experimental parts, not reported before, corresponds to numerals 4 and 5 above. In particular, using the nearest neighbor search, or k-nearest neighbor search when we expect multiple matches for the query polygon is new. In the previous paper, we derived precise bounds, also discussed for a complete presentation, where there was the need for a precise maximum radius. This radius depends on the polygon as well as the amount of noise. Each polygon has associated a maximum allowed noise, posing difficulties for indexing, as convex polygons are more sensitive to noise. We show experimentally in this paper that we can achieve maximum tolerated noise for most polygons by randomly permuting the polygon vertex with a fixed permutation.

1.3 Related Work

Before matching polygons in a natural scene, it is necessary to detect them. In general, reconstructing an arbitrary polygon from partial readings is NP-hard [5]; although some instances can be solvable in practice, such as detecting regular polygons [4,13]. However, shapes and contours can be obtained from other sources, and sometimes the problem consists in measuring if a set of points can be put in correspondence with a nominal polygon; this has applications in manufacturing inspection and city planning [10]. Two arbitrary simple polygons can be compared using several notions of distance [2], being the more general the Fréchet distance as described in [6], which can be computed in polynomial time.

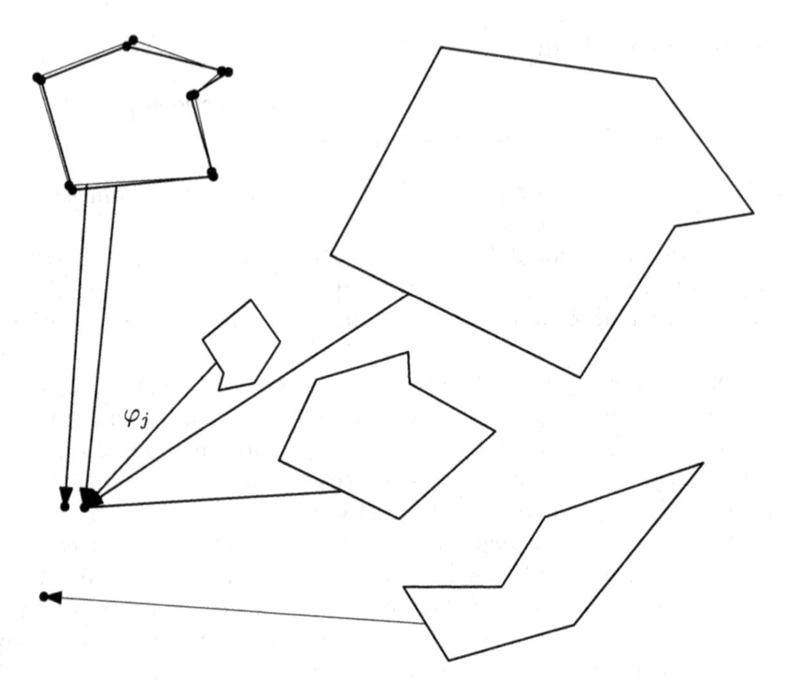

Fig. 1. Application of the invariant φ_j to similar polygons gives the same complex number. For similar polygons plus noise, it gives complex numbers that are close under euclidean distance. For not similar polygons it gives different complex numbers.

Some heuristics have been defined for other simpler realizations of the distance between polygons. One approach is to consider the polygon as a (circular) string with either the edges [17] or the vertices as symbols [11]; this allows efficient comparison of a shape against a nominal polygon allowing insertions and deletions of vertices. Another efficient metric is discussed in [1]. In [15] are discussed algorithms for the specific case of polygon matching upon congruency, including the case of partial matches. A more comprehensive discussion of the problem of shape matching and several efficient approaches are discussed in the survey [18].

To the best of our knowledge, there is no prior attempt to solve the indexed polygon matching discussed in this paper. The metrics mentioned above are designed to compare pairs of polygons and do not contemplate the problem of indexed matching. Moreover, any function φ_j holds more information than a metric because φ_j endows the space of polygons with a two-dimensional coordinate system (complex numbers are two-dimensional). In contrast, a metric can only be considered a one-dimensional coordinate because each polygon is associated with a real number, which is the distance to a fixed polygon.

2 Invariants

2.1 Similarity Invariants for Polygons

In what follows, we shall fix an integer $n \geq 3$.

Definition 1. *For any integer $j = 1, \ldots, \lfloor (n-1)/2 \rfloor$ we consider the function $\varphi_j : \mathbb{C}^n \to \mathbb{C} \cup \{\infty\}$ given by*

$$\varphi_j(z_1, \ldots, z_n) = \frac{\sum_{k=1}^{n} \lambda^{jk} z_k}{\sum_{k=1}^{n} \lambda^{-jk} z_k},$$

where $\lambda = e^{2\pi\sqrt{-1}/n}$ is a nth root of unit.

Proposition 1. *φ_j is invariant under the action of orientation-preserving similarity transformations on polygons with n vertices; that is, if $\alpha, \gamma \in \mathbb{C}$ with $\alpha \neq 0$, then*

$$\varphi_j(\alpha z_1 + \gamma, \alpha z_2 + \gamma, \ldots, \alpha z_n + \gamma) = \varphi_j(z_1, z_2, \ldots, z_n).$$

Proof.

$$\varphi_j(\alpha z_1 + \gamma, \alpha z_2 + \gamma, \ldots, \alpha z_n + \gamma) = \frac{\sum_{k=1}^{n} \lambda^{jk}(\alpha z_k + \gamma)}{\sum_{k=1}^{n} \lambda^{-jk}(\alpha z_k + \gamma)} =$$

$$\frac{\alpha \sum_{k=1}^{n} \lambda^{jk} z_k + \gamma \sum_{k=1}^{n} \lambda^{jk}}{\alpha \sum_{k=1}^{n} \lambda^{-jk} z_k + \gamma \sum_{k=1}^{n} \lambda^{-jk}} = \frac{\alpha \sum_{k=1}^{n} \lambda^{jk} z_k + \gamma \sum_{k=0}^{n-1} \lambda^{jk}}{\alpha \sum_{k=1}^{n} \lambda^{-jk} z_k + \gamma \sum_{k=0}^{n-1} \lambda^{-jk}}$$

$$\frac{\alpha \sum_{k=1}^{n} \lambda^{jk} z_k + \gamma(\lambda^{jn} - 1)/(\lambda^{j} - 1)}{\alpha \sum_{k=1}^{n} \lambda^{-jk} z_k + \gamma(\lambda^{-jn} - 1)/(\lambda^{-j} - 1)} =$$

$$\frac{\alpha \sum_{k=1}^{n} \lambda^{jk} z_k}{\alpha \sum_{k=1}^{n} \lambda^{-jk} z_k} = \frac{\sum_{k=1}^{n} \lambda^{jk} z_k}{\sum_{k=1}^{n} \lambda^{-jk} z_k} = \varphi_j(z_1, z_2, \ldots, z_n).$$

Remarks

Remark 1. The numerator and denominator involved in the definition of φ_j are the coefficients appearing when $Z = (z_1, \ldots, z_n)$ is expressed in certain basis of $\mathbb{C}^n$, namely the basis of star-shaped polygons

$$E_k = ((\lambda^k)^1, (\lambda^k)^2, \ldots, (\lambda^k)^{n-1}, (\lambda^k)^n), \quad k = 1, 2, \ldots, n$$

([8], [14, proof of Proposition 3]). More precisely, if $Z = \sum_{k=1}^{n} x_k E_k$, then

$$\varphi_1 = \frac{x_{n-1}}{x_1}, \quad \varphi_2 = \frac{x_{n-2}}{x_2}, \ldots, \quad \varphi_{(n-1)/2} = \frac{x_{(n+1)/2}}{x_{(n-1)/2}} \quad \text{if } n \text{ is odd},$$

and

$$\varphi_1 = \frac{x_{n-1}}{x_1}, \quad \varphi_2 = \frac{x_{n-2}}{x_2}, \ldots, \quad \varphi_{(n-2)/2} = \frac{x_{(n+2)/2}}{x_{(n-2)/2}} \quad \text{if } n \text{ is even.}$$

All the quotients x_i/x_j satisfy Proposition 1, but only those of the form x_{n-j}/x_j satisfy a more general theorem involving affine transformations as described in [7].

Remark 2. For $Z = (z_1, \ldots, z_n) \in \mathbb{C}^n$ the precise form of the coefficients of the linear combination $Z = \sum_{k=1}^{n} x_k E_k$ is

$$x_k = \frac{1}{n} \sum_{l=1}^{n} \lambda^{-kl} z_l$$

They are precisely the Fourier descriptors or coefficients of the discrete Fourier transform of Z. This is very handy in our experimental construction.

Remark 3. The function φ_j is well-defined except on

$$\mathcal{N}_j = \left\{ (z_1, \ldots, z_n) \in \mathbb{C}^n : \sum_{k=1}^{n} \lambda^{jk} z_k = 0 = \sum_{k=1}^{n} \lambda^{-jk} z_k \right\}.$$

$\mathcal{N}_j$ is a $(n-2)$-dimensional complex linear subspace with measure zero in $\mathbb{C}^n$. According to Remark 1, $\mathcal{N}_j$ is spanned by $\{E_k\}_{k \neq j, n-j}$.

Remark 4. The level sets $\varphi_j^{-1}(c) = \{(z_1, \ldots, z_n) \in \mathbb{C}^n : \varphi_j(z_1, \ldots, z_n) = c\}$ are $(n-1)$-dimensional complex submanifolds with measure zero in $\mathbb{C}^n$ because every point in $\mathbb{C} \cup \{\infty\}$ is a regular value of φ_j, for any j. This follows from a straightforward calculation which shows that $\dfrac{\partial \varphi_j}{\partial z_k} = 0$ implies $\varphi_j = \lambda^{2jk}$. In this sense, the probability that two randomly chosen polygons Z and W satisfy $\varphi_j(Z) = \varphi_j(W)$ is equal to zero.

2.2 Cyclic Shifts and Reversed Labeling

Proposition 2. *The behavior of φ_j under cyclic shift and reversed labeling is given by the formulas*

$$\varphi_j(z_2, z_3, \ldots, z_n, z_1) = \lambda^{-2j} \varphi_j(z_1, z_2, \ldots, z_n),$$
$$\varphi_j(z_n, z_{n-1}, \ldots, z_2, z_1) = \frac{\lambda^{2j}}{\varphi_j(z_1, z_2, \ldots, z_n)},$$

for all $j = 1, \ldots, \lfloor (n-1)/2 \rfloor$. Hence, if Z and W are relabeling of the same polygon, we have by raising to the nth power the equalities

$$\begin{array}{l} \varphi_j(Z)^n = \varphi_j(W)^n \text{ if the labels have the same orientation, and} \\ \varphi_j(Z)^n = \varphi_j(W)^{-n} \text{ if the labels have the opposite orientation.} \end{array} \tag{1}$$

Proof.

$$\varphi_j(z_2, z_3, \ldots, z_n, z_1) = \frac{\sum_{k=1}^{n} \lambda^{jk} z_{k+1}}{\sum_{k=1}^{n} \lambda^{-jk} z_{k+1}} =$$
$$\frac{\lambda^{-j} \sum_{k=1}^{n} \lambda^{j(k+1)} z_{k+1}}{\lambda^{j} \sum_{k=1}^{n} \lambda^{-j(k+1)} z_{k+1}} = \lambda^{-2j} \varphi_j(z_1, z_2, \ldots, z_n),$$

where subscript $n+1$ should be taken as 1. Likewise

$$\varphi_j(z_n, z_{n-1}, \ldots, z_2, z_1) = \frac{\sum_{k=1}^{n} \lambda^{jk} z_{n+1-k}}{\sum_{k=1}^{n} \lambda^{-jk} z_{n+1-k}} =$$
$$\frac{\lambda^{j(n+1)} \sum_{k=1}^{n} \lambda^{-j(n+1-k)} z_{n+1-k}}{\lambda^{-j(n+1)} \sum_{k=1}^{n} \lambda^{j(n+1-k)} z_{n+1-k}} = \frac{\lambda^{2j}}{\varphi_j(z_1, z_2, \ldots, z_n)}.$$

2.3 An Index for Matching Polygons

Exact Matching of Similar Polygons. Assume that a collection of different polygons $Z_1, Z_2, \ldots, Z_m$ of n edges is given. By a preprocessing step we compute pairs $(\ell, \varphi_j(Z_\ell)^n)$. Assume that a query polygon W is given and that the objective is to find all the polygons in the collection such that $W = f(Z_\ell)$ for some unknown similarity transformation f. This corresponds to all the polygons such that $\varphi_j(Z_\ell)^n = \varphi_j(W)^n$ or $\varphi_j(Z_\ell)^n = \varphi_j(W)^{-n}$ (Propositions 1 and 2). Since the probability of collision is zero (Remark 4), all the R polygons mapped to $\varphi_j(Z_\ell)^n$ or $\varphi_j(Z_\ell)^{-n}$ should be similar to the query polygon, and can be found in $O(n+R)$ operations, where R is the number of polygons mapped to $\varphi_j(Z_\ell)^n$ or $\varphi_j(Z_\ell)^{-n}$. Notice that the bound in the running time is time independent of m, the size of the collection.

Matching Similar Polygons Under Noisy Conditions. A slightly more general setup is when there is an unknown noise function at the matching. The image of the query polygon is a similarity transformation f plus noise, namely $W = (f(z_1+\Delta z_1), f(z_2+\Delta z_2), \ldots, f(z_n+\Delta z_n))$. In this case, instead of retrieving just the polygons mapped to $\varphi_j(W)^n$ as above, we retrieve all the polygons within a certain Euclidean distance of the image of the query. That is, if r is the tolerance, then we inspect all the polygons such that $|\varphi_j(W)^n - \varphi_j(Z_\ell)^n|^2 \leq r$.

Proposition 3 below gives a precise bound for the tolerated noise.

Proposition 3. *Let $Z = (z_1, \ldots, z_n) \in \mathbb{C}^n \setminus \{0\}$ be a polygon. Consider an integer $j \in \mathbb{Z} \setminus \frac{n}{2}\mathbb{Z}$ such that $\sum_{l=1}^{n} \lambda^{-jl} z_l \neq 0$. Let ρ be a positive real number such that $n\rho < \mu := |\sum_{l=1}^{n} \lambda^{-jl} z_l|$. Then for any $\Delta Z = (\Delta z_1, \ldots, \Delta z_n)$ with $|\Delta z_k| < \rho$, $k = 1, \ldots, n$, we have*

$$|\varphi_j(z_1, \ldots, z_n) - \varphi_j(z_1 + \Delta z_1, \ldots, z_n + \Delta z_n)| \leq \frac{2n\rho \sum_{l=1}^{n} |z_l|}{\mu(\mu - n\rho)}.$$

Proof.

$$\varphi_j(z_1,\dots,z_n) - \varphi_j(z_1+\Delta z_1,\dots,z_n+\Delta z_n)$$

$$= \frac{\sum_{l=1}^n \lambda^{jl} z_l}{\sum_{l=1}^n \lambda^{-jl} z_l} - \frac{\sum_{k=1}^n \lambda^{jk}(z_k+\Delta z_k)}{\sum_{k=1}^n \lambda^{-jk}(z_k+\Delta z_k)}$$

$$= \frac{\sum_{l=1}^n \lambda^{jl} z_l}{\sum_{l=1}^n \lambda^{-jl} z_l} - \frac{\sum_{k=1}^n \lambda^{jk} z_k + \sum_{k=1}^n \lambda^{jk}\Delta z_k}{\sum_{k=1}^n \lambda^{-jk} z_k + \sum_{k=1}^n \lambda^{-jk}\Delta z_k}$$

$$= \frac{\sum_{l=1}^n \lambda^{jl} z_l \sum_{k=1}^n \lambda^{-jk}\Delta z_k - \sum_{l=1}^n \lambda^{-jl} z_l \sum_{k=1}^n \lambda^{jk}\Delta z_k}{\sum_{l=1}^n \lambda^{-jl} z_l \left(\sum_{k=1}^n \lambda^{-jk} z_k + \sum_{k=1}^n \lambda^{-jk}\Delta z_k\right)} \tag{2}$$

$$= \frac{\sum_{k,l=1}^n (\lambda^{j(l-k)} - \lambda^{j(k-l)}) z_l \Delta z_k}{\sum_{l=1}^n \lambda^{-jl} z_l \left(\sum_{k=1}^n \lambda^{-jk} z_k + \sum_{k=1}^n \lambda^{-jk}\Delta z_k\right)}$$

$$= \frac{2i \sum_{k,l=1}^n \sin\left(\frac{2\pi j(l-k)}{n}\right) z_l \Delta z_k}{\sum_{l=1}^n \lambda^{-jl} z_l \left(\sum_{k=1}^n \lambda^{-jk} z_k + \sum_{k=1}^n \lambda^{-jk}\Delta z_k\right)}.$$

Hence

$$|\varphi_j(z_1,\dots,z_n) - \varphi_j(z_1+\Delta z_1,\dots,z_n+\Delta z_n)|$$

$$\le \frac{2\sum_{k,l=1}^n |z_l|\,|\Delta z_k|}{|\sum_{l=1}^n \lambda^{-jl} z_l| \left(|\sum_{k=1}^n \lambda^{-jk} z_k| - \sum_{k=1}^n |\Delta z_k|\right)} \tag{3}$$

$$< \frac{2n\rho \sum_{l=1}^n |z_l|}{|\sum_{l=1}^n \lambda^{-jl} z_l| \left(\left|\sum_{l=1}^n \lambda^{-jl} z_l\right| - n\rho\right)}.$$

Notice that bounds for the noise depend on the frequency response of the polygon. That is, we cannot input a given tolerance and obtain a proper searching radius. The maximum noise allowed is intrinsic to the polygon.

3 Experiments in Polygon Matching with Noise

This section shows the results obtained when we test the algorithm using an extensive set of polygons of four classes. Figure 2 shows some sample polygons of the four classes we considered, respectively convex, star-shaped, Jordan, and general polygons. We have found that the more complex polygons (e.g., general polygons) are more easily distinguished. We discuss it below.

For evaluating our identification method, we generated a large set of polygons in each of the four considered classes, each class with 100,000 polygons. The polygons were generated with integer coordinates in a grid of size 1024 × 768 using the software kindly provided by Martin Held, according to the heuristics described in [3].

For each polygon Z_ℓ in the collection we computed and stored all $\xi_\ell = \varphi_j(Z_\ell)^n$, $j = 1, \ldots, \lfloor (n-1)/2 \rfloor$, and indexed them using a $2d$-tree. Each polygon was mapped to $\lfloor (n-1)/2 \rfloor$ points in the complex plane. After the mapping, our polygon collection is transformed to a point collection; each point in the collection corresponds to a polygon. The $2d$-tree is used as an inverted index to back-link the points to the original polygons.

Fig. 2. Example polygons. Convex, star shaped, simple (Jordan) and general (with self-intersections).

For querying we took 1,000 polygons in each collection and randomly perturbed *each* vertex with $\pm r$ pixels in the x and y coordinates, with $1 \leq r \leq 25$. The resulting polygons were mapped using the same $\lfloor (n-1)/2 \rfloor$ functions. We searched for the nearest neighbor of each one of the resulting points in the corresponding collection. Figure 3 show the recall as a function of the noise (measured in pixels) for various setups. We first notice that using a single, fixed invariant produces a low recall. Remember that the noise bounds depend on the polygon's response to a frequency. Taking all the invariants φ_j for indexing and requiring any one of them to match the nearest neighbor of the query gives excellent results. Remember, by Remark 4, that false positive matches are improbable, although due to noise, false positives are possible. For the plot, we considered the $\lfloor (n-1)/2 \rfloor$ candidates, one for each invariant, and checked if this list contains the matching polygon.

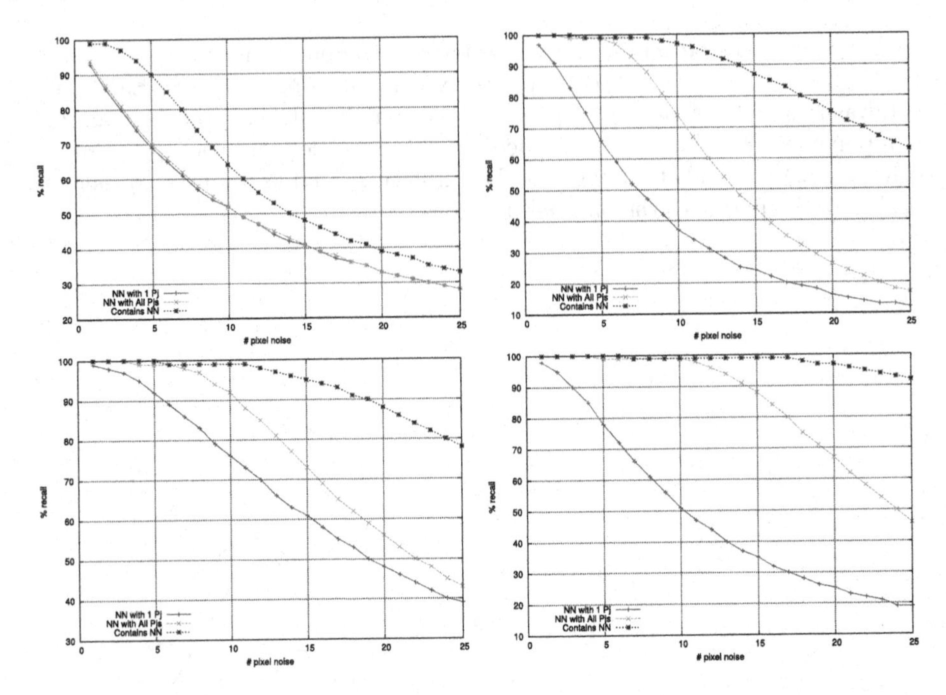

Fig. 3. Searching for noisy polygons. A graph is displayed for each class of polygons. The first class was a set of convex polygons, the second was of star shaped polygons, the third was of simple or Jordan polygons and the fourth class was of polygons without restrictions. The plots show the recall considering *one of the* φ_j (NN with 1Pj), *all of them* (NN with all Pj) and *any of them* (contains NN) respectively. Each point in the plot is the average of 1,000 queries. As the shape is more complex, the identification is easier.

The lowest recall we obtained was for convex polygons, which is consistent with the theoretical results because they have the lowest frequency responses. On the other hand, general polygons have a higher recall because there will be at least one high-frequency response. As per the cyclic shifts, recall Proposition 2, there was no difference in recall when the query was cyclically shifted. We experimented with polygons having between 16 and 32 sides. We saw no significant difference in the plots and only reported the results for 32 sided polygons. The total searching time is negligible, a few milliseconds in a laptop.

3.1 Fixing Recall for Convex Polygons

Observing the disparate results in recall for convex and general polygons and knowing the relationship of the performance and the frequency response of the polygons, we transformed the polygons before storing them. Let Π a random permutation, fixed beforehand. We applied Π to each one of the polygons before computing invariants φ_j, $Z_\Pi = (\Pi(z_1), \Pi(z_2), \ldots, \Pi(z_n))$. At query time, we applied the same permutation to the query. With this change, all the indexed polygons responded equally, obtaining the same recall as generalized polygons.

4 Final Remarks

Polygon matching under similarities is a fundamental problem at the core of many applications. In [9] they define the problem of finding the attitude of a spaceship, that is, finding which star appears in the objective of a camera. Stars are codified as polygons, using as vertices the surrounding stars. More precisely, for each star, the k-nearest stars define a polygon, with the center the target star. The algorithm in [9] is akin to brute-force. They compute thousands of perturbations of each polygon to boost the recall, and the corresponding invariants φ_j defined in [7] are stored with rounded decimals. The query polygon is searched for by exact matching. Hence if it coincides with one of the stored perturbations, a match is reported. The above procedure is wasteful; for each star, there will be a blob of points associated.

Using the heuristics defined herein, we report better recall rates than [9] by using the (k)nearest neighbors instead of exact searching and the random permutation before indexing. We store only one complex number, instead of a blob of points, for each star. The above allows to dramatically reduce space usage for a star index for spatial navigation.

We plan to use polygon indexing as a building block for robust point set retrieval under similarities, with applications to image and multimedia retrieval, computer vision, and robotics.

Acknowledgments. We want to thank David Mount for carefully reading an early version of this manuscript and providing precious suggestions. We are grateful to Tomas Auer and Martin Held [3] who maintain a repository for polygon generation. We used their software to generate polygons of various types for our experiments.

References

1. Arkin, E.M., Chew, L.P., Huttenlocher, D.P., Kedem, K., Mitchell, J.S.B.: An efficiently computable metric for comparing polygonal shapes. IEEE Trans. Pattern Anal. Mach. Intell. **13**(3), 209–216 (1991)
2. Atallah, M.J., Ribeiro, C.C., Lifschitz, S.: Computing some distance functions between polygons. Pattern Recogn. **24**(8), 775–781 (1991)
3. Auer, T., Held, M.: Heuristics for the generation of random polygons. In: Sack, J.-R., Fiala, F., Kranakis, E. (eds.) Proceedings 8th Canadian Conference on Computational Geometry, pp. 38–43. Carleton University Press (1996)
4. Barnes, N., Loy, G., Shaw, D.: The regular polygon detector. Pattern Recogn. **43**(3), 592–602 (2010)
5. Biedl, T., Durocher, S., Snoeyink, J.: Reconstructing polygons from scanner data. Theor. Comput. Sci. **412**(32), 4161–4172 (2011)
6. Buchin, K., Buchin, M., Wenk, C.: Computing the Fréchet distance between simple polygons. Comput. Geom. **41**(1–2), 2–20 (2008). Special Issue on the 22nd European Workshop on Computational Geometry (EuroCG)

7. Chávez, E., Chávez-Cáliz, A.C., López-López, J.L.: Affine invariants of generalized polygons and matching under affine transformations. Comput. Geom. **58**, 60–69 (2016)
8. Chris Ficher, J., Ruoff, D., Shilleto, J.: Perpendicular polygons. Amer. Math. Monthly **92**(1), 23–37 (1985)
9. Antonio Hernández, E., Alonso, M.A., Chávez, E., Covarrubias, D.H., Conte, R.: Robust polygon recognition method with similarity invariants applied to star identification. Adv. Space Res. **59**(4), 1095–1111 (2017)
10. Huang, J.: A new model for general polygon matching problems. Precis. Eng. **33**(4), 534–541 (2009)
11. Kaygin, S., Bulut, M.M.: Shape recognition using attributed string matching with polygon vertices as the primitives. Pattern Recogn. Lett. **23**(1–3), 287–294 (2002)
12. Leszczynski, K., Loose, S.: A polygon matching algorithm and its applications to verification of radiation field placement in radiotherapy. Int. J. Biomed. Comput. **40**(1), 59–67 (1995)
13. Liu, H., Wang, Z.: PLDD: point-lines distance distribution for detection of arbitrary triangles, regular polygons and circles. J. Vis. Commun. Image Represent. **25**(2), 273–284 (2014)
14. López-López, J.L.: The area as a natural pseudo-Hermitian structure on the spaces of plane polygons and curves. Diff. Geom. Appl. **28**(5), 582–592 (2010)
15. McCreath, E.C.: Partial matching of planar polygons under translation and rotation. In: Proceedings of the 20th Canadian Conference on Computational Geometry, pp. 47–50 (2008)
16. Neustupný, E.: Polygons in archaeology. Památky archeologické **87**, 112–136 (1996)
17. Tsai, W.H., Yu, S.S.: Attributed string matching with merging for shape recognition. IEEE Trans. Pattern Anal. Machine Intell. **7**(4), 453–462 (1985)
18. Veltkamp, R.C., Hagedoorn, M.: State of the art in shape matching. In: Lew, M.S. (ed.) Principles of Visual Information Retrieval, pp. 87–119. Springer, London (2001). https://doi.org/10.1007/978-1-4471-3702-3_4

Clustering Adverse Events of COVID-19 Vaccines Across the United States

Ahmed Askar[1,2](✉) and Andreas Züfle[1]

[1] Department of Geography and Geoinformation Science, George Mason University, 4400 University Drive, MS 6C3, Fairfax, VA 22030, USA
aaskar@gmu.edu, azufle@gmu.edu

[2] Food and Drug Administration, Silver Spring, MD 20993, USA

Abstract. We study the similarity of adverse effects of COVID-19 vaccines across different states in the United States. We use data of 300,000 COVID-19 vaccine adverse event reports obtained from the Vaccine Adverse Event Reporting System (VAERS). We extract latent topics from the reported adverse events using a topic modeling approach based on Latent Dirichlet allocation (LDA). This approach allows us to represent each U.S. state as a low-dimensional distribution over topics. Using Moran's index of spatial autocorrelation we show that some of the topics of adverse events exhibit significant spatial autocorrelation, indicating that there exist spatial clusters of nearby states that exhibit similar adverse events. Using Anselin's local indicator of spatial association we discover and report these clusters. Our results show that adverse events of COVID-19 vaccines vary across states which justifies further research to understand the underlying causality to better understand adverse effects and to reduce vaccine hesitancy.

Keywords: Spatial clustering · COVID-19 · Vaccines · Adverse events · Similarity search · Pharmacovigilance · Health geography

1 Introduction

By June 12th, 2021, more than 2.3 billion doses of various brands of COVID-19 vaccines had been administered world-wide with more than 300 million doses administered in the United States [10]. The U.S. Centers for Disease Control and Prevention (CDC) has stated that all U.S. authorized vaccines are safe and efficient [6]. While generally safe, the COVID-19 vaccines have adverse effects, including common side effects such as injection site pain and fever, but also including rare adverse effects that can be more severe. In the United States alone, by June 1st, 2021, a total of 297,410 of adverse events have been reported, collected, and made publicly available by the CDC and the U.S. Food and Drug Administration in a database called the Vaccine Adverse Event Reporting System (VAERS) [14]. As cases of severe symptoms gain public visibility in the news [28], these seemingly contradicting facts of general safety and possibly severe side-effects are a source of confusion leading to vaccine hesitancy among the population [31].

N. Reyes et al. (Eds.): SISAP 2021, LNCS 13058, pp. 307–320, 2021.
https://doi.org/10.1007/978-3-030-89657-7_23

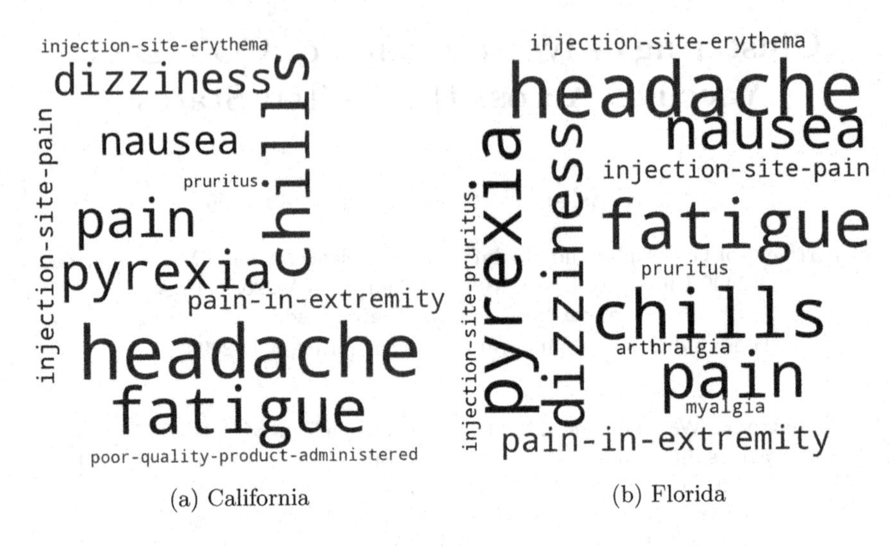

(a) California (b) Florida

Fig. 1. COVID-19 adverse effect clouds per region.

Towards a better understanding of COVID-19 vaccine adverse events we propose a similarity measure to quantify the similarity of sets of adverse events. To illustrate the challenge tackled in this work, Fig. 1 shows word clouds of adverse effects for California (Fig. 1a) and for Florida (Fig. 1b). These word clouds show the font size of the most frequent adverse effects proportional to their relative frequency observed in that state. We observe that common side effects such as headache, pyrexia (fever), and chills appear with similar relative frequency in both states and we also observe that some adverse effects appear more frequently in one region than another. For example, it pyrexia and dizziess are more frequently observed in Florida. Our goal is to measure the (dis-)similarity of the adverse effects observed in different regions. This similarity allows to understand how reported adverse events vary over space, over time, across different vaccine brands, and across different populations. We use our proposed similarity measure to study if we can observe statistically significant clusters of regions exhibiting similar adverse effects using VAERS data for the United States. While our work does not answer the question whether vaccines are safe, we hope that public health researchers and health officials may find our similarity measure useful to better understand adverse events, their variations over space, and the underlying causal factors.

Summarizing our approach, we use a bag-of-words model to describe a set of adverse events, such as reported in a spatial region. We leverage Latent Dirichlet Allocation (LDA) to extract latent topics of adverse effects for each region. LDA has been successfully used to extract domains and research topics from scientific research papers [17] and news topics (such as "Sports", "Politics", "Entertainment") from news articles [29]. To extract latent topics of adverse events, we treat the adverse events reported in a spatial region as documents and individual adverse effects as words. We qualitatively evaluate the modeled topics and show that they are able to represent, for example, adverse events related to "pyrexia/fever" and adverse effects related to "vertigo/dizziness". Then, we describe states of the U.S. by their adverse event topic distribution to evaluate whether topics of vaccine adverse effects vary across the United States. We quantitatively evaluate if this variation exhibits any significant spatial autocorrelation, that is, if spatially close states exhibit similar topics of adverse events.

For this purpose, we first survey existing work in Sect. 2 and formally define an adverse event database in Sect. 3. Our approach to extract latent topics of adverse events using topic modeling is described in Sect. 4. Using these topics as a low-dimensional embedding of adverse events in a spatial region, our approach to quantify spatial autocorrelation and to find spatial clusters of states that exhibit significantly similar (or dissimilar) topics of adverse effects is described in Sect. 5. We explore the global and local spatial autocorrelation of COVID-19 vaccine adverse events in Sect. 7 to discover significant spatial autocorrelation, showing that some topics of adverse events indeed vary in different parts of the United States. Finally, we conclude in Sect. 8 and identify future directions.

2 Related Work

Adverse Effects of Vaccines. Vaccines are, without any doubt, a paramount weapon to fight deadly diseases evident by the fact that "In 1900, for every 1,000 babies born in the United States, 100 would die before their first birthday, often due to infectious diseases" [34]. Furthermore, vaccines not only protect those receiving the vaccines but also vulnerable groups around them, such as new born babies, who may not be able to receive a vaccine [12]. Yet, there are adverse effects [14] including the 300,000 adverse events reported for the COVID-19 vaccines by June 1st, 2021. Understanding and mitigating these adverse events will not only improve the well-being of those receiving the vaccines, but will also decrease fear of vaccines that leads to high vaccine hesitancy as observed during the COVID-19 pandemic [11]. To the best of our knowledge, this is the first study investigating the similarity of adverse effects of COVID-19 vaccines to understand their spatial autocorrelation. We hope that our proposed techniques will find adaption by epidemiologists to improve our understanding of the ecology of past, present, and future infectious diseases.

Topic Modeling of Adverse Events. Topic modeling is an unsupervised learning technique to discover underlying themes of a collection of documents. Latent Dirichlet Allocation (LDA) is one of the more common topic modeling techniques in the literature [4]. In the context of pharmacovigilance, LDA has been used to find potentially unsafe dietary supplements [35], but without the consideration of the spatial distribution of latent topics among adverse effects. In our prior work in [2] we performed a spatio-temporal study on the adverse events of blood thinning drugs and their spatial auto-correlation. This study mainly limited by data availability, having adverse events reported by country only. For this reason, our prior study in [2] used European countries, but most countries had to be removed due to having too few reported adverse events. The wide availability of VAERS COVID-19 vaccine data at United States state level enables us to directly explore the latent adverse event features for spatial auto-correlation.

Pharmacovigilance. The field of pharmacovigilance aims at understanding the occurrence of adverse effects of drugs [18,21]. Existing work has shown that adverse effects of a single drug or multiple combination of drugs may vary over space and time due to racial and ethnic disparities [3,25,27], environment [20, 26], and drug quality [7]. Specifically for vaccines, there is evidence that stress may have an amplifying effect on immune response and adverse events [16]. However, such aspects of understanding the interactions between drugs and other external factors are out of scope of this work. In this work, we investigate the effect of location on adverse effects of the COVID-19 vaccines. While location may be a proxy of other factors (such as stress), this work does not provide or imply any causality between location and adverse events. Yet, we hope that an understanding of the spatial distribution and autocorrleation of adverse events may help experts discover such causalities.

3 Problem Definition

This section formally defines adverse events, adverse effects, and the problem of spatio-temporal clustering of adverse events. First, we provide a definition of adverse effects and events.

Definition 1 (Adverse Effect). *An Adverse Effect is a textual representation of an undesirable experiences associated with the use of a medical product. We let* $\mathcal{A} = \{A_1, ..., A_N\}$ *denote the set of all adverse events and* N *denotes the number of all (possible) adverse effects.*

Data such as collected in the VAERS database is a collection of records each associated with a set of adverse effects, a specific pharmaceutical drug, a location, and time. We call such as record an Adverse Event (AE), formally defined as follows:

Definition 2 (Vaccine Adverse Event Database). *Let* $\mathcal{A}$ *denote a set of adverse effects, let* $\mathcal{S}$ *denote a set of spatial regions, and let* $\mathcal{D}$ *denote a set of vaccine brands. An Adverse Event Report Database* $\mathcal{DB}$ *is a collection of adverse event reports* (s, A, d)*, where* $s \in \mathcal{S}$ *is a spatial region,* $A \subseteq \mathcal{A}$ *is a set of adverse effects, and* $d \in \mathcal{D}$ *is the brand for which the adverse effects are reported. We let* $M := |\mathcal{DB}|$ *denote the number of adverse event reports in* $\mathcal{DB}$

We note that a single adverse event may report multiple adverse effects. As an example, Table 1 shows exemplary adverse events from the VAERS database. The first line in Table 1 implies that "Dizziness", "Injection site pruritus", "Injection site rash", and "Somnolence" are adverse effects reported in Maryland Moderna vaccine.

Table 1. Sample records of Adverse Event Report Database. Each line is an adverse event.

Adverse event ID	Drug	Location	Set of adverse effects
1139067	Moderna	MD	Dizziness, Injection site pruritus, Injection site rash, Somnolence
1004857	Moderna	PA	Nausea, Palpitations, Presyncope, Pyrexia, Tremor
1115746	Moderna	NY	Chills, Headache, Nausea, Pain, Pain in extremity
1148711	Moderna	CA	Axillary pain, Fatigue, Headache, Nausea, Pain in extremity
1240185	Pfizer	IN	Fatigue, Headache, Pain, Pyrexia
1120846	Pfizer	UT	Nausea, Pain in extremity, Sleep disorder, Tinnitus, Vertigo
1104541	Pfizer	GA	Injection site reaction, Rash pruritic
1138693	Pfizer	WI	Eye pruritus, Lip swelling, Nasal pruritus, Swelling face, Urticaria
1200860	Janssen	TX	Headache
1114482	Janssen	MI	Chills, Hyperhidrosis, Pyrexia
1244933	Janssen	IL	Heart rate, Heart rate increased, Pain, Poor quality sleep, Pyrexia
1202067	Janssen	RI	Chills, Injection site erythema, Menstruation irregular, Pyrexia

Our goal is to find clusters of locations that exhibit similar adverse events. Towards this goal, we group adverse events by region.

Definition 3 (Spatial Adverse Events). *Let $\mathcal{DB}$ be an adverse event report database and let $s' \in \mathcal{S}$ be a spatial region. We define*

$$\mathcal{DB}_{s'} := \{(s, A, d) \in \mathcal{DB} | s = s'\}$$

as the set of all adverse events reported in regions$'$.

In the next section, we describe how we obtain latent topics of adverse events to represent each region as a low dimensional topic distribution.

4 Latent Adverse Event Topic Modeling

This section presents our Latent Dirichlet Allocation (LDA) based approach to extract latent topics from adverse events. All our code to access the data and to run the topic modeling can be found at https://github.com/ahmedaskar64/Spatio-Temporal-AEs-Similarity/tree/main.

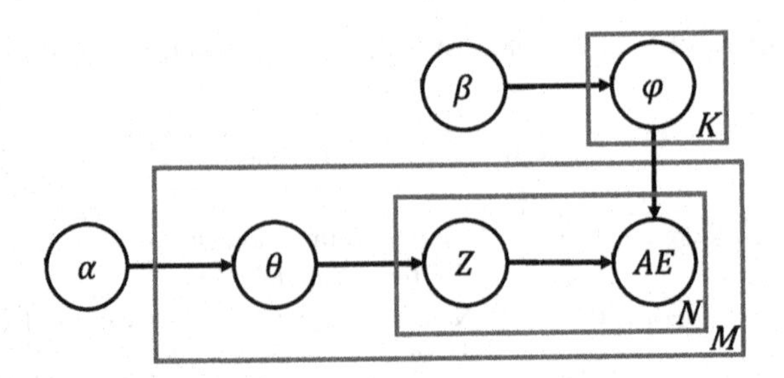

Fig. 2. LDA Topic Modeling of Adverse Events. For each adverse event a topic distribution θ is estimated and for each topic i, an adverse effect distribution φ_i is estimated. Given a topic Z generated from θ, observable adverse effects (AEs) are generated from φ_Z.

A challenge of mining adverse events is the potentially large number of different adverse effects. The FAERS Adverse Event Databases use MedDRA codes [5] and terminology to standardize adverse effects such as using "pyrexia" instead of "heightened temperature" of "fever". Yet, the number of possible adverse effects is too large and the resulting feature space of using bag-of-words semantics to represent adverse effects is too high dimensional. To address this issue, we acknowledge that adverse effects are symptoms of unknown (latent) underlying causes. While one way of identifying causes is involving a medical expert, we propose a data-driven approach to identify underlying topics among adverse events using topic modeling that we interpret as causes. For that, we employ Latent Dirichlet Allocation (LDA) [4] – a generative probabilistic model which assumes that each adverse event is a mixture of underlying (latent) topics, and each topic has a (latent) distribution of more and less likely adverse effects.

A graphical representation of our LDA model using plate notation is shown in Fig. 2. A vector α of length K is used to parameterize the *a priori* distribution of topics. The parameter K corresponds to the number of latent topics used to model adverse events. When an adverse event is created, we assume that its topics are chosen following a *Dirichlet distribution* having parameter α which we use to obtain a topic distribution θ for each of our $M =$ adverse events. Thus, the large plate in Fig. 2 corresponds to a set of M adverse events, each having a topic distribution θ drawn randomly (and Dirichlet distributed) from α.

For each topic, the prior parameter β is used to generate the distribution of adverse effects within a topic. Thus, we assume that a topic generates adverse effects following a Dirichlet distribution having a vector β of length $|\mathcal{A}|$ as parameter, where $\mathcal{A}$ is the set of observed adverse effects (c.f. Definition 1). For each of our K topics, a resulting vector $\varphi_i, 1 \leq i \leq K$ stores the adverse effect distribution of topic K.

To generate the adverse effects of an adverse event, a topic is chosen randomly from the topic distribution θ and, given this topic, a number of N_i adverse effects are generated randomly from the adverse effect distribution φ – where N_i is assumed to be independent from the chosen topic and uniformly distributed.

In Fig. 2, the node AE denotes the (observable) set of all $N = \sum_i N_i$ adverse effects, and Z is a function that maps each word to the topic that generated it. The reason for choosing a Dirichlet distribution rather than a more straightforward uniform or multinomial distribution for the topic and word priors is inspired by research showing that the distribution of words in text can be better approximated using a Dirichlet distribution [23].

To infer the topics of our adverse event database $\mathcal{DB}$, we employ a generative process. Given the observed adverse effects, LDA optimizes the latent variables to maximize the likelihood of matching the observed adverse events and corresponding adverse effects. This generative process works as follows. Adverse events are represented as random mixtures over latent topics, where each topic is characterized by a distribution over all N adverse effects. LDA assumes the following generative process for database $\mathcal{DB}$ consisting of M adverse events, each having a number of N_i adverse effects.

- For each adverse event choose a topic distribution $\theta_m \sim Dir(\alpha), 1 \leq m \leq M$, where $Dir(\alpha)$ is a Dirichlet distribution with prior α. In our experiments, we initially assume each topic to have uniform prior probabilities, having $\alpha_i = \alpha_j$ for $1 \leq i, j \leq K$. This apriori distribution is adapted using Bayesian inference [4] to maximize the likelihood of generating the observed keywords.
- For each topic, choose an adverse effect distribution $\varphi_i \sim Dir(\beta)$, where $1 \leq i \leq K$. For our experiments, we assume each adverse effect to have the same prior probability N^{-1}.
- For each adverse effect ae in adverse event j:
  1. Choose a topic $z \sim Multinomial(\theta_j)$ from the topic distribution of j, and
  2. Choose a word $w \sim Multinomial(\varphi_z)$ from the adverse effect φ_z of topic z.

  Here, $Multinomial(x)$ corresponds to a multinomial distribution drawing from a stochastic vector x.

To describe each adverse event in a latent topic space, we use the adverse event specific topic distributions θ_m which describe each adverse event m as a set of K latent features corresponding to the weight of the respective latent topic. While this topic modeling does not provide us with any semantic of the underlying topics, we know that adverse events having similar latent features also exhibit similar adverse effects. Based on the similarity of latent topics we propose a hierarchical agglomerative clustering approach to find regions that exhibit similar adverse events in Sect. 5 and test these clusters for spatial autocorrelation using Moran's I in Sect. 7.

5 Spatial Clustering of Vaccine Adverse Event Topics

The latent topic modeling of Sect. 4 provides us with a topic distribution θ_i for each adverse event report $d \in \mathcal{DB}$. To describe the topic distribution of a region, we use the average topic distribution of all adverse events reported in the region. To measure similarity between the topics of adverse events of two regions, we use Euclidean distance between these resulting average topic distributions. Formally,

Definition 4 (Region-Wise Adverse Event Distance). *Let $\mathcal{DB}$ be an adverse event database, let $\mathcal{DB}_{s_1}, \mathcal{DB}_{s_2} \subseteq \mathcal{DB}$, let K be a positive integer and let $\theta(ae)$ denote the latent topic distribution of an adverse event $ae \in \mathcal{DB}$ using the LDA model described in Sect. 4, then:*

$$dist(\mathcal{DB}_{s_1}, \mathcal{DB}_{s_2}) := \left\| \frac{\sum_{\mathcal{DB}_{s_1}} \theta(ae)}{|\mathcal{DB}_{s_1}|} - \frac{\sum_{\mathcal{DB}_{s_2}} \theta(ae)}{|\mathcal{DB}_{s_2}|} \right\|_2,$$

where $\|.\|_2$ denotes the Euclidean norm.

To find clusters among regions having similar topics of adverse events we leverage the distance function of Definition 4 and employ a hierarchical agglomerative clustering approach [8]. The advantage of such an approach is that we neither have to guess the number of clusters as often needed for partitioning clustering approaches [22] nor have to define a density threshold as required by density-based clustering algorithms [13,32]. To merge clusters, we employ complete linkage, which defines the distance between two clusters of regions as the maximum pair-wise distance of regions among the clusters.

Figure 3 shows the pair-wise distance (see Definition 3) for each pair of states for the 49 states of the United States excluding Alaska, Puerto Rico, and Hawaii using $K = 10$ adverse event topics. In Fig. 3 darker colors correspond to a higher pair-wise similarity. We observe a large group of mutually similar states having smaller nested clusters of similar states thus explaining our choice for hierarchical clustering. We also observe that is not trivial to delineate clusters due to noise, which explains our choice of complete link clustering to maximize delineation and avoid having clusters "grow together". A high resolution version of Fig. 3 can be found on our project website https://github.com/ahmedaskar64/Spatio-Temporal-AEs-Similarity/tree/main.

6 Spatial Autocorrelation

Given the latent topics of vaccine adverse events as described in Sect. 4 and the clustering approach of Sect. 5, we next investigate if the observed adverse event topics exhibit significant spatial autocorrelation. In other words, can we reject the null hypothesis that topics are independent of location by observing that spatially close regions exhibit similar topics?

For this purpose, we retain all clusters (of all sizes) corresponding to all nodes in the dendrogram excluding clusters of size one and excluding the root of the dendrogram that contains all regions. Given any such cluster of regions that exhibit similar topics of adverse events, we employ Moran's I measure of spatial autocorrelation [24]. Moran's I statistic tests if a variable measured on spatial regions exhibits a significant spatial autocorrelation, either positive (clustered) or negative (dispersed). To measure the spatial autocorrelation of clusters obtained as described in Sect. 5, we use one-hot encoding (or dummy-coding) to encode each individual cluster membership into a binary variable. Thus, for a cluster C,

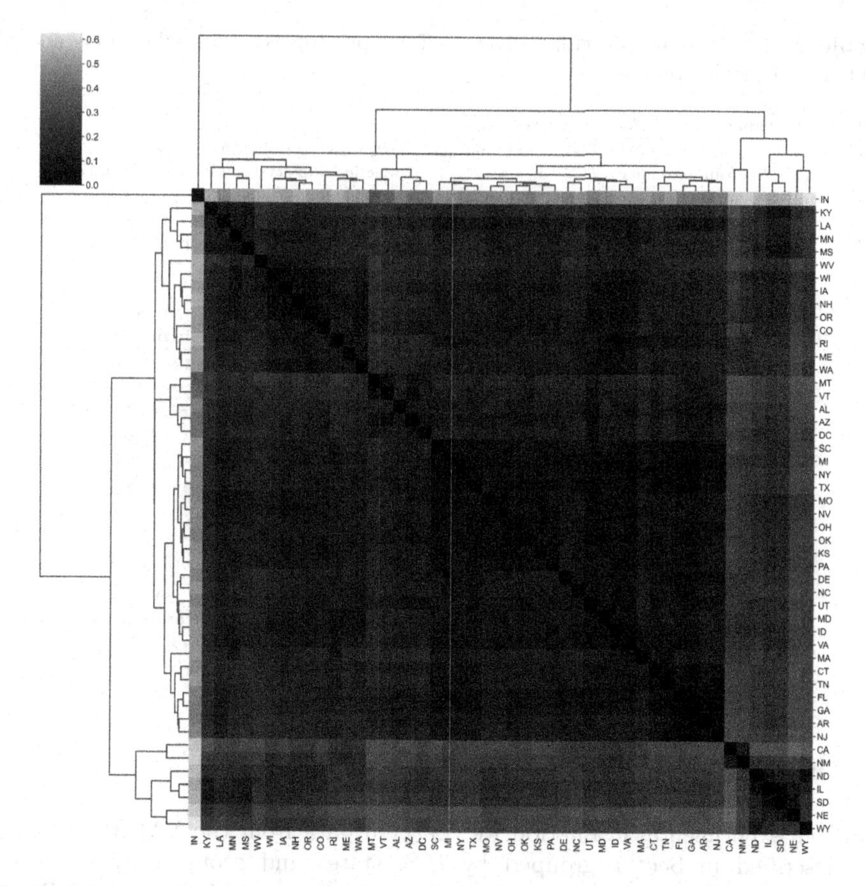

Fig. 3. Pair-wise similarity matrix of latent topics of COVID-19 vaccine adverse events of counties in the United States.

the cluster membership variable of a region r is set to 1 if $r \in C$ and 0 otherwise. Moran's I requires an adjacency metric on regions to assess the similarity between polygonal regions. For this purpose, we employ the *Queen Contiguity* model [15], that is, two regions are considered adjacent if they share boundary. We directly report Moran's I test statistic whose range is in $[-1, -1]$, ranging from strongly dispersed (close to -1) to strongly clustered (close to 1). We also report the p-value of the null-hypothesis that the regions are distributed randomly without any spatial pattern by transforming Moran's I values to z-values and employing a two-tailed z-test [9]. The resulting p-values indicate whether a cluster of regions having similar topics of adverse events are significantly spatially clustered or dispersed. We used the geopandas library for handling spatial attributes and Pysal library for Moran's I test of spatial autocorrelation [19,30].

Table 2. Top-10 most probably adverse effects per topics across all regions and all COVID-19 vaccine brands.

Topic	(Probabilities in %) Adverse effects
1	(4.5)"headache", (3.6)"pyrexia", (3.6)"fatigue", (3.3)"pain", (3.1)"chills", (3.0)"nausea", (2.3)"pain-in-extremity", (1.7)"dizziness", (1.7)"injection-site-erythema", (1.7)"arthralgia"
2	(4.1)"headache", (2.8)"dizziness", (2.6)"pyrexia", (2.6)"pain-in-extremity", (2.5)"fatigue", (2.5)"chills", (2.4)"nausea", (2.4)"pain", (2.1)"injection-site-pain", (1.6)"dyspnoea"
3	(6.9)"headache", (4.1)"pyrexia", (3.8)"fatigue", (3.7)"chills", (3.0)"pain", (2.9)"dizziness", (2.8)"nausea", (1.9)"pain-in-extremity", (1.8)"injection-site-erythema", (1.8)"injection-site-pain"
4	(8.7)"chills", (8.3)"pyrexia", (7.2)"headache", (7.2)"pain", (6.4)"fatigue", (3.9)"nausea", (3.2)"pain-in-extremity", (2.6)"injection-site-pain", (2.2)"myalgia", (2.1)"dizziness"
5	(4.5)"pyrexia", (4.1)"headache", (4.0)"chills", (3.4)"pain", (3.1)"fatigue", (2.5)"nausea", (2.5)"dizziness", (2.1)"injection-site-pain", (2.1)"arthralgia", (2.1)"pain-in-extremity"
6	(3.8)"dizziness", (3.3)"headache", (2.4)"chills", (2.3)"nausea", (2.2)"fatigue", (2.2)"pain", (2.1)"pain-in-extremity", (1.5)"dyspnoea", (1.5)"injection-site-erythema", (1.5)"pyrexia"
7	(6.5)"headache", (5.5)"pyrexia", (5.1)"chills", (4.8)"pain", (4.7)"fatigue", (3.2)"nausea", (2.6)"injection-site-pain", (2.4)"dizziness", (2.0)"injection-site-erythema", (1.7)"pain-in-extremity"
8	(5.7)"headache", (4.4)"fatigue", (4.0)"chills", (3.8)"pain", (3.2)"pyrexia", (3.0)"pain-in-extremity", (2.7)"nausea", (2.1)"injection-site-pain", (1.8)"injection-site-erythema", (1.8)"dizziness"
9	(4.0)"headache", (3.9)"fatigue", (3.6)"pain", (3.2)"chills", (2.9)"nausea", (2.8)"pyrexia", (2.5)"dizziness", (1.9)"pain-in-extremity", (1.9)"injection-site-pain", (1.6)"pruritus"
10	(3.8)"pyrexia", (3.3)"fatigue", (2.9)"headache", (2.8)"pain", (2.6)"chills", (2.4)"dizziness", (2.1)"nausea", (1.9)"pruritus", (1.9)"rash", (1.9)"injection-site-erythema"

7 Experimental Evaluation

For our experimental evaluation we collected data from the VAERS database as described in Sect. 1 grouped by U.S. states and grouped by the three brands of vaccines authorized by 06/14/2021: Janssen, Moderna, and Pfizer. The experiments are conducted on a PC with Intel(R) Xeon(R) CPU $E3$-1240 v6 @3.70 GHz and 32 GB RAM. Windows 10 Enterprise 64-bit is the operating system, and all the algorithms are implemented by Python 3.7. All code, including code to obtain data from the VAERS API, is available at:s https://github.com/ahmedaskar64/Spatio-Temporal-AEs-Similarity/tree/main.

7.1 Qualitative Analysis of Topics

For $K = 10$ latent topics of COVID-19 adverse events Table 2 shows the φ_i vectors of our LDA model which correspond to the adverse effect distribution of the i'th topic. For each topic in Table 2 we show the Top-10 highest probability adverse effects. First, we observe that the resulting ten topics are hard to discriminate, as they all contain common adverse effects such as "headache", "pyrexia" (fever). Yet, we do observe different distributions of these adverse effects. We observe that Topic #4 has high probabilities for common symptoms and consequently low probabilities for rare symptoms. Topic #6 seems to corresponds to light symptoms with a low probability of fever, but higher probability of "dizziness". However, we note that our team does not

Table 3. Moran's I measure of global spatial autocorrelation for each of the $K = 10$ topics of COVID-19 adverse events.

Pattern	p-value	Moran's index	z-score	Topic ID
Clustered	0.0006	0.2756	3.4512	1
Random	0.6214	−0.0635	−0.4938	2
Clustered	0.0966	0.1216	1.6616	3
Random	0.6643	−0.0464	−0.4340	4
Random	0.2054	0.0920	1.2662	5
Random	0.6867	0.0109	0.4033	6
Dispersed	0.0754	−0.1785	−1.7782	7
Clustered	0.0071	0.2149	2.6938	8
Random	0.1988	0.0875	1.2850	9
Clustered	0.0002	0.3163	3.7895	10

include a medical expert, thus we refrain from a deeper analysis of these topics and conclude that our LDA approach has been able to find topics that differ in distribution of adverse effects. We note that due to truncation to only showing the Top-10 most probable adverse effects, we do not show uncommon and rare adverse effects which may define a topic (thus having most of it's probability mass focused within this single topic). The interested reader may find the full list of adverse effect per topic probabilities on our project website (https://github.com/ahmedaskar64/Spatio-Temporal-AEs-Similarity/tree/main), also including the per-topic adverse effect distributions for $K = 3$ and $K = 20$ topics.

7.2 Spatial Anaylsis of COVID-19 Adverse Event Topics

Table 3 shows the degree of spatial autocorrelation of each of the $K = 10$ topics of adverse events. For this purpose, we associated each U.S. state i with it's corresponding φ_{ik} probability of topic $k \in \{1, ..., 10\}$. With each states having it's corresponding probability for topic k, we use Moran's I measure of spatial autocorrelation [24]. Moran's I is a test statistic to test the hypothesis that a spatial phenomenon appears uniformly at random without any spatial pattern. We observe in Table 3 that out of the ten topics, six topics show no spatial autocorrelation (unable to reject the null hypothesis of a random pattern), one topic shows negative spatial autocorrelation (implying a significant dispersed pattern), and three topics exhibit a positive spatial autocorrelation (spatially clustered patterns). First, we note testing ten hypothesis, and at the high p-value of 0.0754 we'd expect one such pattern by chance under the null hypothesis. Accounting for the multiple hypothesis testing problem [33] (for example, using Bonferroni correction [36]), the dispersed pattern of Topic #7 is no significant. However, for the clustered patterns of Topics #1 and #8, and #10 we observe highly significant p-value of 0.0006, 0.0071, and 0.0002, respectively, showing that these

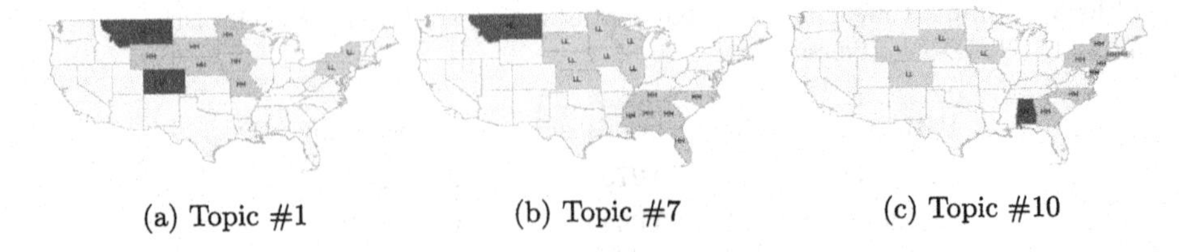

(a) Topic #1 (b) Topic #7 (c) Topic #10

Fig. 4. Local Indicator of Spatial Autocorrelation (LISA). Light red areas correspond to high-high clusters. Light blue areas are low-low clusters. Dark red and dark blue areas corresponds to high-low and low-high outliers. (Color figure online)

three topics of COVID-19 adverse events do exhibit significant spatial autocorrelation. This results shows that some latent topics among the adverse effects of the COVID-19 vaccines indeed depend on location. For a deeper study, we show the Local Indicator of Spatial Autocorrelation (LISA [1]) in Fig. 4, showing the spatial location of clusters of regions that exhibit high (or low) probabilities of the corresponding topic. Using LISA, a cluster is defined as a region having a high (low) value that is surrounded by regions that also have high (low) values. Interestingly, we observe that different parts of the United States exhibit high (low) values in these three significant latent topics. We also observe high-low (low-high) outliers, i.e., regions having high (low) topic probabilities that are surrounded by regions having low (high) topic probabilities. These significant clusters that adverse effects indeed vary locally. The underlying causality warrants further study to understand why certain regions of the United States exhibit different topics of adverse events.

8 Conclusions

In this work, we tackled the problem of measuring (dis-)similarity between adverse events of COVID-19 vaccines observed in different regions. Our measure leverages a topic modeling approach using LDA to map each adverse event from a (textual) set of adverse effects to a latent topic distribution. Using a database of 300,000 adverse event reports of COVID-19 vaccines in the United States, investigate the underlying topics exhibit any spatial autocorrelation to understand if different places exhibit different adverse events. Our results show that some of the latent topics of COVID-19 adverse events show significant positive spatial autocorrelation. Our local analysis of spatial autocorrelation show that certain topics of adverse events have increased (or decreased) likelihood in different parts of the United States.

We hope that teams of medical experts may find this result to investigate the underlying causality. Reasons could be due to vaccine quality issues, storage and cooling issues, or simply due to different brands of vaccines. Our own future work will include looking at the correlation between adverse event topics and different vaccine brands to understand topics and possibly the clusters that we have observed. We will also look into temporal changes of topics to gain an understanding how adverse events may change over time and due to climate.

Finally, we note that all of our implementations, experiments, and results are available at our project website: https://github.com/ahmedaskar64/Spatio-Temporal-AEs-Similarity/tree/main, where we also include additional experiments which we could not fit into this paper.

Acknowledgements. This research was prepared or accomplished by Ahmed Askar in his personal capacity. The opinions expressed in this article are the author's own and do not reflect the view of the U.S. Food and Drug Administration, the Department of Health and Human Services, or the United States government. This research received no external funding.

References

1. Anselin, L.: Local indicators of spatial association—lisa. Geogr. Anal. **27**(2), 93–115 (1995)
2. Askar, A., Züfle, A.: Spatio-temporal clustering of adverse events of post-market approved drugs using latent dirichlet allocation. In: Proceedings of the 17th International Symposium on Spatial and Temporal Databases (2021)
3. Baehr, A., Peña, J.C., Hu, D.J.: Racial and ethnic disparities in adverse drug events: a systematic review of the literature. J. Racial Ethnic Health Disparities **2**(4), 527–536 (2015)
4. Blei, D.M., Ng, A.Y., Jordan, M.I.: Latent dirichlet allocation. J. Mach. Learn. Res. **3**, 993–1022 (2003)
5. Brown, E.G., Wood, L., Wood, S.: The medical dictionary for regulatory activities (meddra). Drug Saf. **20**(2), 109–117 (1999)
6. Centers for disease control and prevention: different COVID-19 vaccines. (https://www.cdc.gov/coronavirus/2019-ncov/vaccines/different-vaccines.html)
7. Chircu, A., Sultanow, E., Saraswat, S.P.: Healthcare RFID in germany: an integrated pharmaceutical supply chain perspective. J. Appl. Bus. Res. (JABR) **30**(3), 737–752 (2014)
8. Day, W.H., Edelsbrunner, H.: Efficient algorithms for agglomerative hierarchical clustering methods. J. Classif. **1**(1), 7–24 (1984)
9. Dixon, W.J., Massey Jr, F.J.: Introduction to statistical analysis (1951)
10. Dong, E., Du, H., Gardner, L.: An interactive web-based dashboard to track COVID-19 in real time. Lancet Infect. Dis. **20**(5), 533-534 (2020). (https://coronavirus.jhu.edu/map.html)
11. Dror, A.A., et al.: Vaccine hesitancy: the next challenge in the fight against Covid-19. Eur. J. Epidemiol. **35**(8), 775–779 (2020)
12. Dushoff, J., et al.: Vaccinating to protect a vulnerable subpopulation. PLoS Med **4**(5), e174 (2007)
13. Ester, M., Kriegel, H.P., Sander, J., Xu, X., et al.: A density-based algorithm for discovering clusters in large spatial databases with noise. In: Kdd, pp. 226–231 (1996)
14. FDA, CDC: vaccine adverse event reporting system. vaers.hhs.gov (2021)
15. Fotheringham, A.S., Brunsdon, C., Charlton, M.: Geographically Weighted Regression: The Analysis of Spatially Varying Relationships. John Wiley & Sons, Hoboken (2003)
16. Glaser, R., Kiecolt-Glaser, J.K., Malarkey, W.B., Sheridan, J.F.: The influence of psychological stress on the immune response to vaccines. Ann. NY Acad. Sci. **840**(1), 649–655 (1998)

17. Griffiths, T.L., Steyvers, M.: Finding scientific topics. Proc. Natl. Acad. Sci. **101**(suppl 1), 5228–5235 (2004)
18. Jeetu, G., Anusha, G.: Pharmacovigilance: a worldwide master key for drug safety monitoring. J. Young Pharmacists **2**(3), 315–320 (2010)
19. Jordahl, K.: Geopandas: python tools for geographic data (2014). https://github.com/geopandas/geopandas
20. Kang, J.H., Kim, C.W., Lee, S.Y.: Nurse-perceived patient adverse events and nursing practice environment. J. Prev. Med. Public Health **47**(5), 273 (2014)
21. Leyens, L., Reumann, M., Malats, N., Brand, A.: Use of big data for drug development and for public and personal health and care. Genet. Epidemiol. **41**(1), 51–60 (2017)
22. Likas, A., Vlassis, N., Verbeek, J.J.: The global k-means clustering algorithm. Pattern Recogn. **36**(2), 451–461 (2003)
23. Madsen, R.E., Kauchak, D., Elkan, C.: Modeling word burstiness using the dirichlet distribution. In: Proceedings of the 22nd international Conference on Machine Learning, pp. 545–552 (2005)
24. Moran, P.A.: Notes on continuous stochastic phenomena. Biometrika **37**(1/2), 17–23 (1950)
25. Okoroh, J.S., Uribe, E.F., Weingart, S.: Racial and ethnic disparities in patient safety. J. Patient Saf. **13**(3), 153–161 (2017)
26. Pereira, F.G.F., Ataíde, M.B.C.D., Silva, R.L., Néri, E.D.R., Carvalho, G.C.N., Caetano, J.Á.: Environmental variables and errors in the preparation and administration of medicines. Rev. Bra. Enfermagem **71**(3), 1046–1054 (2018)
27. Piccardi, C., Detollenaere, J., Bussche, P.V., Willems, S.: Social disparities in patient safety in primary care: a systematic review. Int. J. Equity Health **17**(1), 114 (2018)
28. PolitiFact, The poynter institute: federal VAERS database is a critical tool for researchers, but a breeding ground for misinformation. (https://www.politifact.com/article/2021/may/03/vaers-governments-vaccine-safety-database-critical/)
29. Ramage, D., Hall, D., Nallapati, R., Manning, C.D.: Labeled LDA: a supervised topic model for credit attribution in multi-labeled corpora. In: Proceedings of the 2009 Conference on eempirical Methods in Natural Language Processing, pp. 248–256 (2009)
30. Rey S.J., Anselin L.: PySAL: A python library of spatial analytical methods. In: Fischer M., Getis A. (eds) Handbook of Applied Spatial Analysis. Springer Heidelberg (2010). https://doi.org/10.1007/978-3-642-03647-7_11
31. Sallam, M.: Covid-19 vaccine hesitancy worldwide: a concise systematic review of vaccine acceptance rates. Vaccines **9**(2), 160 (2021)
32. Schubert, E., Sander, J., Ester, M., Kriegel, H.P., Xu, X.: DBSCAN revisited, revisited: why and how you should (still) use DBSCAN. ACM Trans. Database Syst. (TODS) **42**(3), 1–21 (2017)
33. Shaffer, J.P.: Multiple hypothesis testing. Ann. Rev. Psychol. **46**(1), 561–584 (1995)
34. Stratton, K., Ford, A., Rusch, E., Clayton, E.C., et al.: Adverse Effects of Vaccines: Evidence and ccausality. Committee to Review Adverse Effects of Vaccines (2011)
35. Wang, Y., Gunashekar, D.R., Adam, T.J., Zhang, R.: Mining adverse events of dietary supplements from product labels by topic modeling. Stud. Health Technol. Inf. **245**, 614 (2017)
36. Weisstein, E.W.: Bonferroni correction (2004). https://mathworld.wolfram.com/

Similarity Search in Graph-Structured Data

Metric Indexing for Graph Similarity Search

Franka Bause[1(✉)], David B. Blumenthal[2], Erich Schubert[3], and Nils M. Kriege[1]

[1] Faculty of Computer Science, University of Vienna, Vienna, Austria
{franka.bause,nils.kriege}@univie.ac.at
[2] Department Artificial Intelligence in Biomedical Engineering (AIBE), Friedrich-Alexander University Erlangen-Nürnberg (FAU), Erlangen, Germany
david.b.blumenthal@fau.de
[3] Department of Computer Science, TU Dortmund University, Dortmund, Germany
erich.schubert@tu-dortmund.de

Abstract. Finding the graphs that are most similar to a query graph in a large database is a common task with various applications. A widely-used similarity measure is the graph edit distance, which provides an intuitive notion of similarity and naturally supports graphs with vertex and edge attributes. Since its computation is NP-hard, techniques for accelerating similarity search have been studied extensively. However, index-based approaches for this are almost exclusively designed for graphs with categorical vertex and edge labels and uniform edit costs. We propose a filter-verification framework for similarity search, which supports non-uniform edit costs for graphs with arbitrary attributes. We employ an expensive lower bound obtained by solving an optimal assignment problem. This filter distance satisfies the triangle inequality, making it suitable for acceleration by metric indexing. In subsequent stages, assignment-based upper bounds are used to avoid further exact distance computations. Our extensive experimental evaluation shows that a significant runtime advantage over both a linear scan and state-of-the-art methods is achieved.

Keywords: Graphs · Similarity search · Graph edit distance

1 Introduction

Graph-structured data is ubiquitous in many areas such as chemo- and bioinformatics or computer vision. A common task is to search a database containing a large number of graphs for those that are most similar to a given query graph. Such queries are submitted directly by the user or occur as subproblems in downstream

This work has been supported by the Vienna Science and Technology Fund (WWTF) through project VRG19-009, and Deutsche Forschungsgemeinschaft (DFG), project number 124020371, within the Collaborative Research Center SFB 876 "Providing Information by Resource-Constrained Analysis", SFB projects A2 and A6.

N. Reyes et al. (Eds.): SISAP 2021, LNCS 13058, pp. 323–336, 2021.
https://doi.org/10.1007/978-3-030-89657-7_24

machine learning algorithms. A widely accepted concept of graph similarity is the *graph edit distance*, which is the minimum cost for transforming one graph into the other by a sequence of edit operations. A strength of this measure is that it can elegantly be applied to graphs with vertex and edge attributes by defining the costs of edit operations adequately. For example, to compare protein graphs where vertices are annotated by the amino acid sequence of the secondary structure elements they represent, the Levenshtein distance was used [20].

However, the vast majority of efficient methods for similarity search in graph databases are limited to the special case where graphs have categorical labels and the costs of edit operations are uniform (either zero or one) [10,13,16,25,26,28–31]. A fairly recent development in this domain are neural graph embeddings, e.g. [19], which do not return exact similarity search results. For the pairwise computation of the graph edit distance, several exact approaches [10,15] and heuristics such as *bipartite graph matching* based on optimal vertex assignments [20] have been proposed, many of which support the graph edit distance in its full generality [15,20]. Several of these yield lower and upper bounds on the graph edit distance as a byproduct, which have just recently been compared systematically [4]. However, these lower bounds for the general graph edit distance are not yet widely used for similarity search in graph databases. For the methods based on optimal vertex assignments, it has only recently been shown how to derive a distance termed BRANCH that is guaranteed to be a lower bound and proven to satisfy the triangle inequality [2]. BRANCH has been shown to provide an excellent trade-off between tightness and running time [4].

We propose a filter-verification framework for similarity search, which supports the general graph edit distance with arbitrary metric edit costs and is hence suitable for graphs with any attributes comparable with a distance measure. We employ BRANCH as an initial filter accelerated by metric indexing. In the next stages, we derive upper bounds from the optimal assignment and improve them via local search to reduce the candidate set further, before performing verification by exact computation of the graph edit distance. We experimentally evaluate our approach on graphs with attributes and categorical labels showing the effectivity of the filter pipeline. The results show that our approach allows scalable similarity search in attributed graphs with non-uniform edit costs. For the special case of uniform edit costs, where competing methods are available, our approach is shown to outperform the state of the art.

2 Related Work

We discuss approaches for similarity search regarding the graph edit distance and methods for its pairwise exact or approximate computation.

2.1 Similarity Search in Graph Databases

Methods for similarity search in graph databases can be divided into two categories, depending on whether they compare overlapping or non-overlapping substructures. The methods *k-AT* [25], *CStar* [28], *Segos* [26] and *GSim* [30] belong to the first category. These techniques are inspired by the *q-grams* used in the computation of the string edit distance. Either q-grams based on trees [25,26,28] or paths [30] are used. The methods *Pars* [29], *MLIndex* [16], and *Inves* [13] partition the graphs into non-overlapping substructures. They essentially obtain lower bounds based on the observation, that if x non-overlapping substructures of a database graph are not contained in the query graph, the graph edit distance is at least x. *Pars* uses a dynamic partitioning approach to achieve this, while *MLIndex* uses a multi-layered index to manage multiple partitions for each graph. *Inves* is a method used to verify whether the graph edit distance of two graphs is below a specified threshold by first trying to generate enough mismatching non-overlapping substructures. *Mixed* [31] combines the idea of q-grams and graph partitioning. These methods only allow uniform edit costs and are therefore not suitable for graphs with continuous attributes.

The concept of a *median graph* of a set of graphs regarding the graph edit distance has been studied extensively, see [3] and references therein. An application of median graphs is their use as routing objects in hierarchical index structures [3,23]. However, we are not aware of any concrete realization using this concept in a setting comparable to ours.

2.2 Pairwise Computation of the Graph Edit Distance

For computing the exact graph edit distance, both general-purpose algorithms [15] as well as approaches tailored to the verification step in graph databases have been proposed [9], which are usually based on depth- or breadth-first search [9,12], or integer linear programming [15]. As the exact computation of the graph edit distance is not feasible for larger graphs, many heuristics have been proposed, e.g., [2,4,11,14,18,20]. The properties of the dissimilarities obtained from these are in general not well investigated. For heuristics based on optimal vertex assignment [20], which are widely used in practice [24], a variant called Branch was recently studied thoroughly [2]. Branch is a lower bound on the graph edit distance, a pseudo-metric on graphs and supports arbitrary cost models (c.f., Sect. 4.1).

3 Preliminaries

We introduce the required basic concepts of graph theory and discuss database search with a focus on the metric space.

3.1 Graph Theory

A *graph* $G = (V, E, \mu, \nu)$ consists of a set of vertices $V(G)$, a set of edges $E(G) \subseteq V(G) \times V(G)$ between vertices of G, a labeling function for the vertices $\mu\colon V(G) \to L$, and a labeling function for the edges $\nu\colon E(G) \to L$. We discuss only undirected graphs and denote an edge between u and v by uv. The set of neighbors of a vertex $v \in V(G)$ is denoted by $N(v) = \{u \mid uv \in E(G)\}$.

Table 1. Notation for edit costs.

$c_v(u, v)$	Cost of substituting vertex u with vertex v (adjusting the label/attributes)
$c_v(u, \epsilon)$	Cost of deleting the isolated vertex u
$c_v(\epsilon, v)$	Cost of inserting the isolated vertex v
$c_e(uv, wx)$	Cost of substituting edge uv with edge wx (adjusting the label/attributes)
$c_e(uv, \epsilon)$	Cost of deleting the edge uv
$c_e(\epsilon, wx)$	Cost of inserting the edge wx

The set L can be categorical labels or arbitrary attributes including real-valued vectors and complex objects such as strings.

A measure commonly used to describe the dissimilarity of two graphs is the *graph edit distance*, which is the minimum cost for transforming one graph into the other using edit operations. An *edit operation* can be deleting or inserting an isolated vertex or an edge or relabeling any of the two. An *edit path* from graph G_1 to G_2 is a sequence of edit operations $(e_1, e_2, \dots)$ that transforms G_1 into G_2.

Definition 1 (Graph Edit Distance [20]). *Let c be an edit cost function assigning non-negative costs to edit operations. The* graph edit distance *between two graphs G_1 and G_2 is defined as*

$$d_{ged}(G_1, G_2) = \min \left\{ \sum_{i=1}^{k} c(e_i) \mid (e_1, \dots, e_k) \in \Upsilon(G_1, G_2) \right\},$$

where $\Upsilon(G_1, G_2)$ is the set of all possible edit paths from G_1 to G_2.

The costs of the different edit operations can be chosen as required for the specific use case, see Table 1 for our notation. If the edit costs are symmetric, non-negative, and strictly positive for each non-identical edit operation, the graph edit distance is a metric on graphs, treating graph isomorphism as identity [3]. Note that this holds even if the edit costs do not satisfy the triangle inequality (and hence are no metric), because the graph edit distance uses the edit path with minimal cost. In this work, we nonetheless assume that the edit costs respect the triangle inequality, i.e., we assume that the following inequalities hold[1]:

$$c_v(u, w) \le c_v(u, v) + c_v(v, w) \qquad \forall (u, v, w) \in \mathcal{V}^3 \tag{1}$$

$$c_v(u, v) \le c_v(u, \epsilon) + c_v(\epsilon, v) \qquad \forall (u, v) \in \mathcal{V}^2 \tag{2}$$

$$c_e(uv, yz) \le c_e(uv, wx) + c_e(wx, yz) \qquad \forall (uv, wx, yz) \in \mathcal{E}^3 \tag{3}$$

$$c_e(uv, wx) \le c_e(uv, \epsilon) + c_e(\epsilon, wx) \qquad \forall (uv, wx) \in \mathcal{E}^2 \tag{4}$$

Equations (1), (3), and (4) can be enforced via pre-processing without changing the graph edit distance and can hence be assumed to hold w.l.o.g. [5].

[1] For simplicity of notation, we have defined the costs on the vertices and edges instead of their labels. Hence, the sets $\mathcal{V}$ and $\mathcal{E}$ are all possible vertices and edges, respectively.

E.g., if we have $c(u,v)>c(u,w)+c(w,v)$, we can simply substitute $c(u,v)$ with $c(u,w)+c(w,v)$, because a minimum cost edit path cannot contain $c(u,v)$. The only remaining constraint, Equation (2), is met (to the best of our knowledge) in all applications where the graph edit distance is used to address real-world problems [24]. Computing the graph edit distance is NP-hard [29], rendering exact computation possible for small graphs only. There are several heuristics, many of which are based on solving an assignment problem.

Definition 2 (Assignment Problem). *Let A and B be two sets with $|A| = |B| = n$ and $c\colon A \times B \to \mathbb{R}$ a cost function. An* assignment *between A and B is a bijection $f\colon A \to B$. The cost of an assignment f is $c(f) = \sum_{a \in A} c(a, f(a))$. The* assignment problem *is to find an assignment with minimum cost.*

For an assignment instance (A, B, c), we denote the cost of an optimal assignment by $d^c_{\mathrm{oa}}(A, B)$. The assignment problem can be solved in cubic running time using a suitable implementation of the Hungarian method [8].

3.2 Searching in Databases

Databases provide means to store data to be able to retrieve, insert or change it efficiently. In the context of data analysis, retrieval (search) is usually the crucial operation on databases, because it will be performed much more often than updates. We focus on two important types of similarity queries when searching a database DB, the first of which is the *range query* for a radius r:

Definition 3 (Range Query). *A* range query $\mathrm{range}(q, r)$*, with query object q and range r, returns all objects in the database with a distance to the query object not exceeding the range, i.e.,* $\mathrm{range}(q, r) = \{o \in DB \mid d(o, q) \leq r\}$.

The second type of query considered here is the *k-nearest neighbor query.*

Definition 4 (k-Nearest Neighbor Query). *A k-nearest neighbor query (kNN query)* $\mathrm{NN}(q, k)$ *with query object q and parameter k returns the smallest set* $\mathrm{NN}(q, k) \subseteq DB$*, so that* $|\mathrm{NN}(q, k)| \geq k$ *and*

$$\forall o \in \mathrm{NN}(q, k), \forall o' \in DB \setminus \mathrm{NN}(q, k): \;\; d(o, q) < d(o', q).$$

In conjunction with range queries, it is preferable to return all the objects with a distance (including the query object, if part of the database), that does not exceed the distance to the kth neighbor, which may be more than k objects when tied. That yields an equivalency of the results of kNN queries and range queries, i.e., we have $\mathrm{range}(q, r){=}\mathrm{NN}(q, |\mathrm{range}(q, r)|)$ and $\mathrm{NN}(q, k){=}\mathrm{range}(q, r_k)$, where r_k is the maximum distance in $\mathrm{NN}(q, k)$. Provided that the distance used is a metric, both types of queries can be accelerated using metric indices. In our work, we use the *vantage point tree* (*vp-tree*) [27] as a classical method and the more recent *cover tree* [1], because they are available in the ELKI framework [21], but others could also be used. While the vp-tree is a height balanced binary

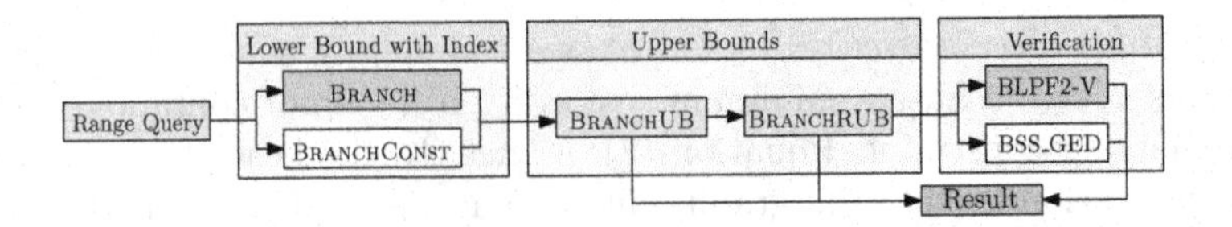

Fig. 1. Overview of the filter pipeline. For general metric edit costs the blue modules are used; yellow modules are more efficient for uniform (edge) edit costs.

tree dividing the data into *near* and *far* halves of the dataset based on the median distance from the vantage point, the cover tree controls the expansion rate by reducing the maximum radius in each level of the tree, branching out if necessary into multiple branches. In both trees, queries are performed top-down by traversing all paths that cannot be dismissed using the routing objects and employing the triangle inequality.

4 Efficient Filtering for the General Graph Edit Distance

We propose a filter pipeline for range queries regarding the graph edit distance following a common paradigm for expensive distances, see e.g. [28]. Here, lower bounds allow to filter out graphs that do not satisfy the query predicate. For the remaining candidates, upper bounds are evaluated to add them immediately to the result set without exact distance computation. Finally, in the verification step, the exact distance is computed for the remaining candidates only. Our approach starts with the optimal assignment based lower bound BRANCH accelerated by metric indexing. From the same optimal assignment, an upper bound is derived (BRANCHUB) and subsequently refined by local search (BRANCHRUB) before the remaining candidates are verified. The pipeline is illustrated in Fig. 1, the individual steps are described in the following.

4.1 Index-Accelerated Lower Bound Filtering

Several lower bounds on the graph edit distance have been proposed or can be derived from known heuristics, see [4]. One of the most effective lower bounds with an excellent trade-off between tightness and runtime is referred to as BRANCH.

Definition 5 (Branch Distance). *For two graphs G_1 and G_2 the* branch distance *is defined as $d_{branch}(G_1, G_2) = d^{c}_{\mathrm{oa}}(V(G_1) \cup \varepsilon_1, V(G_2) \cup \varepsilon_2)$, where ε_i denotes a multiset of ϵ elements, so that $|V(G_i) \cup \varepsilon_i| = |V(G_1) \cup V(G_2)|$ for $i \in \{1, 2\}$, and*

$$c(u,v) = \begin{cases} 0 & \text{if } u = v = \epsilon \\ c_v(u,v) + d_e(u,v) & \text{if } u \neq \epsilon \text{ and } v \neq \epsilon \\ c_v(\epsilon,v) + 1/2 \cdot \sum_{n \in N(v)} c_e(\epsilon, vn) & \text{if } u = \epsilon \text{ and } v \neq \epsilon \\ c_v(u,\epsilon) + 1/2 \cdot \sum_{n \in N(u)} c_e(un, \epsilon) & \text{if } u \neq \epsilon \text{ and } v = \epsilon \end{cases},$$

with $d_e(u,v) = d^{c'}_{\mathrm{oa}}(N(u) \cup \varepsilon_u, N(v) \cup \varepsilon_v)$, where $c'(w,x) = 1/2 \cdot c_e(uw, vx)$.

Remark 1. Note that, by using a customized version of the Hungarian algorithm, BRANCH can also be implemented in a slightly more efficient way, where only one dummy vertex ϵ is added to the vertex sets $V(G_1)$ and $V(G_2)$ (see [7] for details). In this paper, we use the classical implementation employed in [2,20], which corresponds to the characterization provided in Definition 5.

BRANCH has its origin in one of the most successful heuristics for the graph edit distance proposed by Riesen and Bunke [20]. However, in contrast to the original approach, it is guaranteed to underestimate the graph edit distance by dividing all edge costs by two to avoid that the cost of a single edge edit operation is counted twice, once for each endpoint [2]. Since an instance of the assignment problem on the vertices of the two graphs has to be solved, and for each vertex pair an assignment on their edges, BRANCH can be computed in $O(n^2\Delta^3 + n^3)$ time for graphs with n vertices and maximum degree Δ. In the case of uniform edge edit costs, d_e can be computed by multiset intersection of edge labels and the running time reduces to $O(n^3)$. This special case is referred to as BRANCHCONST [2]. It has been shown that, if the edit costs are metric, the branch distance is a pseudo-metric on graphs [2]. This allows to accelerate computing the candidate set w.r.t. this lower bound by employing metric indexing.

4.2 Upper Bound Filtering and Verification

From the solution of the assignment problem of BRANCH, an upper bound can be obtained by deriving the corresponding edit path [20], denoted BRANCHUB here. By definition of the graph edit distance, the cost of every edit path is an upper bound of the graph edit distance. Following [28], we refine the assignment by local search to gain a tighter upper bound. Starting with the assignment obtained for the lower bound, the mapping of two vertex pairs is iteratively swapped, and kept whenever it induces a cheaper edit path, until there is no improvement. We refer to the refined upper bound obtained from the BRANCH assignment as BRANCHRUB.

Eventually, the graphs that were neither filtered out by the lower bound nor approved by the upper bounds are verified by exact graph edit distance computation. We use *BSS_GED* [10] for datasets with discrete labels and uniform costs and *BLPF2-V* otherwise. The latter is based on the integer programming formulation F2 of [15] with the additional constraint that the objective function does not exceed the threshold to allow for early termination.

4.3 Nearest-Neighbor Queries

For kNN queries it is not possible to separate the different steps of the filter pipeline as clearly as shown in Fig. 1. We realize kNN queries using the *optimal multi-step k-nearest neighbor search* algorithm [22]. The database graphs are scanned in ascending order according the lower bound BRANCH regarding the query graph. For each graph, the exact graph edit distance is computed and the k

Table 2. Datasets and their statistics [17]. Some datasets contain graphs with labeled or attributed vertices and edges, as can be seen in the last two columns.

Name	|Graphs|	avg |Vertices|	avg |Edges|	Labels (V/E)	Attributes (V/E)
Cuneiform	267	21.27	44.80	+/+	+/+
Fingerprint	2800	5.42	4.42	−/−	+/+
Letter-high	2250	4.67	4.50	−/−	+/−
Letter-low	2250	4.68	3.13	−/−	+/−
MUTAG	188	17.93	19.79	+/+	−/−
PTC_FM	349	14.11	14.48	+/+	−/−
QM9	129433	18.03	18.63	−/−	+/+

graphs with the smallest exact graph edit distance are maintained. Once we have found at least k objects with an exact distance smaller than the lower bound of all remaining objects, the search can be terminated. This is optimal in the sense that none of the exact distance computations could have been avoided [22]. Accessing the graphs ordered regarding the BRANCH lower bound can be achieved naïvely by sorting all graphs, or by using suitable metric index structures.

5 Experimental Evaluation

In this section, we experimentally address the following research questions:

- Q1 What speed-up in range queries can be achieved when using metric indices compared to a linear scan of the database?
- Q2 How effective are the individual lower and upper bounds in our pipeline?
- Q3 Can the proposed filter pipeline compete with state-of-the-art methods for uniform edit costs?
- Q4 What speed-up in kNN queries can be achieved when using metric indices?
- Q5 Does the proposed filter pipeline scale to a very large dataset?

5.1 Setup

As metric index we chose the *vp-tree* as a classical method and the *cover tree* as a state-of-the-art approach. For both we used the implementation provided by ELKI [21] with a sample size of 5 for the vp-tree and an expansion rate of 1.2 for the cover tree.

For a comparison in databases containing graphs with categorical labels, we used *MLIndex* [16] and *GSim* [30], since the former is considered state-of-the-art, while the latter provided much better results in our experiments. For *MLIndex* the number of partitions was set to threshold + 1 and in *GSim* all provided filters were used. We used the implementations provided by the authors. In addition, we used *CStar* [28], which follows a filter-verification approach related to ours. For verification we used *BSS_GED* [10] and *BLPF2-V* [15] with the Gurobi solver.

We conducted experiments on a wide range of real-world datasets with different characteristics, see Table 2. The costs of inserting, deleting or relabeling

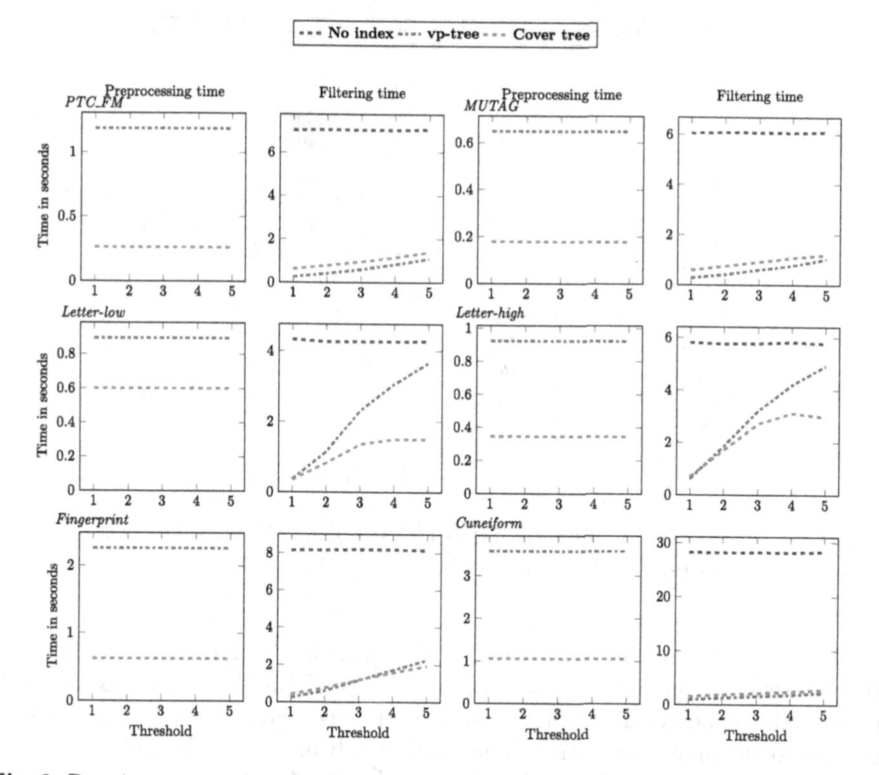

Fig. 2. Runtime comparison for filtering 100 range queries using BRANCH with thresholds 1 to 5 and preprocessing time for constructing the index.

a vertex or edge with a categorical label were set to 1, which is equivalent to the fixed setting in *MLIndex*, *GSim*, and *CStar*. For continuous attributes, the Euclidean distance was used to define the relabeling cost. For simplicity, we did not use domain-specific distances. Continuous attributes were normalized to the range $[0, 1]$ (separately for each dimension), to make distances roughly comparable between different datasets.

5.2 Results

We report on our findings regarding the above research questions.

Q1: Speed-up of range queries through metric indices. We first investigate how much of a speed-up can be achieved by using an index structure when filtering candidates for a range query by a lower bound. We randomly sampled 100 graphs from the dataset to be queries and then performed lower bound filtering without an index, using the cover tree, and the vp-tree.

Figure 2 shows the time needed for filtering 100 range queries (each with thresholds 1 to 5) and additionally the preprocessing time for index construction. The runtime does not depend on the given threshold for a linear scan, but

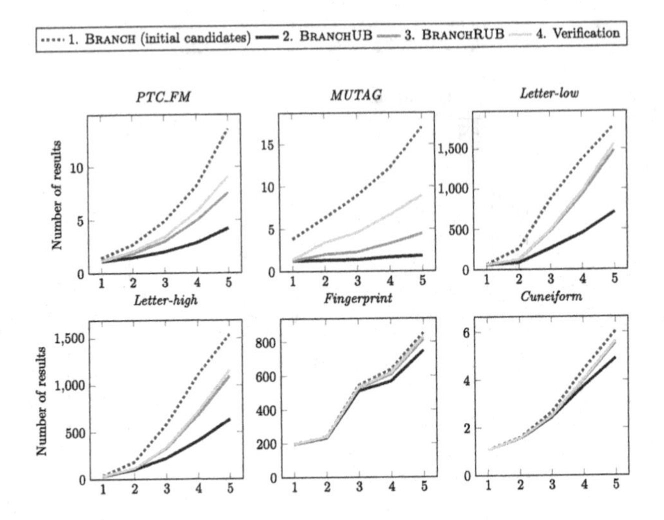

Fig. 3. Average number of initial candidates (dashed) and hits identified in the different stages of the filter pipeline for each threshold.

increases for the metric indices with the threshold, in particular for the *Letter*-datasets. It can be seen that, while on most datasets both index methods provide the same runtime benefit for filtering, the cover tree is much faster in preprocessing than the vp-tree. The runtime advantage on the *Letter*-datasets is quite small for larger thresholds. The runtime of the index structures directly corresponds to the number of BRANCH distance computations. Compared to the cover tree, the vp-tree requires many more distance computations in the preprocessing due to the chosen sample size. In general, the runtime corresponds to the number of candidates, which we investigate in the following.

Q2: Filter pipeline. In this experiment we investigate how the candidate and result set are updated during filtering. Figure 3 shows the average number of candidates for BRANCH and the number of results after each step for 100 range queries. When comparing the size of the candidate sets with the results of the previous experiment, it can be seen, that the runtime for filtering highly depends on the number of candidates. For some datasets almost all candidates remaining after the upper bound filtering are not results. This indicates that improvement is possible with tighter lower bounds. In general, BRANCHRUB manages to report almost all results, except in dataset *MUTAG*.

Q3: Comparison with state-of-the-art methods. Many methods for similarity search in graph databases limited to uniform edit costs have been proposed. We compare to *MLIndex* [16], *CStar* [28] and *GSim* [30]. We used *BSS_GED* [10] for a fast verification in our filter pipeline, as well as in *CStar*. The implementations of *MLIndex* and *GSim* contain their own verification algorithm. Figure 4 shows the runtime for preprocessing and filtering as well as the total query time

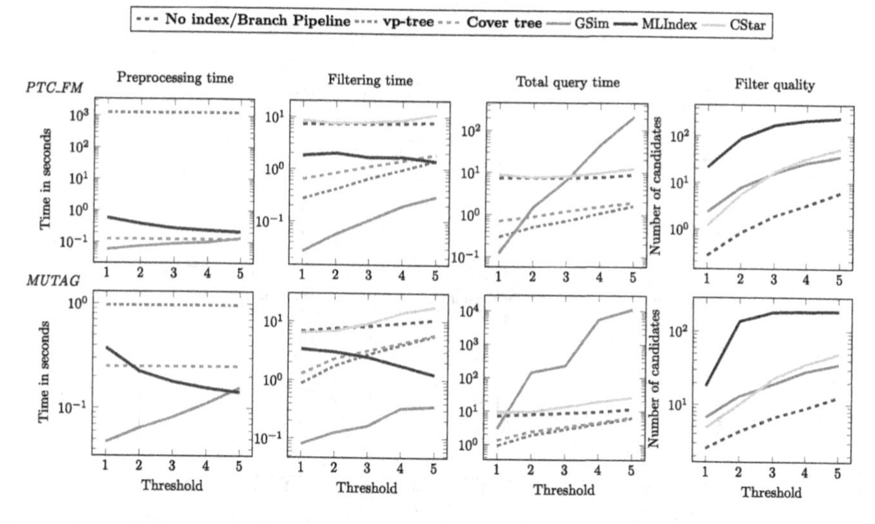

Fig. 4. Runtime for answering 100 range queries and average number of candidates remaining after applying all filters for thresholds 1 to 5. Our approaches are shown with dashed lines and are marked bold in the legend. For *MLIndex* no verification time is given, since it did not finish within the time limit of 2 days.

including filtering and verification for 100 range queries. The average number of candidates remaining after application of all filters in the different methods is also shown. Only these need to be verified by exact graph edit distance computation. For BRANCH only one line is shown, since linear scan, *vp-tree* and the *cover tree* variant apply the same filters and generate the same candidates.

MLIndex produces the largest candidate set, and did not finish the verification process in the time limit. It can be seen that, while *GSim* is quite fast in preprocessing and filtering, the verification step takes a long time. This is due to a combination of a slower verification algorithm and a higher number of candidates that have to be verified. The results indicate that, even when using *BSS_GED* for verification, the approach would not be competitive with the *cover tree* due to the high number of candidates. *CStar* needs much more time for filtering and cannot filter out as many candidates leading also to a higher verification time. Interestingly, the time for verification does not increase proportionally to the number of candidates, which might indicate, that the verification algorithm needs more time to verify certain *difficult* graphs.

Q4: Speed-up of kNN queries through metric indices. We investigate how much of a speed-up can be achieved by using an index structure compared to not using one, when answering kNN queries using the optimal multi-step k-nearest neighbor search, cf. Sect. 4.3. We randomly sampled 20 graphs from the dataset to be queries and then used the cover tree as well as the vp-tree as the underlying metric index to compare them. Figure 5 shows the time needed for answering 20 kNN queries, each with $k \in \{1, \ldots, 5\}$ (excluding preprocessing). Since in the

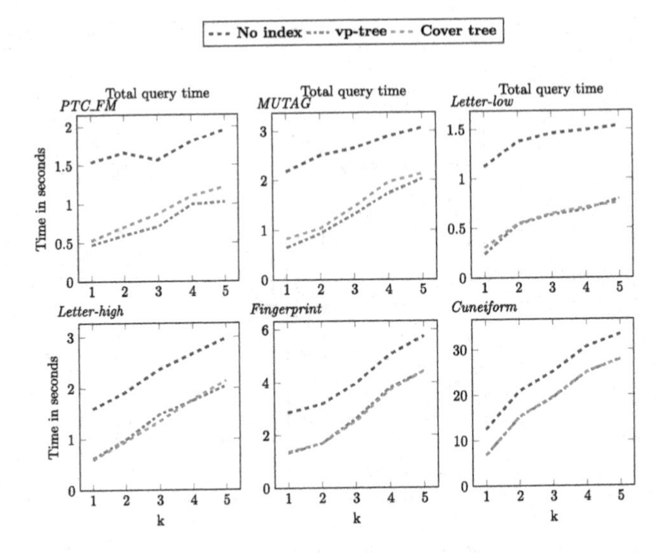

Fig. 5. Runtime comparison for answering 20 kNN queries using BRANCH and $k \in \{1, \ldots, 5\}$.

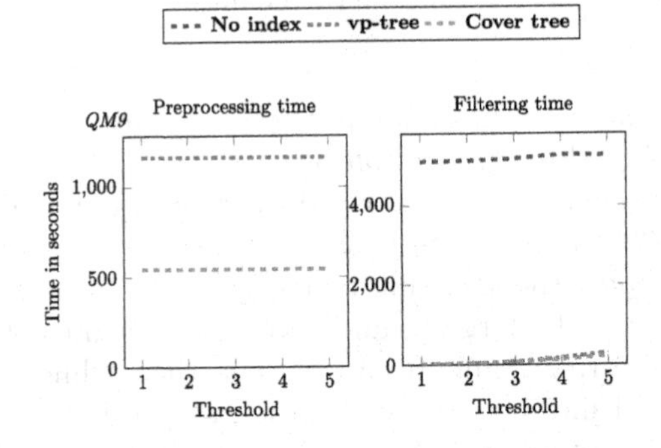

Fig. 6. Runtime comparison for preprocessing and filtering 100 range queries in the dataset *QM9* using BRANCH and thresholds 1 to 5.

optimal multi-step k-nearest neighbor search, the candidates have to be verified during search, before further candidates are explored, the runtime also includes the time needed for verification. It can be seen that, again both index structures provide the same runtime benefit. Taking into account the preprocessing time however, the cover tree has a clear advantage over the vp-tree.

Q5: Similarity search in a large dataset. We investigate the scalability of our approach on the dataset *QM9* with 129 433 graphs with attributed vertices and edges. The results shown in Fig. 6 confirm the high preprocessing time of the vp-tree compared to the cover tree. Both index methods achieve a significant advantage over a linear scan in filtering by reducing the running time by several orders of magnitude depending on the selectivity of the query.

6 Conclusions

We have shown that the recently studied lower and upper bounds on the graph edit distance can be employed to realize scalable graph similarity search in a filter-verification framework accelerated by metric indexing. Our approach supports attributed graphs without restrictions of edit costs. For the extensively studied special case of graphs with discrete labels and uniform edit costs, our approach was shown experimentally to outperform the state-of-the-art methods.

There are several directions of future work to improve the filter-verification pipeline further. Our tightest upper bound was obtained via local search using a straightforward approach. More sophisticated techniques have been proposed recently [6] and can be incorporated to reduce verification. For the verification step, tailored methods that benefit from the already obtained assignment or the upper and lower bound can be developed. A well-known phenomenon of metric trees is that their effectivity decreases with increasing intrinsic dimensionality of the data/distance. Therefore, a suitable lower bound should not only be efficiently computed and tight, but ideally also have a low intrinsic dimensionality. Studying this property for the available lower bounds remains future work. Finally, recent advances in median graph computation [3] suggest to compute routing objects instead of using database graphs. An experimental comparison to such orthogonal approaches remains future work.

References

1. Beygelzimer, A., Kakade, S.M., Langford, J.: Cover trees for nearest neighbor. In: International Conference Machine Learning, ICML, vol. 148, pp. 97–104 (2006)
2. Blumenthal, D.B., Gamper, J.: Improved lower bounds for graph edit distance. IEEE Trans. Knowl. Data Eng. **30**(3), 503–516 (2018)
3. Blumenthal, D.B., Boria, N., Bougleux, S., Brun, L., Gamper, J., Gaüzère, B.: Scalable generalized median graph estimation and its manifold use in bioinformatics, clustering, classification, and indexing. Inf. Syst. **100**, 101766 (2021)
4. Blumenthal, D.B., Boria, N., Gamper, J., Bougleux, S., Brun, L.: Comparing heuristics for graph edit distance computation. VLDB J. **29**(1), 419–458 (2019). https://doi.org/10.1007/s00778-019-00544-1
5. Blumenthal, D.B., Gamper, J.: On the exact computation of the graph edit distance. Pattern Recogn. Lett. **134**, 46–57 (2020)
6. Boria, N., Blumenthal, D.B., Bougleux, S., Brun, L.: Improved local search for graph edit distance. Pattern Recogn. Lett. **129**, 19–25 (2020)
7. Bougleux, S., Gaüzère, B., Blumenthal, D.B., Brun, L.: Fast linear sum assignment with error-correction and no cost constraints. Pattern Recogn. Lett. **134**, 37–45 (2020)
8. Burkard, R.E., Dell'Amico, M., Martello, S.: Assignment Problems. SIAM, Philadelphia (2012)
9. Chang, L., Feng, X., Lin, X., Qin, L., Zhang, W., Ouyang, D.: Speeding up GED verification for graph similarity search. In: International Conference Data Engineering, ICDE, pp. 793–804 (2020)
10. Chen, X., Huo, H., Huan, J., Vitter, J.S.: An efficient algorithm for graph edit distance computation. Knowl. Based Syst. **163**, 762–775 (2019)

11. Gouda, K., Arafa, M., Calders, T.: BFST_ED: a novel upper bound computation framework for the graph edit distance. In: Amsaleg, L., Houle, M.E., Schubert, E. (eds.) SISAP 2016. LNCS, vol. 9939, pp. 3–19. Springer, Cham (2016). https://doi.org/10.1007/978-3-319-46759-7_1
12. Gouda, K., Hassaan, M.: CSI_GED: an efficient approach for graph edit similarity computation. In: International Conference Data Engineering, ICDE, pp. 265–276 (2016)
13. Kim, J., Choi, D., Li, C.: Inves: incremental partitioning-based verification for graph similarity search. In: Extending Database Technology, EDBT, pp. 229–240 (2019)
14. Kriege, N.M., Giscard, P., Bause, F., Wilson, R.C.: Computing optimal assignments in linear time for approximate graph matching. In: ICDM, pp. 349–358 (2019)
15. Lerouge, J., Abu-Aisheh, Z., Raveaux, R., Héroux, P., Adam, S.: New binary linear programming formulation to compute the graph edit distance. Pattern Recogn. **72**, 254–265 (2017)
16. Liang, Y., Zhao, P.: Similarity search in graph databases: A multi-layered indexing approach. In: International Conference Data Engineering, ICDE, pp. 783–794 (2017)
17. Morris, C., Kriege, N.M., Bause, F., Kersting, K., Mutzel, P., Neumann, M.: TUDataset: a collection of benchmark datasets for learning with graphs. In: ICML 2020 Workshop on Graph Representation Learning and Beyond, GRL+ (2020)
18. Neuhaus, M., Riesen, K., Bunke, H.: Fast suboptimal algorithms for the computation of graph edit distance. In: Structural, Syntactic, and Statistical Pattern Recognition. pp. 163–172, August 2006
19. Qin, Z., Bai, Y., Sun, Y.: Ghashing: semantic graph hashing for approximate similarity search in graph databases. In: ACM SIGKDD, pp. 2062–2072 (2020)
20. Riesen, K., Bunke, H.: Approximate graph edit distance computation by means of bipartite graph matching. Image Vis. Comput. **27**(7), 950–959 (2009)
21. Schubert, E., Zimek, A.: ELKI: a large open-source library for data analysis - ELKI release 0.7.5 "Heidelberg". CoRR abs/1902.03616 (2019)
22. Seidl, T., Kriegel, H.: Optimal multi-step k-nearest neighbor search. In: SIGMOD International Conference Management of Data, pp. 154–165 (1998)
23. Serratosa, F., Cortés, X., Solé-Ribalta, A.: Graph database retrieval based on metric-trees. In: SSPR, pp. 437–447 (2012)
24. Stauffer, M., Tschachtli, T., Fischer, A., Riesen, K.: A survey on applications of bipartite graph edit distance. In: GbRPR, pp. 242–252 (2017)
25. Wang, G., Wang, B., Yang, X., Yu, G.: Efficiently indexing large sparse graphs for similarity search. IEEE Trans. Knowl. Data Eng. **24**(3), 440–451 (2012)
26. Wang, X., Ding, X., Tung, A., Ying, S., Jin, H.: An efficient graph indexing method. In: International Conference Data Engineering, ICDE (2012)
27. Yianilos, P.N.: Data structures and algorithms for nearest neighbor search in general metric spaces. In: SODA, pp. 311–321 (1993)
28. Zeng, Z., Tung, A.K.H., Wang, J., Feng, J., Zhou, L.: Comparing stars: on approximating graph edit distance. Proc. VLDB Endow. **2**(1), 25–36 (2009)
29. Zhao, X., Xiao, C., Lin, X., Liu, Q., Zhang, W.: A partition-based approach to structure similarity search. Proc. VLDB Endow. **7**(3), 169–180 (2013)
30. Zhao, X., Xiao, C., Lin, X., Wang, W.: Efficient graph similarity joins with edit distance constraints. In: International Conference Data Engineering, ICDE (2012)
31. Zheng, W., Zou, L., Lian, X., Wang, D., Zhao, D.: Efficient graph similarity search over large graph databases. IEEE Trans. Knowl. Data Eng. **27**(4), 964–978 (2015)

The Minimum Edit Arborescence Problem and Its Use in Compressing Graph Collections

Lucas Gnecco[1(✉)], Nicolas Boria[1], Sébastien Bougleux[2], Florian Yger[1], and David B. Blumenthal[3]

[1] PSL Université Paris-Dauphine, LAMSADE, Paris, France
{lucas.gnecco,nicolas.boria,florian.yger}@dauphine.fr
[2] UNICAEN, ENSICAEN, CNRS, GREYC, Caen, France
sebastien.bougleux@unicaen.fr
[3] Department Artificial Intelligence in Biomedical Engineering (AIBE), Friedrich-Alexander University Erlangen-Nürnberg (FAU), Erlangen, Germany
david.b.blumenthal@fau.de

Abstract. The inference of minimum spanning arborescences within a set of objects is a general problem which translates into numerous application-specific unsupervised learning tasks. We introduce a unified and generic structure called *edit arborescence* that relies on edit paths between data in a collection, as well as the MINIMUM EDIT ARBORESCENCE PROBLEM, which asks for an edit arborescence that minimizes the sum of costs of its inner edit paths. Through the use of suitable cost functions, this generic framework allows to model a variety of problems. In particular, we show that by introducing *encoding size preserving edit costs*, it can be used as an efficient method for compressing collections of labeled graphs. Experiments on various graph datasets, with comparisons to standard compression tools, show the potential of our method.

Keywords: Edit arborescence · Edit distance · Lossless compression

1 Introduction

The discovery of some underlying structure within a collection of data is the main goal of unsupervised learning. Among the different kinds of graph structures available for structure inference, arborescences play an essential role, because they contain the minimal number of edges required to connect all the entries of the collection and induce a meaningful hierarchy within the data. For these reasons, arborescences are widely used in structure inference, in numerous fields ranging from bioinformatics [12] to computational linguistics [19]. For constructing arborescences, distances have to be computed for data objects within the

Supported by Agence Nationale de la Recherche (ANR), projects STAP ANR-17-CE23-0021 and DELCO ANR-19-CE23-0016. FY acknowledges the support of the ANR as part of the "Investissements d'avenir" program, reference ANR-19-P3IA-0001 (PRAIRIE 3IA Institute).

N. Reyes et al. (Eds.): SISAP 2021, LNCS 13058, pp. 337–351, 2021.
https://doi.org/10.1007/978-3-030-89657-7_25

collection. While the computation of distances is trivial for many kinds of simple data (e. g., vectors in Euclidean space), it is often challenging for more complex kinds of data such as strings, trees, or graphs. For such data, *edit distances*— measuring the distance between two objects o_1 and o_2 as the cost of modifications needed to transform o_1 into o_2—provide meaningful measures.

In this work, we propose a unified and generic framework for minimum arborescence computation on collections of structured data for which an edit distance is available. We introduce the concept of *edit arborescence* which generalizes the concept of edit path (a sequence of edit operations, or modifications), and we formalize the Minimum Edit Arborescence Problem (*MEA*). By using appropriate edit cost functions over the edit operations, as well as different sets of allowed edit operations, this generic framework allows to tackle a variety of specific problems, such as event detection in time series [16], morphological forests inference over a language vocabulary [18], or structured data compression [10].

As a proof of concept, we focus on the latter application, and address the problem of compressing a collection of labeled graphs. To the best of our knowledge, this problem has not been addressed in the literature. In graph stores, each graph is encoded individually using space-efficient representations based on different, mainly lossless compression schemes [3,4,8], but without taking into account the other graphs in the store. This is also the case for lossy graph compression schemes [22]. All of these compression schemes are beyond the main focus of this paper, and we refer the reader to the above references.

Contrary to these schemes, our compression method relies heavily on reference-based compression underpinned by an arborescence connecting the graphs of the collection. Intuitively, each graph is represented by an edit path between its parent graph and itself. Each graph can thus be reconstructed recursively up to the root element of the arborescence, which we define as the empty graph. Similar ideas have been proposed for compressing web graphs seen as a temporal graphs with edge insertions and deletions [1], or collections of bitvectors using the Hamming distance [9], recently applied to graph annotations (colors) [2]. While these approaches can be considered as early examples of using *MEA* in compression, our formulation is more general.

We first formalize the concepts of edit distances and arborescences in Sect. 2. In Sect. 3, we introduce edit arborescences and define *MEA*. Section 4 deals specifically with graph data and the graph edit distance, and formalizes the Minimum Graph Edit Arborescence Problem (*MGEA*). Section 5 provides detailed explanations on how to make use of the *MGEA* to address the compression of a set of labeled graphs. In Sect. 6, we report the results of the experimental evaluation. Finally, Sect. 7 concludes the paper and points out to future work.

2 Preliminaries

We consider data (sequence, tree, graph) defined by a combinatorial structure and labels attached to the elements of this structure. Labels may be of any type. Unlabeled and unstructured data are special cases.

Edit Distance. Given a space Ω of all data of a fixed type, an *edit path* is a sequence of elementary modifications (or *edit operations*) transforming an object of Ω into another one. Typical edit operations are the deletion and the insertion of an element of the structure, and the substitution of an attribute attached to an element. Given a cost function $c \geq 0$ defined on edit operations, the *edit distance* $d_c : \Omega \times \Omega \to \mathbb{R}_{\geq 0}$ measures the minimal total cost required to transform $x \in \Omega$ into $y \in \Omega$, up to an equivalence relation: $d_c(x, y) := \min_{P \in \mathcal{P}(x,y)} c(P)$, with $c(P) := \sum_{o \in P} c(o)$ the cost of an edit path P, and $\mathcal{P}(x, y)$ the set of all edit paths transforming x into an element of $[y] := \{z \in \Omega \mid y \sim z\}$, the equivalence class of y for an equivalence relation $\sim$ on Ω. Equality is the equivalence relation usually considered for strings or sequences (Hamming, Levenshtein or discrete time warping distances). Isomorphism is used for trees and graphs.

The set Ω equipped with an edit distance d_c defines an *edit space* (Ω, d_c). We assume that d_c is metric or pseudometric (if $\sim$ is not equality). The set Ω contains a null (or empty) element denoted by $0_\Omega \in \Omega$. Any other element of (Ω, d_c) can be constructed by insertion operations only from 0_Ω.

Arborescences. A directed graph (digraph) is a pair $G := (V, E)$, where $V := \{v_0, .., v_n\}$ is a set of nodes and $E \subseteq V \times V$ is a set of directed edges. Within such a graph, a *spanning arborescence* is a rooted, oriented, spanning tree, i. e., a set of edges that induces exactly one directed path from a root node $r \in V$ to each other node in $V \setminus \{r\}$. By assuming w. l. o. g. that the root element is v_0 and reminding that all other nodes in an arborescence have a unique parent node, an arborescence can be represented by a sequence of node indices $\mathcal{A}$ such that, for all $i \in [1, n]$, $\mathcal{A}[i]$ denotes the index of the unique parent node of node v_i. The set of edges of $\mathcal{A}$ is denoted by $E^{\mathcal{A}}$.

3 The Minimum Edit Arborescence Problem

In this section, we introduce and describe the generic MINIMUM EDIT ARBORESCENCE PROBLEM (*MEA*), a versatile problem. An instance of *MEA* is a finite dataset X living in an edit space (Ω, d_c). *MEA* asks for a minimum-cost edit arborescence rooted at the null element.

Given a set $X := \{x_0, x_1, ..., x_n\} \subset \Omega$ such that $x_0 := 0_\Omega$, we define an *edit arborescence* as a pair $(\mathcal{A}, \Psi)$, where $\mathcal{A}$ is a sequence of n indices that defines an arborescence rooted at the index 0, such that for all $i \in [1, n]$, $\mathcal{A}[i]$ is the parent-index of i. $\Psi := (P_1, ..., P_n)$ is a sequence of edit paths, such that $P_i \in \mathcal{P}(x_{\mathcal{A}[i]}, x_i)$ holds for all $i \in [1, n]$, i. e., P_i is an edit path between x_i and its parent in $\mathcal{A}$. $\mathbb{A}(X)$ is the set of all edit arborescences on X.

Definition 1 (*MEA*). *Given a finite set $X \subset \Omega$ and an edit cost function c, the* MINIMUM EDIT ARBORESCENCE PROBLEM *(MEA) asks for an edit arborescence $(A^\star, \Psi^\star)$ on $X \cup \{0_\Omega\}$, which is rooted at the null element $0_\Omega \in \Omega$ and has a minimum cost $c(\Psi^\star)$ among all $(\mathcal{A}, \Psi) \in \mathbb{A}(X)$, with $c(\Psi) := \sum_{P \in \Psi} c(P)$.*

By definition, it holds that $c(\Psi^\star) = \min_{(\mathcal{A},\Psi)\in\mathbb{A}_c(X)} \sum_{P_i\in\Psi} d_c(x_{\mathcal{A}[i]}, x_i)$, where $\mathbb{A}_c(X)$ is the set of edit arborescences in (Ω, d_c), i. e., edit arborescences with edit paths restricted to minimal-cost edit paths w. r. t. c. This generic definition can translate into various optimization problems, with different characteristics in terms of complexity and/or approximability, depending on the edit space.

Exact Solver. Whenever exact edit distances and corresponding edit paths can be computed, the following procedure produces an optimal solution for *MEA*:

1. Construct the complete directed weighted graph on the set $X \cup \{0_\Omega\}$, denoted by $\mathcal{K}(X, d_c) := (V^{\mathcal{K}}, E^{\mathcal{K}}, w)$, with node set $V^{\mathcal{K}} := X \cup \{0_\Omega\}$ and edge weights $w(u, v) := d_c(u, v)$ for all $(u, v) \in E^{\mathcal{K}}$. Note that any edge entering the root can be removed.
2. Solve the MINIMUM SPANNING ARBORESCENCE PROBLEM (*MSA*) on $\mathcal{K}(X, d_c)$ with 0_Ω as root node.

For a connected weighted directed graph G and a root node r in G, the MINIMUM SPANNING ARBORESCENCE PROBLEM (*MSA*) asks for a spanning arborescence $A^\star$ on G, which is rooted in r and has minimum weight $w(A^\star)$, where $w(A) := \sum_{(u,v)\in A} w(u, v)$ [13]. *MSA* can be solved in polynomial time, e. g., in $O(|V^G|^2)$ time with Tarjan's implementation [23] of Edmonds' algorithm [13]. Hence, the main difficulty of the problem consists in computing the edge weights in $\mathcal{K}$, i. e., the edit distances between elements of X.

Lemma 1. *As long as the edit space (Ω, d_c) allows for a polynomial time computation of minimum-cost edit paths, the corresponding version of MEA belongs to the complexity class $\mathcal{P}$.*

Proof. By assumption, d_c is computed in polynomial time by some algorithm `ALG-DIST` that is called $O(n^2)$ times with complexity $O_{\texttt{ALG-DIST}}$ in order to generate the complete graph $\mathcal{K}(X, d_c)$. So, *MEA* can be solved in $O(n^2 O_{\texttt{ALG-DIST}} + (n+1)^2)$ time complexity by using Tarjan's implementation of Edmond's algorithm. □

A Heuristic for Non-polynomial Cases. We adapt the algorithm described above to cases where the edit distance is not solvable in polynomial time. The method is based on approximations or heuristics to estimate the edit distance, and allows the user to choose the desired balance between computation time and accuracy. Also, the algorithm takes advantage of prior knowledge over the data (such as relevant candidate couples of elements) which is often available in practical cases. Given a set $X \subset \Omega$, Algorithm 1 computes a *low-cost* edit arborescence $(\mathcal{A}, \Psi) \in \mathbb{A}(X)$ based on approximations or heuristics to estimate the edit distance. It starts by constructing a size-reduced auxiliary digraph $\mathcal{K}$ (lines 1 to 3) that connects 0_Ω to each element $x_i \in X$, and each x_i to $k \leq |X| - 1$ randomly selected elements of $X \setminus \{x_i\}$. If some promising edges are known *a priori* (e. g., if X has an implicit internal structure), they are added to $\mathcal{K}$. Then, Algorithm 1 computes optimal or low-cost edit paths whose costs provide weights

Algorithm 1. A generic heuristic for *MEA*.

Require: A finite set X of elements from an edit space (Ω, d) with origin 0_Ω, a parameter $k \in [0, |X - 1|]$, two edit distance heuristics `ALG-1` and `ALG-2`.
Ensure: A low-cost edit arborescence $(\mathcal{A}, \Psi)$ for the *MEA* problem.
1: Set $x_0 := 0_\Omega$ and initialize auxiliary graph $\mathcal{K}(X \cup x_0, E^\mathcal{K}, w)$ with $E^\mathcal{K} := \{x_0\} \times X$.
2: **for** $x \in X$ **do** Sample k children $\tilde{X} \in \binom{X \setminus \{x\}}{k}$ and set $E^\mathcal{K} := E^\mathcal{K} \cup (\{x\} \times \tilde{X})$.
3: **if** prior information available **then** Add promising edges to $E^\mathcal{K}$.
4: **for** $(x_i, x_j) \in E^\mathcal{K}$ **do**
5: **if** $i = 0$ **then** Analytically compute the edit path P_{ij}.
6: **else if** identifiers available **then** Compute edit path P_{ij} induced by identifiers.
7: **else** Call `ALG-1` to compute low-cost edit path P_{ij}.
8: Set $w(x_i, x_j) := c(P_{ij})$.
9: Run Edmonds' algorithm on $\mathcal{K}$ to obtain $\mathcal{A}$.
10: **if** tightening **then for** $i \in [1, n]$ **do** Call `ALG-2` to compute tighter edit path $P_{\mathcal{A}[i]i}$.
11: **for** $i \in [1, n]$ **do** Set $\Psi[i] := P_{\mathcal{A}[i]i}$.
12: **return** $(\mathcal{A}, \Psi)$

for the edges of $\mathcal{K}$ (lines 4 to 8). For 0_Ω' out-edges, optimal edit paths can be computed analytically (insertions only). If identifying attributes are available for all elements of X (e. g., unique node labels if X is a set of graphs), it is sometimes possible to compute optimal edit paths from these identifiers. Otherwise, low-cost edit paths are computed by calling a polynomial edit distance heuristic `ALG-1`. Once all edge weights for $\mathcal{K}$ have been computed, an optimal arborescence $\mathcal{A}$ on $\mathcal{K}$ is constructed by Edmonds' algorithm (line 9). Optionally, a tighter edit distance heuristic `ALG-2` can be called to shorten the paths in $\mathcal{A}$ before returning the edit arborescence (line 10). The more precise, and thus potentially more costly heuristic `ALG-2` is called only n times.

4 Minimum Graph Edit Arborescence Problem

In the remainder of the paper, we will focus on the specific case of *MEA* where the space Ω is a space of labeled graphs.

Graphs. We assume that graphs are finite, simple, undirected, and labeled. However, all presented techniques can be straightforwardly adapted to directed or unlabeled graphs. A labeled graph G is a four-tuple $G := (V^G, E^G, \ell_V^G, \ell_E^G)$, where V^G and E^G are sets of nodes and edges, while $\ell_V^G : V^G \to \Sigma_V$ and $\ell_E^G : E^G \to \Sigma_E$ are labeling functions that annotate nodes and edges with labels from alphabets Σ_V and Σ_E, respectively. $\mathbb{G}(\Sigma_V, \Sigma_E)$, or $\mathbb{G}$ for short, denotes the set of all graphs for fixed alphabets Σ_V and Σ_E. $0_\mathbb{G}$ denotes the empty graph (the null element of $\mathbb{G}$). Two graphs $G, H \in \mathbb{G}$ are *isomorphic*, denoted by $G \simeq H$, if and only if there is a bijection between V^G and V^H that preserves both edges and labels.

Edit Operations and Edit Paths. We consider the following elementary edit operations, where ϵ is a dummy node and ϵ_ℓ is a dummy label:

- Node deletion (nd): (v, ϵ_ℓ), with $v \in V^G$ isolated.
- Edge deletion (ed): (e, ϵ_ℓ), with $e \in E^G$.
- Node relabeling (nr): $(v, \ell) \in V^G \times (\Sigma_V \setminus \{\ell_V^G(v)\})$.
- Edge relabeling (er): $(e, \ell) \in E^G \times (\Sigma_E \setminus \{\ell_E^G(e)\})$.
- Node insertion (ni): (ϵ, ℓ), with $\ell \in \Sigma_V$.
- Edge insertion (ei): $(e, \ell) \in (\binom{V^G}{2} \setminus E^G) \times \Sigma_E$.

For each edit path P composed of such operations, there are many equivalent edit paths with a same edit cost, obtained just by reordering the operations in P. In particular, as the deletion of a node assumes that its incident edges have been previously deleted, these operations can be replaced by node-edge deletions (ned): delete all the edges incident to a node and then delete this node. So we can distinguish two different types of edge deletions: implied edge deletion (i-ed), i. e., an edge deletion in a node-edge deletion, and non-implied edge deletion (ni-ed), i. e., an edge deletion between two nodes that are not deleted by P. The cost of an edit path P can thus be rewritten as $c(P) = \sum_{t \in T} \sum_{o \in P^t} c^t(o)$, where P^t is the (possibly empty) set of all edit operations of type $t \in T$, with $T := \{\text{ni-ed}, \text{i-ed}, \text{nd}, \text{nr}, \text{er}, \text{ni}, \text{ei}\}$, and c^t is an edit cost function for type t.

Remark 1. Any concatenation $\sigma_{\text{ni-ed}}(P^{\text{ni-ed}}) \sqcup \sigma_{\text{i-ed}}(P^{\text{i-ed}}) \sqcup \ldots \sqcup \sigma_{\text{ei}}(P^{\text{ei}})$ of edit operations, with σ_t a permutation on P^t, defines an edit path equivalent to P.

Remark 2. $c(P) = \sum_{t \in T} c^t |P^t|$ if c_t is a constant for each type of operation t.

Node Maps and Induced Edit Paths. A *node map* (or *error-correcting bipartite matching*) between a graph G and a graph H is a relation $\pi \in (V^G \cup \{\epsilon\}) \times (V^H \cup \{\epsilon\})$ such that the following two conditions hold:

- For each node $u \in V^G$, there is exactly one node $v \in V^H \cup \{\epsilon\}$ such that $(u, v) \in \pi$. We denote this node v by $\pi(u)$.
- For each node $v \in V^H$, there is exactly one node $u \in V^G \cup \{\epsilon\}$ such that $(u, v) \in \pi$. We denote this node u by $\pi^{-1}(v)$.

Let $\Pi(V^G, V^H)$ be the set of all node maps. Each node map $\pi \in \Pi(G, H)$ can be transformed into an edit path, denoted by $P[\pi]$ (*induced edit path*), such that, for each $(u, v) \in \pi$, there is a corresponding edit operation (u, ℓ): u is deleted if $v = \epsilon$, it is relabeled if $(u, v) \in V^G \times V^H$ and $\ell_V^G(u) \neq \ell_V^H(v)$, or a new node is inserted if $u = \epsilon$. Operations on edges are induced by the operations on nodes, i. e., from the pairs $((u, \pi(u)), (v, \pi(v))$ with $u, v \in V^G \cup \{\epsilon\}$. For details, we refer to [6]. What is important here is that any type of edit operation is taken into account by a node map. In particular, implied and non-implied edge deletions can be distinguished with a specific cost for each type.

Graph Edit Distance. The cost of an optimal edit path from a graph G to a graph $H' \simeq H$ defines the graph edit distance (GED) from G to H (d_c with graph isomorphism as equivalence relation): $\text{GED}(G, H) := \min_{P \in \mathcal{P}(G,H)} c(P)$. GED is hard to compute and approximate, even when restricting to simple special cases [5,24]. However, many heuristics are able to reach tight upper and/or lower

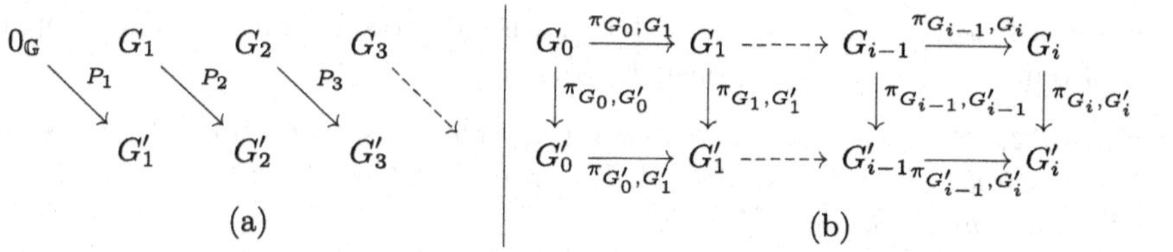

Fig. 1. (a) Paths in a non-reconstructible edit arborescence. (b) Composition of node maps used to construct reconstructible edit arborescences.

bounds. They are based on a reformulation of GED as an ERROR-CORRECTING GRAPH MATCHING PROBLEM: $\mathrm{GED}(G, H) = \min_{\pi \in \Pi(V^G, V^H)} c(P[\pi])$, which is equivalent to the above definition under mild assumptions on the edit cost function c. We refer to [6,20] for an overview.

Problem Formulation and Hardness. We can now define *MGEA*:

Definition 2 (*MGEA*)**.** *The* MINIMUM GRAPH EDIT ARBORESCENCE PROBLEM *(MGEA) is a MEA problem with* $\Omega := \mathbb{G}, X := \{G_1, ..., G_n\}$ *and* $d_c :=$ GED.

As the problem of computing GED is $\mathcal{NP}$-hard, Lemma 1 does not apply here.

Theorem 1. *MGEA is* $\mathcal{NP}$*-hard.*

The proof is omitted here due to space constraints, we refer the reader to [15] for a detailed proof, based on a reduction from the Hamiltonian cycle problem.

5 Arborescence-Based Compression

In this section, we show how to leverage *MGEA* for compressing a set of labeled graphs. For this, we introduce reconstructible and non-reconstructible edit arborescences, formulate the ARBORESCENCE-BASED COMPRESSION PROBLEM (*ABC*), and present an encoding for induced edit paths.

Reconstructible Edit Arborescence. When $d_c :=$ GED, since the definition of GED is based on graph isomorphism, applying an induced edit path $P[\pi_{G,H}]$ to a graph G yields a graph $H' \simeq H$. When using such edit paths within an edit arborescence $(\mathcal{A}, \Psi) \in \mathbb{A}(X)$, with $X \subset \mathbb{G}$ a set of graphs, the configuration described in Fig. 1(a) occurs. Namely, the edit paths in Ψ may be disjoint due to the isomorphism relation between the source graphs G_i and target graphs G'_i. Thus, the arborescence does not allow the reconstruction of any graph that is not directly connected to the root element. In this sense, only specific edit arborescences allow to reconstruct all graphs in X up to isomorphism.

Definition 3 (Reconstructability). *An edit arborescence* $(\mathcal{A}, \Psi) \in \mathbb{A}(X)$ *is* reconstructible *if and only if each graph* $G_i \in X$ *can be constructed up to isomorphism by applying the sequence of edit paths* $P := (P_{s^1}, P_{s^2}, \ldots, P_i)$ *to the empty graph* $0_{\mathbb{G}}$*, where* $(0_{\mathbb{G}}, G_{s^1}, G_{s^2}, \ldots, G_i)$ *is the path from* $0_{\mathbb{G}}$ *to* G_i *in* $\mathcal{A}$.

By composition of node maps (Fig. 1(b)), it is easy to show the following property (proof omitted due to space constraints).

Lemma 2. *For any edit arborescence $(\mathcal{A}, \Psi)$ on a set of graphs X, there is a reconstructible edit arborescence $(\mathcal{A}, \Psi')$ on a set X', such that X' is isomorphic to X and it holds that $c(\Psi') = c(\Psi)$.*

By Definition 3, the set of graphs X can be reconstructed up to graph isomorphism based on the encoding of a reconstructible edit arborescence.

Problem Formulation. For compressing a finite set of graphs $X \subset \mathbb{G}$, we are interested in finding a reconstructible edit arborescence $(\mathcal{A}, \Psi) \in \mathbb{A}(X)$ with a small encoding size $|\mathrm{C}(\mathcal{A}, \Psi)|$, where $\mathrm{C}(\cdot)$ denotes the encoding (a binary string) for the code C. Ideally we would like to minimize this size over $\mathbb{A}(X)$ and all possible codes. To encode an edit arborescence, we encode both of its elements, i. e., the arborescence $\mathcal{A}$ defined as a sequence of indices, and the sequence Ψ of edit paths induced by the node maps. In order to derive a useful expression for the code length function $|\mathrm{C}(\cdot)|$, the edit path encodings are concatenated. Thus, the encoding size to optimize is given by $|\mathrm{C}(\mathcal{A}, \Psi)| = |\mathrm{C}(M_{\mathcal{A}})| + |\mathrm{C}(\mathcal{A})| + \sum_{P \in \Psi} |\mathrm{C}(P)|$, where $M_{\mathcal{A}}$ is the overhead for decoding the different parts of $\mathrm{C}(\mathcal{A}, \Psi)$. In order to optimize this size , we must define an *encoding size preserving* cost function which forces the encoding sizes of edit paths to coincide with their edit cost.

Definition 4 (Encoding Size Preservation). *Let* C *be a code for edit paths. An edit cost function c is* encoding size preserving *w. r. t. code* C *if and only if there is a constant γ such that $|\mathrm{C}(P)| = c(P) + \gamma$ holds for any edit path P. Put differently, an encoding size preserving cost function assigns to each edit operation the space required in memory to encode the operations with code* C.

Assuming that a code C and an encoding size preserving cost function c w. r. t. C exist, the encoding size for any edit arborescence $(\mathcal{A}, \Psi) \in \mathbb{A}(X)$ can be rewritten as $|\mathrm{C}(\mathcal{A}, \Psi)| = |\mathrm{C}(M_{\mathcal{A}})| + |\mathrm{C}(\mathcal{A})| + c(\Psi) + \gamma|X|$. Since the encoding size for $\mathcal{A}$ depends only on the number of nodes, the problem of minimizing $|\mathrm{C}(\mathcal{A}, \Psi)|$ amounts to minimizing $c(\Psi)$. Consequently, finding a compact encoding of a set of graphs X reduces to a *MGEA* problem as introduced in Sect. 4.

Definition 5 (*ABC*)**.** *Let $X := \{G_1, \ldots, G_n\} \subset \mathbb{G}$ be a finite set of graphs,* C *be a code for edit paths, and c be an encoding size preserving edit cost function for* C*. Then, the* Arborescence-Based Compression Problem *(ABC) asks for a minimum weight reconstructible edit arborescence $(\mathcal{A}, \Psi')$ on some set of graphs $X' := \{G'_1, \ldots, G'_n\}$ such that, for all $i \in [1, n]$, $G'_i \simeq G_i$.*

We stress that, thanks to the use of encoding size preserving edit costs, the value that is optimized by *ABC* corresponds to the length of the code $\mathrm{C}(\mathcal{A}, \Psi')$ up to a constant. In other words, solving *ABC* produces the most compact arborescence-based representation of X. Given the simple correspondence between reconstructible edit arborescences and their non-reconstructible counterparts, *ABC* reduces to *MGEA* by restricting to encoding size preserving edit costs. Since *MGEA* is $\mathcal{NP}$-hard (Theorem 1), we propose to heuristically compute a low-cost edit arborescence as detailed below.

Algorithm 2. *ABC* encoding of graph collections.

Require: A set of graphs X, a code C, and an edit cost function c.
Ensure: Encoding $\mathrm{C}(\mathcal{A}, \Psi')$ of a reconstructible edit arborescence $(\mathcal{A}, \Psi')$ on X.
1: Compute $(\mathcal{A}, \Psi)$ with Algorithm 1.
2: Initialize list $L := [(0_{\mathbb{G}}, \pi_{\text{id}})]$, where π_{id} is the identity.
3: Initialize encoding $\mathrm{C}(\mathcal{A}, \Psi') := \mathrm{C}(M_{\mathcal{A}})\mathrm{C}(E^{\mathcal{A}})$.
4: **while** $L \neq \emptyset$ **do**
5: Pop an element $(G_i, \pi_{G_i,G_i'})$ from L.
6: **for all** children G_j of G_i in $\mathcal{A}$ **do**
7: Get $\pi \in \Pi(G_i, G_j)$ with $P[\pi] = \Psi[j]$ and initialize node ID $v' := 1$.
8: Initialize $\pi' \in \Pi(G_i', G_j')$ as node map of insertions and deletions only.
9: **for all** $v \in V^{G_i}$ **if** $\pi(v) \neq \epsilon$ **then** Set $\pi'(\pi_{G_i,G_i'}(v)) := v'$ and increment v'.
10: Concatenate $\mathrm{C}(P[\pi'])$ to $\mathrm{C}(\mathcal{A}, \Psi')$.
11: **if** G_j is no leaf in $\mathcal{A}$ **then** Append $(G_j, \pi' \circ \pi_{G_i,G_i'} \circ \pi^{-1})$ to L.
12: **return** $\mathrm{C}(\mathcal{A}', \Psi')$

Heuristic Solver for ABC. Algorithm 2 sketches our strategy to tackle the *ABC* problem. Given a set X of graphs, it first uses Algorithm 1, which outputs a non-reconstructible edit arborescence $(\mathcal{A}, \Psi)$ on X. After initializing the code, it starts encoding a reconstructible edit arborescence by going through the arborescence in BFS order (line 4). For each new node G_j with parent node G_i, a node map π' from $G_i' \simeq G_i$ to $G_j' \simeq G_j$ is reconstructed (lines 7 to 9), and the code of its induced edit path is added to the code of the arborescence (line 10). If G_j is not a leaf, the node map representing the isomorphism between G_j and G_j' is computed for later use (line 11).

Remark 3 (Star Ratio) In the worst case, we obtain a star $\mathcal{S} \in \mathbb{A}(X)$, which connects the empty graph $0_{\mathbb{G}}$ to all the graphs in X. This yields the upper bound $|\mathrm{C}(\mathcal{A}, \Psi)| \leq |\mathrm{C}(\mathcal{S})|$ on the encoding size of the obtained arborescence. Since encoding a graph from the empty graph by insertion operations only is similar to encoding the graph itself, the encoding size for the star is close to the encoding size for X, for a similar encoding strategy. Consequently, the *star ratio* $|\mathrm{C}(\mathcal{A})|/|\mathrm{C}(\mathcal{S})|$ provides a good indicator for the compression quality.

Remark 4 The encoded structure is designed to allow a straightforward decompression of any graph G_i, which, starting from the empty graph, simply consists in consecutively applying the edit paths along the path from the root to G_i in $\mathcal{A}$.

A Code for Induced Edit Paths. We show that there is a code C for edit paths and an edit cost function c such that c is encoding size preserving w. r. t. C. Using the notations introduced in Sect. 4, we encode an edit path as the concatenated string $\mathrm{C}(P[\pi]) := \mathrm{C}(M_P)\mathrm{C}(P^{\text{ni-ed}})\mathrm{C}(P^{\text{nd}})\mathrm{C}(P^{\text{nr}})\mathrm{C}(P^{\text{er}})\mathrm{C}(P^{\text{ni}})\mathrm{C}(P^{\text{ei}})$, where M_P denotes the overhead for decoding each string $\mathrm{C}(P^t)$. Note that the set $P^{\text{i-ed}}$ is not encoded, since implied edge deletions can be implicitly represented by node deletions. Similarly, we encode a set of edit operations P^t as $\mathrm{C}(P^t) := \mathrm{C}(o_1^t)\mathrm{C}(o_2^t)\cdots\mathrm{C}(o_{|P^t|}^t)$, with $o_i^t \in P^t$. Any edit operation $o := (a, \ell)$ is encoded as $\mathrm{C}(o) := \mathrm{C}(a)\mathrm{C}(\ell)$, with $\mathrm{C}(a) := \emptyset$ if $a = \epsilon$, and $\mathrm{C}(\ell) := \emptyset$ if $\ell = \epsilon_\ell$. That is, the dummy elements ϵ and ϵ_ℓ in deletion operations and node insertions are not encoded. Ultimately, the encoding size for $P[\pi]$ hence depends on how the nodes, edges, and their labels are encoded in the codes of the edit operations.

We consider fixed-length codes for nodes, edges, and their labels (other codes will be studied in future works). For a set $X \in \mathbb{G}$, nodes are encoded as integers on β_V bits, edges are encoded as a pairs of integers on $2\beta_V$ bits, and node or edge labels are encoded on, respectively, β_{Σ_V} and β_{Σ_E} bits. Dictionaries can be used for the labels and encoded in the overhead $M_\mathcal{A}$ or known *a priori*. In order to decode each set of edit operation P^t, M_P must contain their sizes $|P^t|$. They are encoded on β_P bits for each edit path. We obtain $|\mathrm{C}(P[\pi])| = \beta_P + \sum_{t \in T} c^t|P^t|$, where $c^{\mathrm{nr}} := \beta_V + \beta_{\ell_V}$, $c^{\mathrm{nd}} := \beta_V$, $c^{\mathrm{ni}} := \beta_{\Sigma_V}$, $c^{\mathrm{er}} := c^{\mathrm{ei}} := 2\beta_V + \beta_{\Sigma_E}$, $c^{\text{ed-ni}} := 2\beta_V$, and $c^{\text{ed-i}} := 0$. With these constant costs, the pair (C, c) defined above is encoding size preserving (Remark 2 and Definition 4) with constant β_P for any node map π, i.e., $|\mathrm{C}(P[\pi])| = c(P[\pi]) + \beta_P$. Therefore, the encoding size for a spanning edit arborescence $(\mathcal{A}, \Psi) \in \mathbb{A}(X)$ reduces to $|\mathrm{C}(\mathcal{A}, \Psi)| = |\mathrm{C}(M_\mathcal{A})| + |\mathrm{C}(\mathcal{A})| + c(\Psi) + \beta_P|X|$, which implies that minimizing $|\mathrm{C}(\cdot)|$ is an *ABC* problem.

6 Experiments

We performed an empirical evaluation of our compression method in the context of data archiving. Since no dedicated algorithms for compressing graph collections exist in this context, we compared it to the generic tar.bz compression, sufficient to highlight the potential of our method. Other generic compression tools such as zip or tar.gz yielded worse compression ratios than tar.bz in initial tests.

Datasets. We used eight different datasets (Table 1). The datasets AIDS and MUTA from the IAM Graph Database Repository [21], and ACYCL, PAH, and MAO from GREYC's Chemistry Dataset[1] contain graphs modeling chemical compounds. We also tested on time-evolving minimum spanning trees (MSTs) induced by the pairwise correlations of a large-scale U.S. stocks time series dataset.[2] Such MSTs are widely used for detecting critical market events such as financial crises [11,17]. We constructed three versions of the MSTs with the code in [17]: STOCKS-F (edge labels are floating-point stocks correlations), STOCKS-I (the correlations are rounded to integers), and STOCKS-N (no edge label). For all datasets, graphs were initially stored in GXL format.[3]

Parameters and Implementation. We tested two versions of our *ABC* method (Algorithm 2)— with and without additional tar.bz compression of the obtained codes. For both versions, the out-degree k of all nodes in $\mathcal{K}$ was varied across $\{0.1 \cdot |X|, 0.2 \cdot |X|, \ldots, 1.0 \cdot |X|\}$, and we did 5 repetitions for each value. For the experiments reported in Table 2, we performed 10 repetitions for

[1] https://brunl01.users.greyc.fr/CHEMISTRY/index.html.
[2] https://www.kaggle.com/borismarjanovic/price-volume-data-for-all-us-stocks-etfs.
[3] https://userpages.uni-koblenz.de/~ist/GXL/index.php.

Table 1. Number of graphs $|X|$, maximum and average number of nodes $|V|$, as well as node and edge label alphabet sizes $|\Sigma_V|$ and $|\Sigma_E|$ for all datasets.

Dataset	$\|X\|$	max $\|V\|$	avg $\|V\|$	$\|\Sigma_V\|$	$\|\Sigma_E\|$	Dataset	$\|X\|$	max $\|V\|$	avg $\|V\|$	$\|\Sigma_V\|$	$\|\Sigma_E\|$
ACYCL	183	11	8.15	3	1	MAO	68	27	18.38	3	4
MUTA	4337	417	30.32	14	3	STOCKS-N	1600	213	212.99	213	0
AIDS	1500	95	15.72	∞	3	STOCKS-I	1600	213	212.99	213	100
PAH	94	28	20.7	1	1	STOCKS-F	1600	213	212.99	213	∞

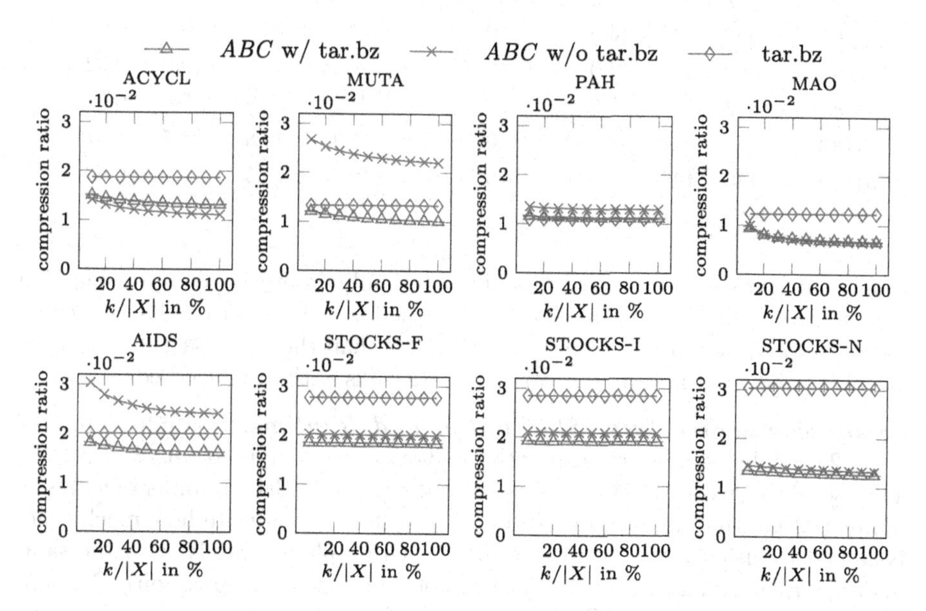

Fig. 2. Mean compression ratios w. r. t. out-degree k for tar.bz and *ABC* w/ or w/o tar.bz. For STOCKS, the values for $k = 0$ in the plots correspond to the setting where the auxiliary graph $\mathcal{K}$ only contains temporal edges.

each dataset. For STOCKS, we also added all temporal edges to $\mathcal{K}$ and always used the node maps induced by the stock identities across time (cf. lines 3 and 6 in Algorithm 1). On the other datasets, the node maps were computed and refined using the GED heuristics `ALG-1` := `BRANCH-UNIFORM` [25] and `ALG-2` := `IPFP` [6]. All algorithms were implemented in C++ using the GEDlibrary GEDLIB [7] and the *MSA* library MSArbor [14].[4] Tar.bz compressions were performed at the default compression level (9, *i.e.* the highest compression). Tests were run on a Linux system with an Intel Haswell CPU (24 cores, 2.4 GHz each) and 19 GB of main memory.

Compression Ratio. Figure 2 shows that, for all datasets except PAH, *ABC* with tar.bz significantly outperformed tar.bz compression alone and led to smaller compression ratios than *ABC* w/o tar.bz for all datasets except ACYCL.

[4] https://github.com/lucasgneccoh/gedlib.

Table 2. Mean compression and decompression times (in sec.), and standard deviations, of ABC with tar.bz for $k = 0.4 \cdot |X|$, as well as mean depths, star ratios, numbers of leafs $|\mathcal{L}|$ and inner nodes $|\mathcal{I}|$ of the computed arborescences.

Dataset	$\lvert\mathcal{L}\rvert$	$\lvert\mathcal{I}\rvert$	Avg. depth	Star ratio	Compression	Decompression
ACYCL	65	118	19.5	0.37	8 ± 0.6	0.3 ± 0.02
MUTA	1751	2586	95.4	0.47	16052 ± 2385.0	14.6 ± 1.34
AIDS	641	859	46	0.63	673 ± 59.0	5.5 ± 0.31
PAH	35	59	13.8	0.38	16 ± 1.3	0.3 ± 0.04
MAO	23	45	13.3	0.17	13 ± 1.6	0.2 ± 0.02
STOCKS-N	133	1467	75.9	0.51	1662 ± 31.9	15.0 ± 0.61
STOCKS-I	153	1446	441.5	0.71	2095 ± 46.8	18.7 ± 0.37
STOCKS-F	148	1452	426.3	0.71	2166 ± 28.5	18.9 ± 0.43

Using out-degrees $k > 0.4 \cdot |X|$ only marginally improved compression. For STOCKS, using only temporal edges ($k = 0$) led to very good results. Moreover, STOCKS-N can be more compressed with ABC than the other STOCKS datasets, as cheaper edit paths can be computed for graphs with unlabeled edges.

Arborescence Structure, Star Ratio, and Runtime. Columns 2 to 4 of Table 2 provide statistics regarding the arborescences computed with $k = 0.4 \cdot |X|$. They seem to have a good balance between depth and width (number of leaves vs. number of internal nodes). The star ratios (column 5) indicate how much space is gained by using ABC w. r. t. encoding each graph separately with the same underlying encoding scheme (a star ratio of 1 means no compression). Columns 6 to 9 summarize the ABC compression and decompression times. The most important observation is that, although ABC is much slower than tar.bz, the runtimes are still acceptable in application scenarios where a data holder wants to offer compressed graph datasets for download (compressing the largest dataset MUTA took about four to five hours). Indeed, unlike compression, decompression is fast even on the largest datasets (a couple of seconds). Runtime variations w. r. t. k are detailed in Fig. 3 for four datasets. As expected, the time required for computing the arborescences increases linearly with k, and the runtime of the refinement phase is independent of k. As the refinement algorithm IPFP is randomized, the runtimes of the refinement phase have a higher variability than the runtimes of the arborescence phase.

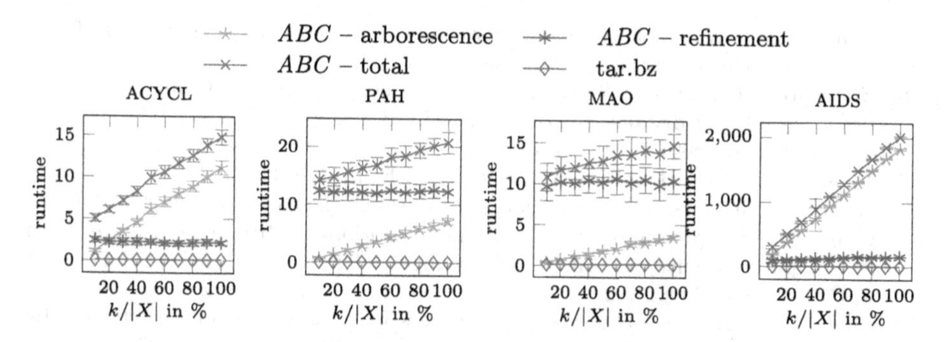

Fig. 3. Means and standard deviations of runtimes (in sec.) for *ABC* with tar.bz and its subroutines, and tar.bz alone. *ABC*-total includes the final tar.bz step.

7 Conclusions

In this paper, we have proposed the concept of an *edit arborescence* and have introduced the Minimum Edit Arborescence Problem (*MEA*). *MEA* yields a generic framework for inferring hierarchies in finite sets of complex data objects such as graphs or strings, which can be compared via *edit distances*. We have shown how to leverage *MEA* for the lossless compression of collections of labeled graphs — a task, for which no dedicated algorithms are available to date. Experiments on eight datasets show that our approach *ABC* clearly outperforms standard compression tools in terms of compression ratio and that it achieves reasonable compression and decompression times. More precisely, the experiments showed that (1) on seven out of eight datasets, our *ABC* method clearly outperformed tar.bz compression in terms of compression ratio; (2) compressing with *ABC* is computationally expensive but still reasonable in settings where the compression is carried out by an institutional data holder; (3) decompression is much faster and only takes a couple of seconds even for the largest test datasets.

References

1. Adler, M., Mitzenmacher, M.: Towards compressing web graphs. In: Proceedings of the Data Compression Conference, p. 203. IEEE Computer Society (2001). https://doi.org/10.5555/882454.875027
2. Almodaresi, F., Pandey, P., Ferdman, M., Johnson, R., Patro, R.: An efficient, scalable, and exact representation of high-dimensional color information enabled using de bruijn graph search. J. Comput. Biol. **27**(4), 485–499 (2020). https://doi.org/10.1089/cmb.2019.0322
3. Besta, M., Hoefler, T.: Survey and taxonomy of lossless graph compression and space-efficient graph representations. CoRR arXive:1806.01799 [cs.DS] (2018)
4. Besta, M., et al.: Demystifying graph databases: analysis and taxonomy of data organization, system designs, and graph queries. CoRR arXive:1910.09017 [cs.DB] (2019)

5. Blumenthal, D.B.: New Techniques for Graph Edit Distance Computation. Ph.D. thesis, Free University of Bozen-Bolzano (2019)
6. Blumenthal, D.B., Boria, N., Gamper, J., Bougleux, S., Brun, L.: Comparing heuristics for graph edit distance computation. VLDB J. **29**(1), 419–458 (2020). https://doi.org/10.1007/s00778-019-00544-1
7. Blumenthal, D.B., Bougleux, S., Gamper, J., Brun, L.: GEDLIB: A C++ library for graph edit distance computation. In: Conte, D., Ramel, J.Y., Foggia, P. (eds.) Graph-Based Representations in Pattern Recognition. GbRPR 2019. Lecture Notes in Computer Science, **11510**, 14–24. Springer, Cham (2019). https://doi.org/10.1007/978-3-030-20081-7_2
8. Boldi, P., Vigna, S.: The webgraph framework i: Compression techniques. In: WWW 2004, pp. 595–602 (2004). https://doi.org/10.1145/988672.988752
9. Bookstein, A., Klein, S.: Compression of correlated bit-vectors. Inf. Syst. **16**(4), 387–400 (1991). https://doi.org/10.1016/0306-4379(91)90030-D
10. Chwatala, A.M., Raidl, G.R., Oberlechner, K.: Phylogenetic comparative methods. J. Math. Model. Algorithms **8**, 293–334 (2009). https://doi.org/10.1007/s10852-009-9109-1
11. Coelho, R., Gilmore, C.G., Lucey, B., Richmond, P., Hutzler, S.: The evolution of interdependence in world equity markets-evidence from minimum spanning trees. Physica A **376**, 455–466 (2007). https://doi.org/10.1016/j.physa.2006.10.045
12. Cornwell, W., Nakagawa, S.: Phylogenetic comparative methods. Curr. Biol. **27**(9), R333–R336 (2017). https://doi.org/10.1016/j.cub.2017.03.049
13. Edmonds, J.: Optimum branchings. J. Res. Natl. Bur. Stand. B **71**(4), 233–240 (1967). https://doi.org/10.6028/jres.071b.032
14. Fischetti, M., Toth, P.: An efficient algorithm for the min-sum arborescence problem on complete digraphs. INFORMS J. Comput. **5**(4), 426–434 (1993). https://doi.org/10.1287/ijoc.5.4.426
15. Gnecco, L., Boria, N., Bougleux, S., Yger, F., Blumenthal, D.B.: The minimum edit arborescence problem and its use in compressing graph collections [extended version] (2021). https://arxiv.org/abs/2107.14525
16. Guralnik, V., Srivastava, J.: Event detection from time series data. In: Fayyad, U.M., Chaudhuri, S., Madigan, D. (eds.) SIGKDD 1999, pp. 33–42. ACM (1999). https://doi.org/10.1145/312129.312190
17. Liu, T., Coletti, P., Dignös, A., Gamper, J., Murgia, M.: Correlation graph analytics for stock time series data. In: EDBT 2021 (2021). https://edbt2021proceedings.github.io/docs/p173.pdf
18. Luo, J., Narasimhan, K., Barzilay, R.: Unsupervised learning of morphological forests. Trans. Assoc. Comput. Linguist. **5**, 353–364 (2017)
19. Moschitti, A., Pighin, D., Basili, R.: Semantic role labeling via tree kernel joint inference. In: CoNLL 2006, pp. 61–68. ACL (2006)
20. Riesen, K.: Structural Pattern Recognition with Graph Edit Distance: Approximation, Algorithms and Applications. Advances in Computer Vision and Pattern Recognition. Springer, Heidelberg (2016). https://doi.org/10.1007/978-3-319-27252-8
21. Riesen, K., Bunke, H.: IAM graph database repository for graph based pattern recognition and machine learning. In: da Vitoria Lobo, N., et al. (eds.) Structural, Syntactic, and Statistical Pattern Recognition. SSPR /SPR 2008. Lecture Notes in Computer Science, **5342**, 287–297. Springer, Berlin, Heidelberg (2008). https://doi.org/10.1007/978-3-540-89689-0_33
22. Sourek, G., Zelezny, F., Kuzelka, O.: Lossless compression of structured convolutional models via lifting. CoRR (2021). arXiv:2007.06567 [cs.LG]

23. Tarjan, R.E.: Finding optimum branchings. Networks **7**(1), 25–35 (1977). https://doi.org/10.1002/net.3230070103
24. Zeng, Z., Tung, A.K.H., Wang, J., Feng, J., Zhou, L.: Comparing stars: on approximating graph edit distance. Proc. VLDB Endow. **2**(1), 25–36 (2009). https://doi.org/10.14778/1687627.1687631
25. Zheng, W., Zou, L., Lian, X., Wang, D., Zhao, D.: Efficient graph similarity search over large graph databases. IEEE Trans. Knowl. Data Eng. **27**(4), 964–978 (2015). https://doi.org/10.1109/TKDE.2014.2349924

Graph Embedding in Vector Spaces Using Matching-Graphs

Mathias Fuchs[1(✉)] and Kaspar Riesen[1,2]

[1] Institute of Computer Science, University of Bern, 3012 Bern, Switzerland
mathias.fuchs@inf.unibe.ch, kaspar.riesen@fhnw.ch

[2] Institute for Informations Systems, University of Applied Sciences and Arts Northwestern Switzerland, 4600 Olten, Switzerland

Abstract. An evergrowing amount of readily available data and the increasing rate at which it can be acquired leads to fast developments in many fields of intelligent information processing. Often the underlying data is complex, making it difficult to represent it by vectorial data structures. This is where graphs offer a versatile alternative for formal representations. Actually, quite an amount of graph-based methods for pattern recognition and related fields have been proposed. A considerable part of these methods rely on graph matching. In our recent work we propose a novel encoding of specific graph matching information. The idea of this encoding is to formalize the stable cores of specific classes by means of graphs (called matching-graphs). In the present paper we propose to use these matching-graphs to create a vectorial representation of a given graph. The basic idea is to produce hundreds of matching-graphs first, and then represent each graph g as a binary vector that shows the occurrence of each matching-graph in g. In an experimental evaluation on three data sets we show that this graph embedding is able to improve the classification accuracy of two reference systems with statistical significance.

Keywords: Graph matching · Matching-graphs · Graph edit distance

1 Introduction and Related Work

Pattern recognition is a major field of research which aims at solving various problems like the recognition of facial expressions [1], the temporal sorting of images [2], or enhancing weakly lighted images [3], to name just a few examples. The field of pattern recognition can be divided in two main approaches. *Statistical approaches*, which rely on *feature vectors* for data representation and *structural approaches*, which use *strings*, *trees*, or *graphs* for the same task. Since graphs are able to encode more information than merely an ordered list of numbers, they offer a compelling alternative to vectorial approaches. Hence, they are widely used and adopted in various pattern recognition tasks that range

Supported by Swiss National Science Foundation (SNSF) Project Nr. 200021_188496.

N. Reyes et al. (Eds.): SISAP 2021, LNCS 13058, pp. 352–363, 2021.
https://doi.org/10.1007/978-3-030-89657-7_26

from predicting the demand of medical services [4], over skeleton based action recognition [5], to the automatic recognition of handwriting [6]. The main drawback of graphs is, however, the computational complexity of basic operations, which in turn makes graph based algorithms often slower than their statistical counterparts.

A large amount of graph based methods for pattern recognition have been proposed from which many rely on *graph matching* [7]. Graph matching is typically used for quantifying graph proximity. *Graph edit distance* [8,9], introduced about 40 years ago, is recognized as one of the most flexible graph distance models available. In contrast with many other distance measures (e.g. *graph kernels* [10] or *graph neural networks* [11]), graph edit distance generally offers more information than merely a dissimilarity score, viz. the information which subparts of the underlying graphs actually match with each other (known as *edit path*).

In a recent paper [12], the authors of the present paper propose to explicitly exploit the matching information of graph edit distance. This is done by encoding the matching information derived from *graph edit distance* into a data structure, called *matching-graph*. The main contribution of the present paper is to propose and research another employment of these matching-graphs. In particular, we use these matching-graphs to embed graphs into a vector space by means of subgraph isomorphism. That is, each graph g is represented by a vector of length of the number of matching-graphs available, where each entry in the vector equals 1 if the matching-graph occurs in g and 0 otherwise.

The proposed process of creating vector space embedded representations based on found substructures is similar in spirit to approaches like frequent substructure based approaches [14], subgraph matching kernels [15] or graphlet approaches [16]. The common idea is to first generate a set of subgraphs and treat them as features. In [14] a graph g is represented by a vector that counts the number of times a certain subgraph occurs in g. The subgraphs that are used for embedding are derived via *FSG algorithm* [14]. Related to this in [15] a *Subgraph Matching Kernel (SMKernel)* is proposed. This kernel is derived from the common subgraph isomorphism kernel and counts the number of matchings between subgraphs of fixed sizes in two graphs. Another related approach uses *graphlets* for embedding [16]. Graphlets are small induced subgraphs of fixed size that contain a given set of nodes including all edges.

The major principle of our approach is similar to that of [14–16]. However, the main difference of the above mentioned approaches to our proposal lies in the creation of the subgraphs. We employ graph edit distance to create matching-graphs as basic substructures. These matching-graphs offer a natural way of defining significant and large sets of subgraphs that can be readily used for embedding.

The remainder of this paper is organized as follows. Section 2 makes the paper self-contained by providing basic definitions and terms used throughout this paper. Next, in Sect. 3 the general procedure for creating a matching-graph is explained together with a detailed description of the vector space embedding for graphs. Eventually, in Sect. 4, we empirically confirm that our approach is able to improve the classification accuracy of two reference systems. Finally, in Sect. 5, we conclude the paper and discuss some ideas for future work.

2 Graphs and Graph Edit Distance – Basic Definitions

2.1 Graph and Subgraph

Let L_V and L_E be finite or infinite label sets for nodes and edges, respectively. A *graph* g is a four-tuple $g = (V, E, \mu, \nu)$, where

- V is the finite set of nodes,
- $E \subseteq V \times V$ is the set of edges,
- $\mu : V \rightarrow L_V$ is the node labeling function, and
- $\nu : E \rightarrow L_E$ is the edge labeling function.

A part of a graph, called a *subgraph*, is defined as follows. Let $g_1 = (V_1, E_1, \mu_1, \nu_1)$ and $g_2 = (V_2, E_2, \mu_2, \nu_2)$ be graphs. Graph g_1 is a subgraph of g_2, denoted by $g_1 \subseteq g_2$, if

- $V_1 \subseteq V_2$,
- $E_1 \subseteq E_2$,
- $\mu_1(u) = \mu_2(u)$ for all $u \in V_1$, and
- $\nu_1(e) = \nu_2(e)$ for all $e \in E_1$.

Obviously, a subgraph g_1 is obtained from a graph g_2 by removing some nodes and their incident edges, as well as possibly some additional edges from g_2.

Two graphs g_1 and g_2 are considered *isomorphic* if there is a matching part for each node and edge of g_1 in g_2 (and vice versa). In this regard it is also required that the labels on the nodes and edges exactly correspond (if applicable).

In close relation to graph isomorphism is *subgraph isomorphism.* Intuitively speaking a subgraph isomorphism states whether a graph is contained in another graph. More formally a graph g_1 is subgraph isomorphic to a graph g_2 if there exists a subgraph $g \subseteq g_2$ that is isomorphic to g_1. The concept of subgraph isomorphism is one of the building blocks used in our embedding framework (see Sect. 3).

2.2 Graph Matching

When graphs are used to formally represent objects or patterns, a measure of distance or similarity is usually required. Over the years several dissimilarity measures for graphs have been proposed. Some of the most prominent ones would be *graph kernels* [10], *spectral methods* [17], or *graph neural networks* [18].

A kernel is a function that implicitly maps data to a feature space, by representing it in the form of pairwise comparisons [19]. Intuitively, graph kernels measure the similarity between pairs of graphs and thus provide an embedding in a – typically unknown – feature space. Graphs can also be represented in the form of their *laplacian matrix*. The eigenvalues and eigenvectors of these matrices are known to contain information about the branching and clustering of the nodes and can be used for the definition of various similarity measures [17]. Another emerging graph matching method makes use of *deep neural networks*. Some approaches use neural networks to map the graphs into an Euclidean space [11], while other approaches directly take pairs of graphs as input and output a similarity score [18].

A further prominent graph matching method, which is actually employed in the present paper, is *graph edit distance* [8,9]. One of the main advantages of graph edit distance is its high degree of flexibility, which makes it applicable to virtually any kind of graphs.

Given two graphs g_1 and g_2, the general idea of graph edit distance is to transform g_1 into g_2 using some *edit operations*. A standard set of edit operations is given by *insertions*, *deletions*, and *substitutions* of both nodes and edges. Sometimes, in other applications additional operations like merging and splitting are used. We denote the substitution of two nodes $u \in V_1$ and $v \in V_2$ by $(u \rightarrow v)$, the deletion of node $u \in V_1$ by $(u \rightarrow \varepsilon)$, and the insertion of node $v \in V_2$ by $(\varepsilon \rightarrow v)$, where ε refers to the empty node. For edge edit operations we use a similar notation.

A set $\{e_1, \ldots, e_t\}$ of t edit operations e_i that transform a source graph g_1 completely into a target graph g_2 is called an *edit path* $\lambda(g_1, g_2)$ between g_1 and g_2. Let $\Upsilon(g_1, g_2)$ denote the set of all edit paths transforming g_1 into g_2 while c denotes the cost function measuring the strength $c(e_i)$ of edit operation e_i. The graph edit distance between $g_1 = (V_1, E_1, \mu_1, \nu_1)$ and $g_2 = (V_2, E_2, \mu_2, \nu_2)$ can now be defined as follows.

$$d_{\lambda_{\min}}(g_1, g_2) = \min_{\lambda \in \Upsilon(g_1, g_2)} \sum_{e_i \in \lambda} c(e_i) \quad , \tag{1}$$

Optimal algorithms for computing the edit distance are computationally demanding, as they rely on combinatorial search procedures. In order to counteract this problem we use the often used approximation algorithm BP [20]. The approximated graph edit distance between g_1 and g_2 computed by algorithm BP is termed $d_{\mathrm{BP}}(g_1, g_2)$ from now on.

3 Graph Embedding by Means of Matching-Graphs

The general idea of the proposed approach is to embed a given graph into a vector space by means of *matching-graphs*. These matching-graphs are built by extracting information on the matching of pairs of graphs and by formalizing and encoding this information in a data structure. Matching-graphs can be interpreted as denoised core structures of their respective class. The idea of matching-graphs emerged in [12] where this data structure is employed for the first time for improving the overall quality of graph edit distance. In the next subsection we first formalize the graph embedding, and in Subsection 3.2 we then describe in detail the creation of the matching-graphs.

3.1 Graph Embedding Using Matching-Graphs

Let g be an arbitrary graph stemming from a given set of graphs. Using a set $\mathcal{M} = \{m_1, \ldots, m_N\}$ of matching-graphs, we embed g as follows

$$\varphi(g) = (sub(m_1, g), \ldots, sub(m_N, g)),$$

where

$$sub(m_i, g) = \begin{cases} 1 & \text{if } m_i \subseteq g \\ 0 & \text{else} \end{cases}$$

That is, for our embedding we employ subgraph isomorphism that provides us with a binary similarity measure which is 1 or 0 for subgraph-isomorphic and non-subgraph-isomorphic graphs, respectively. There are various algorithms available that can be applied to the subgraph isomorphism problem. Namely various tree search based algorithms [21,22], as well as decision tree based techniques [23]. In the present paper we employ the VF2 algorithm which makes use of efficient heuristics to speed up the search [22].

Obviously, our graph embedding produces binary vectors with a dimension that is equal to the number of matching-graphs actually available. This specific graph embedding is similar in spirit to the frequent substructure approaches [14], the subgraph matching kernel [15], or graphlet kernel [16] reviewed in the introduction of the present paper. However, the special aspect and novelty of our approach is the employment of matching-graphs for embedding.

3.2 Creating Matching-Graphs

In order to produce the N matching-graphs for embedding, we pursue the following procedure. We consider a pair of graphs g_i, g_j for which the graph edit distance is computed. Resulting from this a (suboptimal) edit path $\lambda(g_i, g_j)$ can be obtained. For each edit path $\lambda(g_i, g_j)$, two matching-graphs $m_{g_i \times g_j}$ and $m_{g_j \times g_i}$ are now built (one for the source graph g_i and one for the target graph g_j). These matching-graphs contain all nodes of g_i and g_j that are substituted according to edit path $\lambda(g_i, g_j)$. All nodes that are deleted in g_i or inserted in g_j are not considered in either of the two matching-graphs.

We observe isolated nodes in some experiments. Graph edit distance can handle isolated nodes. However we still decide to remove isolated nodes from our matching-graphs because we aim at building as small as possible cores of the graphs that are actually connected. Note that we also remove incident edges of nodes that are not included in the resulting matching-graphs.

An edge $u_1, u_2 \in E_i$ that connects two substituted nodes $u_1 \rightarrow v_1$ and $u_2 \rightarrow v_2$, is added to the matching-graph $m_{g_i \times g_j}$, if, and only if, there is an edge $(v_1, v_2) \in E_j$ available.

Using the described procedure for creating a matching-graph out of two input graphs, we now employ a simplified version of an iterative algorithm [13] that builds a set of matching-graphs. The algorithm takes as input k sets of graphs

$G_{\omega_1}, \ldots, G_{\omega_k}$ with graphs from k different classes $\omega_1, \ldots, \omega_k$ as well as the number of matching-graphs kept from one iteration to another (see Algorithm 1).

Algorithm 1: Algorithm for iterative matching-graph creation.

```
input : sets of graphs from k different classes 𝒢 = {G_ω1, ..., G_ωk}, the maximum number
        n of matching-graphs to keep in each iteration
output: sets of matching-graphs for each of the k different classes ℳ = {M_ω1, ..., M_ωk}
1  Initialize ℳ as the empty set: ℳ = {}
2  foreach set of graphs G ∈ 𝒢 do
3      Initialize M as the empty set: M = {}
4      foreach pair of graphs g_i, g_j ∈ G × G with j > i do
5          M = M ∪ {m_{g_j×g_i}, m_{g_i×g_j}}
6      end
7      do
8          M' = a subset of n random elements of M
9          foreach pair of graphs m_i, m_j ∈ M' × M' with j > i do
10             M = M ∪ {m_{m_j×m_i}, m_{m_i×m_j}}
11         end
12     while M has changed in the last iteration
13     ℳ = ℳ ∪ M
14 end
```

The algorithm iterates over all k sets (classes) of graphs from $\mathcal{G}$ (main loop of Algorithm 1 from line 2 to line 14). For each set of graphs G and for all possible pairs of graphs g_i, g_j stemming from the current set G, the initial set of matching-graphs M is produced first (line 3 to 6). Eventually, we aim at iteratively building matching-graphs out of pairs of existing matching-graphs. The motivation for this procedure is to further reduce the size of the matching-graphs and to find small core-structures that are often available in the corresponding graphs. Due to computational limitations, we have to randomly select a subset of size n from the current matching-graphs in M (line 8)[1]. Based on this selection, the next generation of matching-graphs is built. This process is continued until no more changes occur in set M. Finally, set $\mathcal{M}$ – actually used for graph embedding – is compiled as the union of all matching-graphs individually produced for all available classes.

The dimension of the created vectors directly depends on the number of matching-graphs. Hence, our method might result in feature vectors that are initially very large. In order to reduce potential redundancies and select informative matching-graphs, we apply a recursive feature elimination based on feature weights of random forests on our graph embeddings [24].

[1] Qualitative results of our research show, that the finally created matching-graphs do not differ substantially regardless the initial random set of graphs. Hence, the process of creating matching-graphs is not executed in multiple iterations.

4 Experimental Evaluation

4.1 Experimental Setup

From an experimental point of view we aim at answering the following question. Can the created feature vectors (based on our novel matching-graphs) be used to improve the classification accuracy of existing procedures where the graph matching distances are directly used as a basis for classification? In order to answer this question, we compare our embedding with two reference systems in a classification experiment.

The first reference system is a k-nearest-neighbor classifier (k-NN) that directly operates on d_{BP} (denoted as k-NN(d_{BP})). The second reference system is a Support Vector Machine (denoted as SVM($-d_{\mathrm{BP}}$)) that exclusively operates on a similarity kernel $\kappa(g_i, g_j) = -d_{\mathrm{BP}}(g_i, g_j)$ [25]. For classifying the embedded graphs, we also employ an SVM that operates on the embedding vectors (using standard kernel functions for feature vectors). We denote our novel approach as $\mathrm{SVM}_{\mathrm{vec}}$.

We chose the above mentioned classifiers as a baseline, because our goal is to leverage the power of graph edit distance to build a novel graph representation. That is, we decide to compare our novel method with these classifiers that are often used in conjunction with graph edit distance.

The proposed approach is evaluated on three different data sets representing molecules. The first two sets stem from the IAM graph repositoy [26][2] (AIDS and Mutagenicity) and the third originates from the National Cancer Institute [27][3](NCI1).

Each data set consists of two classes. The AIDS data set consists of two classes that represent molecules with activity against HIV or not. Mutagenicity consists of molecules with or without the *mutagen* property, whereas the NCI1 data set consists of chemical compounds that contain activity against non-small cell lung cancer or not. For all data sets the nodes contain a discrete label (symbol of the atom) whereas the edges have no labels.

For the experimental evaluation each data set is split in to three predefined random disjoint sets for training, validation, and testing. Details about the size of the individual splits can be found in Table 1.

4.2 Validation of Metaparameters

For the BP algorithm that approximates the graph edit distance the following parameters are commonly optimized. The costs for node and edge deletions, as well as a weighting parameter $\alpha \in [0, 1]$ that is used to trade-off the relative importance of node and edge costs. However, for the sake of simplicity we employ unit cost of 1.0 for both deletions and insertions of both nodes and edges and optimize the weighting parameter α only.

---

[2] www.iam.unibe.ch/fki/databases/iam-graph-database.

[3] https://ls11-www.cs.tu-dortmund.de/staff/morris/graphkerneldatasets.

Table 1. We show the total number of graphs for each data set as well as the corresponding number of graphs in the training, validation, and test sets.

Data set	Total	Training	Validation	Test
AIDS	2,000	250	250	1,500
Mutagenicity	4,337	1,500	500	2,337
NCI1	4,110	2,465	822	823

For the creation of the matching-graphs – actually also dependant on graph edit distance – the same cost parameters are employed. For the iterative matching-graph creation process (Algorithm 1) we set the number of matching-graphs considered for the next iteration to $n = 200$ for all data sets. The stop criterion of the iterative process checks whether or not the last iteration resulted in a change of the underlying set M. Hence, the final number of matching-graphs to be employed for graph embedding is self-controlled by the algorithm.

As discussed in Sect. 3.1 the dimension of the created vectors initially refers to the number of matching-graphs. As our method might generate thousands of matching-graphs, the dimension of the resulting vectors can be very large. Hence, we apply the feature selection process as discussed in Sect. 3.2.

In Fig. 1 we can see the cross validation accuracy as a function of the number of features after each step of the recursive feature elimination process. It is clearly visible that if the dimension of the vectors becomes too small, the validation accuracy drops by a large margin. However, before this significant drop the accuracy remains relatively stable. In Table 2 we compare the number of selected features and the total amount of available features for all data sets. On AIDS and Mutagenicity about 4% of the originally available features are selected, while on NC1 about 13% of the features are finally used.

Table 2. The amount of features created for each data set and the final amount used after feature selection.

	AIDS	Mutagenicity	NCI1
Total features	4,955	86,752	4,544
Selected features	199	4,139	618

For the optimization of the SVM that operates on our embedded graphs we evaluate three different standard kernels for feature vectors, viz. the Radial Basis Function (RBF), Linear kernel, and a Sigmoid kernel [28]. For all functions we optimize parameter C, which is the trade off between margin maximization and error minimization. In case of RBF and Sigmoid kernel also parameter γ is optimized (all optimizations are conducted by means of an exhaustive grid search).

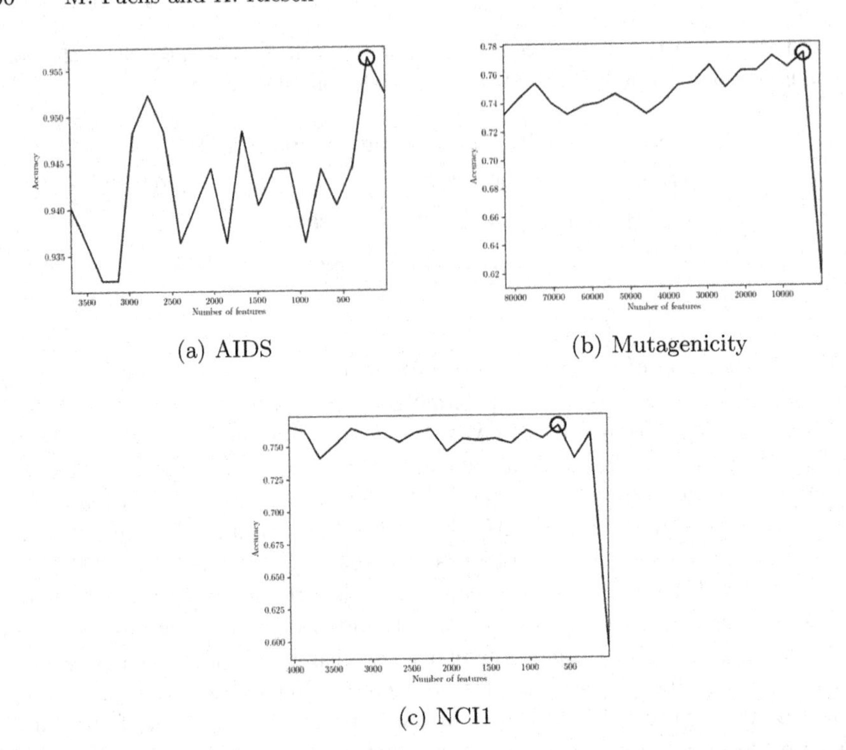

(a) AIDS

(b) Mutagenicity

(c) NCI1

Fig. 1. Cross validation accuracy as a function of the number of features during the recursive feature elimination process. The global optimum is indicated with a small circle.

4.3 Test Results and Discussion

In Table 3 we show the classification accuracies of both reference systems, viz. k-NN(d_{BP}) and SVM($-d_{\mathrm{BP}}$), as well as the results of our novel approach SVM$_{\mathrm{vec}}$ on all data sets.

We observe that our approach achieves better classification results compared to both baseline classifiers on all data sets. On the Mutagenicity data set our approach outperforms both reference systems with statistical significance. On AIDS and NCI1 we achieve a statistically significant improvement compared with the first and second reference system, respectively. The statistical significance is based on a Z-test using a significance level of $\alpha = 0.05$.

A more detailed analysis of the validation and test results on the Mutagenicity data set brings to light the following interesting result (see Table 4). While for both reference systems the validation and test accuracies are more or less stable, we observe a massive overfitting of our novel approach. That is, the classification accuracy drops from 88.2% on the validation set to 76.3% on the test set. Note that this effect is visible on this specific data set only and needs further investigations in future work.

Table 3. Classification accuracies of two reference systems (a k-NN classifier that operates on the original edit distances (k-NN(d_{BP})) and an SVM that uses the same edit distances as kernel values (SVM($-d_{BP}$))) and our proposed system (an SVM using the embedded graphs (SVM$_{vec}$)). Symbol ∘/∘ indicates a statistically significant improvement over the first and second system, respectively. We are using a Z-test at significance level $\alpha = 0.05$.

	Reference Systems		Ours
Data Set	k-NN(d_{BP})	SVM($-d_{BP}$)	SVM$_{vec}$
AIDS	98.6	99.4	99.6 ∘/-
Mutagenicity	72.4	69.1	76.3 ∘/∘
NCI1	74.4	68.6	76.7 -/∘

Table 4. The difference of validation and test accuracies using the different classifiers on the Mutagenicity data set. The effect of overfitting of our novel system is clearly observable.

	Reference Systems				Ours	
	k-NN(d_{BP})		SVM($-d_{BP}$)		SVM$_{vec}$	
Data Set	va	te	va	te	va	te
Mutagenicity	74.8	72.4	69.8	69.1	88.2	76.3

5 Conclusions and Future Work

In the present paper we propose to use matching-graphs – small, pre-computed core structures extracted from a training set – to build vector representations of graphs. The matching-graphs are based on the edit path between pairs of graphs. In particular, the resulting matching-graphs contain only nodes that are actually substituted via graph edit distance. First, we build a relatively large set of N small matching-graphs by means of an iterative procedure. Eventually, we embed our graphs in an N-dimensional vector space such that the i-th dimension corresponds to the i-th matching-graph. More formally, each entry of the resulting vector represents whether or not the corresponding matching-graph occurs as a subgraph in the graph to be embedded. Finally, we reduce the dimension of the created vectors by means of a standard feature selection. Hence we follow the paradigm of overproducing and selecting features.

By means of an experimental evaluation on three graph data sets, we empirically confirm that our novel approach is able to statistically significantly improve the classification accuracy when compared to classifiers that directly operate on the graphs.

For future work we see several rewarding paths to be pursued. First, we aim at evaluating the procedure on additional data sets and in this regard apply it also on data sets with continuous labels. Furthermore it could be interesting to employ the vectorized representation in conjunction with other classifiers and compare our approach with other subgraph or graphlet based kernels.

References

1. Jain, N., Kumar, S., Kumar, A., Shamsolmoali, P., Zareapoor, M.: Hybrid deep neural networks for face emotion recognition. Pattern Recogn. Lett. **115**, 101–106 (2018)
2. Padilha, R., Andaló, F.A., Lavi, B., Pereira, L.A., Rocha, A.: Temporally sorting images from real-world events. Pattern Recogn. Lett. **147**, 212–219 (2021)
3. Li, C., Guo, J., Porikli, F., Pang, Y.: LightenNet: a convolutional neural network for weakly illuminated image enhancement. Pattern Recogn. Lett. **104**, 15–22 (2018)
4. Jin, R., Xia, T., Liu, X., Murata, T., Kim, K.S.: Predicting emergency medical service demand with bipartite graph convolutional networks. IEEE Access **9**, 9903–9915 (2021)
5. Feng, D., Wu, Z., Zhang, J., Ren, T.: Multi-scale spatial temporal graph neural network for skeleton-based action recognition. IEEE Access **9**, 58256–58265 (2021)
6. Fischer, A., Suen, C.Y., Frinken, V., Riesen, K., Bunke, H.: A fast matching algorithm for graph-based handwriting recognition. In: Kropatsch, W.G., Artner, N.M., Haxhimusa, Y., Jiang, X. (eds.) Graph-Based Representations in Pattern Recognition. GbRPR 2013. Lecture Notes in Computer Science, **7877**, 194–203. Springer, Berlin, Heidelberg (2013). https://doi.org/10.1007/978-3-642-38221-5_21
7. Conte, D., Foggia, P., Sansone, C., Vento, M.: Thirty years of graph matching in pattern recognition. Int. J. Pattern Recogn. Artif. Intell. **18**(03), 265–298 (2004)
8. Bunke, H., Allermann, G.: Inexact graph matching for structural pattern recognition. Pattern Recogn. Lett. **1**(4), 245–253 (1983)
9. Sanfeliu, A., Fu, K.S.: A distance measure between attributed relational graphs for pattern recognition. IEEE Trans. Syst. Man Cybern. **3**, 353–362 (1983)
10. Vishwanathan, S.V.N., Schraudolph, N.N., Kondor, R., Borgwardt, K.M.: Graph kernels. J. Mach. Learn. Res. **11**, 1201–1242 (2010)
11. Scarselli, F., Gori, M., Tsoi, A.C., Hagenbuchner, M., Monfardini, G.: The graph neural network model. IEEE Trans. Neural Netw. **20**(1), 61–80 (2008)
12. Fuchs, M., Riesen, K.: Matching of matching-graphs - a novel approach for graph classification. In: Proceedings of the 25th International Conference on Pattern Recognition, ICPR 2020, Milano, Italy, 10–15 January 2021
13. Fuchs, M., Riesen, K.: Iterative creation of matching-graphs - finding relevant substructures in graph sets. In: Proceedings of the 25th Iberoamerican Congress on Pattern Recognition, CIARP25 2021, Porto, Portugal, 10–13 May 2021
14. Deshpande, M., Kuramochi, M., Wale, N., Karypis, G.: Frequent substructure-based approaches for classifying chemical compounds. IEEE TKDE **17**(8), 1036–1050 (2005)
15. Kriege, N., Mutzel, P.: Subgraph matching kernels for attributed graphs (2012). arXiv preprint: arXiv:1206.6483
16. Shervashidze, N., Vishwanathan, S.V.N., Petri, T., Mehlhorn, K., Borgwardt, K.: Efficient graphlet kernels for large graph comparison. In: Artificial Intelligence and Statistics, pp. 488–495. PMLR (2009)
17. Kannan, N., Vishveshwara, S.: Identification of side-chain clusters in protein structures by a graph spectral method. J. Mol. Biol. **292**(2), 441–464 (1999)

18. Li, Y., Gu, C., Dullien, T., Vinyals, O., Kohli, P.: Graph matching networks for learning the similarity of graph structured objects. In: International Conference on Machine Learning, pp. 3835–3845. PMLR (2019)
19. Cristianini, N., Shawe-Taylor, J.: An Introduction To Support Vector Machines and Other Kernel-based Learning Methods. Cambridge University Press, Cambridge (2000)
20. Riesen, K., Bunke, H.: Approximate graph edit distance computation by means of bipartite graph matching. Image Vision Comput. **27**(7), 950–959 (2009)
21. Ullmann, J.R.: An algorithm for subgraph isomorphism. J. ACM (JACM) **23**(1), 31–42 (1976)
22. Cordella, L.P., Foggia, P., Sansone, C., Vento, M.: A (sub) graph isomorphism algorithm for matching large graphs. IEEE Trans. Pattern Anal. Mach. Intell. **26**(10), 1367–1372 (2004)
23. Messmer, B.T., Bunke, H.: A decision tree approach to graph and subgraph isomorphism detection. Pattern Recogn. **32**(12), 1979–1998 (1999)
24. Darst, B.F., Malecki, K.C., Engelman, C.D.: Using recursive feature elimination in random forest to account for correlated variables in high dimensional data. BMC Genet. **19**(1), 1–6 (2018)
25. Neuhaus, M., Bunke, H.: Bridging the Gap Between Graph Edit Distance And Kernel Machines, **68**. World Scientific, Singapore (2007)
26. Riesen, K., Bunke, H.: IAM graph database repository for graph based pattern recognition and machine learning. In: da Vitoria Lobo, N., et al. (eds.) Structural, Syntactic, and Statistical Pattern Recognition. SSPR/SPR 2008. Lecture Notes in Computer Science, **5342**, 287–297. Springer, Berlin, Heidelberg (2008). https://doi.org/10.1007/978-3-540-89689-0_33
27. Wale, N., Karypis, G.: Comparison of descriptor spaces for chemical compound retrieval and classification. In: Sixth International Conference on Data Mining (ICDM 2006), pp. 678–689 (2006). https://doi.org/10.1109/ICDM.2006.39
28. Lin, H.T., Lin, C.J.: A study on sigmoid kernels for SVM and the training of non-PSD kernels by SMO-type methods. Neural Comput. **3**(1–32), 16 (2003)

An A*-algorithm for the Unordered Tree Edit Distance with Custom Costs

Benjamin Paaßen(✉)

Institute of Informatics, Humboldt-University of Berlin,
Rudower Chaussee 25, 12489 Berlin, Germany
benjamin.paassen@hu-berlin.de

Abstract. The unordered tree edit distance is a natural metric to compute distances between trees without intrinsic child order, such as representations of chemical molecules. While the unordered tree edit distance is MAX SNP-hard in principle, it is feasible for small cases, e.g. via an A* algorithm. Unfortunately, current heuristics for the A* algorithm assume unit costs for deletions, insertions, and replacements, which limits our ability to inject domain knowledge. In this paper, we present three novel heuristics for the A* algorithm that work with custom cost functions. In experiments on two chemical data sets, we show that custom costs make the A* computation faster and improve the error of a 5-nearest neighbor regressor, predicting chemical properties. We also show that, on these data, polynomial edit distances can achieve similar results as the unordered tree edit distance.

Keywords: Unordered tree edit distance · A* algorithm · Tree edit distance · Chemistry

1 Introduction

Tree structures occur whenever data follows a hierarchy or a branching pattern, like in chemical molecules [6,10], in RNA secondary structures [11], or in computer programs [9]. To perform similarity search on such data, we require a measure of distance over trees. A popular choice is the tree edit distance, which is defined as the cost of the cheapest sequence of deletions, insertions, and relabelings that transforms one tree to another [3,14,16]. Unfortunately, the edit distance becomes MAX SNP-hard for unordered trees, like tree representations of chemical molecules [15]. Still, for smaller trees, we can compute the unordered tree edit distance (UTED) exactly using strategies like A* algorithms [7,13]. Roughly speaking, an A* algorithm starts with an empty edit sequence and then successively extends the edit distance such that a heuristic lower bound for the cost of the edit sequence remains as low as possible. The tighter our lower bound h, the more we can prune the search and the faster the A* algorithm becomes. Horesh et al. have proposed a heuristic based on the

Funding by the German Research Foundation (DFG) under grant number PA 3460/2-1 is gratefully acknowledged.

N. Reyes et al. (Eds.): SISAP 2021, LNCS 13058, pp. 364–371, 2021.
https://doi.org/10.1007/978-3-030-89657-7_27

histogram of node degrees [7] and Yoshino et al. have improved upon this heuristic by also considering label histograms and by re-using intermediate values via dynamic programming [13]. However, both approaches assume unit costs, i.e. that deletions, insertions, and relabelings all have a cost of 1, irrespective of the content that gets deleted, inserted, or relabeled. This is unfortunate because, in many domains, we have prior knowledge that suggests different costs or we may wish to learn costs from data [9]. Accordingly, most tree edit distance algorithms are general enough to support custom deletion, insertion, and replacement costs, as long as they conform to metric constraints [3,14,16].

In this paper, we develop three novel heuristics for the A* algorithm which all support custom costs. The three heuristics have linear, quadratic, and cubic complexity, respectively, where the slower heuristics provide tighter lower bounds. Based on these novel heuristics, we investigate three research questions:

RQ1: Which of the three heuristics is the fastest? And how do they compare against the state-of-the-art by Yoshino et al. [13]?
RQ2: Do custom edit costs actually contribute to similarity search?
RQ3: How does UTED compare to polynomial edit distances in similarity search?

We investigate these research questions on two example data sets of chemical molecules, both represented as unordered trees. To answer RQ2 and RQ3, we consider a regression task where we try to predict the chemical properties of a molecule (boiling point and stability, respectively) via a nearest-neighbor regression. We begin our paper with more background and related work before we describe our proposed A* algorithm, present our experiments, and conclude.

2 Background and Related Work

Let Σ be an arbitrary set which we call *alphabet*. Then, we define a *tree* over Σ as an expression of the form $\hat{x} = x(\hat{y}_1, \ldots, \hat{y}_K)$, where $x \in \Sigma$ and where $\hat{y}_1, \ldots, \hat{y}_K$ is a list of trees over Σ, which we call the *children* of $\hat{x}$. If $K = 0$, we call $x()$ a *leaf*. We denote the set of all trees over Σ as $\mathcal{T}(\Sigma)$.

In this paper, we are concerned with similarity search on trees. In the literature, there are three general strategies to compute similarities on trees. First, we can construct a feature mapping $\phi : \mathcal{T}(\Sigma) \to \mathbb{R}^n$, which maps an input tree to a feature vector, and then compute a (dis-)similarity between features, e.g. via $d(\hat{x}, \hat{y}) = \|\phi(\hat{x}) - \phi(\hat{y})\|$. For example, we can represent trees by *pq*-grams [2], by counts of typical tree patterns [5], or by training a neural network [6,8]. The second strategy are tree kernels k, i.e. functions that directly compute inner products $k(\hat{x}, \hat{y}) = \phi(\hat{x})^T \cdot \phi(\hat{y})$ without the need to explicitly compute ϕ [1,5].

In this paper, we focus on a third option, namely tree edit distances [3]. Let Σ be an alphabet with $- \notin \Sigma$. Roughly speaking, a tree edit distance $d(\hat{x}, \hat{y})$ between two trees $\hat{x}$ and $\hat{y}$ from $\mathcal{T}(\Sigma)$ is the cost of the cheapest sequence of deletions, insertions, and relabelings of nodes in $\hat{x}$ such that we obtain $\hat{y}$

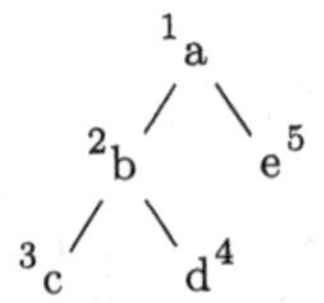

$M_1 = \{(1,1),(2,0),(3,4),(4,3),(5,2)\}$ $d_1^{\text{UTED}}(\hat{x},\hat{y}) = 1$

$M_2 = \{(1,1),(2,0),(3,4),(4,3),(5,0),(0,2)\}$ $d_1^{\text{CUTED}}(\hat{x},\hat{y}) = 3$

$M_3 = \{(1,1),(2,0),(3,2),(4,3),(5,4)\}$ $d_1^{\text{TED}}(\hat{x},\hat{y}) = 3$

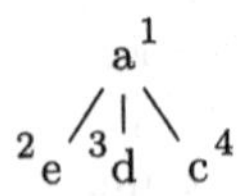

Fig. 1. An illustration of mappings according to the unordered tree edit distance [15] (top), the constrained unordered tree edit distance [14] (center), and the ordered tree edit distance [16] (bottom) between the same two trees. The distances assume unit costs. Numbers in superscript show the depth first order.

[3,14,16]. More precisely, let $x_1, \ldots, x_m$ be the nodes of $\hat{x}$[1] and $y_1, \ldots, y_n$ be the nodes of $\hat{y}$ in depth-first-search order. Then, we define a *mapping* between $\hat{x}$ and $\hat{y}$ as a set of tuples $M \subset \{0,1,\ldots,m\} \times \{0,1,\ldots,n\}$ such that each $i \in \{1,\ldots,m\}$ occurs exactly once on the left and each $j \in \{1,\ldots,n\}$ occurs exactly once on the right. Figure 1 illustrates three example mappings between the trees $a(b(c,d),e)$ (left) and $a(e,d,c)$ (right), namely M_1, M_2, and M_3 (center left). Each mapping M can be translated into a sequence of edits by deleting all nodes x_i where $(i,0) \in M$, by replacing nodes x_i with y_j where $(i,j) \in M$ and $x_i \neq y_j$, and by inserting all nodes y_j where $(0,j) \in M$. We denote the set of all mappings between $\hat{x}$ and $\hat{y}$ as $\mathcal{M}(\hat{x},\hat{y})$. Next, we define a *cost function* as a metric $c : (\Sigma \cup \{-\}) \times (\Sigma \cup \{-\}) \to \mathbb{R}$, and we define the cost of a mapping M as $c(M) = \sum_{(i,j)\in M} c(x_i, y_j)$ where $x_0 = y_0 = -$. A typical cost function is $c_1(x,y) = 1$ if $x \neq y$ and $c_1(x,y) = 0$ if $x = y$, which we call *unit costs*. Finally, we define the tree edit distance $d_c : \mathcal{T}(\Sigma) \times \mathcal{T}(\Sigma) \to \mathbb{R}$ according to cost function c as the minimum $d_c(\hat{x},\hat{y}) = \min_{M\in\mathcal{M}(\hat{x},\hat{y})} c(M)$.

We obtain different edit distances depending on the additional restrictions we apply on the set of mappings $\mathcal{M}(\hat{x},\hat{y})$. The unordered tree edit distance (UTED) requires that mappings respect the ancestral ordering, i.e. if $(i,j) \in M$, then descendants of i can only be mapped to descendants of j [3]. A cheapest example mapping according to unit costs is M_1 (Fig. 1, top). The constrained unordered tree edit distance (CUTED) [14] additionally requires that a deletion/insertion of a node implies either deleting/inserting all of its siblings or all of its children but one. This forbids M_1 and M_3, where b is deleted but both its sibling and more than one child are maintained. M_2 is a cheapest mapping according to CUTED with unit costs. The ordered tree edit distance (TED) [16] requires that the ancestral ordering and the depth-first ordering is maintained. Accordingly, neither M_1 nor M_2 are permitted because they swap the order of c and d. $M_3$is a cheapest mapping according to TED with unit costs. Note that UTED is MAX-SNP hard. However, CUTED and TED are both polynomial [14,16] via dynamic programming and we consider them as baselines in our experiments.

[1] Note that we use 'node' and 'label' interchangeably in this paper. To disambiguate between two nodes with the same label, we use the index.

3 Method

In this section, we explain our proposed A* algorithm for the unordered tree edit distance (UTED). We begin with the general scheme, which we adapt from Yoshino et al. [13], and then introduce three heuristics to plug into the A* algorithm.

Algorithm 1. The A* algorithm to compute the unordered tree edit distance $d_c^{\text{UTED}}(\hat{x}, \hat{y})$ between two trees $\hat{x}$ and $\hat{y}$, depending on a cost function c and a heuristic h.

1: **function** ASTAR_UTED(trees $\hat{x}$ and $\hat{y}$, cost function c, heuristic h)
2: Initialize a priority queue Q with the partial mapping $M = \{(1, 1)\}$
3: and value $c(x_1, y_1) + h(\{2, \ldots, m\}, \{2, \ldots, n\})$.
4: **while** Q is not empty **do**
5: Poll partial mapping M with lowest value f from Q.
6: $i \leftarrow \min\{1, \ldots, m+1\} \setminus I_M$.
7: **if** $i = m + 1$ **then**
8: **return** $c(M \cup \{(0, j) | 1 \leq j \leq n, j \notin J_M\})$.
9: **end if**
10: Retrieve $(k, l) \in M$ with largest k such that x_k is ancestor of x_i and $l > 0$.
11: $h^p \leftarrow h\big(\{1, \ldots, m\} \setminus (\mathcal{X}_k \cup I_M), \{1, \ldots, m\} \setminus (\mathcal{Y}_l \cup J_M)\big)$.
12: $M_0 \leftarrow M \cup \{(i, 0)\}$
13: $h_0 \leftarrow h\big(\mathcal{X}_k \setminus I_{M_0}, \mathcal{Y}_l \setminus J_{M_0}\big) + h^p$.
14: **for** $j \in \mathcal{Y}_l \setminus J_M$ **do**
15: Let $y'_0, \ldots, y'_t$ be the path from y_l to y_j in $\hat{y}$ with $y'_0 = y_l$ and $y'_t = y_j$.
16: $M_j \leftarrow M \cup \{(i, j), (0, y'_1), \ldots, (0, y'_{t-1})\}$.
17: $h_j \leftarrow h\big(\mathcal{X}_i \setminus I_{M_j}, \mathcal{Y}_j \setminus J_{M_j}\big) + h\big(\mathcal{X}_k \setminus (\mathcal{X}_i \cup I_{M_j}), \mathcal{Y}_l \setminus (\mathcal{Y}_j \cup J_{M_j})\big) + h^p$.
18: **end for**
19: Put M_j with value $c(M_j) + h_j$ onto Q for all $j \in \{0\} \cup (\mathcal{Y}_l \setminus J_M)$.
20: **end while**
21: **end function**

A Algorithm:* We first introduce a few auxiliary concepts that we require for the A* algorithm. First, let M be some subset of $\{0, \ldots, m\} \times \{0, \ldots, n\}$. Then, we denote with I_M the set $\{i > 0 | \exists j : (i, j) \in M\}$ and with J_M the set $\{j > 0 | \exists i : (i, j) \in M\}$, i.e. the set of left-hand-side and right-hand-side indices of M. Next, let $\hat{x}$ and $\hat{y}$ be trees with nodes $x_1, \ldots, x_m$ and $y_1, \ldots, y_n$, respectively. Then, we define $\mathcal{X}_i$ and $\mathcal{Y}_j$ as the index sets of all descendants of x_i and y_j, respectively. Finally, let c be a cost function. Then, we define a *heuristic* as a function $h : \mathcal{P}(\{1, \ldots, m\}) \times \mathcal{P}(\{1, \ldots, n\}) \to \mathbb{R}$, such that for any $I \subseteq \{1, \ldots, m\}$ and $J \subseteq \{1, \ldots, n\}$ it holds

$$h(I, J) \leq \min_{M \in \mathcal{M}^{\text{UTED}}(\hat{x}, \hat{y})} \sum_{(i,j) \in M : i \in I, j \in J} c(x_i, y_j). \quad (1)$$

Algorithm 1 shows the pseudocode for the A* algorithm. We initialize a partial mapping $M = \{(1,1)\}$ which maps the root of $\hat{x}$ to the root of $\hat{y}$. If this is undesired, all input trees can be augmented with a placeholder root node. Next, we initialize a priority queue Q with M and its lower bound. Now, we enter the main loop. In each step, we consider the current partial mapping M with the lowest lower bound f (line 5). If I_M already covers all nodes in $\hat{x}$, we complete M by inserting all remaining nodes of $\hat{y}$ and return the cost of the resulting mapping (lines 7–9)[2]. Otherwise, we extend M by mapping the smallest non-mapped index i either to zero (lines 12–13), or to j for some available node y_j from $\hat{y}$ (lines 14–18). In the latter case, we need to maintain the ancestral ordering of the tree $\hat{y}$. Accordingly, we first retrieve the lowest ancestor x_k of x_i such that $(k,l) \in M$ with $l > 0$ and only permit i to be mapped to descendants $\mathcal{Y}_l$. Note that a $(k,l) \in M$ with $l > 0$ must exist because we initialized M with $\{(1,1)\}$. Further, if we map i to a non-direct descendant of y_l, we make sure to insert all nodes on the ancestral path $y'_0, \ldots, y'_t$, first. We generate lower bounds h_j for all extensions M_j and put them back onto the priority queue (line 19).

Note that the space complexity of this algorithm can be polynomially limited by representing the partial mappings in a tree structure. However, the worst-case time complexity remains exponential because the algorithm may need to explore combinatorially many possible mappings. Generally, though, the tighter the lower bound provided by h, the fewer partial mappings need to be explored before we find a complete mapping. To further cut down the time complexity, we tabulate the lower bounds h^p for the ancestor mappings (k,l) (line 11), as recommended by Yoshino et al. [13].

Heuristics: The final ingredient we need is the actual heuristic h. We define three heuristics in increasing relaxation and decreasing time complexity. First, $h_3(I,J) = \min_{M \subseteq \mathcal{M}(I,J)} \sum_{(i,j) \in M} c(x_i, y_j)$, where $\mathcal{M}(I,J)$ denotes the set of all mappings between I and J, irrespective of ancestral ordering. Accordingly, Inequality 1 is trivially fulfilled because any mapping that respects ancestral ordering is also in $\mathcal{M}(I,J)$. Importantly, this relaxation can be solved in $\mathcal{O}((m+n)^3)$ via the Hungarian algorithm [4]. While polynomial, this appears rather expensive for a heuristic. Therefore, we also consider further relaxations. Without loss of generality, let $|I| \geq |J|$, otherwise exchange the roles of $\hat{x}$, $\hat{y}$, I and J. Then, we define

$$h_2(I,J) = \min_{I' \subseteq I : |I'| = |I| - |J|} \Big(\sum_{i \in I'} c(x_i, -) \Big) + \Big(\sum_{i \in I \setminus I'} \min_{j \in J} c(x_i, y_j) \Big). \tag{2}$$

Note that this is a lower bound for $h_3(I,J)$ because we expand the class of permitted mappings M to one-to-many mappings, which is a proper superset of $\mathcal{M}(I,J)$. Further, h_2 can be solved in $\mathcal{O}(m \cdot n)$ because we can evaluate $c_i := \min_{j \in J} c(x_i, x_j)$ for all i in $|I| \cdot |J|$ steps and we can solve the outer minimization by finding the $|I| - |J|$ smallest terms according to $c(x_i, -) - c_i$ and using those

[2] Strictly speaking, this is only valid if the lower bound f is exact for insertions. This is the case for all heuristics considered in this paper.

as I', which is possible in $O(|I|)$. In case even $\mathcal{O}(m \cdot n)$ is too expensive, we relax further to $h_1(I, J) = \min_{I' \subseteq I : |I'| = |I| - |J|} \sum_{i \in I'} c(x_i, -)$. This is obviously a lower bound for h_2 and can be solved in $\mathcal{O}(\max\{m, n\})$.

4 Experiments

We evaluate our three research questions on two data sets from Chemistry, namely the Alkanes data set of 150 alkane molecules by Micheli et al. [6] and the hundred smallest molecules from the ZINC molecule data set of Kusner et al. [8]. In the former case, the molecules are directly represented as trees (with 8.87 nodes on average) with hydrogen counts as node labels. In the latter case, we use the syntax tree of the molecule's SMILES representation (with 13.82 nodes on average) [12], where nodes are labeled with syntactic blocks. Note that this is a lossy representation because we cut aromatic rings to obtain trees.

Regarding RQ1, we compute all pairwise UTED values using the three heuristics h_1, h_2, and h_3, both with unit costs and with custom costs. As custom cost function c, we use the difference in hydrogen count between two carbon atoms for the alkanes data set. For the ZINC data set, we use the difference in electron count. For further reference, we also compare to the heuristic of Yoshino et al. [13] for unit costs. We execute all computations in Python on a consumer desktop PC with Intel core i9-10900 CPU and 32 GB RAM, and measure time using Python's `time` function. All experimental code is available at https://gitlab.com/bpaassen/uted.

Table 1. The average runtime in milliseconds (top) and the number of partial mappings searched (bottom) per distance computation for each heuristic.

	Data set	Unit				Custom		
		h_1	h_2	h_3	h_{yoshino}	h_1	h_2	h_3
Runtime	Alkanes	8.70	12.15	10.72	9.52	**7.34**	8.21	9.92
	ZINC	549.38	277.15	192.97	266.66	130.62	75.53	**68.12**
Search size	Alkanes	376	348	260	279	318	302	**246**
	ZINC	24586	9164	4158	6781	6643	2655	**1379**

Table 1 shows the average runtime in milliseconds (top) for each heuristic on both data sets. On alkanes, h_1 is fastest and on ZINC, h_3 is fastest. All heuristics get faster for custom costs. Surprisingly, h_{yoshino} is not the fastest for unit costs, even though it is optimized for this setting. This may just be due to an unfavourable constant factor, though: h_{yoshino} is successful in reducing the size of the search space almost to the same level as h_3 (see Table 1, bottom). Further, Fig. 2 displays a linear regression for the runtime versus tree size in a log-log plot, indicating that h_{yoshino} and h_3 have the lowest slopes/best scaling behavior for large trees.

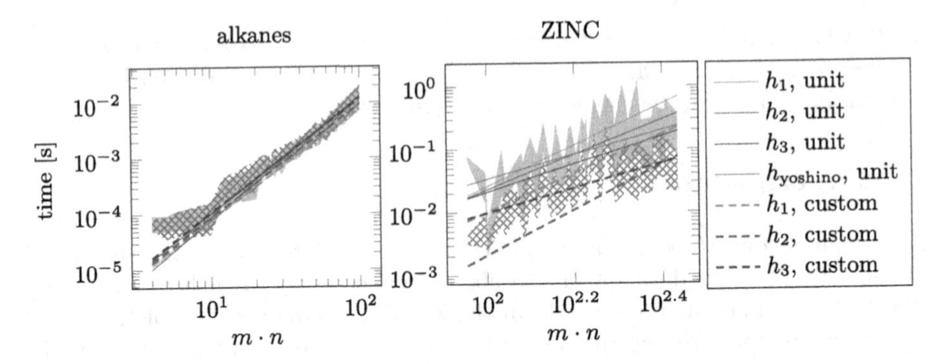

Fig. 2. A log-log regression of the runtime needed for computing UTED for all four heuristics (indicated by color) on the alkanes data (left) and the ZINC data (right). Shading indicates region between 25th and 75th percentile of the runtimes for h_{yoshino} (orange, solid), and h_3 with custom costs (purple, crosshatch), respectively. (Color figure online)

Table 2. Average RMSE (± std.) of a 5-NN regressor across 15 cross validation folds for UTED, CUTED, and TED with unit and custom costs.

Data set	Unit			Custom		
	UTED	CUTED	TED	UTED	CUTED	TED
Alkanes	0.27 ± 0.24	0.27 ± 0.24	0.27 ± 0.24	$\mathbf{0.25 \pm 0.24}$	$\mathbf{0.25 \pm 0.24}$	$\mathbf{0.25 \pm 0.24}$
ZINC	1.33 ± 0.85	1.31 ± 0.86	1.36 ± 0.84	$\mathbf{1.24 \pm 0.87}$	1.26 ± 0.87	1.29 ± 0.86

Regarding RQ2 and RQ3, we perform a 5-nearest neighbor regression[3] to predict the boiling point of alkanes and the chemical stability measure of [8] for ZINC molecules, respectively. Table 2 shows the prediction error for both data sets in 15-fold cross validation. For reference, we do not only evaluate UTED with unit and custom costs, but also CUTED and TED. We observe that all methods perform better with custom costs compared to unit costs. For alkanes, there is no measurable difference between UTED, CUTED, and TED. For ZINC, TED performs worst, CUTED performs better than UTED for unit costs, and UTED performs better than CUTED for custom costs.

5 Conclusion

We proposed three novel heuristics to compute the unordered tree edit distance via an A* algorithm. In contrast to prior work, our heuristics can accommodate custom costs, not only unit costs. Our three heuristics provide different trade-offs of time complexity (linear, quadratic, cubic) versus how much they prune the A* search.

[3] We also tested lower K, which achieved worse results for all methods.

In our experiments on two chemical experiments, we observed that this trade-off works in favor of the linear heuristic for small trees but that the cubic heuristic takes over for larger trees. Interestingly, the cubic heuristic compared favorably even to the current state-of-the-art heuristic. When applying custom costs, all our heuristics became faster thanks to the disambiguation provided by the custom cost function.

Regarding similarity search, we investigated the performance of a 5-nearest neighbor regressor, predicting chemical properties. We observed that custom costs lowered the regression error. However, we also saw that a similar performance can be achieved with a polynomial, restricted edit distance. Future work might investigate further tree data set to check whether these results generalize beyond chemistry.

References

1. Aiolli, F., Da San Martino, G., Sperduti, A.: An efficient topological distance-based tree kernel. IEEE TNNLS **26**(5), 1115–1120 (2015)
2. Augsten, N., Böhlen, M., Gamper, J.: The pq-gram distance between ordered labeled trees. ACM TDS **35**(1), 1–36 (2008)
3. Bille, P.: A survey on tree edit distance and related problems. Theoret. Comput. Sci. **337**(1), 217–239 (2005)
4. Bougleux, S., Brun, L., Carletti, V., Foggia, P., Gaüzère, B., Vento, M.: Graph edit distance as a quadratic assignment problem. Pat. Rec. Lett. **87**, 38–46 (2017)
5. Collins, M., Duffy, N.: New ranking algorithms for parsing and tagging: kernels over discrete structures, and the voted perceptron. In: Proceedings of the ACL, pp. 263–270 (2002)
6. Gallicchio, C., Micheli, A.: Tree echo state networks. Neurocomputing **101**, 319–337 (2013)
7. Horesh, Y., Mehr, R., Unger, R.: Designing an A* algorithm for calculating edit distance between rooted-unordered trees. J. Comp. Bio. **13**(6), 1165–1176 (2006)
8. Kusner, M.J., Paige, B., Hernández-Lobato, J.M.: Grammar variational autoencoder. In: Proceedings of the ICML, pp. 1945–1954 (2017)
9. Paaßen, B., Gallicchio, C., Micheli, A., Hammer, B.: Tree edit distance learning via adaptive symbol embeddings. In: Proceedings of the ICML, pp. 3973–3982 (2018)
10. Rarey, M., Dixon, J.S.: Feature trees: a new molecular similarity measure based on tree matching. J. Comput. Aided Mol. Design **12**(5), 471–490 (1998)
11. Shapiro, B.A., Zhang, K.: Comparing multiple RNA secondary structures using tree comparisons. Bioinformatics **6**(4), 309–318 (1990)
12. Weininger, D.: Smiles, a chemical language and information system. 1. Introduction to methodology and encoding rules. J. Chem. Inf. CS **28**(1), 31–36 (1988)
13. Yoshino, T., Higuchi, S., Hirata, K.: A dynamic programming A* algorithm for computing unordered tree edit distance. In: Proceedings of the IIAI-AAI, pp. 135–140 (2013)
14. Zhang, K.: A constrained edit distance between unordered labeled trees. Algorithmica **15**(3), 205–222 (1996)
15. Zhang, K., Jiang, T.: Some MAX SNP-hard results concerning unordered labeled trees. Inf. Proc. Lett. **49**(5), 249–254 (1994)
16. Zhang, K., Shasha, D.: Simple fast algorithms for the editing distance between trees and related problems. SIAM Comput. **18**(6), 1245–1262 (1989)

FIMSIM: Discovering Communities by Frequent Item-Set Mining and Similarity Search

Jakub Peschel[1(✉)], Michal Batko[1], Jakub Valcik[2], Jan Sedmidubsky[1], and Pavel Zezula[1]

[1] Masaryk University, Brno, Czech Republic
{jpeschel,batko,sedmidubsky,pzezula}@mail.muni.cz
[2] Konica Minolta Global R&D, Brno, Czech Republic
Jakub.Valcik@konicaminolta.cz
https://www.muni.cz, https://research.konicaminolta.com

Abstract. With the growth of structured graph data, the analysis of networks is an important topic. Community mining is one of the main analytical tasks of network analysis. Communities are dense clusters of nodes, possibly containing additional information about a network. In this paper, we present a community-detection approach, called FIMSIM, which is based on principles of frequent item-set mining and similarity search. The frequent item-set mining is used to extract cores of the communities, and a proposed similarity function is applied to discover suitable surroundings of the cores. The proposed approach outperforms the state-of-the-art DB-Link Clustering algorithm while enabling the easier selection of parameters. In addition, possible modifications are proposed to control the resulting communities better.

Keywords: Community mining · Frequent item-set mining · Similarity search · Network analysis

1 Introduction

In recent years, the type of data has dramatically changed. The data is becoming more and more context-dependent, and, therefore, it is necessary to analyze it with respect to the context of a target application. An example of this kind of data is structured network data, such as internet pages, social interactions, protein-to-protein interactions, and many others. With the growing amount of structured network data, there is also a rising need to analyze it efficiently. One of the most important tasks in network data analysis is *discovering communities*, also referred to as the community-mining or community-detection task.

Community mining is a process of uncovering hidden relationships among the elements of network data. These relations lead to the creation of community structures that represent densely packed clusters of network elements. The discovery of such communities can help better understand graph dynamics and

N. Reyes et al. (Eds.): SISAP 2021, LNCS 13058, pp. 372–383, 2021.
https://doi.org/10.1007/978-3-030-89657-7_28

an organizational network structure and can be used as an improvement of recommendation systems, police investigation, business reorganization, and many others.

To discover communities in network-based data, we consider a general data representation in the form of a graph consisting of *nodes* that are interconnected by *edges*. We further consider only undirected and unweighted graphs without self-loops over the nodes. The nodes are not required to contain any additional information, nor any additional network knowledge is taken into account for the purpose of community mining. It is also worth noting that the analyzed graph has to be sparse so that meaningful communities can be discovered. If the graph starts to be too much dense, most of the nodes are becoming the candidates of the community, which can easily degrade to the pathological case when all the nodes belong to one community.

A *community* itself is generally understood as a group of nodes that are more interconnected between themselves in comparison with external nodes [7,8,15]. However, there is no generally accepted definition of a community. For example, Radicchi et al. proposes two categories of definitions: strong communities and weak communities [15]. The strong communities consist of nodes with a majority of their respective neighbours as a part of the community, while the weak ones simplify the condition that a total number of connections (edges) between the community members must be higher than the number of edges connecting community members with the others. Newman et al. consider the community as some sort of hierarchy that can be gradually built from smaller communities, and a given level of hierarchy with a suitable community granularity can be selected [11]. This can be achieved by hierarchical clustering, where strongly connected nodes are gradually grouped together to form a community, and a dendrogram of such groupings then reflects the hierarchical representation. Nevertheless, the lack of a formal definition and the universality of abstract definition often leads to the approaches that strictly prefer disjoint partitioning of the graph instead of overlapping. To solve this issue, we consider the following main assumptions:

- Communities are clusters of highly interconnected nodes;
- Communities have some level of hierarchical structure;
- Nodes can belong to multiple communities.

In this paper, we propose to combine the principles of *frequent item-set mining* and *similarity searching* for the two-step discovery of overlapping communities. The proposed method offers a new perspective on solving the community detection problem as well as outperforms a traditional method for the discovery of overlapping communities.

2 Related Work

Community mining is an NP-complete problem due to the combinatorial nature of graph subsets [16]. In this paper, we focus primarily on three main approaches for the discovery of overlapping communities: clique percolation, local expansion and link clustering.

2.1 Clique Percolation

Clique percolation methods are based on discovering small clique-like structures that are then based on their overlap merged together. For the purpose of better clique searching, k-cliques are used. These are completely connected subgraphs consisting of k nodes. Unlike maximal clique searching, k-clique searching for one fixed k is a polynomial problem. After the discovery of a set of all k-cliques, their overlap is checked. If two k-cliques share k–1 nodes, they are merged together as a new community. If two communities share a contained k-clique, they are connected together into one bigger community until no such merging is possible.

This approach was first proposed by Palla et al. [13] and led to the development of CFinder [1]. Kumpula et al. then improved the clique percolation method by defining sequential order for edges added into a graph to detect k-cliques [9].

The problem of the clique percolation approach is in its chaining of cliques. There exist a possibility that nodes of resulting communities can have a high distance (number of hops) between themselves.

2.2 Local Expansion

This approach uses randomly picked nodes as seeds for the community and greedily expand the community to maximise a fitness function that evaluates the quality of a resulting community.

An example of such an approach is LFM [10].

$$f(c) = \frac{k_{in}^c}{(k_{in}^c + k_{out}^c)^\alpha} \tag{1}$$

The algorithm picks a random node as a seed and expands it to maximize function in Eq. 1 where $k_{in/out}^c c$ is external/internal degree of community c and α is resolution parameter controlling the size of the community. After the maximisation is finished, LFM picks a new random node from unassigned nodes as a seed for the next community.

The random seed selection leads to situations where the overlapping community is not detected because its nodes are already assigned to different communities. This approach also does not restrict the diameter of communities, which mean that nodes inside communities may be distanced.

2.3 Link Clustering

Another approach to community mining is to detect overlapping communities by clustering links. The idea is, that node can belong to multiple communities, but the link between nodes defines the relationship between these two objects [3]. By grouping similar edges into clusters, respective nodes can be assigned into communities defined by these clustered edges. There exist several link clustering methods such as OCMiner [5] and DB Link Clustering [18].

The latter mentioned uses six steps to identify communities:

1. An index of the edges is generated from the graph.
2. An incidence matrix between edges and nodes is created from the adjacency matrix.
3. A similarity matrix between each pair of edges is computed. Modified Jaccard coefficient is used as a distance function.
4. For each yet unassigned edge in the graph, if this edge is a core edge, a new cluster is created, and neighbouring edges connected to the core edges are assigned to this cluster. Edge is core edge if the cardinality of the set of similar edges in the neighbourhood of studied edge is higher than core size parameter.
5. Unassigned edges that are not part of the core are checked, whether they can be assigned to some existing cluster of edges to get final link partitioning.
6. Final link partitioning is transformed into the node communities by collecting incident nodes.

From the selected approaches, the link clustering results in community structures that are the most coherent. The diameter is controlled by the usage of core links that need to be adjacent to each other and, as such, does not tend to grow too much. Because of this, the link clustering approach leads to the most structurally cohesive community structure.

3 Community Mining Process

We propose a two-step approach to discover cohesive communities within a reasonably sparse graph. First, we detect possible community candidates, so-called *cores*, which are densely interconnected sets of nodes. To find such cores, we take inspiration in the problem of frequent item-set mining, for which many different algorithms can be used. Second, we possibly enrich each discovered candidate community with its *surrounding* – a set of nodes that do not need to be mutually interconnected but are densely connected to the core. To find such surroundings, we define a core-to-surrounding distance function and search for suitable surrounding candidates by evaluating range queries, for which many different similarity-search algorithms exist. The enriched communities are finally refined to retain only communities with reasonable size and inter-community overlaps.

3.1 Preliminaries

We consider structured network data as a graph G consisting of a pair of sets (V, E), where V is a set of nodes and E is a set of undirected and unweighted edges. We define a community as a structure consisting of two types of nodes: *core* and *surrounding* nodes.

The core is a group of heavily interconnected nodes of sufficient size. The ideal core of the community is a fully connected subgraph, otherwise known as *clique*. Such core ensures that in the resulting community, the diameter is maximally three hops: one from a starting node to core, one across the core, and the last one from the core to an end node. Although the clique is ideal core, this requirement is too strict; thus, the selection of the right relaxation can dramatically improve the detection of communities over the whole graph.

Although cores of the communities are cohesive groups, the extension of such structure can result in better grouping. This can be caused by imperfections in the process of capturing relations between network nodes. As an example, two members of the community do not use the same communication platform as with the rest of the community. This is the reason why each core is extended by its surrounding. The surrounding is defined as a set of nodes that have "sufficient" interconnection into the core of the community but do not meet the criteria to become a core.

3.2 FIMSIM: Community Mining Algorithm

For the purpose of processing, a graph is assumed in the form of a list of neighbours. The first step is the detection of cores. To detect the suitable core of the community, we use the task of mining frequent item sets, which was originally introduced by Agrawal et al. [2]. This task is often referred to as a market basket analysis and served as a way to detect items commonly bought together. The frequent item-set mining searches a database containing sets of unique items for subsets. Any subset occurring in the database more frequently than *minimal threshold* θ is returned as a frequent set.

In the list of neighbours representation, each node can be viewed as a market basket and a set of neighbours as the bought items. We further extend the list of neighbours of a node by adding the node itself. Frequent sets FS obtained from this extended representation are a superset of the cliques and densely connected clusters provided with reasonable parameters. By selecting the parameter θ, we will obtain all the cliques of size θ and higher.

To prove this assumption, let C be set of nodes of the clique of size k and F_C set of nodes from graph, such as each set of neighbours of selected node $\forall f_C \in F_C$ contains all nodes of clique $C \subseteq f_C$. Then every node of clique C contributes to a frequency of C by one, and thus frequency must be at least $|C| = k$. If $k \geq \theta$, the clique will be part of frequent sets FS. There is one specific pathological case (as illustrated in Fig. 1), where a set of nodes is referenced by third party nodes frequently enough to appear as heavily connected. This case is finally eliminated in the refinement step.

After frequent sets are obtained, it is necessary to eliminate undesirable ones. Sets are filtered based on two parameters; their common occurrence must be higher than their size to eliminate part of cores referenced from third party nodes, and their size must be at least θ. This second condition removes small cores that are often the result of the bottom-up approach of frequent item-set mining.

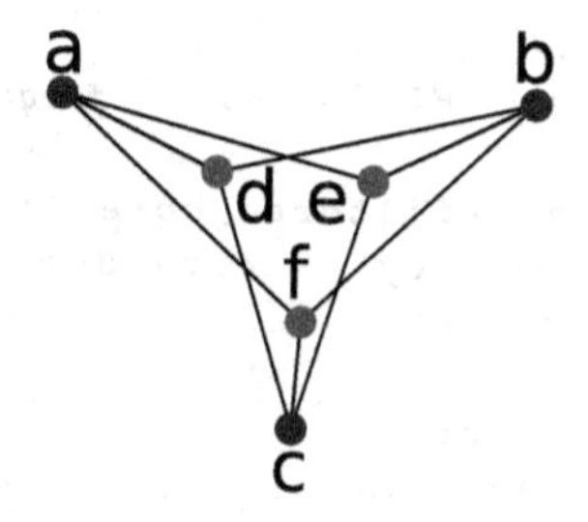

Fig. 1. Pathological case resulting in frequent sets {a, b, c} and {d, e, f} without an existing connection between nodes.

After obtaining the cores of the communities, a similarity range query is applied with the core as a query. For each node, there is then sequentially measured similarity to each core, and if the node is sufficiently similar, the node is marked with the respective core. There is a possibility to replace the range query with a K-NN query, which will allow better control over the resulting size of the community. The similarity is measured against the set of nodes of the core for each node in the neighbourhood of the core.

There is a number of standard distance functions for sets like the Jaccard coefficient that can be found in [17]. However, for comparison of node's neighbour list A with core B, we need a distance function that would result in zero in these two special cases:

1. $A \subseteq B$ – All neighbours of the node A are members of core B;
2. $B \subseteq A$ – Neighbour list of node A contains whole core B.

For this purpose, we propose to use the following distance function:

$$1 - max\left(\frac{|A \cap B|}{|A|}, \frac{|A \cap B|}{|B|}\right). \tag{2}$$

One parameter of the range query is distance radius r. The radius has a major impact on the density of the resulting communities. When distance is smaller, more edges are connected to the core of the communities. Standard measures often prefer this density, so values between 0 and 0.2 are preferred.

As a result of the range query, all nodes are marked with respective cores. By aggregating them into groups, raw communities are obtained. It is possible that the result contains duplicates and products of the pathological cases of a core. These can be eliminated by checking if the core is part of the community. Nodes from the pathological case have very low to zero common neighbours present in the core set and thus are not selected into the community by range query. After filtration, duplicates are removed.

After obtaining the communities, the results can be used as the new cores and search of the surrounding can be started again. This leads to communities with a potentially bigger distance between nodes and thus it is necessary that such approach is used only for suitable use cases. The pseudocode of the whole algorithm is depicted in Fig. 2.

```
 1 def FIMSIM(G, θ, r):
 2     unfiltered_cores = FIM(G, θ); # frequent item-set mining
 3     cores = {};
 4     foreach (core in unfiltered_cores):
 5         if (core.frequency >= core.size and core.size >= θ):
 6             cores += core;
 7     assignments = {};
 8     foreach (core in cores):
 9         assignments += simsearch(G, core, r);
10     aggregations = aggregate(assignments)
11     duplicit_communities = {};
12     foreach (candidate in aggregations):
13         foreach (core in cores):
14             if(candidate contains core):
15                 duplicit_communities += candidate;
16     communities = deduplications(duplicit_communities)
17     return communities;
```

Fig. 2. Pseudocode of the proposed FIMSIM algorithm for discovering communities.

4 Experiments

In the experimental part, we compared the proposed approach with the DB-Link Clustering. Both of these approaches share similar parameter space and results in similar graph structures, as can be seen in Fig. 3.

The prototype implementation of FIMSIM is developed with the analytical framework ADAMiSS [14], and for evaluation of similarity, MESSIF framework [4] is used. The former framework allows for a potential optimization based on the density of the analyzed graph.

4.1 Dataset

To evaluate whether this approach is eligible for community mining, we created a collaboration network from a pseudonymised dataset consisting of logged interactions of users with documents on a shared drive provided by Konica Minolta. Typically each document is created, modified and read by various users; thus, the dataset captures active collaboration of users on the document's content and passive interactions of users who only accessed the document.

The input data are in form of tuples: *date user_id*, *action_type*, *document_id*, *document_type*. The *action_type* consisted of six categories: *Download*, *Previewed*, *Edit*, *Uploaded*, *Created* and *Item Shared*.

We created the user-user interaction network from this data, where two users are connected when they cooperated on the creation process of at least one document. Thus if there is an access log with *action_type Created*, *Uploaded* or *Edit* for both users with same *document_id*. The resulting network consists of 128 users with an average degree of 8.531.

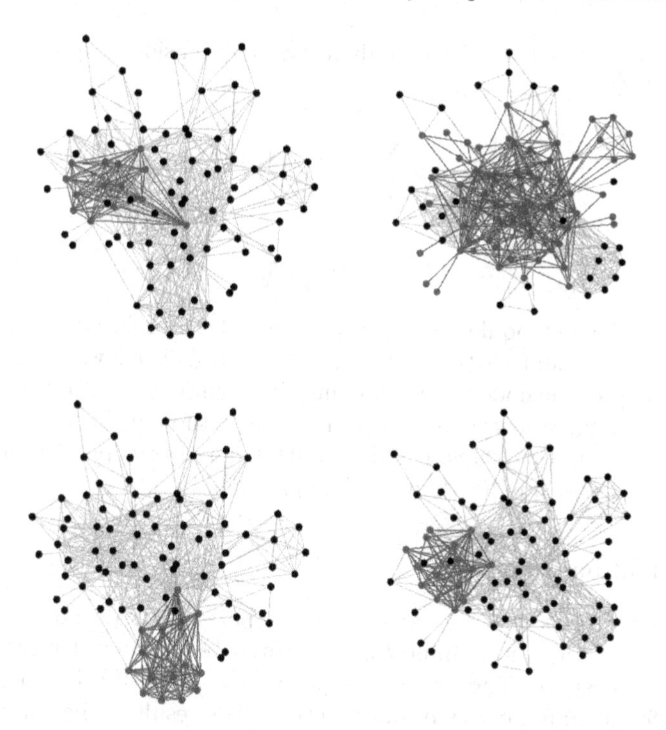

Fig. 3. Example of discovered communities: FIMSIM (Left) and DB-Link Clustering (Right). The example contains two biggest communities obtained from the best runs of both the algorithms. In particular, DB-Link Clustering uses $\theta = 6$ and $r = 0.4$, while FIMSIM uses $\theta = 12$ and $r = 0.1$.

This data represents a collaboration of workers in the organisation and, as such, can be analysed by managers. The discovered communities may function as decision support for managers to create and validate the matrix structure of the organisation.

4.2 Evaluation Criteria

The traditional way of evaluating the quality of graph partitioning is by *modularity* [12].

$$Q = \frac{1}{2 \cdot |E|} \sum_{c \in C} \sum_{i,j \in V} \delta_{ci} \delta_{cj} \left(A_{i,j} - \frac{k_i k_j}{2 \cdot |E|} \right) \tag{3}$$

Modularity measures the number of edges that connects nodes in the same partitioning reduced by the expected amount of edges in a randomly wired network. This metric is primarily used for the evaluation of exclusive partitioning and thus penalizes overlapping communities.

Because of limitations of the modularity, an extension proposed by Chen et al. was used [6].

$$Q = \frac{1}{2 \cdot |E|} \sum_{c \in C} \sum_{i,j \in V} \alpha_{ci} \alpha_{cj} \left(A_{i,j} - \frac{k_i k_j}{2 \cdot |E|} \right), \tag{4}$$

$$\alpha_{ci} = \frac{k_{ci}}{\sum_{c_2 \in C} k_{c_2 i}} \tag{5}$$

In this version of modularity, the Kronecker delta is replaced with a coefficient of how many communities the node is involved in. This allows for a decrease in the importance of the node involved in multiple communities and thus is not that penalising for higher density on the overlap. Even though this approach tries to solve the problem of overlaps, it still evaluates non-overlapping communities as having much higher quality than overlapping ones.

4.3 Evaluation

Although algorithms share parameter space, parameters do not match one to one. Because of that, it is important to watch best-achieved results over the whole searched space. The experiment showed that FIMSIM outperforms the quality of found communities in both metrics. The result of the comparison can be seen in Figs. 4 and 5.

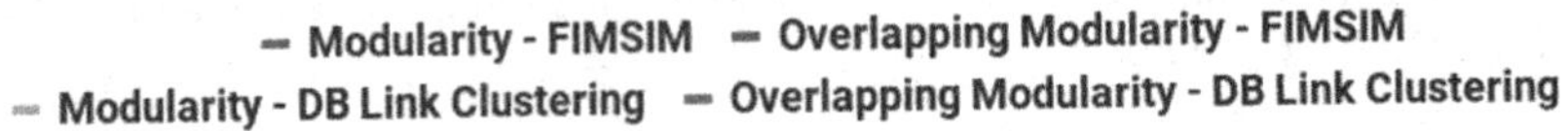

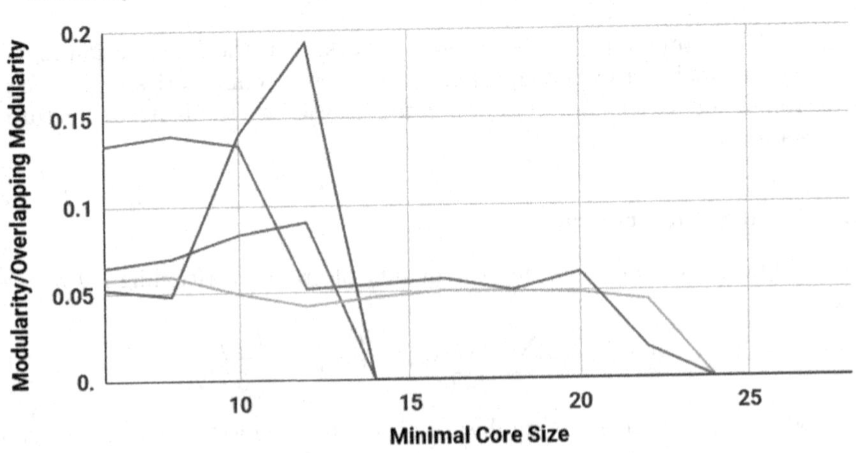

Fig. 4. Quality comparison

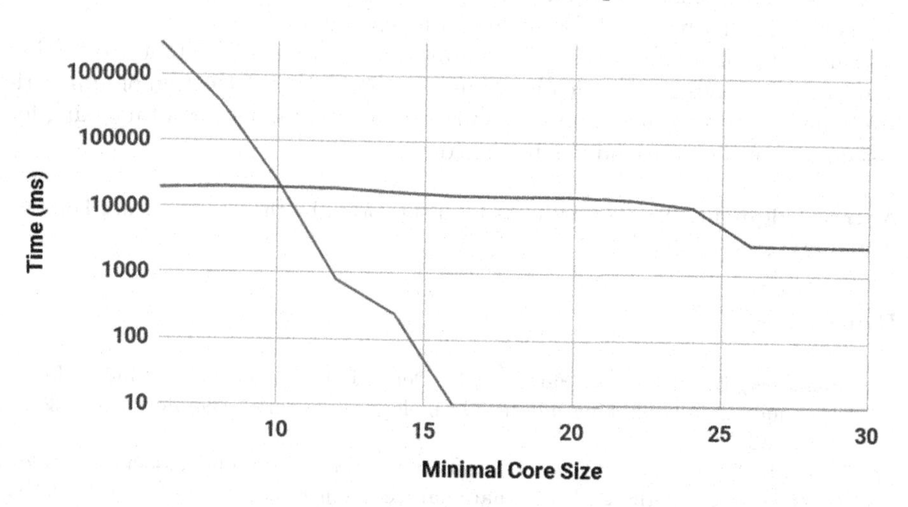

Fig. 5. Time comparison

The experiment also showed that our approach can be better used without knowledge of the correct parameters than DB Link clustering. Because of the nature of frequent item-set mining, when the size of the core is too huge, the algorithm stops almost immediately. Because of that, it can be beneficial to start from higher numbers and lower the core size parameter until a suitable amount of cores is found.

The disadvantage of our approach is that once the core size parameter is too small, the discovery of cores is computationally challenging. Our experiments showed that higher quality of communities is achieved at the higher value of the core size parameter.

5 Conclusion

In this paper, we proposed a different approach to community mining based on a combination of frequent item-set mining and similarity searching. The proposed method uses a two-step process of finding core community candidates and assigning surroundings to them. We proposed a distance function for the selection of suitable surroundings based on two extreme cases.

In cooperation with Konica Minolta, a collaboration network was created as a representation of real-world data, and a new approach was tested. With the usage of modularity and overlapping modularity, we showed that the proposed approach achieves higher-quality results than the state-of-the-art DB-Link Clustering approach that discovers a similar type of communities.

In several steps, we discussed the possibility of modifying the approach to achieve better flexibility in terms of the size of the community as well as the quality of the surrounding of the cores. We also discussed an iterative approach, where after obtaining communities, these are taken as the new cores and with the new range, surrounding for them can be chosen.

Due to the modular nature of the approach, there is a possibility to further improve the proposed approach in terms of optimizing the selection of appropriate frequent item-set algorithms, as well as employing some sort of indexing for assignment of the surroundings to the cores.

Acknowledgment. This research has been supported by the Czech Science Foundation project No. GA19-02033S.

References

1. Adamcsek, B., Palla, G., Farkas, I.J., Derényi, I., Vicsek, T.: CFinder: locating cliques and overlapping modules in biological networks. Bioinformatics **22**(8), 1021–1023 (2006)
2. Agrawal, R., Srikant, R., et al.: Fast algorithms for mining association rules. In: Proceedings of the 20th International Conference on Very Large Data Bases, VLDB, vol. 1215, pp. 487–499 (1994)
3. Ahn, Y.Y., Bagrow, J.P., Lehmann, S.: Link communities reveal multiscale complexity in networks. Nature **466**(7307), 761–764 (2010)
4. Batko, M., Novak, D., Zezula, P.: MESSIF: metric similarity search implementation framework. In: Thanos, C., Borri, F., Candela, L. (eds.) Digital Libraries: Research and Development. DELOS 2007. LNCS, vol. 4877, pp. 1–10. Springer, Heidelberg (2007). https://doi.org/10.1007/978-3-540-77088-6_1
5. Bhat, S.Y., Abulais, M.: OCMiner: a density-based overlapping community detection method for social networks. Intell. Data Anal. **19**(4), 917–947 (2015)
6. Chen, D., Shang, M., Lv, Z., Fu, Y.: Detecting overlapping communities of weighted networks via a local algorithm. Physica A: Stat. Mech. Appl. **389**(19), 4177–4187 (2010)
7. Fortunato, S.: Community detection in graphs. Phys. Rep. **486**(3–5), 75–174 (2010)
8. Girvan, M., Newman, M.E.: Community structure in social and biological networks. Proc. Natl. Acad. Sci. **99**(12), 7821–7826 (2002)
9. Kumpula, J.M., Kivelä, M., Kaski, K., Saramäki, J.: Sequential algorithm for fast clique percolation. Phys. Rev. E **78**(2), 026109 (2008)
10. Lancichinetti, A., Fortunato, S., Kertész, J.: Detecting the overlapping and hierarchical community structure in complex networks. New J. Phys. **11**(3), 033015 (2009)
11. Newman, M.E.: Communities, modules and large-scale structure in networks. Nat. Phys. **8**(1), 25–31 (2012)
12. Newman, M.E., Girvan, M.: Finding and evaluating community structure in networks. Phys. Rev. E **69**(2), 026113 (2004)
13. Palla, G., Derényi, I., Farkas, I., Vicsek, T.: Uncovering the overlapping community structure of complex networks in nature and society. Nature **435**(7043), 814–818 (2005)
14. Peschel, J., Batko, M., Zezula, P.: Techniques for complex analysis of contemporary data. In: Proceedings of the 2020 International Conference on Pattern Recognition and Intelligent Systems, pp. 1–5 (2020)
15. Radicchi, F., Castellano, C., Cecconi, F., Loreto, V., Parisi, D.: Defining and identifying communities in networks. Proc. Natl. Acad. Sci. **101**(9), 2658–2663 (2004)

16. Schaeffer, S.E.: Graph clustering. Comput. Sci. Rev. **1**(1), 27–64 (2007)
17. Zezula, P., Amato, G., Dohnal, V., Batko, M.: Similarity Search: The Metric Space Approach, vol. 32. Springer, Heidelberg (2006)
18. Zhou, X., Liu, Y., Wang, J., Li, C.: A density based link clustering algorithm for overlapping community detection in networks. Physica A: Stat. Mech. Appl. **486**, 65–78 (2017)

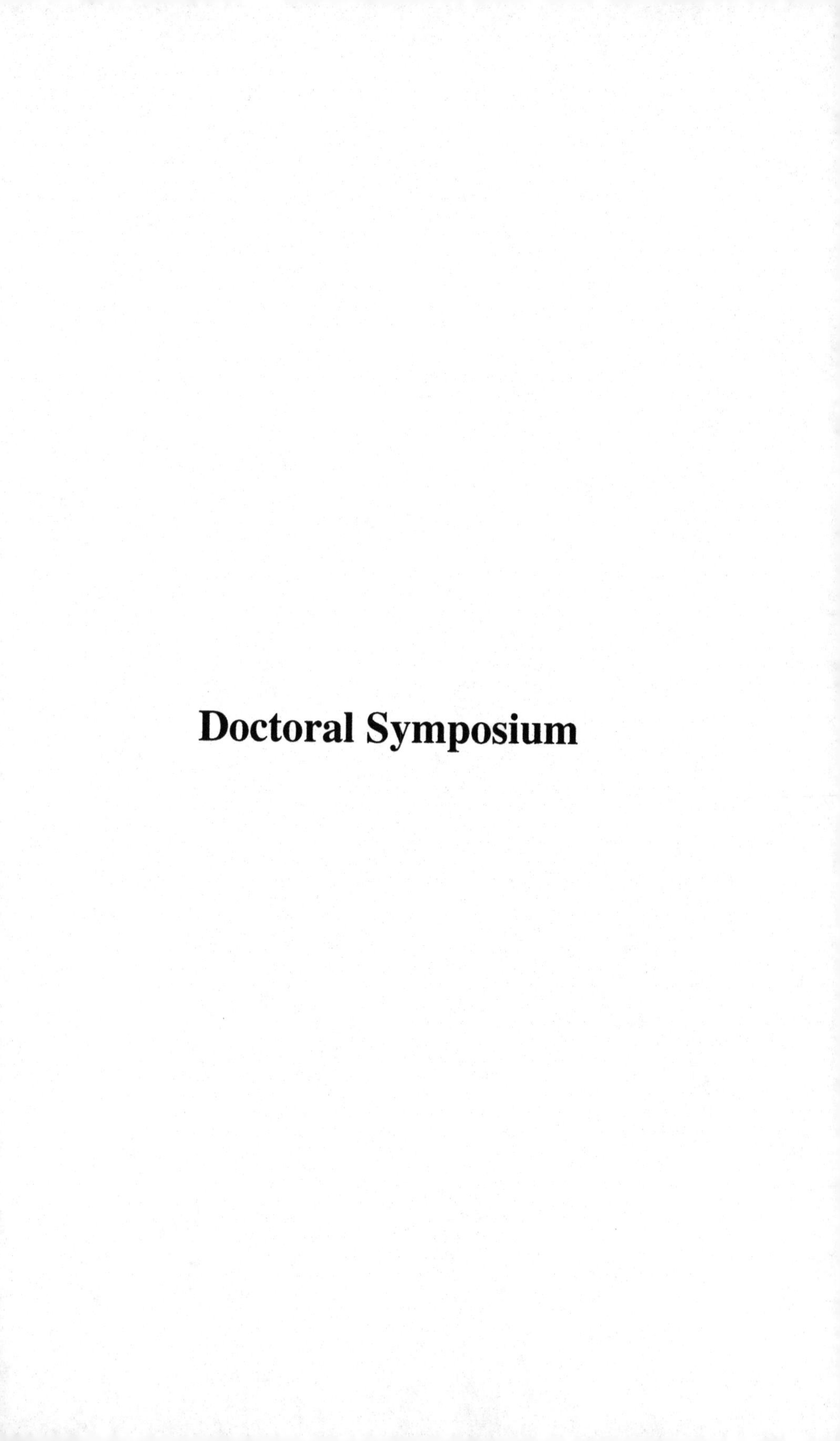

Doctoral Symposium

Towards an Italian Healthcare Knowledge Graph

Marco Postiglione(✉)

University of Naples Federico II, Naples, Italy
marco.postiglione@unina.it

Abstract. Electronic Health Records (EHRs), Big Data, Knowledge Graphs (KGs) and machine learning can potentially be a great step towards the technological shift from the *one-size-fit-all* medicine, where treatments are based on an equal protocol for all the patients, to the *precision* medicine, which takes count of all their individual information: lifestyle, preferences, health history, genomics, and so on. However, the lack of data which characterizes low-resource languages is a huge limitation for the application of the above-mentioned technologies. In this work, we will try to fill this gap by means of transformer language models and few-shot approaches and we will apply similarity-based deep learning techniques on the constructed KG for downstream applications. The proposed architecture is general and thus applicable to any low-resource language.

Keywords: Knowledge Graphs · Electronic Health Records · Transformer language models

1 Introduction

The Big Data paradigm has become a reality thanks to the availability of an ever-growing quantity of data and the synergetic progress in computing infrastructures and data analysis techniques [6]. To the current state, Big Data solutions are already in use to support us in our daily life for safety [26], entertainment [25] and healthcare [3] inter alia.

In the healthcare industry, the recent progress made in *Electronic Health Records (EHRs)* has enabled the collection of huge quantities of data related to the medical histories of patients (e.g. laboratory measurements, radiology imaging, clinical notes). Being closer to the actual practice of medicine as compared with the idealized information presented in textbooks and journals, EHRs provide the possibility to (1) identify possible causal relations between healthcare entities (e.g. symptoms, diseases, measurements) which are not even written in books [30] and (2) suggest personalized treatments.

The heterogeneous information of EHRs can be organized in graph data structures, a.k.a. *Knowledge Graphs (KGs)*, which capture the relationships between different entities by linking them through edges. Once the KG is constructed, not only can data be easily and interactively visualized and explored

Supported by Oracle Labs.

N. Reyes et al. (Eds.): SISAP 2021, LNCS 13058, pp. 387–394, 2021.
https://doi.org/10.1007/978-3-030-89657-7_29

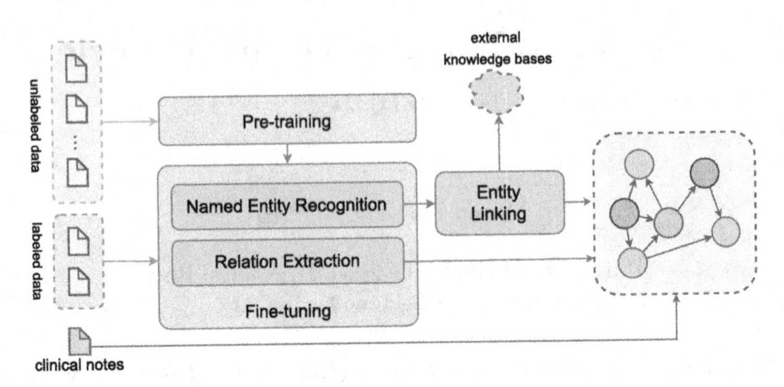

Fig. 1. Overview of the planned methodology.

by analysts and physicians, but they can also be analyzed with machine and deep learning techniques to solve complex tasks, such as providing personalized therapies. For example, KGs have been effectively used for the prediction of adverse drug reactions in patients [46] and for drug repurposing for the treatment of COVID-19 [40] especially thanks to embedding methods which allow to represent the KG entities in an Euclidean space and thus to exploit distance and similarity-based metrics to analyze relations between nodes.

In this work, we will try to pave the way towards the application of healthcare KGs in low-resource languages, where all the advances detailed above—in EHRs, KGs and analytics techniques—cannot be fully exploited due to the lack of data. Specifically, the overall project contributions are summarized as follows:

1. Pre-training of a *transformer* language model [4,8,29] based on Italian biomedical corpora
2. Definition of few-shot learning approaches to use the pre-trained model to recognize entities and relations from clinical notes
3. Entity linking to external knowledge bases with similarity-based approaches
4. Smart navigation and analysis of the constructed KG with deep learning similarity-based techniques.

2 Planned Methodology

Figure 1 shows an outline of the planned methodology, which will be detailed in the remainder of this section.

2.1 Language Model: Pre-training and Fine-Tuning

The automatic understanding and processing of clinical notes is a challenging task due to several peculiarities, i.e. negations, synonyms, alternate spelling of entities, non-standard abbreviations, polysemous words [7,43]. Thanks to their effectiveness in leveraging both words and their contexts, transformer language

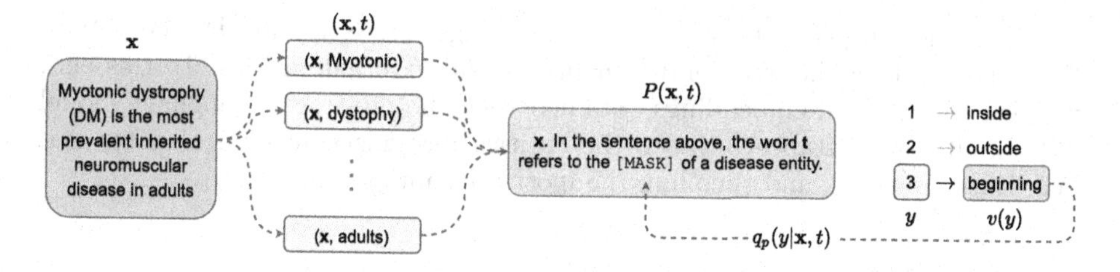

Fig. 2. Example of PETER application.

models (e.g. BERT [8], GPT-3 [4], T5 [29]) have proven to be a valuable solution for this challenge. They are first *pre-trained* on huge quantities of unlabeled text data and then *fine-tuned* with labeled data to solve downstream tasks (e.g. sentence classification, part-of-speech labeling). Inspired by recent works in biomedical language understanding which have shown that performance of downstream tasks can be strongly improved by pre-training on biomedical text data (e.g. papers, clinical notes) [2,20,22], we will pre-train an Italian biomedical language model and fine-tune it to build our KG by detecting entities (*Named Entity Recognition*, a.k.a. NER [14]) and extracting unknown relational facts (*Relation Extraction* [28,35,44,45]) from clinical notes.

The lack of annotated data characterizing low-resource languages imposes the use of few-shot learning approaches [12,19,27,33,36,42]. *Pattern-Exploiting Training (PET)* [32] has been proved to be an effective technique to fine-tune language models for few-shot classification tasks. Hence, we developed PETER (**P**attern-**E**xploiting **T**raining for Named **E**ntity **R**ecognition), a slight adaptation which allows to use PET for NER and we intuitively describe it in Fig. 2. Given a sequence of tokens $\mathbf{x}$ (i.e. a sentence) a "pattern" is applied to each token to generate input examples containing a *mask* token which will be replaced by the model with the appropriate label (which indicates if the token is at the *beginning*, *inside* or *outside* of an entity mention). In this way, the language model leverages the knowledge it has acquired during the pre-training phase to solve the downstream task, hence requiring few samples to obtain satisfactory performance.

2.2 Entity Linking

The knowledge retrieved from EHRs will be extended with external knowledge bases containing additional useful information. For example, WikiData [37] allows to link diseases with their corresponding *International Classification of Diseases (ICD)* code and to enrich nodes with suggested drugs, therapies, health specialty, and so on. The *Entity Linking* task consists in annotating mentions with their corresponding identifier in an external knowledge base. It involves candidate-entity *generation* and *ranking* [1], i.e. retrieving all the possible entities which may be linked to an entity mention and returning the most likely one. Current literature shows the effectiveness of semantic similarity-based approaches which use neural networks and word embeddings [11,18] to capture the semantic correspondence between entity mentions and external entities. To this end, we plan to use word embeddings obtained with the pre-trained language model to compute the distance between the entities recognized in clinical notes and WikiData concepts, and thus link the most appropriate one.

2.3 Knowledge Graph Analysis

Despite the construction of the KG being an important step in our project, its smart navigation and analysis is essential to effectively use it as a supportive tool in the actual practice of medicine. To this end, we will leverage similarity measures which take count of not only the informative content of nodes (i.e. node properties) but also the topological graph structure (e.g. path length and depth). More specifically, we will leverage *knowledge representation learning* techniques, which aim to learn low-dimensional embedding of nodes and relations.

Embeddings have to guarantee the possibility to define *scoring functions* [10], which are used to measure the plausibility of facts, i.e. $(head, relation, tail)$ triples. As an example, a scoring function f can be exploited to return the probability that a patient suffers from a disease given all its attributes (e.g. age, sex, lifestyle) and laboratory exams $\rightarrow^{e.g.}$ $f(Alice, hasDisease, Diabetes) = 0.6$.

We plan to employ similarity-based functions, which use semantic matching to calculate the semantic similarity between entities [39,41,47]. To this end, neural networks have been proven to effectively encode the semantic matching principle by feeding entities or relations or both into deep networks to compute a similarity score [17]. In particular, *Graph Convolutional Neural networks (GCNs)* have been proven to be effective in leveraging the attributes associated with nodes [15]. Node structures, attributes and relation types can be integrated in weighted GCN models [34], which treat multi-relational KGs as multiple single-relational graphs and learn weights when combining GCN embeddings for each subgraph and node. The output of the l-th layer for the node v_i can be thus computed as:

$$h_i^{l+1} = \sigma\Big(\sum_{j \in N_i} \alpha_t^l g(h_i^l, h_j^l)\Big), \tag{1}$$

where: h_i^l and h_j^l are the input vectors for nodes v_i and v_j, respectively, and v_j is a node in the neighborhood N_i of v_i; α_t is a learnable parameter specifying the strength of the relation type t between two adjacent nodes; σ is an activation function; g incorporates neighboring information with a coefficient matrix.

3 Early Results

Our research activities have so far been focused on the first steps of our research plan, i.e. data collection, pre-training and the definition of few-shot learning techniques. In particular, the hospital *Azienda Ospedaliera Universitaria (AOU) Federico II* has provided a database with information about hospitalizations in

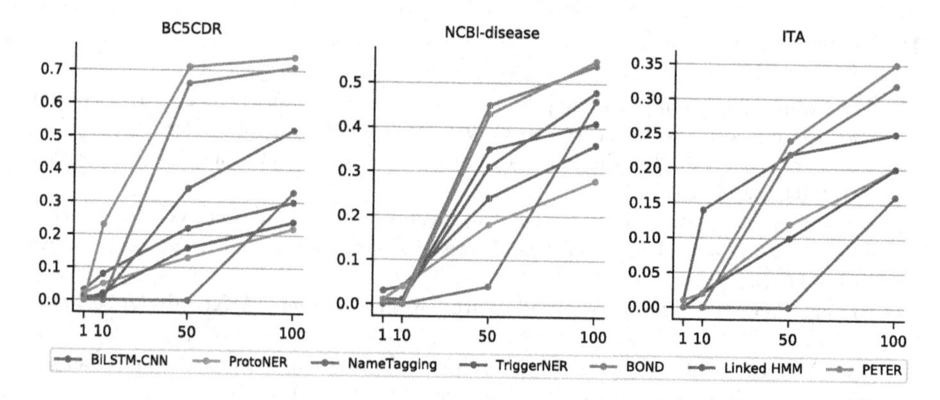

Fig. 3. PETER comparison with the state-of-the-art in terms of F1 scores.

their cardiological departments. With reference to the pre-training phase, all the clinical notes included in the above-mentioned database (646, 774 sentences) have been collected and integrated with information collected from the forum *Medicitalia*[1] (14, 484, 684 sentences) and *DBpedia* [21] (7, 129 sentences).

Furthermore, a team of 8 biomedical engineers has labeled a subset of the clinical notes to allow the fine-tuning of NER models. More in detail, the dataset contains 6186 *disease* and 4918 *symptom* mentions annotated.

Finally, we compared PETER with state-of-the-art few-shot NER techniques [5,13,16,23,24,31] on three datasets: BC5CDR [38], NCBI-disease [9], and the Italian NER dataset described above. Results in Fig. 3 show that PETER obtains higher results w.r.t. the other techniques in terms of F1 scores (y-axis) in several few-shot contexts (x-axis). While models fine-tuned on BC5CDR and NCBI-Disease are initialized with BioBERT [20], i.e. a biomedical transformer, the Italian model has been initialized with GilBERTo[2], which is trained on general text corpora. The resulting poor performance shows the crucial importance of the pre-training step we are currently working on.

4 Conclusion and Future Work

In this paper, we have described the planned research methodology for the construction and analysis of an Italian KG in healthcare. We have so far (1) collected the required data to train the language model and (2) defined the few-shot approach to be used. In future work, we will link medical entities to external knowledge bases and analyze the developed KG with similarity-based techniques which take count of both the topological graph structure and the information content of nodes.

---

[1] https://www.medicitalia.it/.
[2] https://github.com/idb-ita/GilBERTo.

References

1. Al-Moslmi, T., Ocaña, M.G., Opdahl, A.L., Veres, C.: Named entity extraction for knowledge graphs: a literature overview. IEEE Access **8**, 32862–32881 (2020)
2. Alsentzer, E., et al.: Publicly available clinical BERT embeddings. ArXiv abs/1904.03323 (2019)
3. Amendola, S., Lodato, R., Manzari, S., Occhiuzzi, C., Marrocco, G.: RFID technology for IoT-based personal healthcare in smart spaces. IEEE Internet Things J. **1**, 144–152 (2014)
4. Brown, T.B., et al.: Language models are few-shot learners. ArXiv abs/2005.14165 (2020)
5. Cao, Y., Hu, Z., Chua, T.S., Liu, Z., Ji, H.: Low-resource name tagging learned with weakly labeled data. ArXiv abs/1908.09659 (2019)
6. Chen, M., Mao, S., Liu, Y.: Big data: a survey. Mob. Netw. Appl. **19**(2), 171–209 (2014)
7. Dalloux, C., et al.: Supervised learning for the detection of negation and of its scope in French and Brazilian Portuguese biomedical corpora. Nat. Lang. Eng. **27**, 181–201 (2020)
8. Devlin, J., Chang, M.W., Lee, K., Toutanova, K.: BERT: pre-training of deep bidirectional transformers for language understanding. In: NAACL-HLT (2019)
9. Dogan, R., Leaman, R., Lu, Z.: NCBI disease corpus: a resource for disease name recognition and concept normalization. J. Biomed. Inform. **47**, 1–10 (2014)
10. Ebisu, T., Ichise, R.: TorusE: knowledge graph embedding on a lie group. In: AAAI (2018)
11. Francis-Landau, M., Durrett, G., Klein, D.: Capturing semantic similarity for entity linking with convolutional neural networks (2016)
12. Fries, J., Wu, S., Ratner, A., Ré, C.: SwellShark: a generative model for biomedical named entity recognition without labeled data. arXiv:1704.06360 [cs] (2017). http://arxiv.org/abs/1704.06360
13. Fritzler, A., Logacheva, V., Kretov, M.: Few-shot classification in named entity recognition task. In: Proceedings of the 34th ACM/SIGAPP Symposium on Applied Computing (2019)
14. Grishman, R., Sundheim, B.: Message understanding conference-6: a brief history. In: COLING (1996)
15. Hamilton, W.L., Ying, Z., Leskovec, J.: Inductive representation learning on large graphs. In: NIPS (2017)
16. Hofer, M., Kormilitzin, A., Goldberg, P., Nevado-Holgado, A.: Few-shot learning for named entity recognition in medical text. ArXiv abs/1811.05468 (2018)
17. Ji, S., Pan, S., Cambria, E., Marttinen, P., Yu, P.S.: A survey on knowledge graphs: representation, acquisition and applications. IEEE Trans. Neural Netw. Learn. Syst., 1–21 (2021)
18. Karadeniz, I., Özgür, A.: Linking entities through an ontology using word embeddings and syntactic re-ranking. BMC Bioinform. **20**, 1–12 (2019)
19. Kim, S., Toutanova, K., Yu, H.: Multilingual named entity recognition using parallel data and metadata from Wikipedia. In: ACL (2012)
20. Lee, J., et al.: BioBERT: a pre-trained biomedical language representation model for biomedical text mining. Bioinformatics **36**, 1234–1240 (2020)
21. Lehmann, J., et al.: DBpedia - a large-scale, multilingual knowledge base extracted from Wikipedia. Semant. Web **6**, 167–195 (2015)

22. Lewis, P., Ott, M., Du, J., Stoyanov, V.: Pretrained language models for biomedical and clinical tasks: understanding and extending the state-of-the-art. In: CLINICALNLP (2020)
23. Liang, C., et al.: BOND: BERT-assisted open-domain named entity recognition with distant supervision. In: Proceedings of the 26th ACM SIGKDD International Conference on Knowledge Discovery & Data Mining (2020)
24. Lin, B.Y., et al.: TriggerNER: learning with entity triggers as explanations for named entity recognition. In: ACL (2020)
25. Lippell, H.: Big data in the media and entertainment sectors. In: New Horizons for a Data-Driven Economy (2016)
26. Mannering, F., Bhat, C., Shankar, V., Abdel-Aty, M.: Big data, traditional data and the tradeoffs between prediction and causality in highway-safety analysis. Anal. Methods Accid. Res. **25**, 100113 (2020)
27. Mintz, M.D., Bills, S., Snow, R., Jurafsky, D.: Distant supervision for relation extraction without labeled data. In: ACL/IJCNLP (2009)
28. Nguyen, T., Grishman, R.: Relation extraction: perspective from convolutional neural networks. In: VS@HLT-NAACL (2015)
29. Raffel, C., et al.: Exploring the limits of transfer learning with a unified text-to-text transformer. ArXiv abs/1910.10683 (2020)
30. Rotmensch, M., Halpern, Y., Tlimat, A., Horng, S., Sontag, D.: Learning a health knowledge graph from electronic medical records. Sci. Rep. **7**, 1–11 (2017)
31. Safranchik, E., Luo, S., Bach, S.H.: Weakly supervised sequence tagging from noisy rules. In: AAAI (2020)
32. Schick, T., Schütze, H.: Exploiting cloze-questions for few-shot text classification and natural language inference. In: EACL (2021)
33. Schmidhuber, J.: On learning how to learn learning strategies (1994)
34. Shang, C., Tang, Y., Huang, J., Bi, J., He, X., Zhou, B.: End-to-end structure-aware convolutional networks for knowledge base completion (2018)
35. Shen, Y., Huang, X.: Attention-based convolutional neural network for semantic relation extraction. In: COLING (2016)
36. Thrun, S., Pratt, L.Y.: Learning to learn (1998)
37. Vrandecic, D., Krötzsch, M.: Wikidata: a free collaborative knowledgebase. Commun. ACM **57**, 78–85 (2014)
38. Wei, C.H., et al.: Assessing the state of the art in biomedical relation extraction: overview of the BioCreative V chemical-disease relation (CDR) task. Database: J. Biol. Databases Curation **2016** (2016)
39. Xue, Y., Yuan, Y., Xu, Z., Sabharwal, A.: Expanding holographic embeddings for knowledge completion. In: NeurIPS (2018)
40. Yan, V.K.C., et al.: Drug repurposing for the treatment of COVID-19: a knowledge graph approach. Adv. Ther. **4** (2021)
41. Yang, B., tau Yih, W., He, X., Gao, J., Deng, L.: Embedding entities and relations for learning and inference in knowledge bases. CoRR abs/1412.6575 (2015)
42. Yarowsky, D., Ngai, G.: Inducing multilingual POS taggers and NP bracketers via robust projection across aligned corpora. In: NAACL (2001)
43. Yoon, W., So, C.H., Lee, J., Kang, J.: CollaboNet: collaboration of deep neural networks for biomedical named entity recognition. BMC Bioinform. **20**(S10), 55–65 (2019). https://doi.org/10.1186/s12859-019-2813-6
44. Zeng, D., Zhao, C., Quan, Z.: CID-GCN: an effective graph convolutional networks for chemical-induced disease relation extraction. Front. Genet. **12**, 115 (2021)
45. Zeng, X., He, S., Liu, K., Zhao, J.: Large scaled relation extraction with reinforcement learning. In: AAAI (2018)

46. Zhang, F., Sun, B., Diao, X., Zhao, W., Shu, T.: Prediction of adverse drug reactions based on knowledge graph embedding. BMC Med. Inform. Decis. Making **21**, 1–11 (2021)
47. Zhang, W., Paudel, B., Zhang, W., Bernstein, A., Chen, H.: Interaction embeddings for prediction and explanation in knowledge graphs. In: Proceedings of the Twelfth ACM International Conference on Web Search and Data Mining (2019)

Progressive Query-Driven Entity Resolution

Luca Zecchini(✉)

University of Modena and Reggio Emilia, Modena, Italy
luca.zecchini@unimore.it

Abstract. *Entity Resolution (ER)* aims to detect in a dirty dataset the records that refer to the same real-world entity, playing a fundamental role in data cleaning and integration tasks. Often, a data scientist is only interested in a portion of the dataset (e.g., *data exploration*); this interest can be expressed through a query. The traditional *batch* approach is far from optimal, since it requires to perform ER on the whole dataset before executing a query on its *cleaned* version, performing a huge number of useless comparisons. This causes a waste of time, resources and money. Proposed solutions to this problem follow a *query-driven* approach (perform ER only on the useful data) or a *progressive* one (the entities in the result are emitted as soon as they are solved), but these two aspects have never been reconciled. This paper introduces `BrewER` framework, which allows to execute *clean queries on dirty datasets* in a *query-driven* and *progressive* way, thanks to a preliminary filtering and an iteratively managed sorted list that defines emission priority. Early results obtained by first `BrewER` prototype on real-world datasets from different domains confirm the benefits of this combined solution, paving the way for a new and more comprehensive approach to ER.

Keywords: Entity resolution · Data integration · Data cleaning

1 Introduction

Entity Resolution (ER) is a fundamental task for data integration [11], aiming to detect in a dirty dataset the records (*duplicates* [10]) that represent the same real-word object (*entity*). The duplicates are detected by applying a *matching function* (e.g., a trained binary classifier) to each possible pair of records (or to the pairs formed by records appearing in the same block, if a *blocking function* [12] is applied), in order to determine if they refer or not to the same entity (in the first case, they are referred to as *matches*).

Once a cluster of matches is found, its records are merged to create a single consistent record representing the entity (*data fusion* [7]), removing dataset redundancy. The matches often present missing, wrong or conflicting values; data fusion is performed through the application of a *resolution function*, which determines the value to be assigned to each attribute of the entity according to the *aggregation function* (e.g., maximum/minimum value, majority voting, etc.) defined for it by the data scientist.

N. Reyes et al. (Eds.): SISAP 2021, LNCS 13058, pp. 395–401, 2021.
https://doi.org/10.1007/978-3-030-89657-7_30

In concrete situations (e.g., *data exploration*), a data scientist is often only interested in a specific portion of the dataset; this interest can be expressed through a query. Since performing a query on the dirty dataset may lead to an inconsistent result, it is necessary to perform ER on the dataset; then, the query can be executed on the obtained *cleaned* version. As shown in Fig. 1a, all the entities appearing in the result are returned to the data scientist at the end of this pipeline.

When computational resources are limited and/or time is a critical component, this approach (called *batch*) is far from optimal, wasting time, resources and money (e.g., in the case for *pay-as-you-go* contracts, widely used by cloud providers) to perform comparisons which are guaranteed to be useless. These comparisons are required to generate entities with no chance of appearing in the result (e.g., the query in Fig. 2 returns only `Canon` cameras, so each comparison performed to retrieve an entity whose brand is `Nikon` is useless and should be avoided) and cause performance degradation, as the data scientist can only run the query after all comparisons have been performed.

In order to overcome the described problems, an innovative approach (Fig. 1b) must be able to perform *clean queries on dirty datasets* and it is required to be both *query-driven* (i.e., to perform ER only on the portion of data effectively useful to answer the query, according to the `WHERE` clauses of the query itself) and *progressive* (i.e., to emit the entities appearing in the result as soon as they are solved, following the ordering expressed by the `ORDER BY` clause). This is exactly the aim of `BrewER`.

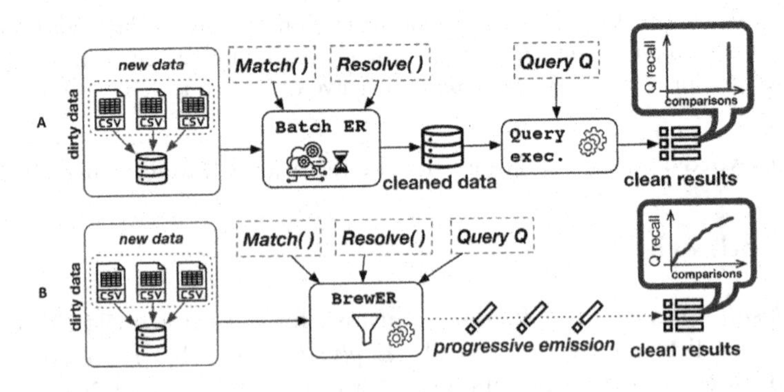

Fig. 1. The traditional *batch* pipeline and the one proposed by `BrewER`.

2 Related Work

Solutions in literature [4–6] propose a *Query-Driven Approach (QDA)* to ER, executing clean queries on dirty datasets performing comparisons only on the portion of data relevant for the executed query; however, the adopted techniques are not suitable for supporting the progressive emission of the results (e.g., the `ORDER BY` clause is not managed). On the other hand, even progressive solutions have been presented [13,15,16], but neither of them considers the possibility of an integration with QDA principles, which is far from trivial. A draft of combined approach based on a graph structure has been presented in [14], but it is limited to approximate solutions for a keyword-search scenario. Therefore, BrewER approach represents a novelty in literature.

3 BrewER: A Progressive Query-Driven ER Framework

BrewER framework for ER reconciles the described approaches, executing clean queries on dirty datasets in a query-driven and progressive way.

The implementation of the query-driven approach consists of a preliminary filtering of the blocks, which aims to keep only the blocks that could generate an entity appearing in the result. BrewER approach to the eventual blocking function adopted by the user is agnostic; if no blocking is performed, the whole dataset can be interpreted as a single block. Furthermore, blocks are transitively closed. If the WHERE clauses of the query are in OR (at the moment, only conjunctive and disjunctive queries are supported), it is checked that at least one record in the block satisfies at least one of the defined clauses; on the other hand, if the clauses are in AND, it is verified that all the clauses, each considered by itself, are satisfied by at least one record in the block. The records which satisfy at least one of the clauses are called *seed records*. The aggregation function for each attribute is defined by the user; MIN, MAX, AVG, VOTE (majority voting) and RANDOM are supported, with SUM to be implemented. In case of numeric attributes, the described filtering is not applied using functions that can generate new values (i.e., AVG and SUM).

The progressive emission is obtained through an iteratively managed sorted list (Fig. 2) called *Ordering List (OL)*. The records appearing in the blocks that pass the filtering are marked as *unsolved* (ER not yet performed) and inserted in OL, each one with a list containing the identifiers of its *neighbours* (i.e., the records in the same block). At the beginning of each iteration, the elements in OL are sorted according to the ordering mode and the attribute, called *Ordering Key (OK)*, expressed by the ORDER BY clause of the query. Then, the first element (i.e., the one with the highest emission priority) is checked. If it is marked as *unsolved* (2.1a), it is compared with its neighbours (even for the matching function BrewER adopts an agnostic approach); once identified the cluster of matches, all the matching elements are removed from OL, while a single element representing the cluster (with the aggregated OK value), marked as *solved*, is inserted (2.1b). If it is marked as *solved* (2.2a), the resolution function is applied on the represented cluster: if the obtained entity satisfies the query, it is emitted; otherwise, it is discarded. Comparisons involve seed neighbours first: if a non-seed does not match any seed, it can be discarded. Iterations can run until OK is empty or can be stopped after the emission of k entities (TOP(K) queries).

Optimizations. In case of *discordant ordering* (MIN/DESC or MAX/ASC), it is possible to optimize the described algorithm by inserting in OL only the seed records, while the non-seed records in their blocks only appear in their lists of neighbours (fewer comparisons). This is possible because if a matching non-seed neighbour (whose OK value is not therefore sorted in OL) alters the OK value for the first element, the generated *solved* record priority is updated by changing its position when sorting OL at the beginning of the next iteration (*delayed emission*), guaranteeing the correctness of the ordering (while in MAX/DESC or MIN/ASC cases this variant could alter the correct emission ordering).

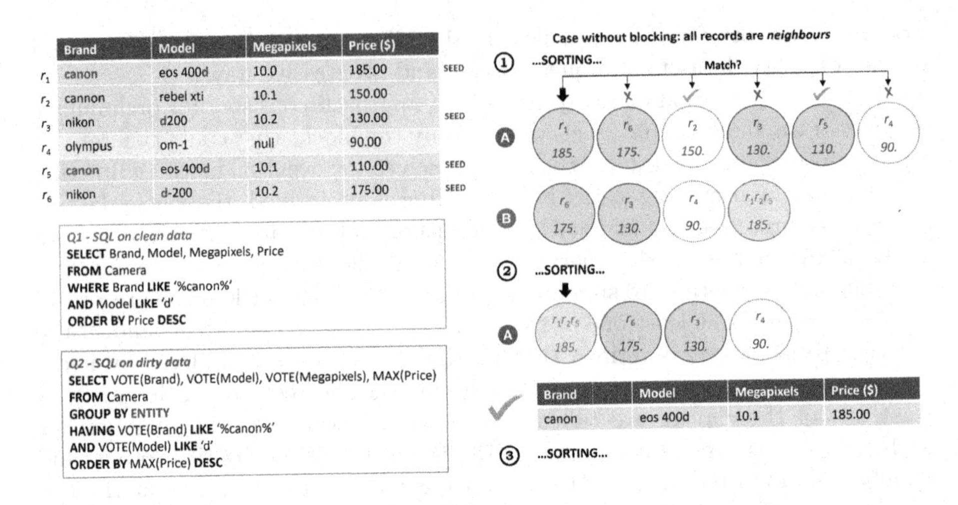

Fig. 2. BrewER in action.

4 Evaluation

The evaluation of BrewER, whose implementation is realized in Python, is performed on real-world datasets from different domains (Table 1) with known ground truth: SIGMOD20 [2,8] (camera specifications from e-commerce websites, pre-processed using a variant of the algorithm described in [17]), SIGMOD21 [3] (USB stick specifications from e-commerce websites) and its superset (both provided by Altosight [1]), Funding [9] (organizations presenting financing requests). All strings are put in lowercase and OK values cast to float, filtering out the records whose OK is null, since they do not alter the emission ordering.

Plots in Figs. 3 and 4 show the early results, in terms of progressive *query recall* (number of emitted entities/size of the result set) after x performed comparisons (ground truth as matching function), obtained computing mean values on batches of 20 queries (both for conjunctive and disjunctive case), selected as the ones emitting most entities out of wider batches of at least 50 queries. Values for the considered attributes are randomly selected from lists containing the most common ones. Figure 3 clearly shows the progressive nature of *BrewER* and highlights its potential in anticipating emissions, considering as batch baseline an adapted version of QDA [5]. When progressiveness is lower because of delayed emissions (discordant case), optimized algorithm generates a further significant reduction of the comparisons (Fig. 4; analogous plots for disjunctive case).

Table 1. Characteristics of the selected datasets.

Name	Records	Duplicates	Entities (Mean size)	Attributes	Ordering key
SIGMOD20 [8,17]	13.58k	12.01k	3.06k (4.439)	5	Megapixels
SIGMOD21	1.12k	1.08k	190 (5.879)	5	Price
Altosight	12.47k	12.44k	453 (27.534)	5	Price
Funding [9]	17.46k	16.70k	3.11k (5.609)	18	Amount

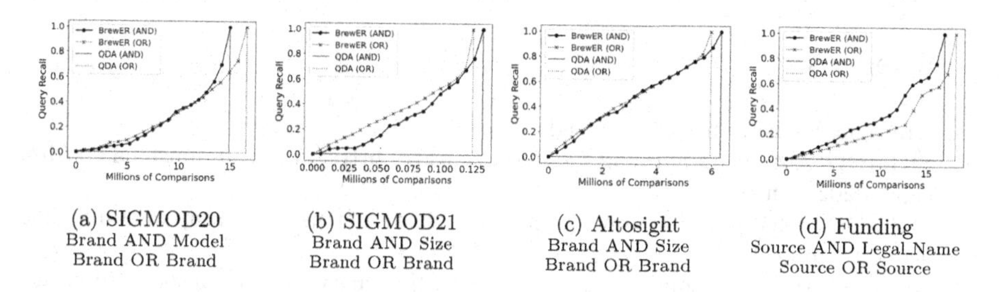

(a) SIGMOD20 Brand AND Model Brand OR Brand
(b) SIGMOD21 Brand AND Size Brand OR Brand
(c) Altosight Brand AND Size Brand OR Brand
(d) Funding Source AND Legal_Name Source OR Source

Fig. 3. Progressive query recall in `MAX/DESC` and `MIN/ASC` cases (no blocking).

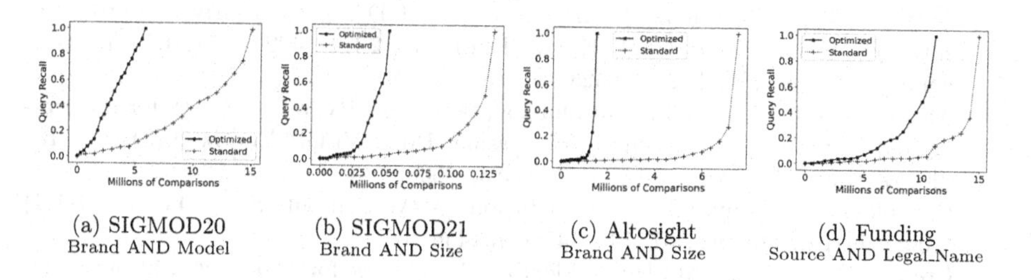

(a) SIGMOD20 Brand AND Model
(b) SIGMOD21 Brand AND Size
(c) Altosight Brand AND Size
(d) Funding Source AND Legal_Name

Fig. 4. Progressive query recall in `MAX/ASC` and `MIN/DESC` cases (no blocking).

5 Conclusions and Next Steps

Early results confirm the benefits of the approach adopted by `BrewER`, in terms both of reduction of performed comparisons and of progressive emission of the results, paving the way for new and more comprehensive solutions to ER tasks.

`BrewER` has a lot of room for improvement (the implementation itself needs to be optimized, even considering the migration to a faster language), with significant scenarios to be deepened and integrated in the framework; as for discordant case and non-seed record comparisons, it is fundamental to find out and avoid all situations causing useless comparisons. Benefits have to be evaluated in case of blocking: the actual agnostic approach is considered as a strength, since it allows to combine `BrewER` with the most innovative solutions in this field (the same happens for matching functions), but even the possibility of including blocking itself in the progressive pipeline has to be investigated. Cases to be studied are that of `TOP(K)` queries (supposed to take full advantage from this approach) and that of temporal series, while missing value imputation, together with the possibility of keeping track of executed queries, can turn `BrewER` into a powerful data preparation tool, leading to a progressive cleaning of the dataset. Furthermore, since ER can be seen as a case of binary classification, `BrewER` impact is not strictly bound to this field, and it is important to study how to extend the presented techniques to other classification tasks.

`BrewER` is going to be presented and further explored in a dedicated research paper, containing the formalized algorithm and new experiments covering some of these cases; the code will be made available at the time of its publication.

References

1. Altosight Website. https://altosight.com/
2. SIGMOD 2020 Programming Contest Website. http://www.inf.uniroma3.it/db/sigmod2020contest/
3. SIGMOD 2021 Programming Contest Website. https://dbgroup.ing.unimo.it/sigmod21contest/
4. Altwaijry, H., Kalashnikov, D.V., Mehrotra, S.: Query-driven approach to entity resolution. Proc. VLDB Endow. **6**(14), 1846–1857 (2013). https://doi.org/10.14778/2556549.2556567
5. Altwaijry, H., Kalashnikov, D.V., Mehrotra, S.: QDA: a query-driven approach to entity resolution. IEEE Trans. Knowl. Data Eng. **29**(2), 402–417 (2017). https://doi.org/10.1109/TKDE.2016.2623607
6. Altwaijry, H., Mehrotra, S., Kalashnikov, D.V.: QuERy: a framework for integrating entity resolution with query processing. Proc. VLDB Endow. **9**(3), 120–131 (2015). https://doi.org/10.14778/2850583.2850587
7. Bleiholder, J., Naumann, F.: Data fusion. ACM Comput. Surv. **41**(1), 1:1-1:41 (2008). https://doi.org/10.1145/1456650.1456651
8. Crescenzi, V., et al.: Alaska: A Flexible Benchmark for Data Integration Tasks. CoRR abs/2101.11259 (2021). https://arxiv.org/abs/2101.11259
9. Deng, D., et al.: Unsupervised string transformation learning for entity consolidation. In: ICDE 2019, pp. 196–207. IEEE (2019). https://doi.org/10.1109/ICDE.2019.00026
10. Elmagarmid, A.K., Ipeirotis, P.G., Verykios, V.S.: Duplicate record detection: a survey. IEEE Trans. Knowl. Data Eng. **19**(1), 1–16 (2007). https://doi.org/10.1109/TKDE.2007.250581
11. Papadakis, G., Ioannou, E., Thanos, E., Palpanas, T.: The four generations of entity resolution. Synth. Lect. Data Manag. Morgan & Claypool Publishers (2021). https://doi.org/10.2200/S01067ED1V01Y202012DTM064
12. Papadakis, G., Skoutas, D., Thanos, E., Palpanas, T.: Blocking and filtering techniques for entity resolution: a survey. ACM Comput. Surv. **53**(2), 31:1-31:42 (2020). https://doi.org/10.1145/3377455
13. Papenbrock, T., Heise, A., Naumann, F.: Progressive duplicate detection. IEEE Trans. Knowl. Data Eng. **27**(5), 1316–1329 (2015). https://doi.org/10.1109/TKDE.2014.2359666

14. Pietrangelo, A., Simonini, G., Bergamaschi, S., Naumann, F., Koumarelas, I.K.: Towards progressive search-driven entity resolution. In: SEBD 2018, CEUR Workshop Proceedings, vol. 2161. CEUR-WS.org (2018). http://ceur-ws.org/Vol-2161/paper16.pdf
15. Simonini, G., Papadakis, G., Palpanas, T., Bergamaschi, S.: Schema-agnostic progressive entity resolution. IEEE Trans. Knowl. Data Eng. **31**(6), 1208–1221 (2019). https://doi.org/10.1109/TKDE.2018.2852763
16. Whang, S.E., Marmaros, D., Garcia-Molina, H.: Pay-as-you-go entity resolution. IEEE Trans. Knowl. Data Eng. **25**(5), 1111–1124 (2013). https://doi.org/10.1109/TKDE.2012.43
17. Zecchini, L., Simonini, G., Bergamaschi, S.: Entity resolution on camera records without machine learning. In: DI2KG@VLDB 2020, CEUR Workshop Proceedings, vol. 2726. CEUR-WS.org (2020). http://ceur-ws.org/Vol-2726/paper3.pdf

Discovering Latent Information from Noisy Sources in the Cultural Heritage Domain

Fabrizio Scarrone(✉)

Università degli studi di Torino, Turin, Italy

Abstract. Today, there are many publicly available data sources, such as online museum catalogues, Wikipedia, and social media, in the cultural heritage domain. Yet, the data is heterogeneous and complex (diverse, multi-modal, sparse, and noisy). In particular, availability of *social media (such as Twitter messages) is both a boon and a curse* in this domain: social media messages can potentially provide information not available otherwise, yet such messages are short, noisy, and are dominated by grammatical and linguistic errors. The key claim of this research is that the availability of publicly available information related to the cultural heritage domain can be improved with tools capable of signaling to the various classes of users (such as the public, local governments, researchers) the entities that make up the domain and the relationships existing among them. To achieve this goal, I focus on developing novel algorithms, techniques, and tools for leveraging multi-modal, sparse, and noisy data available from multiple public sources to integrate and enrich useful information available to the public in the cultural heritage domain. In particular, research aims to develop novel models that take advantage of multi-modal features extracted by deep neural models to improve the performance for various underlying tasks.

Keywords: Multi-modal · Information extraction · Information integration and latent information discovery · Entity recognition · Neural-networks · Attention models · Cultural heritage domain

1 Introduction

Today, there are many publicly available data sources, such as online museum catalogues(e.g. [17]), Wikipedia (e.g. [18]), and social media (e.g. [19]), in the cultural heritage domain. The key premise of the research is that the availability of publicly available information related to the cultural heritage domain can be significantly improved with tools capable of signaling to the various classes of users (such as the public, local governments, researchers) the entities that make up the domain and the relationships existing among them. Yet, realizing this idea is non-trivial due to the complexity (multi-modality, sparsity, and noise) of the data available in this domain (Fig. 1).

Results presented in this paper were obtained using the Chameleon testbed.

N. Reyes et al. (Eds.): SISAP 2021, LNCS 13058, pp. 402–408, 2021.
https://doi.org/10.1007/978-3-030-89657-7_31

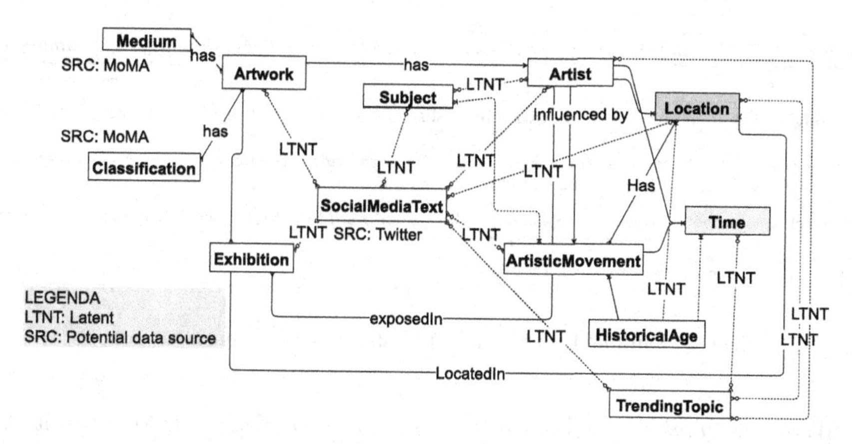

Fig. 1. A simplified ER-diagram outlining some of the data sources and their contributions to the information available in the cultural heritage domain – here any edge marked as "LTNT" represent a latent relationship that needs to be extracted

Problem Statement. Given the above context, in this thesis, I focus on developing novel algorithms, techniques, and tools for leveraging multi-modal, sparse, and noisy data available from multiple public sources to integrate and enrich useful information available to the public in the cultural heritage domain.

2 Methodology

Figure 1 includes an ER-diagram outlining some of the data sources and their contributions to the information available in the cultural heritage domain. For example, one can observe that a number of entities (such as *artwork*, *artist*, *exhibition*, *artistic movement*) are available from various sources. Some of the relationships among this entities can be explicit in these sources (such as an *artwork* is being associated to an *artist* or an *artist* being included in an *exhibition*), while some other relationships, such as an *artist* being influenced by an *artistic movement* or an *artist* being interested on a particular *subject* may be latent – in Fig. 1, such latent relationships are highlighted with edges marked as "LTNT" and they represent areas of interest for this research.

2.1 Social Media to Our Help (?)

Availability of *social media (such as Twitter messages) is both a boon and a curse* in this domain:

- *How can social media help in this domain?:* Social media messages can potentially provide information not available otherwise: many museums or collections post regular Twitter messages about *artists*, *artworks*, or *exhibitions*; in addition, there may be online communities that discuss a particular *artistic movement*. These messages may not only provide data points that are not available otherwise, but can also help contextualize available data or help discover latent relationships among entities in the domain of interest.

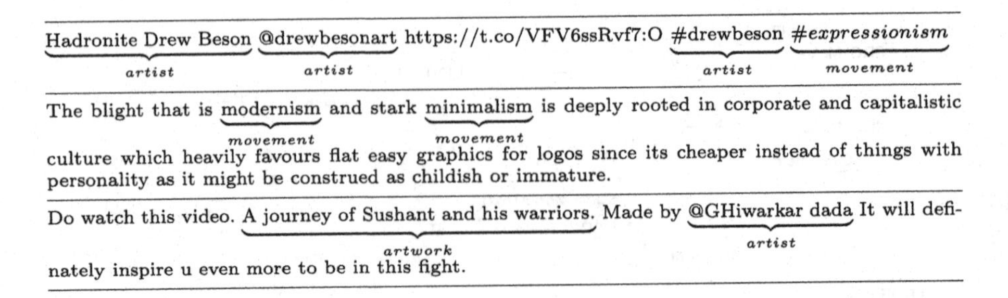

Fig. 2. Sample Twitter messages and associated entity types

– *Why is leveraging social media in this domain challenging?:* Yet, despite the opportunities they provide, leveraging social media data in this context is not trivial: Twitter messages are short, noisy, and are largely dominated by text full of grammatical and linguistic errors. Moreover, many messages in this domain are automatically generated through APIs and their structures are not linguistically based. Being extremely short, they often lack context to help interpret their content. In fact, our experience with Mechanical Turk [20], has shown that even manually labeling portions of the messages for a supervised methodology is difficult due to the underlying ambiguities.

2.2 A Multi-modal Approach

It is important to note that social media, in this context, is often multi-modal in that many messages in this domain are accompanied by visuals – our experience has shown that roughly 30% of the messages have one or more associated images. Recently, joint learning from sources with different underlying modalities has become an emerging research interest and, with the rise of deep learning techniques, such approaches have found increasing use in diverse domains, such image understanding/annotation [16,21], and natural language processing [5].

While in theory, these visuals can also help provide context necessary to better interpret these social media messages for effective information extraction and integration, in practice, these visuals can be very diverse (such as the visual representation of the artwork, picture of an artist, a snapshot of the exhibit venue, or an announcement flyer) and they themselves lack descriptive labels. Consequently, leveraging such visual data to implement a multi-modal approach is not possible with existing techniques. We next illustrate this with a specific learning task: *entity and entity type recognition.*

2.3 A Specific Task: Entity and Entity Type Recognition

Consider the task of recognizing where entities of various types, such as *artists* (A), *artworks* (W), *movements* (M), or *venues* (V), occur in a given Twitter as visualized in Fig. 2. As the examples in this figure illustrate, this is a highly difficult task, primarily because, the tweets are short, many times poorly organized, and they lack context to help identify that entities they may contain.

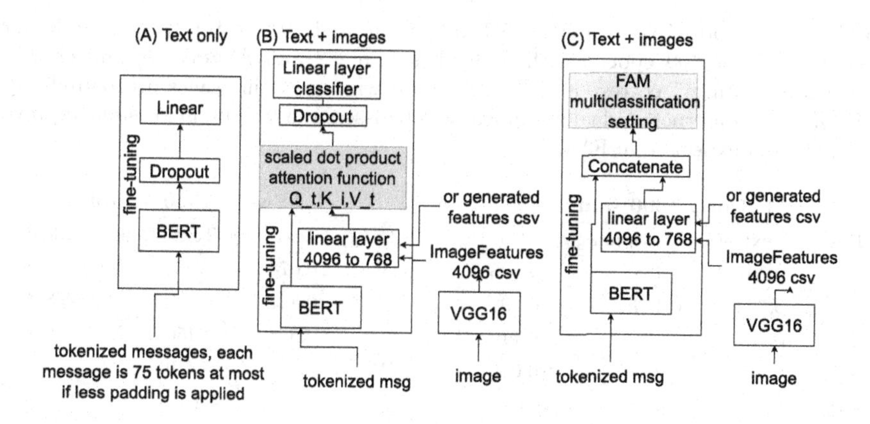

Fig. 3. Three different models

The closest literature to this task is on the *named entity recognition* (NER) problem, where one aims to discover entities, such as persons, locations, that are referred to by their names in a given text document corpus [9–11]. Existing approaches to this problem are often studied in four major groups: 1) rule-based approaches rely on hand-crafted rules [12]; unsupervised learning approaches [22] attempt to solve the problem without the help of hand-labeled training examples; 3) feature-based supervised learning techniques rely on with careful feature engineering; and 4) more recent deep-learning based approaches aim to automatically discover latent representations needed for the classification and/or detection task from the raw input [1,2,5,6,15]. This latter approach can benefit from non-linear transformations and eliminate (or reduce) the efforts needed for designing hand crafted features for this task. However, existing deep-learning based approaches require significant amount of contextual information to help interpret the text and perform poorly when provided out-of-context and short Twitter messages as input data. Therefore, I am developing novel, multi-modal approach to addressing this extremely difficult task.

Data Set. Using the Twitter API, we are building a Cultural Heritage dataset collecting related tweets with their associated images. The entities in the text messages are manually labeled by Mechanical Turkers, using the BIO tagging scheme [7][1] Latent embeddings for the textual part of the messages are obtained using BERT [1] a model Trained on a large corpus of unlabelled text including the entire Wikipedia and Book Corpus. Bert generates context related embedding at sub-word level. Latent embeddings for the images associated the to messages, on the other hand, are obtained from VGG16 [3] a convolutional model trained using ImageNet 15 million labeled images in 22,000 categories.

[1] We will make this data set publicly available to the research community.

Table 1. F1 score (75% training, 15% validation, and 10% testing); each model has been trained for 600 epochs with a batch size of 32 and AdamW optimizer [8] – here "ideal features" corresponds to a scenario where text messages are paired with synthetically-constructed ideal visual features to assess the maximum possible improvements we can expect from FAM

F1 Score, 1000 samples

Entity	Text only	Text+images	Ideal fts
A	0,597	0,458	0,598
M	0,734	0,449	0,792
V	0,667	0,071	0,714
W	0,612	0,255	0,619

F1 Score, 1500 samples

Entity	Text only	Text+images	Ideal fts
A	0,678	0,545	0,728
M	0,827	0,675	0,895
V	0,538	0,456	0,690
W	0,621	0,449	0,708

Novel Multi-modal Attention Mechanisms. As we mentioned above, a particular challenge in addressing the entity and entity type recognition challenge is the lack of context when interpreting the messages. I argue that multi-modal attention, if used effectively, can help alleviate this problem. Attention as a mechanism by which a network can capture the interaction among elements of a given data object (or across multiple data objects) to discover features weights and use this weighting to help improve the network performance [2,13,14]. Since our goal is to leverage visual information to help us provide context to recognize entities in short text messages, we are especially interested in cross-attention mechanisms, such as [21]. Figure 3 depicts, the possible models: Model (A) uses only the text messages and applies a linear layer after the Bert model in order to classify the tokens in the available classes. Model (B) uses text messages and images embeddings – Bert embeddings and visual features are combined in the attention module and the result goes through a linear model for classification. The attention applied in this case is similar to that the one present in [2].

I am, however, proposing a novel cross-attention mechanism, depicted as Model (C): in this model textual and visual embeddings are passed to a novel factorizing attention module (FAM) module for a multi-classification setting, inspired by factorization machines [4] – FAM accounts for the second order interactions within and across textual and visual features while analyzing the combined sparse feature space for feature weights that will inform the search for entities in short text. As shown in Fig. 1, FAM has the potential to provide significant improvements in F1 scores for all entity types when provided idealized features. While the results are less accurate when provided real images, the difference quickly improves as the number of samples increases from 1000 to 1500, which indicates that with a reasonable training corpus, the provided FAM will surpass the text-only accuracies.

3 Conclusions and Future Work

Here, I provided an overview of my PhD work on improving information availability in the cultural heritage domain. My research is focusing on novel algorithms and techniques these can include novel multi-modal, cross-attention mechanism but also instance and n-shot learning, for leveraging multi-modal, sparse, and noisy data available from multiple public sources to improve information quality. Future work will include building a sufficiently large corpus to train and evaluate algorithms and tackle latent information extraction tasks outlined in Fig. 1.

References

1. Devlin, J., Chang, M., Lee, K., Toutanova, K.: Bert: pre-training of deep bidirectional transformers for language understanding, NAACL-HLT (2019)
2. Vaswani, A., et al.: Attention is all you need. In: Advances in Neural Information Processing Systems, pp. 6000–6010 (2017)
3. Simonyan, K., Zisserman, A.: Very Deep Convolutional Networks for Large-Scale Image Recognition. ICLR (2015)
4. Rendle, S.: Factorization machines. In: 2010 IEEE International Conference on Data Mining (2010)
5. Asgari-Chenaghlu, M., Feizi-Derakhshi, M.R., Farzinvash, L., Balafar, M.A., and Motamed, C.: A multimodal deep learning approach for named entity recognition from social media, CoRR (2020)
6. Zhang, Q., Fu, J., Liu, X., Huang, X.: Adaptive co-attention network for named entity recognition in tweets. In: AAAI (2018)
7. https://en.wikipedia.org/wiki/Inside_outside_beginning_(tagging)
8. Loshchilov, I., and Hutter, F., Decoupled Weight Decay Regularization, ICLR (2019)
9. Li, J., Sun, A., Han, J., Li, C.: A survey on deep learning for named entity recognition, CoRR (2018)
10. Yadav, V., Bethard, S.: A survey on recent advances in named entity recognition from deep learning models, CoRR (2019)
11. Grishman, R., Sundheim, B.: Message understanding conference-6: a brief history. In: COLING 1996 Volume 1: The 16th International Conference on Computational Linguistics, vol. 1 (1996)
12. Sekine, S., Nobata, C.: Definition, dictionaries and tagger for extended named entity hierarchy. In: LREC, pp. 1977–1980 (2004)
13. Bahdanau, D., Cho, K., Bengio, Y.: Neural machine translation by jointly learning to align and translate, CoRR (2014)
14. Luong, T., Pham, H., Manning, C.D.: Effective approaches to attention-based neural machine translation. In: Proceedings of the 2015 Conference on Empirical Methods in Natural Language Processing (2015). http://dx.doi.org/10.18653/v1/d15-1166
15. Collobert, R., Weston, J., Bottou, L., Karlen, M., Kavukcuoglu, K., Kuksa, P.: Natural language processing (almost) from scratch. J. Mach. Learn. Res. **12**, 2493–2537 (2011)
16. Baltrusaitis, T., Ahuja, C., Morency, L.-P.: Multimodal machine learning: a survey and taxonomy. IEEE Trans. Pattern Anal. Mach. Intell. **41**, 423–443 (2019)
17. MuseumofModernArt, GitHub repository. https://github.com/MuseumofModernArt/collection
18. Wikipedia, The Free Encyclopedia, "List of art movements". https://en.wikipedia.org/wiki/List_of_art_movements. Accessed 21 July 2021

19. MoMA The Museum of Modern Art [@@MuseumModernArt]. Tweets [Twitter profile]. Twitter. https://twitter.com/museummodernart. Accessed 21 July 2021
20. Amazon Mechanical Turk. https://www.mturk.com/. Accessed 21 July 2021
21. Wei, X., Zhang, T., Li, Y., Zhang, Y., Wu, F.: Proceedings of the IEEE/CVF Conference on Computer Vision and Pattern Recognition (CVPR), pp. 10941–10950 (2020)
22. Nadeau, D., Turney, P.D., Matwin, S.: Unsupervised named entity recognition: generating gazetteers and resolving ambiguity. In: CSCSI, pp. 266–277 (2006)

Author Index

Printed in the United States
by Baker & Taylor Publisher Services